AF772803

Physical and Numerical Models in Knot Theory

Including Applications to the Life Sciences

Series on Knots and Everything — Vol. 36

Physical and Numerical Models in Knot Theory

Including Applications to the Life Sciences

Editors

Jorge A. Calvo
Ave Maria University, USA

Kenneth C. Millett
University of California, Santa Barbara, USA

Eric J. Rawdon
Duquesne University, USA

Andrzej Stasiak
Université de Lausanne, Switzerland

World Scientific

NEW JERSEY • LONDON • SINGAPORE • BEIJING • SHANGHAI • HONG KONG • TAIPEI • CHENNAI

Published by

World Scientific Publishing Co. Pte. Ltd.
5 Toh Tuck Link, Singapore 596224
USA office: 27 Warren Street, Suite 401-402, Hackensack, NJ 07601
UK office: 57 Shelton Street, Covent Garden, London WC2H 9HE

Library of Congress Cataloging-in-Publication Data
Physical and numerical models in knot theory : including applications
to the life sciences / editors, Jorge A. Calvo ... [et al.].
p. cm. (series on knots and everything ; v. 36)
Includes bibliographical references.
ISBN 981-256-187-0 (alk. paper)
1. Knot theory. I. Calvo, Jorge Alberto, 1971– II. Title. III. Series.

QA612.2 .P64 2005
514'.2242--dc22

2005049743

British Library Cataloguing-in-Publication Data
A catalogue record for this book is available from the British Library.

About the cover: the contact points of the ideal trefoil as architectures – by Ben Laurie

Copyright © 2005 by World Scientific Publishing Co. Pte. Ltd.
All rights reserved. This book, or parts thereof, may not be reproduced in any form or by any means, electronic or mechanical, including photocopying, recording or any information storage and retrieval system now known or to be invented, without written permission from the Publisher.

For photocopying of material in this volume, please pay a copying fee through the Copyright Clearance Center, Inc., 222 Rosewood Drive, Danvers, MA 01923, USA. In this case permission to photocopy is not required from the publisher.

Printed in Singapore by B & JO Enterprise

To our families and our teachers.

PREFACE

The physical properties of knotted and linked configurations in space have long been of interest to mathematicians. More recently, these properties have become significant to biologists, physicists, and engineers among others. Their depth of importance and breadth of application are now widely appreciated, and valuable progress continues to be made each year. Nevertheless, many fundamental and basic questions in the general domain of modeling physical knotting remain completely unanswered or at least are not fully or satisfactorily answered.

This volume consists primarily of chapters by researchers who spoke at two conferences: a special session on "Numerical Methods, Calculations, and Simulations in Knot Theory and its Applications" held at the Western Sectional meeting of the AMS in San Francisco in May 2003, and a conference on "Knots, Random Walks and Biomolecules" sponsored by the Bernoulli Centre of the Swiss Federal Institute of Technology in Lausanne and held in Les Diablerets, Switzerland, in July 2003.

The special session produced several contributions from researchers who are using computers to study problems that would otherwise be intractable. While computations have long been used to analyze problems, formulate conjectures, and search for special structures in knot theory, increased computational power has made them a staple in many facets of the field. From properties of knot invariants, to knot tabulation, studies of hyperbolic structures, knot energies, and the exploration of spaces of knots, computers have opened the doors to problems that would have otherwise been too difficult to do by hand.

The conference in Switzerland produced chapters concentrating on the models that researchers use to understand knotting, linking, and entanglement in physical and biological systems. Topics range from knotted umbilical cords, to studies of DNA knots, knots in proteins, and the structure of tight knots.

Together, the chapters in this volume present a unique arrangement of treatises exploring, in broad terms, four major themes: physical knot theory, knot theory in the life sciences, computational knot theory, and geometric knot theory.

Part 1. Physical Knot Theory

Physical knot theory encompasses the study of knotting in structures composed of "real" materials, as well as the mathematical modeling of such structures. Knot energy, ropelength, and thickness (all of which measure the compactness of an embedding), and in particular the configurations that optimize these quantities, play a central role in this area. Chapters in this section discuss the dynamics of knots slipping off the end of a metal chain, the energy spectrum of tightly-knotted Quantum Chromodynamic (QDC) flux tubes, and the search for ideal configurations of knots. We begin with an article which, despite being one of the first papers ever written on physical knots, remains essentially unknown.

1. **On the Theory of Solid Knots**
   by Otto Krötenheerdt and Sigrid Veit; translated by Ted Ashton.

   Originally published in 1974 in the *Wissenschaftliche Beiträge der Martin-Luther-Universität Halle-Wittenberg* and translated by Ted Ashton, this appears to be the earliest treatise on thick knots and knots formed from identical beads.

2. **A Tutorial on Knot Energies**
   by E. J. J. van Rensburg.

   This chapter looks at different properties of knot energies defined for lattice knots, polygonal knots, and smooth knots. Several energies are analyzed with respect to these properties.

3. **Universal Energy Spectrum of Tight Knots and Links in Physics**
   by R. Buniy and T. Kephart.

   A system of tightly knotted, linked, or braided flux tubes can be given a universal mass-energy spectrum as the lengths of the fixed radius tubes depend only on the topology of the configuration. This proposal is motivated by examples from plasma physics, but the principal focus is on the model of glueballs as knotted QCD flux tubes.

4. **Knot Dynamics in a Driven Hanging Chain: Experimental Results**
by A. Belmonte.

The author investigates the formation and motion of knots in a linear hanging chain shaken vertically with varying amplitude and frequency. An interesting phase diagram is presented that defines the optimal ratios between the amplitude and frequency required to generate knots. In addition, the complicated dynamics of the unknotting of different knots are analyzed.

5. **Biarcs, Global Radius of Curvature, and the Computation of Ideal Knot Shapes**
by M. Carlen, B. Laurie, J. Maddocks, and J. Smutny.

The authors introduce and exploit a $C^{1,1}$ biarc discretization to compute approximations to the ideal shapes of the trefoil and figure-eight knots with a tight control of numerical error. Geometric and parametric contact and approximate contact sets of thickened configurations are rigorously defined and carefully analyzed on the two approximately ideal shapes that are computed.

Part 2. Knot Theory in the Life Sciences

The chapters in this section explore knotting, linking, and folding in a biological setting. The issues addressed by these chapters include the dynamics of DNA gel-electrophoresis, DNA elasticity and packing, statistical models for protein folding, and the entangling of umbilical cords in the womb.

6. **Knotted Umbilical Cords**
by Alain Goriely.

In about 1% of newborn babies, the umbilical cord is knotted. The author reviews the medical literature on knotted umbilical cords, starting with the 1609 work of Louise Bourgois, and then offers a detailed explanation of the conditions required for the formation of knots in umbilical cords.

7. **Modelling DNA as a Thick Polymer: Application to DNA Elasticity and Packaging Thermodynamics**
by C. Micheletti and D. Marenduzzo.

The authors consider a thick polymer model to provide a coarse-grained description for double-stranded DNA. They use the model to gain insight into how the intrinsic thickness of DNA affects the behavior of the biomolecule subject to compaction.

8. **Monte Carlo Simulations of Gel-Electrophoresis of DNA Knots**
by C. Weber, M. Fleurant, P. De Los Rios, and G. Dietler.

Circular DNA molecules of the same size but forming different knot types can be separated by gel electrophoresis according to their knot type. The authors present a numerical simulation that mimics the physical behavior of knotted DNA molecules during the process of gel electrophoresis.

9. **Atomic Force Microscopy of Complex DNA Knots**
by F. Valle, M. Favre, J. Roca, and G. Dietler.

This chapter analyzes DNA knots formed by circularization within intact phage particles in order to answer the long-standing question of how DNA is packed in phage heads. The authors apply atomic force microscopy imaging to investigate such DNA knots.

10. **Protein Folds, Knots and Tangles**
by W. Taylor.

The author introduces some interesting topological concepts to describe the structure and topology of properly folded protein chains. Applying one of these concepts, it is possible to detect whether or not a linear protein chain is knotted.

11. **Tying Down Open Knots: A Statistical Method For Identifying Open Knots With Applications To Proteins**
by K. C. Millett and B. M. Sheldon.

A new strategy of describing the nature of knotting that occurs in open random walks or proteins of varying lengths is investigated. Random walk and protein data are presented in a novel manner that gives deeper spatial information about the associated knotting spectrum and suggest new directions of research.

12. **Scaling of the Average Crossing Number in Equilateral Random Walks, Knots and Proteins**
by A. Dobay, J. Dubochet, A. Stasiak, and Y. Diao.

The authors compare configurations of properly folded proteins with those of random walks. They observe that, with respect to scaling behavior, small proteins (up to 300 amino acids) resemble self-attracting random walks while large proteins behave like ideal random walks in which independent segments neither attract nor repel each other.

13. **Folding Complexity in a Random-Walk Copolymer Model**
by G. Arteca.

Physical properties of polymers are frequently approximated using a rather simple model of equilateral chains. More complex models are needed to describe correctly the effect of varying segment lengths in natural polymers like polysaccharides and proteins. The author investigates, both analytically and numerically, the potential effects of varying the segment lengths on selected physical properties of modeled polymers.

Part 3. Computational Knot Theory

This section highlights computational aspects of knot theory. These chapters utilize statistical methods (such as Monte Carlo explorations) and deterministic algorithms to measure the probability of knotting, the average crossing number in different types of random polygons, the ropelength of a knot, and to analyze properties of polymers modeled as random walks.

14. **Universal Characteristics of Polygonal Knot Probabilities**
by K. Millett and E. Rawdon.

The authors explore several functions for fitting probability graphs of different knotting models with the hopes of finding a single function that works for both the unknot and non-trivial knots. Of particular interest is the finite-range of knotting in the models.

15. **The Average Crossing Number of Gaussian Random Walks and Polygons**
by Y. Diao and C. Ernst.

Prior results on the average crossing number of equilateral random walks and polygons are extended to Gaussian random walks and polygons. The mean average crossing number, ACN, for Gaussian random walks and polygons of length n is of the form $\frac{1}{2\pi} n \ln n + O(n)$.

16. **Ropelength of Tight Polygonal Knots**
by J. Baranska, P. Pieranski, and E. Rawdon.

The authors analyze polygonal knots provided by the SONO algorithm arriving at upper bounds for the minimum smooth ropelength of a few knots. The analysis suggests another formula allowing one to estimate the minimum ropelength of smooth knots from the data obtained for polygons with fewer edges.

17. **A Fast Octree-Based Algorithm for Computing Ropelength** by T. Ashton and J. Cantarella.

    This chapter presents an algorithm using octrees for computing the polygonal ropelength in (expected) $O(n \log n)$ time. This improves on the standard algorithm, which runs in $O(n^2)$ time. Some performance results for the authors' implementation are included.

18. **Topological Entropic Force Between a Pair of Random Knots Forming a Fixed Link** by T. Deguchi.

    The author explores the probability that two unknotted components are linked as a function of the distance between their centers of mass. Physical arguments are presented to derive an analytic expression for the cases of the trivial and Hopf link.

19. **Under-Knotted and Over-Knotted Polymers: 1. Unrestricted Loops** by N. Moore, R. Lua, and A. Grosberg.

    The authors present computer simulations to examine the probability distributions of the gyration radius for zero-thickness random equilateral knots. Of particular interest is the dependence of the distributions on length and topology.

20. **Under-Knotted and Over-Knotted Polymers: 2. Compact Self-Avoiding Loops** by R. Lua, N. Moore, and A. Grosberg.

    The authors explore compact random knots, namely those forming Hamiltonian closed paths on the 3-dimensional cubic lattice. In particular, they analyze the effect of topology and the size of the knotted regions.

21. **On the Mean Gyration Radius and the Radial Distribution Function of Ring Polymers With Excluded Volume Under a Topological Constraint** by M. Shimamura and T. Deguchi.

    The competition between topological effects and those of excluded volume on the average size of ring polymers is studied in this chapter. In addition, radial distribution functions of segments of ring polymers with fixed knotting are discussed. Numerical results suggest that topological constraints on ring polymers lead to entropic repulsion among polymer segments.

22. **Thermodynamics and Topology of Disordered Knots. Correlations in Trivial Lattice Knot Diagrams**
by S. Nechaev and O. Vasilyev.

In this investigation of the statistical properties of random lattice knots, the focus is on the "knottedness" of densely packed knots as measured by the Jones-Kauffman polynomial. The results of this study provide support for a conjecture that collapsed closed polymer chains form systems of densely packed mutually segregated folds.

23. **Generating Large Random Knot Projections**
by Y. Diao, C. Ernst, and Uta Ziegler.

This chapter addresses the difficult problem of randomly generating knot projections with a large number of crossings. Two approaches to generate diagrams of large knots are presented, and even though they both have the drawback of lack of control over the resulting probability distribution, they are of undeniable interest.

Part 4. Geometric Knot Theory

The central questions of the chapters in this final section address a variety of topics, all of which focus on the geometry of knotting and folding. These include knot enumeration, flat ribbons, quadrisecants, writhe of non-closed curves, ropelength-critical clasps, hyperbolic invariants, and thick surfaces.

24. **Minimal Flat Knotted Ribbons**
by L. Kauffman.

The author obtains upper bounds for the minimal length required to make a flat knotted ribbon by examining the configurations which naturally arise when a flat ribbon is tied as a trefoil or figure-eight knot and pulled tight. In the case of a trefoil, the conjectured optimal configuration takes on the shape of a regular pentagon; in the case of the figure-eight knot, it has a hexagonal shape.

25. **Quadrisecants of Knots with Small Crossing Number**
by G. T. Jin.

This chapter examines the set of quadrisecants for several embeddings of the knots with five crossings or fewer. It also conjectures that the "quadrisecant approximation" of a knot K – that is, the piecewise-linear knot whose vertices are the intersection points of K with each of its quadrisecants – has the same knot type as K.

26. **On the Writhe of Non-Closed Curves**
by E. Starostin.

The author defines the writhe of an open space curve and explores its computation. The approach is based on closing the tangent indicatrix with a geodesic. A relationship connecting the writhe with the Gauss integral over the open curve is also studied.

27. **On a Mathematical Model for Thick Surfaces**
by P. Strzelecki and H. von del Mosel.

The authors of this paper extend the concept of thick knots to surfaces by using a quantity – known as the global radius of curvature $\Delta[X]$ – which is defined over a large class of non-smooth parametric surfaces. The main result is that a surface X is in fact a continuously differentiable manifold and can be "thickened" by a constant magnitude of θ whenever $\Delta[X] \geq \theta > 0$.

28. **Some Ropelength-Critical Clasps**
by J. Sullivan and N. Wrinkle.

The Gehring ropelength problem asks to minimize the length of a link whose components stay at a unit distance from each other; a balance criterion describes which configurations are critical. Several explicit configurations of clasped ropes which are balanced, and hence critical, are presented.

29. **Remarks on Some Hyperbolic Invariants of 2-Bridge Knots**
by J. Hoste and P. Shanahan.

The authors describe a recursive technique suitable for computing several hyperbolic invariants of a class of 2-bridge knots. These include the representation and character varieties, the trace and cusp fields, and the A-polynomial. Applications and generalizations of this computational technique are also discussed.

30. **Conjectures on the Enumeration of Alternating Links**
by P. Zinn-Justin.

This chapter parallels the enumeration of alternating links and tangles with statistical models on random lattices. The insight provided by this parallelism leads to conjectures on asymptotic counting as the number of crossings becomes large.

We believe that this collection provides the reader with an excellent entry to the foundations of work in these areas and to the broad spectrum of new questions, techniques, and research progress. Without the extraordinary efforts of the large number of researchers whose work is presented in the collection, we would never have been able to bring the project to a timely conclusion. The editors thank all the authors and referees for their dedication to this project.

Jorge Alberto Calvo
Department of Mathematics
Ave Maria University
Naples, FL 34119, USA
Email: jorge.calvo@avemaria.edu

Kenneth C. Millett
Department of Mathematics
University of California
Santa Barbara, CA 93105, USA
Email: millett@math.ucsb.edu

Eric J. Rawdon
Department of Mathematics and
Computer Science
Duquesne University
Pittsburgh, PA 15282, USA
Email: rawdon@mathcs.duq.edu

Andrzej Stasiak
Laboratoire d'Analyse Ultrastructurale
Bâtiment de Biologie
Université de Lausanne
CH-1015 Lausanne-Dorigny,
Switzerland
Email: andrzej.stasiak@unil.ch

CONTENTS

CHAPTER 1

ON THE THEORY OF SOLID KNOTS

Otto Krötenheerdt

Sigrid Veit

Translated by Ted Ashton

*Otto Krötenheerdt and Sigrid Veit, Mathematics Section, Martin Luther University Halle-Wittenberg**

1. Problem Statement and Results

In the past, chemical research has provided many applications for Knot Theory. For example, knots and links have been found in the molecular structure of certain chemical compounds. A cycloparaffin[a] molecule with over 50 carbon atoms can exist, for example, as a simple ring or as a knotted ring resembling a trefoil[b] (see figure 1). H. L. FRISCH and E. WASSERMANN, in their work[1], call these "topological isomers" and note that they differ in some, predominately physical, characteristics. G. SCHILL and C. ZÜRCHER report in [5] on mechanically linked molecules — Catenanes and Rotaxanes. For example, the simplest Catenane is represented by two linked rings (see figure 2). In classical knot theory, a *knot* is the image of S^1 in $\mathbb{R}^3$ and thus one-dimensional; *links* are combinations of knots and hence also one-dimensional. However the previously-mentioned chemical examples are of knotted and linked three-dimensional objects and classical Knot Theory is only partially applicable to these. So we would like to call these objects

*This paper was first published in German on 30 December 1974 in the *Wissenschaftliche Beiträge der Martin-Luther-Universität Halle-Wittenberg*. It was translated into English in 2004 by Ted Ashton at the University of Georgia, Athens.

[a]Also called "cycloalkane". Note that all footnotes were added by the translator.

[b]German: *Kleeblattschlinge*, literally "cloverleaf loop".

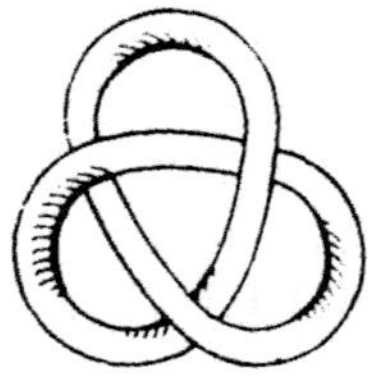

Fig. 1.

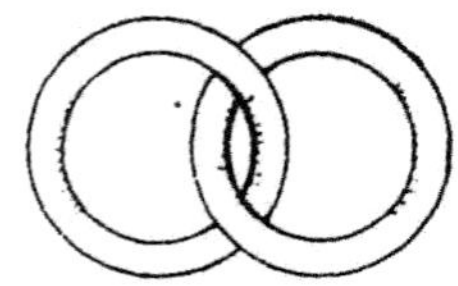

Fig. 2.

"solid knots".[c]

The study of the geometric characteristics of solid knots and links will involve, apart from topological questions (such as the question of knot type or isotopy class), also "metric questions."[d] So, for example, a solid knot constructed of congruent components must have some minimum number of those components for its core[e] to form a given knot (for instance a trefoil or a particular rosette knot[f] R_n^m).

In what follows we will focus on two special kinds of solid knots, namely solid knots built from congruent components, in section 3, and solid knots of uniform thickness[g], in section 4. Section 2 will prepare the way for those via a lemma regarding certain polygonal knots (and here we use the term in the classical sense). We will restrict our attention to certain torus knots, those which in [2] and [3] are called "rosette knots of second order" R_2^m ($m =$

[c]German: *massive Knoten.* As will soon be seen, Krötenheerdt and Veit include in this definition what are currently termed "thick knots" — the "solid knots of uniform thickness d" of section 4 — and also knots formed from congruent components, like pearl necklaces.

[d]German: *metrische Fragen.* One presumes these to be questions of distance and size and so forth. The sentence is perhaps better translated, "The study of the characteristics of solid knots and links involves not only topological questions but also geometric ones."

[e]Here the authors use the term *Seele*, meaning "soul" or "spirit" and define it as a knot in the classical sense imagined to be inside the solid knot.

[f]German: *Rosettenknoten.* In [2], Krötenheerdt defines the term "*Rosettenknoten* R_n^m of order n" to mean closed braids of n strands of the form

$$Z = \begin{cases} (s_1 s_2^{-1} s_3 s_4 - 1 \cdots s_{n-2}^{-1} s_{n-1})^m & \text{for } n \text{ even,} \\ (s_1 s_2^{-1} s_3 s_4 - 1 \cdots s_{n-2} s_{n-1}^{-1})^m & \text{for } n \text{ odd,} \end{cases}$$

where $n > 1$ (naturally), $m > 1$ (or the knot is the unknot) and $(m, n) = 1$ (or the knot becomes a link).

[g]Now commonly called "thick knots".

$3, 5, \dots$); the most basic representative of these second order rosette knots[h] is the trefoil, R_2^3. If we denote by $n_0(R_2^m)$ the minimum number of edges which are necessary to form a polygonal R_2^m knot, then the lemma will give us that $n_0(R_2^m)$ is bounded above by $m + 3$ and in the $m = 3$ case it is actually $n_0(R_2^3) = 6$.

We define as "Solid Knots made of Congruent Components" finite unions of congruent bodies B_i $(i = 1, \dots, n)$ in $\mathbb{R}^3$ with the following characteristics:

(a) There exists a polygonal knot K — the core — which intersects the interior of each B_i.
(b) The B_i — the "components" — are convex and bounded.
(c) No two components share an interior point, but the two components B_i and B_{i+1} share at least one boundary point (defining B_{n+1} to be B_1).
(d) K intersects the boundary of each B_i in exactly two points, of which one is also in the boundary of B_{i-1} and the other in the boundary of B_{i+1} (defining B_0 as B_n).
(e) For each B_i, the part of K contained in B_i, along with the shortest curve on the boundary of B_i necessary to make a closed curve, forms a trivial knot.

For those solid knots we use the symbol $\mathbf{K}(K, B)$ where B is any representative of the B_i (which will be referred to as "components of the type B").

(Although the part of (c) which says that B_i and B_{i+1} share a point can be derived from (d), it was stated here explicitly for the sake of later emphasis on the geometry of the situation).

We denote by $n_0(K, B)$ the minimum number of congruent components of type B needed to form a solid knot $\mathbf{K}(K, B)$ for a given knot type K. The following can be shown:

1. If $K := R_2^m$ $(m = 3, 5, \dots)$ and $B := \mathcal{C}$ (cube) then $n_0(R_2^m, \mathcal{C}) \leq 6m + 6$ (this estimate also applies for the Archimedean solids[i] $(3, 8, 8)$, $(4, 6, 6)$, $(4, 6, 8)$, $(3, 4, 3, 4)$, $(3, 4, 4, 4)$).

---

[h] Specifically the $T(m, 2)$ torus knots.

[i] German: *halbregulären Polyeder*, literally "halfregular polyhedra". In the order listed these are: the truncated cube, the truncated octahedron, the truncated cuboctahedron (or great rhombicuboctahedron), the cuboctahedron, and the (small) rhombicuboctahedron. The symbols are vertex configurations, sometimes called "C&R symbols" and

2. If $K := R_2^m$ $(m = 3, 5, \dots)$ and $B := \mathcal{B}$ (ball) then $n_0(R_2^m, \mathcal{B}) \leq 4m + 4$.
3. If $K := R_2^m$ $(m = 3, 5, \dots)$ and $B := \mathcal{K}_a$ (*Kalotte*[j] – a special portion of a ball which has significant applications and is better described in section 3) then $n_0(R_2^m, \mathcal{K}_a) \leq 13m + 9$.

We define as "Solid Knots of Uniform Thickness *d*" the unions of congruent balls, $\mathcal{B}_t$ $(0 \leq t \leq 1)$, in $\mathbb{R}^3$ with the following characteristics:

(a) There is a knot K – the core of the solid knot – on which the centers, C_t, of the $\mathcal{B}_t$ lie.
(b) The $\mathcal{B}_t$ have diameter d where d is the thickness of the solid knot.
(c) As t varies montonically along the interval 0 to 1, C_t varies montonically along the knot K and at $\mathcal{B}_1$ has come back around to C_0.
(d) Each ball $\mathcal{B}_t$ intersects the border of the union of the $\mathcal{B}_t$ exactly in the points of a great circle.

For these solid knots we use the symbol $\mathbf{K}(K, d)$ and occasionally shorten it to $\mathbf{K}$.

(In the above definition, it does not matter whether the $\mathcal{B}_t$ are understood to be solid balls or simply spherical surfaces. Due to characteristic (d), the core K cannot have corners and cannot penetrate the surface of $\mathbf{K}(K, d)$.)

We denote by $l_0(K, d)$ the minimum length of the core K of a thick knot $\mathbf{K}(K, d)$ given a particular knot type for K and a thickness d. It can be show that with $K := R_2^m$ we have

$$l_0(R_2^m, d) \leq \left(\frac{1 + 7m}{2} + \frac{11 + 5m}{2} \cdot \frac{\pi}{4}\right) \cdot d \quad (m = 3, 5, \dots).$$

From this it follows, for example, that for the trefoil R_2^3

$$l_0(R_2^3, d) \leq \left(11 + 13 \cdot \frac{\pi}{4}\right) \cdot d = d \cdot 21.210 \dots$$

In sections 3 and 4 some results of previous investigations into solid links are mentioned; further remarks about them are planned for a future paper.

---

describe the faces which meet at any given vertex (cf. the parenthetical remark on page 8 regarding the corollary). The truncated octahedron, for example, has a square and two hexagons meeting at every vertex.

[j]The word *Kalotte* can be translated "calotte" (the skullcap worn by clergy in the Roman Catholic church), "crown," or "roof" among other things. As the shape is defined in this paper and does not appear to have given name in the mathematical realms, the translator has chosen to leave the German word but to treat it henceforth as an English noun with plural "kalottes".

Further investigations could consider other types of components, find better bounds on $n_0(K, B)$ and $l_0(K, d)$ and, for particular knot types, accurate values for $n_0(K, B)$. Also, it seems possible and desirable to carry over results already known about rosette knots of order n and/or other knot classes.

2. A Lemma about Polygonal Knots

There exists a minimum number, $n_0(K)$, of edges needed to represent a given knot K as a polygonal knot in regular projection. In particular

Lemma 1: *Every polygonal knot of the type $R_2^m (m = 3, 5, \dots)$ can be formed with $m + 3$ edges; thus*

$$n_0(R_2^m) \le m + 3. \tag{1}$$

Proof. A trefoil R_2^3 is representable by six edges, as one can see from the standardized regular projection given in figure 3. (The points A_i and B_i in the following text and in figures 3 and 4 have indices $3, 5, \dots$ from the m of the rosette knot R_2^m). Now we can form an R_2^5 with eight edges by replacing the edge A_3B_3 by the chain $A_3A_5B_5B_3$, with A_3A_5 running under DE and B_5B_3 crossing over DE and the edge A_5B_5 (in the standardized regular projection of the knot R_2^5) lying parallel to DE outside the triangle DES, not crossing any other edge. So R_2^5 can be formed with $6 - 1 + 3 = 5 + 3$ edges while R_2^3 can be formed with $3 + 3$ edges. By induction it follows that

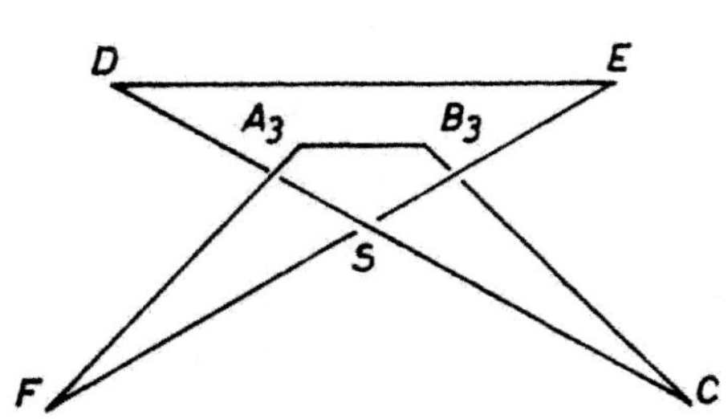

Fig. 3.

we can form an R_2^{m+2} knot from

$$(m + 3) - 1 + 3 = (m + 2) + 3$$

edges by proceeding from the standardized regular projection of an R_2^m (consisting of $m+3$ edges) and replacing the edge A_mB_m with the chain $A_mA_{m+2}B_{m+2}B_m$ in such a way that

(a) in the cases $m = 3, 7, \ldots$, the edge A_mA_{m+2} crosses under the edge DE and $B_{m+2}B_m$ crosses over DE, while the edge $A_{m+2}B_{m+2}$ in the standard regular projection of the knot R_2^{m+2} lies parallel to DE, outside the triangle DES, and crosses no other edge of R_2^{m+2};

(b) in the cases $m = 5, 9, \ldots$, the edge A_mA_{m+2} crosses over the edge DE and $B_{m+2}B_m$ crosses under DE, while the edge $A_{m+2}B_{m+2}$ in the standard regular projection of the knot R_2^{m+2} lies parallel to DE, inside the triangle DES, and crosses no other edge of R_2^{m+2} (see figure 4).

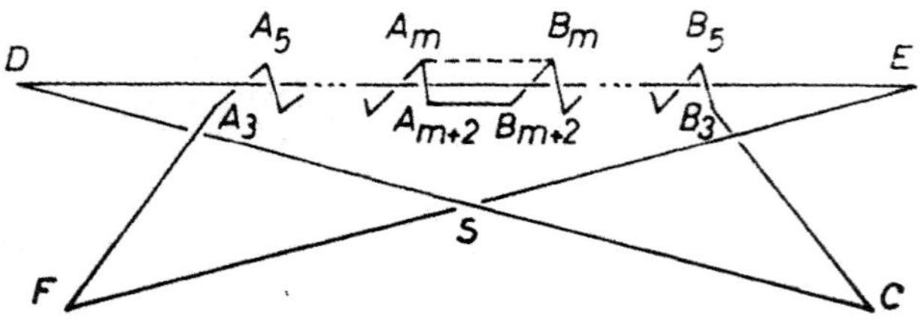

Fig. 4.

Thus we have shown that any polygonal knot of the type $R_2^m (m = 3, 5, \ldots)$ can be formed with $m+3$ edges.

Remark 1 For the trefoil we have exactly

$$n_0(R_2^3) = 6,$$

since any polygonal knot other than the unknot requires at least 6 edges (there are no knotted pentagons or squares).

Remark 2 For each $m = 2, 4, \ldots$, R_2^m is a concatenation of two trivial knots. The estimate (1) applies also to $m = 4, 6, \ldots$, as one can show by an appropriate construction procedure like the one in the proof of the above lemma; in this case $n_0(R_2^m)$ is understood to be the sum of the number of edges in the two trivial knots which form the R_2^m. In the case $m = 2$ we have

$$n_0(R_2^2) = 6$$

(both of the link components are chains of three edges each).

3. Solid Knots Made of Congruent Components

In this section we will turn to the investigation of solid knots formed from congruent components. If we strengthen the requirement that two succesive components must share a point and instead insist that successive components with flat faces must share an entire face, then it seems that, e.g., from the regular polyhedra, cubes would be suitable components.

For solid knots made of congruent cubes we prove the following theorem.

Theorem 2: Any solid knot made of congruent cubes $\mathcal{C}$ with core $K = R_2^m$ $(m = 3, 5, \dots)$ can be formed with $6m + 6$ cubes; so for $\mathbf{K}(R_2^m, \mathcal{C})$ we have

$$n_0(R_2^m, \mathcal{C}) \leq 6m + 6 \quad (m = 3, 5, \dots). \tag{2}$$

Proof Any solid knot $\mathbf{K}(R_2^3, \mathcal{C})$ can be built from $24 = 6 \cdot 3 + 6$ cubes (see figure 5a; figure 5b shows the core with the centers of the cubes marked). The realization of the solid knot can be accomplished with $\mathbf{K}(R_2^5, \mathcal{C})$, $24 + 12 = 36 = 6 \cdot 5 + 6$ cubes (see figure 6; the centers of the additional cubes are shown by small circles). If one uses the same basic construction as for

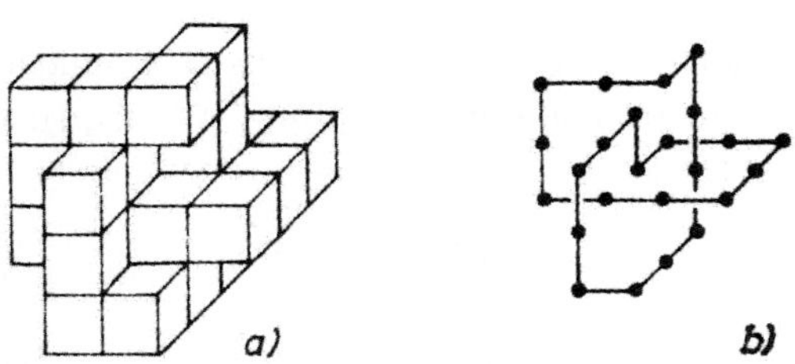

Fig. 5.

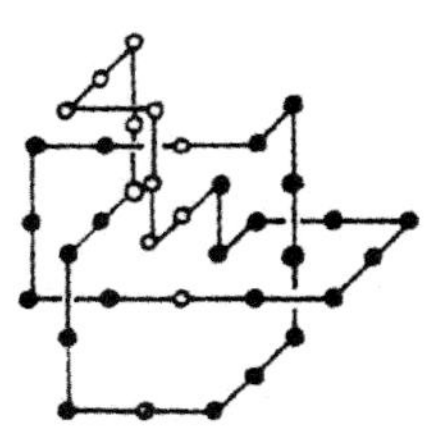

Fig. 6.

the polygonal knots (see section 2), then each solid knot $\mathbf{K}(R_2^{m+2}, \mathcal{C})$ can be created from the solid knot $\mathbf{K}(R_2^m, \mathcal{C})$ in the same way as the solid knot $\mathbf{K}(R_2^5, \mathcal{C})$ was created from $\mathbf{K}(R_2^3, \mathcal{C})$, that is, by inserting 12 additional cubes. Given that

$$n_0(R_2^m, \mathcal{C}) \leq 6m + 6,$$

it then follows that

$$n_0(R_2^{m+2}, \mathcal{C}) \leq 6m + 6 + 12 = 6(m+2) + 6.$$

Hence by the Principle of Mathematical Induction we have the desired result.

Corollary 3: *The bound* (2) *remains valid if the cubes* $\mathcal{C}$ *are replaced with any of the Archimedean solids* $(3,8,8)$, $(4,6,6)$, $(4,6,8)$, $(3,4,3,4)$ *or* $(3,4,4,4)$.

(The symbol $(n_1, n_2, \ldots, n_i)$ means that at every corner of the polyhedron, an n_1-gon, an n_2-gon, . . ., an n_i-gon meet in that order.) The corollary follows directly simply by noticing that the Archimedean solids mentioned all result from trimming cubes in various ways so every one of the cube faces contains a face of the Archimedean solid.

Remark Theorem 2, as stated, applies only if the intersection of two neighbouring cubes is exactly one face. It is easy to see that the bound (2) applies even more if we drop the stronger requirement and only demand that neighboring components possess at least one common point.

If we use balls as components, we get

Theorem 4: Any solid knot formed from congruent balls $\mathcal{B}$ around the core $K = R_2^m$ $(m = 3, 5, \ldots)$ can be realised using $4m + 4$ balls; so for $\mathbf{K}(R_2^m, \mathcal{B})$ we have

$$n_0(R_2^m, \mathcal{C}) \leq 4m + 4 \quad (m = 3, 5, \ldots). \tag{3}$$

Proof Any solid knot $\mathbf{K}(R_2^3, \mathcal{B})$ can be formed with $16 = 4 \cdot 3 + 4$ balls (see figure 7). As explanation, imagine the centers of eight of the balls lying in one plane p; the centers of these balls are A, F, G_3, H_3, I_3, J, K and N_3 (the index on certain of the points coming from the m in R_2^m). The centers of six additional balls lie in a plane parallel to p which is "under" p in figure 7; the centers of these balls are B, C, D and E and

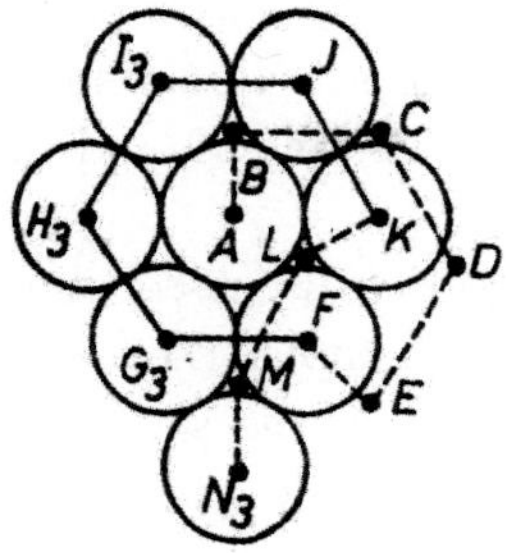

Fig. 7.

also L and M, and the distance between the two planes is fixed so that A, I_3, J and B form a regular tetrahedron. With two more appropriately-placed balls with centers O_3 and P_3 "over" p, we can make a connection between the ball centered at N_3 and the one centered at A. The closed path $ABCDEFG_3H_3I_3JKLMN_3O_3P_3A$ is a polygonal knot of type R_2^3 and with the associated balls forms the solid knot $\mathbf{K}(R_2^3, \mathcal{B})$.

A solid $\mathbf{K}(R_2^5, \mathcal{B})$ can be formed using $16 + 8 = 24 = 4 \cdot 5 + 4$ balls. In order to see this, we supplement the existing balls of the $\mathbf{K}(R_2^3, \mathcal{B})$ with four new balls in the plane p whose centers are the points G_5, H_5, I_5 and T_5 (see figure 8). The path $G_3H_3I_3$ is replaced by the path $G_3G_5H_5I_5I_3$, and the point H_3 is renamed Q_5. Then the points O_3 and P_3 are shifted "over" p into suitable points O_5 and P_5 so that the balls around these points form a connection between the ball centered at N_3 and the one centered at Q_5. With two additional balls around suitable points R_5 and S_5 "under" p we

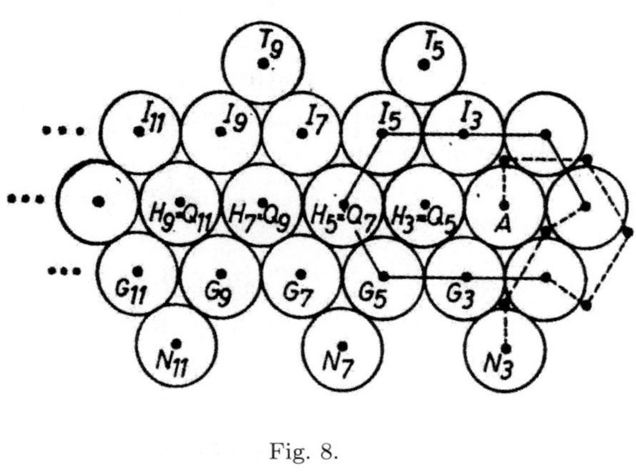

Fig. 8.

can connect the ball centered at Q_5 with the one at T_5; and finally, two further new balls around appropriate points U_5 and V_5 "over" p can form a connection between the ball around T_5 and the one at A.

If one uses the previously-discussed method of constructing the R_2^m ($m = 3, 5, \ldots$) polygonal knots, then every solid knot $\mathbf{K}(R_2^{m+2}, \mathcal{B})$ can be built from a $\mathbf{K}(R_2^m, \mathcal{B})$ in the same fashion as the $\mathbf{K}(R_2^5, \mathcal{B})$ was created from a $\mathbf{K}(R_2^3, \mathcal{B})$ by carefully adding eight balls:

(a) In the cases $m = 3, 7, 11, \ldots$ we supplement the original balls of the $\mathbf{K}(R_2^m, \mathcal{B})$ with four new balls, whose centers, G_{m+2}, H_{m+2}, I_{m+2} and T_{m+2} lie in the plane p (see figure 8). Then we replace the path $G_m H_m I_m$ with the path $G_m G_{m+2} H_{m+2} I_{m+2} I_m$ and rename the point H_m to Q_{m+2}. Next we shift the points O_m and P_m "over" p into suitable positions O_{m+2} and P_{m+2} so that the balls around these points form a connection between the ball around N_m and the one around Q_{m+2}. Then we can add two more balls around selected points R_{m+2} and S_{m+2} to form a connection "under" p between the ball around Q_{m+2} and the one around T_{m+2} and similarly we can add two balls "over" p to make a connection between the ball at T_{m+2} and the one around Q_m (where $Q_3 := A$).

(b) In the cases $m = 5, 9, 13, \ldots$ we supplement the original balls of the $\mathbf{K}(R_2^m, \mathcal{B})$ by four new balls with their centers, G_{m+2}, H_{m+2}, I_{m+2} and N_{m+2}, lying in the plane p (see figure 8). So the path $G_m H_m I_m$ gets replaced by the path $G_m G_{m+2} H_{m+2} I_{m+2} I_m$ and we rename H_m to Q_{m+2}. Next we move the points R_m and S_m "under" p over to suitable points R_{m+2} and S_{m+2} so that the balls around these points form a connection between the ball around Q_{m+2} and the ball around T_m. Two more balls centered at the appropriate points Q_{m+2} and P_{m+2} "over" p suffice to form a connection between the balls around N_{m+2} and Q_{m+2}; a final two balls at L_{m+2} and M_{m+2} "under" p will connect the ball around Qm with the one around N_{m+2}.

So given

$$n_0(R_2^m, \mathcal{B}) \leq 4 \cdot m + 4 \quad (m = 3, 5, \ldots)$$

it follows that

$$n_0(R_2^{m+2}, \mathcal{B}) \leq 4 \cdot m + 4 + 8 = 4(m + 2) + 4.$$

Hence by the Principle of Mathematical Induction we have Theorem 4.

Of particular importance — particularly regarding applications in Chemistry — is the investigation of solid knots formed from special congruent ball pieces — the so-called "kalottes". These are formed by taking a ball, $\mathcal{B}_1$ of radius r_1 and slicing it with two planes, p_1 and p_2 which are at an angle of $\frac{\pi}{3}$ to each other and are both tangent planes of a ball, $\mathcal{B}_2$, which is concentric with $\mathcal{B}_1$ and has radius $r_2 = \frac{\sqrt{3}}{3}r_1$. These shapes model the situation when the diameters of the normal atomic spheres are larger than the distances between centers of neighboring atoms in certain bonds so that neighboring atomic spheres intersect. As the original concept of the kalotte is one of intersecting balls, we will assume in what follws that two neighboring kalottes intersect in a disk. Then for solid knots formed from these kalottes, $\mathcal{K}_a$, we have

Theorem 5: Any solid knot built from congruent kalottes around a core $K = R_2^m$ $(m = 3, 5, \ldots)$ can be formed with $13m + 9$ kalottes; thus we have the bound

$$n_0(R_2^m, \mathcal{K}_a) \leq 13m + 9 \quad (m = 3, 5, \ldots). \tag{4}$$

Proof The construction of solid knots from the above congruent kalottes can be based on the construction of solid knots formed from congruent balls. If the core of a solid knot makes an angle of $\frac{2\pi}{3}$ within a ball of radius r centered at M, we can replace the ball with three kalottes (see figure 9). But we cannot simply multiply the number of balls of the solid

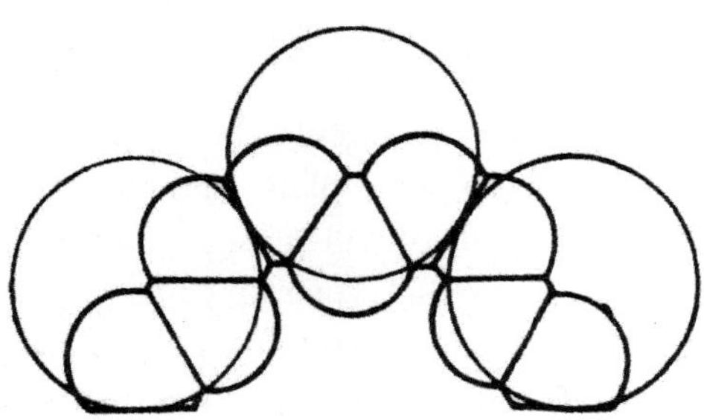

Fig. 9.

knot $\mathbf{K}(R_2^3, \mathcal{B})$ of figure 7 by 3 to get an upper bound for $n_0(R_2^3, \mathcal{K}_a)$ since the core R_2^3 doesn't form an angle of $\frac{2\pi}{3}$ at B, E, K and at least one of N_3, P_3, O_3 or A. So we select for the polygonal knot of type R_2^3 the new arrangement given in figure 10.

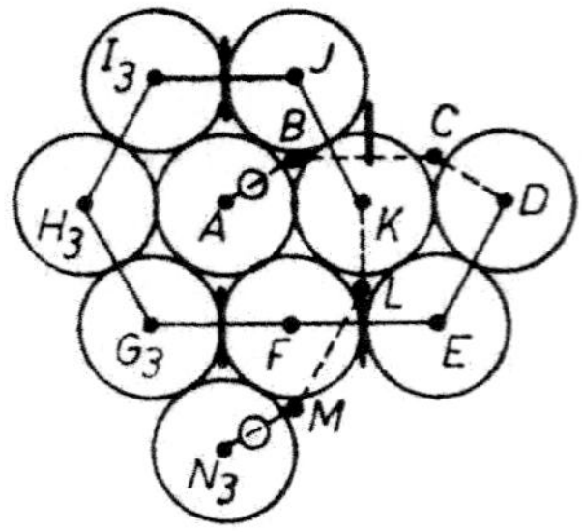

Fig. 10.

We place three kalottes in each of the balls around J, K, L, M and B in figure 9; we can do this because the angles at corners J, K, L, M and B of the polygonal knot each have an angle of $\frac{2\pi}{3}$. Four kalottes, in the arrangement shown in figure 11a, replace the ball around F. Since the knot makes an angle of $\frac{\pi}{2}$ at D, we remove the balls around C, D and E and replace them by a "bridge" formed from nine kalottes as in figure 11b; that is possible because the point B' (the former point of contact between around B and the ball around C) and the point E' (the former point of contact between the ball around E and the one around F) are separated by a distance $|B'E'| = r \cdot 2 \cdot \sqrt{2} = r \cdot 2.8\ldots$ and the "bridge" can adjust to any length from $2 \cdot \sqrt{3} \cdot r$ through $\frac{2}{3} \cdot \sqrt{3} \cdot r$ to meet up with B'' and E'' by suitable turns of kalottes 1 and 2 around the axis a_2 and kalottes 8 and 9 around axis a_1.

Then at A' (the point of contact betweeen the ball around A and the one around B) and at M' (the point of contact between the balls at M and N_3) the respective kalottes have a circular face lying parallel to the plane p; furthermore, they lie in the same plane and the distance between the centers of those faces — which we will term $\bar{A}'$ and $\bar{M}'$ respectively — is $|\bar{A}'\bar{M}'| = |AN_3| = 2 \cdot r \cdot \sqrt{3} = r \cdot 3.4\ldots$ so we connect the kalotte at A' and the kalotte at M' with a "bridge" built from eleven kalottes as in figure 11c; that is possible because in this "bridge" the distance between $\bar{A}''$ and $\bar{M}''$ can be set to any value between $r \cdot 4 \cdot \frac{2}{3} \cdot \sqrt{3}$ and $r \cdot 2 \cdot \frac{2}{3} \cdot \sqrt{3}$ by suitable turns of kalottes 1 and 2 around axis a_2 and kalottes 10 and 11 around axis a_1.

Although the knot forms an angle of $\frac{2\pi}{3}$ at each of the corners G_3, H_3 and I_3 and we could place the usual three kalottes in each place, we instead replace those three balls with a "bridge" built of seven kalottes as in figure 11d. That is possible because the points F'' and I_3'' of the "bridge" naturally

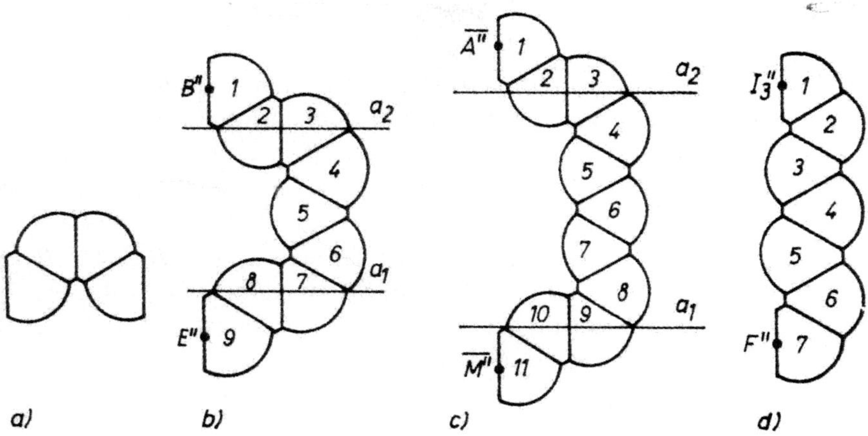

Fig. 11.

have a distance $r \cdot 2 \cdot \sqrt{3}$ from the seven kalottes and because the point F' (the former point of contact between the ball around F and the ball around G) and the point I_3' (which was where the ball around I_3 met up with the ball around J) were also $r \cdot 2 \cdot \sqrt{3}$ apart.

With the procedure described we have arranged

$$5 \cdot 3 + 4 + 9 + (2 + 11) + 7 = 48 = 13 \cdot 3 + 9$$

congruent kalottes in such a way that every kalotte shares a circular face with each of two other kalottes. If one imagines the center of each kalotte connected by an edge with the center of the two neighbor kalottes which share a face with it, then one has a closed path of 48 edges forming a polygonal R_2^3 knot. This forms the solid knot $\mathbf{K}(R_2^3, \mathcal{K}_a)$ and so we have $n_0(R_2^3, \mathcal{K}_a) \leq 13 \cdot 3 + 9$.

In forming a $\mathbf{K}(R_2^5, \mathcal{K}_a)$ from a $\mathbf{K}(R_2^3, \mathcal{K}_a)$, we first (as instructed in the proof of Theorem 4) add four balls to the $\mathbf{K}(R_2^3, \mathcal{B})$ shown in figure 10 around the points G_5, H_5, I_5 and T_5. Then we replace the balls around G_3 and I_3 each with the four kalottes of figure 11a. The balls around G_5, H_5 and I_5 are again replaced by the "bridge" of seven kalottes found in figure 11d. The balls around O_5, P_5, U_5 and V_5 can be placed in such a way that the polygonal R_2^5 has an angle of $\frac{2\pi}{3}$ at N_3, O_5, P_5, U_5 and A so each of these six balls can be replaced with three kalottes as desired. For the connection between the ball around Q_5 and the ball around T_5 we use the same "bridge construction" underneath the plane p as with the connection between A and N_3 above p in building $\mathbf{K}(R_2^3, \mathcal{K}_a)$, i.e., we need two kalottes and also a "bridge" of eleven kalottes. The remaining balls of the $\mathbf{K}(R_2^5, \mathcal{B})$ are replaced with kalottes in the same way as during the building of the solid knot $\mathbf{K}(R_2^3, \mathcal{K}_a)$. So we added $2 \cdot 4 + 6 \cdot 3$ kalottes to the 48 kalottes of the solid knot $\mathbf{K}(R_2^3, \mathcal{K}_a)$ in order to get the solid knot $\mathbf{K}(R_2^5, \mathcal{K}_a)$. Since $48 + 2 \cdot 4 + 6 \cdot 3 = 48 + 2 \cdot 13$ we have $n_0(R_2^5, \mathcal{K}_a) \leq 13 \cdot 5 + 9$.

The construction given in the proof of Theorem 4 lets a $\mathbf{K}(R_2^{m+2}, \mathcal{B})$ be built from a $\mathbf{K}(R_2^m, \mathcal{B})$ (for $m = 3, 5, \ldots$) by the addition of 8 balls, but for each m those eight balls are situated as above, causing $2 \cdot 4 + 6 \cdot 3 = 2 \cdot 13$ kalottes to be added in the transition from a $\mathbf{K}(R_2^m, \mathcal{K}_a)$ to a $\mathbf{K}(R_2^{m+2}, \mathcal{K}_a)$, i.e.,

$$n_0(R_2^m, \mathcal{K}_a) \leq 13 \cdot m + 9$$

implies that

$$n_0(R_2^{m+2}, \mathcal{K}_a) \leq 13 \cdot (m + 2) + 9$$

for each $m = 3, 5, \ldots$ and so Theorem 5 is established.

For a "solid trefoil" from congruent kalottes — the name we want to use for $\mathbf{K}(R_2^3, \mathcal{K}_a)$ — it follows from Theorem 5 that 48 kalottes are sufficient. (In [1], H. L. Frisch and E. Wassermann determine that a Cycloparaffin must consist of at least 50 carbon atoms to form a "trefoil". Here "trefoil" means the same as our $\mathbf{K}(R_2^3, \mathcal{K}_a)$. The 50 carbon atoms mentioned are understood to partially intersect, so the kalotte shape described is appropriate to the situation.)

The considerations and construction procedures for solid knots made of congruent components which we have so far can be transferred over to links formed from congruent-component solid knots. For cubes, for balls and for the described ball pieces (kalottes), some such links have been studied — ones where the "rosette links of second order," R_2^m ($m = 2, 4, \ldots$), form the core; the simplest representative of the "Rosette links of second order" is R_2^2 — two linked rings. Regarding links formed from congruent-component solid knots we can also ask the question about the minimum number of components necessary for a given "link core" of type V and congruent components of kind B. We designate this minimum number $n_0(V, B)$; it is worth noting that each link consists of several knots which do not all have to contain the same number of components. We include here, without proof, the following bounds for $V := R_2^m$ $(m = 2, 4, \ldots)$, $B := \mathcal{C}$ (cubes), $B := \mathcal{B}$ (balls) and $B := \mathcal{K}_a$ (kalottes — the shape described above):

$$n_0(R_2^m, \mathcal{C}) \leq 6 \cdot m + 4 \quad (m = 2, 4, \ldots), \tag{5}$$

$$n_0(R_2^m, \mathcal{B}) \leq 4 \cdot m + 2 \quad (m = 4, 6, \ldots), \tag{6}$$

$$n_0(R_2^m, \mathcal{K}_a) \leq 13 \cdot m + 4 \ (m = 4, 6, \ldots). \tag{7}$$

(Each of the links of congruent-component solid knots above consists of two pieces; during the constructions a particularly simple representation was chosen for each piece.)

Remark 1 In the case $m = 2$, instead of the bound (6) we have the actual value

$$n_0(R_2^2, \mathcal{B}) = 12.$$

Supposing that we have five balls arranged so that each ball touches two others and their centers are at the corners of a regular pentagon, is it possible to place a sixth ball so that it touches, inside the pentagon, the plane containing the centers of the other balls? Well, the height h of a pyramid whose base is a regular pentagon and whose sides are equilateral triangles with sidelengths a is

$$h = \frac{a}{2}\sqrt{2\left(1 - \frac{\sqrt{5}}{5}\right)} > \frac{a}{2}.$$

Remark 2 It is possible to build a trival solid knot (i.e. a solid knot whose core is equivalent to a circle) with 18 kalottes (see figure 9) and to link this with a second similar knot. In the work mentioned [1], it turns out the simplest catenane in the form of two linked rings (see figure 2) needs at least 20 carbon atoms per ring. It is also mentioned that it is possible for two 18-unit carbon rings to link with each other but due to the inter-atomic repulsions two more atoms must be inserted in each ring. Furthermore it turns out in [1] that at least 66 $(= 20 + 26 + 20)$ carbons atoms are required for a chain of three carbon rings. If one imagines such a chain from three trivial congruent-kalotte knots, it is easy to see that the minimum number of kalottes necessary is not more than 62 $(= 18 + 26 + 18)$.

4. Solid Knots of Uniform Thickness d

We now turn our attention to solid knots of uniform thickness d. We get an estimate for the value of $l_0(R_2^m, d)$ from examining solid knots formed from congruent cubes. It is possible to embed into any solid knot $\mathbf{K}(R_2^m, \mathcal{C})$ which was designed according to the plan in the proof of Theorem 2 a solid knot of uniform thickness d, where d is the edge length of the cubes. The R_2^m core of the $\mathbf{K}(R_2^m, \mathcal{C})$ we designed consists of $6m+6$ edges each of length d. This core, in the case $m = 3$, forms a right angle at 13 of its vertices and and a straight angle at the other 11. The method used to produce a $\mathbf{K}(R_2^{m+2}, \mathcal{C})$ from a $\mathbf{K}(R_2^m, W)$ $(m = 3, 5, \dots)$ involved adding twelve cubes. The core R_2^{m+2} which arose from the core R_2^m in this way had twelve new edges which met at five of the new vertices in right angles and at seven of them in straight angles. So the core of a $\mathbf{K}(R_2^m, \mathcal{C})$ $(m = 3, 5, \dots)$ makes

$$\text{a right angle at } 13 + \frac{m-3}{2} \cdot 5 \text{ vertices}$$

and

$$\text{a straight angle at } 11 + \frac{m-3}{2} \cdot 7 \text{ vertices.}$$

In each cube containing a vertex at which the core knot is straight, we simply keep the core as it is. In the cubes where the core makes a right angle, we replace the portion of the core in the cube with a circular elbow of radius $\frac{d}{2}$ in such a fashion that the resulting closed curve forms the core of a $\mathbf{K}(R_2^m, d)$ embedded in the $\mathbf{K}(R_2^m, \mathcal{C})$. For a solid knot of uniform thickness d thus constructed, we have now found that

$$l_0(R_2^m, d) \leq \left(11 + \frac{m-3}{2} \cdot 7\right) \cdot d + \left(13 + \frac{m-3}{2} \cdot 5\right) \cdot \frac{\pi}{4} \cdot d \qquad (m = 3, 5, \dots).$$

So then we immediately have the following theorem:

Theorem 6: For any given solid knot of uniform thickness d with core knot R_2^m $(m = 3, 5, \dots)$,

$$l_0(R_2^m, d) \leq \left(\frac{1+7m}{2} + \frac{11+5m}{2} \cdot \frac{\pi}{4}\right) \cdot d. \tag{8}$$

Hence in the $m = 3$ case we get a bound for the "solid trefoil" of uniform thickness d of

$$l_0(R_2^3, d) \leq \left(11 + 13 \cdot \frac{\pi}{4}\right) \cdot d,$$

i.e.

$$l_0(R_2^3, d) \leq d \cdot 21.210\ldots$$

We wish to examine in the following, as well, the solid knots of greatest possible uniform thickness d which can be embedded in a $\mathbf{K}(R_2^m, \mathcal{K}_a)$ developed according to the construction given in the proof to Theorem 5. Every such $\mathbf{K}(R_2^m, \mathcal{K}_a)$ consists of $13m+9$ kalottes and the core R_2^m makes an angle of $\frac{2\pi}{3}$ in the center of each kalotte. In each kalotte we replace the portion of the core within it with a $\frac{\pi}{3}$-radian circular arc[k] with radius $\sqrt{3} \cdot r_2$ (the *kalotte* under consideration, according to the definitions in section 3 is a ball $\mathcal{B}_1$ with center C_1 of radius r_1 truncated by two planes, p_1 and p_2, which meet at an angle of $\frac{\pi}{3}$ and are tangential to a ball $\mathcal{B}_2$ around C_1 of radius $r_2 := \frac{1}{3} \cdot \sqrt{3} \cdot r_1$); the radius $\sqrt{3} \cdot r_2$ was selected so that it is equal to the distance from the intersection of planes p_1 and p_2 to the centers of the circular faces. If the arcs are inserted into each kalotte in such a way that they meet the kalotte's circular faces perpendicularly in their centers, then the $13m+9$ arcs form a closed curve with knot type R_2^m. Since the circular faces of each kalotte have radius $\rho = \sqrt{{r_1}^2 - {r_2}^2} = \frac{1}{3} \cdot \sqrt{6} \cdot r_1$, we have thus found a solid $\mathbf{K}(R_2^m, d)$ knot of uniform thickness $d = \frac{2}{3} \cdot \sqrt{6} \cdot r_1$, which is embedded in $\mathbf{K}(R_2^m, \mathcal{K}_a)$; with the given embedding, a larger thickness d is not possible. The length of l of the core knot of the $\mathbf{K}(R_2^m, d)$ built in this fashion obeys

$$\begin{aligned} l &= (13m+9) \cdot \frac{2\pi}{6} \cdot \sqrt{3} \cdot r_2 = (13m+9) \cdot \frac{\pi}{3} \cdot \sqrt{3} \cdot \frac{1}{3}\sqrt{3} \cdot r_1 \\ &= (13m+9) \cdot \frac{\pi}{3} \cdot r_1 = (13m+9) \cdot \frac{\pi}{3} \cdot \frac{1}{4}\sqrt{6} \cdot d. \end{aligned}$$

This is a worse bound for $l_0(R_2^m, d)$ than the one given in Theorem 6. In the case $m = 3$ — the "solid trefoil" of uniform thickness d — this gives a length for the core of

$$l = 4 \cdot \pi \cdot \sqrt{6} \cdot d = d \cdot 30.786\ldots \text{[sic]}$$

The results found were appplied also to links composed of knots with uniform thickness d whose core are the rosette links of second order R_2^m ($m = 2, 4, \ldots$). We state without proof that from the embedding of a link of solid knots $\mathbf{K}(R_2^m, d)$ of uniform thickness d into a link of solid

[k] German *Sechstelkreisbogen*. *Sechstel* means "one-sixth" and *Kreisbogen* is a circular arc.

knots $\mathbf{K}(R_2^m, \mathcal{C})$ of congruent cubes, one can get the bound

$$l_0(R_2^m, d) \leq \left(\frac{7m}{2} + \frac{8+5m}{2} \cdot \frac{\pi}{4}\right) \cdot d \qquad (m = 4, 6, \dots) \tag{9}$$

In the case $m = 2$ simple considerations lead directly to

$$l_0(R_2^2, d) = 2 \cdot \pi \cdot d.$$

References

1. H. L. Frisch and E. Wasserman. Chemical Topology. *J. Amer. Chem. Soc*, 83:3789–3795, 1961.
2. O. Krötenheerdt. Über einen speziellen Typ alternierender Knoten. *Math. Ann.*, 153:270–284, 1964.
3. O. Krötenheerdt. Zur Lösung des Isotopieproblems der Rosettenknoten. *Beiträge zur Algebra und Geometrie*, 1:19–31, 1971.
4. O. Krötenheerdt. Probleme der Lagerung von Figuren. *Mitteil. Math. ges. DDR*, Heft 3/4:65–74, 1975.
5. G. Schill and C. Zürcher. Mechanisch verknüpfte Moleküle: Catenane und Rotaxane. *Die Naturwissenschaft*, 58:Jan., 1971.

CHAPTER 2

A TUTORIAL ON KNOT ENERGIES

E.J. Janse van Rensburg
Department of Mathematics and Statistics
York University, Toronto, Ontario, M3J 1P3, Canada
E-mail: rensburg@yorku.ca

In this paper I give a tutorial of knots energies from a mathematical point of view. A wide variety of knot energies that appeared in the literature is reconsidered, and put into context by discussing their properties. These energies include lattice knot, polygonal knot, and energies of C^1 and C^2 knots. In this paper I discuss the properties of a variety of knot energies proposed in the literature.

1. Introduction to Knot Energies

A *knot energy* is a non-negative real function defined on the space of conformations of an embedded simple closed curve in three space. The *knotting* of such a curve is a topological problem, but in this tutorial I shall primarily focus on the *energy* of conformations of simple closed curves; this is a geometrical question. There has been significant progress in the understanding of knot energies since their introduction by Fukuhara in 1988.[23] A systematic study of knot energies by Diao *et. al.* appeared in *Topology and Geometry in Polymer Science*.[15] A similar study on the energies of polygonal knots can be found in a subsequent paper.[16]

The scientific motivation for the study of knot energies is found in the study of knots in DNA and polymers,[2] the study of electrophoretic mobility of knotted DNA,[35,64] and the knotted flow and field lines in fluids and electromagnetism.[41,43] The role that tangled and knotted DNA plays in cell division, and the manipulation of knotted DNA by topoisomerases, showed that knots and tangles are an important part in the study of DNA.[2,10,65]

Progress on understanding the separation of knotted DNA by electrophoresis [64] was made by modeling knot energies and relating the results to the electrophoretic gel velocity.[62] Further studies into the electrophoresis of knotted polymers can be found in the literature,[35] and knots in charged polymers continues to receive attention in the literature.[21]

Energies of knots were also considered from a fluid mechanics and electromagnetic point of view.[41] The ideas leading to knot energies from this point of view are originally due to Kelvin's vortex atom theory, and is traceable to work by Maxwell and Gauss.[56]

More recently, implementations of graphic tools on the computer to visualize knots (such as KnotPlot and Evolver) made it possible to endow knotted loops with an energy. Gradient flows along the energy landscape to local minima were supposed to produce canonical conformations of the knots. A variety of energy functions were introduced for this purpose, including the *normal energy* and the *symmetric energy*.[3,59]

In the context of the minimization of knot energies as in the Evolver and in KnotPlot, one may consider "canonical conformations" of knotted curves, or "ideal knots", as knotted curves at local minima of a knot energy. These conformations are quite distinct from those projected in standard texts on knot theory.[57] The local minima of knot energies could lead to new knot invariants.[61,62] However, a "canonical conformation" of a knot is still an undefined concept, and it is not even clear what the properties of an "ideal knot" should be. Knot energies have illuminated this question somewhat. However, I shall not address the issue in this tutorial, and focus instead on the properties of knot energies. Putting these on a firm mathematical footing could create the circumstances for addressing ideal knots.

In Section 2 a number of Examples of knots energies are discussed. I focus in particular on the energies of polygonal knots, since those can be studied computationally. However, curvature energies, the Gromov distortion and Möbius energy, as well as thickness energies, all energies of C^2 or C^1 curves, are also reviewed. In Section 3 the general properties of polygonal knot energies are studied. The definitions of "good" knot energies (in this case they are "semi-ideal" or "ideal") are given. Examples of knot energies that either fail or pass as semi-ideal or as ideal are discussed. In Section 4 the notion of thickness of a curve is considered somewhat closer. This issue will be first considered in Section 2, and remains one of the more interesting and important of knot energies, and it (still!) deserves further study.

2. Examples of Knot Energies

Let $\mathcal{S}$ be the set of simple closed curves, or let $\mathcal{S}$ be a set of polygons (equilateral or otherwise). A function f on $\mathcal{S}$ is scale invariant if for any non-negative real number r one has $f(r\,S) = f(S)$ for any $S \in \mathcal{S}$. A *knot energy* is a scale invariant function f from $\mathcal{S}$ into non-negative real numbers:

$$f : \mathcal{S} \to \mathbb{R}^+, \tag{1}$$

where $\mathbb{R}^+$ is the set of all non-negative real numbers. The energy of the curve $S \in \mathcal{S}$ is $f(S)$.

Generally, each curve $S \in \mathcal{S}$ has a well-defined topological knot type $\mathcal{K}$. Let $\mathcal{S}_\mathcal{K}$ be the set of all curves in $\mathcal{S}$ with knot type $\mathcal{K}$. The energy of the knot type $\mathcal{K}$ is the infimum of f over all curves of knot type $\mathcal{K}$; by abusing notation, this is defined by

$$f(\mathcal{K}) = \inf\{f(S) \,|\, \text{ for all } S \in \mathcal{S}_\mathcal{K}\ \}. \tag{2}$$

An "ideal knot" with respect to the energy f exists if the infimum is realised on a certain curve $C_\mathcal{K} \in \mathcal{S}_\mathcal{K}$. In other words, if there exists a curve $C_\mathcal{K}$ of knot type $\mathcal{K}$ such that

$$f(C_\mathcal{K}) = \min\{f(S) \,|\, \text{ for all } S \in \mathcal{S}_\mathcal{K}\ \}, \tag{3}$$

then $C_\mathcal{K}$ is an ideal knot of knot type $\mathcal{K}$ with respect to the knot energy f.

Example 1: [Crossover Energy] Define $c_r(S)$ to be the minimum crossover number of the curve S over all its possible regular projections. Then $c_r(S)$ is an energy function of the curve S. The crossing energy of the knot type $\mathcal{K}$ is defined by

$$c_r(\mathcal{K}) = \min\{c_r(S) \,|\, S \text{ is a simple closed curve of knot type } \mathcal{K}\}. \tag{4}$$

The energy of a knot type $\mathcal{K}$ is its minimal crossing number. For example, for the unknot ϕ, $c_r(\phi) = 0$, while for the trefoil knot-type, $c_r(3_1) = 3$. Every simple closed curve of knot type $\mathcal{K}$ that admits a regular projection with crossing number equal to the crossover number of $\mathcal{K}$ is an ideal knot. Standard knot tables usually order knot types according to $c_r(\mathcal{K})$. So for example the knot type $\mathcal{K}$ with designation N_ℓ is the ℓ-th knot in the table with $N = c_r(\mathcal{K})$. △

Example 2: [Stick Energy] Let $\mathcal{P}$ be the set of all polygonal closed curves. Let $p \in \mathcal{P}$, and define $s(p)$ to be equal to the number of edges (or sticks) in p. The stick energy of knot-type $\mathcal{K}$ is the minimal number of edges of a

Table 1. Upper Bounds on Stick Numbers

$s(\mathcal{K})$	$\mathcal{K}$
3	ϕ^*
6	3_1^*
7	4_1^*
8	$5_1^*, 5_2^*, 6_1^*, 6_2^*, 6_3^*, 8_{19}^*, 8_{20}^*, (3_1^+\#3_1^+)^*, (3_1^+\#3_1^-)^*$
9	$7_1^*, 7_2^*, 7_3^*, 7_4^*, 7_5^*, 7_6^*, 7_7^*, 8_{16}^*, 8_{17}^*, 8_{18}^*, 8_{21}^*, 9_{29}^*, 9_{34}^*$
	$9_{40}^*, 9_{41}^*, 9_{42}^*, 9_{44}^*, 9_{46}^*, 9_{47}^*, 9_{49}^*$
10	$8_1, 8_2, 8_3, 8_4, 8_5, 8_6, 8_7, 8_8, 8_9, 8_{10}, 8_{11}, 8_{12}, 8_{13}, 8_{14}, 8_{15}$
	$9_1, 9_4, 9_5, 9_7, 9_8, 9_9, 9_{10}, 9_{12}, 9_{13}, 9_{14}, 9_{16}, 9_{17}, 9_{19}, 9_{20}, 9_{22}$
	$9_{24}, 9_{26}, 9_{28}, 9_{30}, 9_{31}, 9_{32}, 9_{33}, 9_{35}, 9_{37}, 9_{38}, 9_{39}, 9_{43}$
	$9_{45}, 9_{48}, (3_1\#4_1), (3_1^+\#3_1^+\#3_1^\pm)^*$
11	$9_2, 9_3, 9_6, 9_{11}, 9_{15}, 9_{18}, 9_{21}, 9_{23}, 9_{25}, 9_{27}, 9_{36}, (3_1^+\#5_1^+)$
	$(3_1^+\#5_1^-), (3_1^+\#5_2^+), (3_1^+\#5_2^-), (4_1\#4_1), (3_1^+\#6_1^+)$
	$(3_1^+\#6_1^-), (3_1^+\#6_2^+), (3_1^+\#6_2^-), (3_1\#6_3)$
12	$(4_1\#5_1), (4_1\#5_2), (4_1\#6_1), (4_1\#6_2), (4_1\#6_3)$

polygon that can be tied to have knot type $\mathcal{K}$. It is known that $s(\phi) = 3$ and $s(3_1) = 6$.[49,50,51] Further investigations showed that $s(4_1) = 7$,[39] and yet more upper bounds on stick numbers are listed in Table 1. These are due to Rawdon and Scharein,[55] and coordinates for such minimal knots can be found in that paper. The known stick numbers in Table 1 are labeled by an asterisk; these are all due to the work of Calvo and Millett[8] and Rawdon and Scharein.[55] The other estimates are all upper bounds on the stick numbers. These upper bounds could be made sharp by ruling out the possibility that a given knot can be realised in a polygon with a smaller number of edges.

Realisations of octagonal knots in Table 1 are due to Calvo and Millett[8] and Calvo ruled out other octagonal knots except 8_{18},[5,7] and later showed that $s(8_{18}) = 9$. Nine-stick knots were constructed by Meissen[38] and Scharein.[58]

For a knot type $\mathcal{K}$, the stick number have also been bound by the crossing number:[42,24,48]

$$\frac{1}{2}\left(5 + \sqrt{9 + 8c_r(\mathcal{K})}\right) \leq s(\mathcal{K}) \leq c_r(\mathcal{K}) + 2. \tag{5}$$

The lower bound was improved by Calvo to $(7 + \sqrt{1 + 8c_r(\mathcal{K})})/2$ Calvo.[6] In the case of torus knots there also are the bounds

$$\begin{aligned} 2\min\{2p, q\} &\leq s(T_{p,q}) \leq p\left\lceil \frac{2q}{p} \right\rceil; \\ 6r + 2 &\leq s(T_{2r+1,3r+1}) \leq 6r + 3, \end{aligned} \tag{6}$$

due to Jin and Kim.[32] Any embedding of a polygon that realises a polygonal

Table 2. Upper Bounds on Equilateral Stick Numbers

$s_e(\mathcal{K})$	$\mathcal{K}$
3	ϕ^*
6	3_1^*
7	4_1^*
8	$5_1^*, 5_2^*, 6_1^*, 6_2^*, 6_3^*, 8_{20}^*$
9	$7_1^*, 7_2^*, 7_3^*, 7_4^*, 7_5^*, 7_6^*, 7_7^*, 8_{16}^*, 8_{17}^*, 8_{18}^*, 8_{19}, 8_{21}^*, 9_{34}^*$ $9_{40}^*, 9_{41}^*, 9_{42}^*, 9_{44}^*, 9_{46}^*, 9_{47}^*, 9_{49}^*, (3_1^+\#3_1^+)$
10	$8_1, 8_2, 8_3, 8_4, 8_5, 8_6, 8_7, 8_8, 8_9, 8_{10}, 8_{11}, 8_{12}, 8_{13}, 8_{14}, 8_{15}$ $9_1, 9_4, 9_5, 9_7, 9_8, 9_9, 9_{10}, 9_{12}, 9_{13}, 9_{14}, 9_{16}, 9_{17}, 9_{19}, 9_{20}, 9_{22}$ $9_{24}, 9_{26}, 9_{28}, 9_{29}, 9_{30}, 9_{31}, 9_{32}, 9_{33}, 9_{35}, 9_{37}, 9_{38}, 9_{39}, 9_{43}$ $9_{45}, 9_{48}, (3_1^+\#3_1^-)$
11	$9_2, 9_3, 9_6, 9_{11}, 9_{15}, 9_{18}, 9_{21}, 9_{23}, 9_{25}, 9_{27}, 9_{36}, (3_1\#4_1)$

knot of type $\mathcal{K}$ with stick energy equal to $s(\mathcal{K})$ is an ideal conformation with respect to the stick energy. Coordinates can be found in Rawdon and Scharein.[55] △

Example 3: [Equilateral Stick Energy] Consider again the set of polygonal knots, but this time only that subset of equilateral polygonal knots $\mathcal{P}_e$. Let $s_e(p)$ be the number of edges in a polygon $p \in \mathcal{P}_e$. The *equilateral stick energy* $s_e(\mathcal{K})$ of knots of knot type $\mathcal{K}$ is the minimum number of edges in an equilateral polygon that may be tied into a knot of type $\mathcal{K}$. It is known that $s_e(\emptyset) = 3$ and $s_e(3_1) = 6$. Notice that it is thought that $s(3_1^+\#3_1^+) = 8 < 9 = s_e(3_1^+\#3_1^+)$, and also that[55] $s(8_{19}) = 8 < 9 = s_e(8_{19})$. More estimates and results[8] are listed in Table 2. Equilateral stick numbers marked with an asterisk are known to be exact, the remaining numbers are upper bounds. In the case of general (non-equilateral) polygons there are better bounds for stick numbers in Rawdon and Scharein's data.[55] In particular, the knots 8_{19}, 9_{29}, 10_{16}, 10_{79}, 10_{107}, 10_{119}, and 10_{147} have estimates for $s(\mathcal{K})$ smaller than the corresponding estimates for $s_e(\mathcal{K})$. There are conjectures that $s(\mathcal{K}) < s_e(\mathcal{K})$ for some knots types, but these remain unproven. Much less are known about composite knots, and fewer bounds are known for the equilateral compounded stick knots than for the general case, if one compares Table 2 with 1. An equilateral stick knot is ideal with respect to the equilateral stick energy if it is realized with minimum $s_e(\mathcal{K})$. The coordinates of such conformations are given in Rawdon and Scharein.[55] △

Example 4: [Lattice Edge Energy] Consider the space of polygons $\mathcal{P}_L$ with edges and vertices in the cubic lattice $\mathbb{Z}^3$. Injecting these polygons into $\mathbb{R}^3$ define their knot types unambiguously. The (cubical) lattice edge

Table 3. Upper Bounds on Square Lattice Edge Numbers

$m(\mathcal{K})$	$\mathcal{K}$
4	ϕ^*
24	3_1^*
30	4_1
34	5_1
36	5_2
40	$6_1, 6_2, 6_3$
42	8_{19}
44	$7_1, 7_3, 7_4, 7_7, 8_{20}$
46	$7_2, 7_5, 7_6, 8_{21}$
48	$8_3, 8_7, 9_{42}$
50	$8_1, 8_2, 8_4, 8_5, 8_6, 8_9, 8_{13}, 8_{16}, 9_{47}, 10_{161}$
52	$8_8, 8_{10}, 8_{11}, 8_{12}, 8_{14}, 8_{15}, 8_{17}, 8_{18}$
54	9_1
56	9_2
60	10_1

energy $m(p)$ of lattice polygon $p \in \mathcal{P}_L$ is the number of edges in p. The lattice edge energy of the knot type $\mathcal{K}$ is the minimum number of edges of any lattice knot with knot type $\mathcal{K}$. It is known that $m(\phi) = 4$, and $m(3_1) = 24$.[13]

In Table 3 some numerical estimates of the lattice edge energy of some cubical lattice knots are listed. Known results are marked by an asterisk. It is known that $m(\mathcal{K}) > 4$ if $\mathcal{K} \neq \phi$, and $m(\mathcal{K}_1 \# \mathcal{K}_2) \leq m(\mathcal{K}_1) + m(\mathcal{K}_2)$.[30]

It is also known that[30]

$$m(\mathcal{K}) \geq (3 + \sqrt{9 + 8c_r(\mathcal{K})})/2. \tag{7}$$

An upper bound is also known[20]

$$m(\mathcal{K}) \leq 136\,(c_r(\mathcal{K}))^{3/2} + 84c_r(\mathcal{K}) + 22\sqrt{c_r(\mathcal{K})} + 11. \tag{8}$$

Furthermore, if $q(m)$ is the number of knot types of lattice edge energy equal to m (for example, $q(4) = 1$, and $q(24) = 1$), and $Q(n) = \sum_{m \leq n} q(m)$ is the number of knot types of minimal lattice energy at most n, and $Q(n)$ grows at most exponentially with n. Indeed, the limit

$$\lim_{n\to\infty} [Q(n)]^{1/n} = \mathcal{Q} \tag{9}$$

exists, and $Q(n) \leq \mathcal{Q}^n$.[30]

If $\mathcal{K}^n = \mathcal{K}\#\mathcal{K}\#\ldots\#\mathcal{K}$ (the connected sum of n lattice knots of knot type $\mathcal{K}$ each), then the *minimal edge index* $\mu_\mathcal{K}$ is defined by

$$\mu_\mathcal{K} = \lim_{n\to\infty} \frac{1}{n} m(\mathcal{K}^n) \tag{10}$$

Table 4. Upper Bounds on Square Lattice Curvature

$c_r(\mathcal{K})$	$\mathcal{K}$
4	ϕ^*
12	3_1^*
14	4_1
17	$5_1, 5_2$
18	$6_2, 6_3, 9_{47}^*$
19	$6_1, 7_7, 8_{20}$
20	$7_4, 7_6, 8_{18}, 8_{19}, 8_{21}$
21	$7_2, 8_8, 8_{13}, 8_{17}, 9_{42}$
22	$7_1, 7_3, 8_2, 8_5, 8_7, 8_9, 8_{11}, 10_{161}$
23	$7_5, 8_4, 8_6, 8_{10}, 8_{13}, 8_{16}$
24	$8_1, 8_3, 8_{12}, 8_{15}$
26	9_2
27	9_1
29	10_1

and this limit always exists.[30] The best upper bound is $\mu_{3_1} \leq 17$. A good lower bound is not known, but it can be shown that $\mu_{\mathcal{K}} > 0$ for any non-trivial knot type $\mathcal{K}$.[29] △

Example 5: [Lattice Curvature Energy] The *lattice curvature energy* of a lattice knot K is the number $k(K)$ of right angles in K. The *minimal lattice curvature* energy of the knot type $\mathcal{K}$, denoted by $k(\mathcal{K})$, is the least number of right angles in any realisation of a lattice knot of type $\mathcal{K}$ in the cubic lattice:

$$k(\mathcal{K}) = \min\{k(K) \mid \text{ for all lattice polygons of knot-type } \mathcal{K}\}. \tag{11}$$

Clearly, $k(\phi) = 4$, while $k(3_1) = 12$.[31] It is known that

$$k(\mathcal{K}) \geq \left(3 + \sqrt{9 + 8c_r(\mathcal{K})}\right), \tag{12}$$

where $c_r(\mathcal{K})$ is the crossing energy of the knot type $\mathcal{K}$. It is known that $c_r(\mathcal{K}) \geq 12$ if $\mathcal{K} \neq \phi$, while $c_r(\mathcal{K}_1 \# \mathcal{K}_2) \leq c_r(\mathcal{K}_1) + c_r(\mathcal{K}_2)$.[31] If $b(\mathcal{K})$ is the bridge index of the knot type $\mathcal{K}$, then $c_r(\mathcal{K}) \geq 6b(\mathcal{K})$,[31] and a corollary of this is that $c_r(9_{47}) = 18$ (since 9_{47} is the first three bridge knot in the knot tables, and one may realise 9_{47} with 18 right angles as a lattice knot).

The limit

$$\nu_{\mathcal{K}} = \lim_{n\to\infty} \frac{1}{n} c_r(\mathcal{K}^n) \tag{13}$$

exists, where ν_K is the *minimal lattice curvature index* of the knot type $\mathcal{K}$.[31] The best upper bound on ν_{3_1} is $\nu_{3_1} \leq 7.36$.[29] It can be shown that

$\nu_{\mathcal{K}} > 0$ if $\mathcal{K} \neq \phi$. The estimated minimal lattice curvature energies of some knot types are in Table 4. Known values are indicated by an asterisk.

Lattice knots with minimal lattice curvature are ideal representations of knots with respect to this energy. △

Example 6: [Curvature Energies] Consider the space C^2 of differentiable simple and closed space curves. If $\alpha(t)$ is a unit-length curve in C^2 parameterized by the unit speed (or arc length) parameter $t \in I = [0,1)$, then $\alpha'(t) = 1$ and the curvature of $\alpha(t)$ at $t = t_0$ is given by $\kappa(t_0) = |\alpha''(t_0)|$. The total curvature κ_α of α is

$$\kappa_\alpha = \int_0^1 \kappa(t)dt \tag{14}$$

and one may define *curvature moments* by

$$\kappa_\alpha^{(p)} = \int_0^1 [\kappa(t)]^p \, dt; \qquad \text{for } p > 0 \text{ real.} \tag{15}$$

A *curvature energy* of the knot type $\mathcal{K}$ may be defined by the infimum

$$C_p(\mathcal{K}) = \inf_\alpha \left\{ \kappa_\alpha^{(p)} \,\middle|\, \alpha \in C^2 \text{ has knot type } \mathcal{K} \right\}. \tag{16}$$

Another curvature energy may be defined by

$$C_m(\mathcal{K}) = \inf_\alpha \left\{ \max_{t \in [0,1)} \{\kappa(t)\} \right\} \tag{17}$$

where the infimum is over all C^2 curves α of knot type $\mathcal{K}$.

The total curvature of a space curve of knot type $\mathcal{K}$ is related to the bridge number $b(\mathcal{K})$ of $\mathcal{K}$. In particular, $C_1(\mathcal{K}) = 2\pi b(\mathcal{K})$,[40] a fact easily seen by "lifting" a $b(\mathcal{K})$-bridge projection by its bridges into three dimensions. △

Example 7: [The Gromov Distortion] Suppose that C^1 is the space of differentiable closed space curves, and that $\alpha(t)$ is an arc-length parameterized unit-length curve in C^1. Define

$$g(s,t) = \frac{|s-t|}{\|\alpha(s) - \alpha(t)\|} \tag{18}$$

where $0 < |s-t| \leq 1/2$ and $\|\alpha(s) - \alpha(t)\|$ is the Euclidean distance between the points $\alpha(s)$ and $\alpha(t)$. It follows that $\lim_{|s-t| \to 0^+} g(s,t) = 1$, and so continuously extend $g(s,t)$ to $0 \leq |s-t| \leq 1/2$ by defining $g(s,s) = 1$. Define the *Gromov distortion*[26,27] of $\alpha(t)$ by

$$D_s(\alpha) = \sup_{\substack{s,t \in [0,1] \\ |s-t| \leq 1/2}} g(s,t). \tag{19}$$

This is scale invariant, and so is a knot energy functional. The infimum of $D_s(\alpha)$ can be used to define the Gromov distortion of a knot type:

$$D_s(\mathcal{K}) = \inf_{\alpha}\{D_s(\alpha) \,\big|\, \text{for all } C^1 \text{ closed curves of knot type } \mathcal{K}\}. \tag{20}$$

The Gromov distortion of a circle is $\pi/2$, and it can be shown that $D_s(\alpha) > \pi/2$ if α is not a geometric circle.[28] Therefore, $D_s(\phi) = \pi/2$, in the case of the unknot, and $D_s(\mathcal{K}) > \pi/2$ for any non-trivial knot type $\mathcal{K}$. There also exists infinitely many knot types with Gromov distortion bounded by a large enough finite fixed number M.[23] The Gromov distortion was also studied in references.[34,43] △

Example 8: [The Möbius energy] Suppose that $\alpha(t)$ is the parametric equation of a simple closed C^2 curve. Define

$$E_M(\alpha) = \int_\alpha \int_\alpha \left(\frac{1}{\|\alpha(s)-\alpha(t)\|^2} - \frac{1}{D^2(\alpha(s),\alpha(t))} \right) \|\alpha'(s)\| \|\alpha'(t)\| ds\, dt \tag{21}$$

where $D(\alpha(s), \alpha(t))$ is the shortest distance along α between the points $\alpha(s)$ and $\alpha(t)$. $E_M(\alpha)$ is the *Möbius energy*, [22,33,43,44] of the curve α. $E_M(\alpha)$ can be generalized to the O'Hara functional

$$E_j^p(\alpha) = \frac{1}{j} \left[\int_0^1 \int_0^1 \left(\frac{1}{\|\alpha(s)-\alpha(t)\|^j} - \frac{1}{D^j(\alpha(s),\alpha(t))} \right)^p ds\, dt \right]^{1/p} \tag{22}$$

where $\alpha(t)$ is now an arc-length parameterization of the curve. It can be shown that $E_M(\alpha) = 2\,E_2^1(\alpha) - 4$. It is known that most of this family of energies are (globally) minimized by circles.[1] The limit $\lim_{j\to 0^+} \lim_{p\to\infty} E_j^p$ is the Gromov distortion of the curve.[44]

The Möbius energy of a knot type $\mathcal{K}$ is the infimum of E_M over all simple closed C^2 curves α of knot type $\mathcal{K}$:

$$E_M(\mathcal{K}) = \inf_{\alpha\in\mathcal{K}} \int_0^1 \int_0^1 \left(\frac{1}{\|\alpha(s)-\alpha(t)\|} - \frac{1}{D(\alpha(s),\alpha(t))} \right) \|\alpha'(s)\| \|\alpha'(t)\|\, ds\, dt. \tag{23}$$

One may similarly define the energy of knot types $E_j^p(\mathcal{K})$.

It is known that $E_M(\alpha) = 4$ if and only if α is a circle, and that $E_m(\mathcal{K}) \geq 2\pi c_r(\mathcal{K}) + 4$,[22] where $c_r(\mathcal{K})$ is the crossing number or crossing energy of $\mathcal{K}$ as defined in Example 1. △

Example 9: [The Open Rope Energy] Let $\alpha(t)$ be an open smooth curve of length L with arc-length parameterization $t \in [0, L]$, and let $\mathcal{K}$ be a given knot type. Suppose furthermore that $\alpha(t)$ has the following properties:

- The endpoints of $\alpha(t)$ are at the points $(0,0,0)$ and $(0,0,\ell)$.
- The set $\alpha \setminus \{\alpha(0), \alpha(L)\}$ is strictly contained in the space bounded by the planes $z = 0$ and $z = \ell$.
- The tangents $\alpha'(t)$ are parallel to the z-axis at $t = 0$ and $t = L$.
- If B is the solid cylinder of radius L and axis containing the points $(0,0,0)$ and $(0,0,\ell)$, and with circular faces in the planes $z = 0$ and $z = \ell$, then (α, B) is a knotted arc-ball pair of knot type $\mathcal{K}$.

The ϵ-neighbourhood of α and B will be denoted by α_ϵ and B_ϵ. If $x = \alpha(t)$ is a point in α, define the plane Σ_x to contain x and be normal to $\alpha'(t)$. If ϵ is small enough, then the component of $\Sigma_x \cap \alpha_\epsilon$ containing x is a meridian disk of α_ϵ. ϵ is *nice* if (α, B) is homotopic to (α_s, B_s), and if every component of $\Sigma_x \cap \alpha_s$ containing x is a meridian disk, for all $0 < s < \epsilon$. The *open thickness* of α is defined by

$$t^o(\alpha) = \sup\{\epsilon \,|\, s \text{ is nice for all } 0 < s < \epsilon \text{ and } \epsilon < L\}. \tag{24}$$

The *open rope thickness* of the knot type $\mathcal{K}$ is defined by

$$t^o(\mathcal{K}) = \sup_\alpha \{t^o(\alpha) \,|\, \alpha \text{ has knot type } \mathcal{K}\}. \tag{25}$$

The open rope energy of $\mathcal{K}$ is defined by

$$E^o(\mathcal{K}) = \frac{[L-\ell]}{t^o(\mathcal{K})} \tag{26}$$

and note that the unit is *rope radii*. $E^o(\mathcal{K})$ may be interpreted as the excess rope needed to tie a knot of type $\mathcal{K}$ in a flexible unit radius rope stretched between parallel planes. The open rope energies have been estimated experimentally;[17] the results are listed in Table 5. The uncertainties in brackets are standard deviations. For example, one notes that a flexible unit radius rope of excess length 20.7 ± 0.3 would be enough to tie a trefoil between two planes. The uncertainties in the measurements are included in brackets; the digit states the uncertainty in the last digit in the measured average.

The general trends in Table 5 are similar to those obtained for the minimal edge numbers in Table 3. The non-alternating eight crossing knots can be tied with less rope, consistent with the minimal edge numbers. The exceptional nature of these knots are also demonstrated by other energies, and one could surmise that crossing number is not the most natural quantity for tabulating knots. Rigorous upper bounds on the rope energies can be obtained by considering the shortest lattice knots, and replacing their corners by arcs (quarter circles) of radius 1/2. For example, one may show

Table 5. Estimates of Open Rope Energies

n	3_n	4_n	5_n	6_n	7_n	8_n
1	20.7(3)	29.3(4)	37.4(4)	46.4(4)	55.8(9)	62.5(6)
2			37.4(3)	47.2(3)	55.2(5)	63.8(6)
3				46.9(3)	56.7(6)	63.3(9)
4					55.4(4)	62.7(6)
5					55.8(5)	63.3(6)
6					56.0(8)	63.7(5)
7					57.5(6)	62.7(7)
8						62.9(4)
9						63.0(5)
10						62.4(5)
11						63.5(8)
12						60.6(4)
13						62.5(3)
14						63.2(4)
15						64.7(7)
16						66.3(5)
17						66.3(1.0)
18						69.0(1.5)
19						45.0(8)
20						48.4(7)
21						51.6(6)

that $E^o(3_1) \leq 30.1$, not a very good bound, but a bound nevertheless. Numerical simulations have been done by Pieranski *et.al.*[47] on open knot energies. △

Example 10: [Closed Thickness Energies] A more natural knot energy would be derived from the thickness of closed curves.[4] Thicknesses for C^2 curves can be defined in terms of double critical points and curvatures of the curve, see Example 26 and Theorem 12 with the accompanying citations. This thickness is often referred to as a *Disk Thickness*, since a C^2 curve can be thickened along normal disks. A formal definition is as follows:

Definition 11: (Disk Thickness) Let α be a unit length C^2 curve. A number $r > 0$ is *nice* if for $x \in \alpha$ and $y \in \alpha$ we define the normal disks of radius r at x and y, $D_x(r)$ and $D_y(r)$, and it is the case that $D_x(r) \cap D_y(r) = \emptyset$, for all pairs $x \neq y$. The *disk-thickness* $t_D(\alpha)$ is the supremum over all such nice r:

$$t_D(\alpha) = \sup \left\{ r \,\middle|\, r \text{ is nice.} \right\}. \tag{27}$$

◇

If $0 < r < t_D(\alpha)$, then the r-neighbourhood $\alpha(r)$ of α has a circle bundle structure $D^2 \times S^1$. Since $D_x(r)$ is disjoint from $D_y(r)$ if $x \neq y$, $\alpha(r)$ can be retracted onto α by retracting each $D_x(r)$.

The disk thickness is in fact closely related to the curvature and to "double critical chords". In particular, let $c(x, y)$ be a chord in a knot α between the points x and y. The chord $c(x, y)$ is double critical if $x \neq y$ and if $c(x, y)$ is normal to α at both the points x and y. Denote the infimum of all double critical chords in α by $D(\alpha)$, and suppose that κ is the maximum curvature of α. Then it is known that[37]

Theorem 12: For a unit length C^2 curve α, $t_D(\alpha) = \min\{D(\alpha)/2, 1/\kappa\}$. ♢

Since $t_D(\alpha)$ is explicitly dependent on the curvature κ, it may become undefined if α is not a C^2 curve. In particular, the supremum of $t_D(\alpha)$, for α constrained to have knot type $\mathcal{K}$, could be limiting to a curve that is not C^2. One may therefore wish to define thicknesses for a more general class of curves; in that case the major problem becomes the strong deformation retract of $\alpha(r)$ onto α; this is ensured in the C^2 case by the disk-bundle structure. This problem has also been considered for C^1 curves.[9,18,25] △

3. Properties of Knot Energies: Polygonal Knots

Since any scale invariant functional $f : \mathcal{P} \to I\!R^+$ from the space of knotted curves or polygons to non-negative real-numbers is an energy, there seems to be a bewildering array of knot energies.

However, closer examination shows that most arbitrary defined energies are not interesting, and that care is needed to find an energy with "good" properties (for example, finding energies that are simple to work with, but approaches infinity if a knot approaches a self-intersecting curve, and have local minima on knot types that distinguishes between knot types).

Polygonal knots (that is, embeddings of polygons in three space) are an important class of knots. They are particularly important in the computational study of knots so that many energy functionals have been defined for the class of polygonal knots. Certain energy functionals are also derived from the energies of C^k knots as a result of a polygonal approximation to the knot. Examples include the curvature energies and the Möbius energy, and these were defined as discrete approximations in order to estimate knot energies computationally, see for example reference.[33]

A systematic study of knot energies for equilateral polygonal knots[16] focused on knots that are embeddings of equilateral polygons in three space.

Let P_n be a polygonal curve composed of n edges $\langle e_1, e_2, \ldots, e_n \rangle$.

Definition 13: (Energy Functional of Equilateral Polygonal Knots) Let $f : \mathcal{P}_n \to I\!R^+$ be a real-valued function from the space of equilateral polygonal curves of length n edges, $\mathcal{P}_n$, into non-negative real numbers such that f is scale-invariant: That is, for any $P_n \in \mathcal{P}_n$ and $\alpha > 0$,

$$f(\alpha P_n) = f(P_n). \tag{28}$$

Then f is an energy function on n-step equilateral polygonal knots.

Let $\mathcal{P}_n(\mathcal{K})$ be the class of equilateral n-gons embedded into three space with knot type $\mathcal{K}$. The energy function of a knot type $\mathcal{K}$ within $\mathcal{P}_n(\mathcal{K})$ is an infimum over all equilateral n-gons in $\mathcal{P}_n(\mathcal{K})$:

$$f_n(\mathcal{K}) = \inf \left\{ f(P_n) \,\middle|\, \text{for all } P_n \in \mathcal{P}_n(\mathcal{K}) \right\}. \tag{29}$$

The energy of a knot type $\mathcal{K}$ over all equilateral polygons is the infimum

$$f(\mathcal{K}) = \inf_{n>0} \left(\inf \left\{ f(P_n) \,\middle|\, \text{for all } P_n \in \mathcal{P}_n(\mathcal{K}) \right\} \right). \tag{30}$$

$f(\mathcal{K})$ is independent of the number of steps in the polygonal representation of the knot type. If a knot type $\mathcal{K}$ cannot be represented by a polygon of length n, then its energy in polygons of length n is infinite: $f_n(\mathcal{K}) = \infty$ if there is no polygon P_n of length n such that $P_n \in \mathcal{K}$. ♢

The properties of knot energies of equilateral polygonal knots are set out in the next definition:

Definition 14: Let $f : \mathcal{P}_n \to I\!R^+$ be an energy function on equilateral polygons. Then

(1) f is a basic energy function if $f(P_n)$ is the absolute minimum within $\mathcal{P}_n$ if and only if P_n is the regular polygon R_n.
(2) f is a strong energy function if for any given positive number $a > 0$ there are only finitely many knot types $\mathcal{K}$ such that $f(\mathcal{K}) < a$.
(3) f is a charge energy function if $f(P_n) \to \infty$ whenever P_n approaches a polygonal curve with self-intersections (or a singular polygon).
(4) f is a semi-ideal energy function if it satisfies conditions (1), (2) and (3).

♢

There are more to energy functions of polygonal knots. Knot energies of smooth knots, such as a curvature energy, or a Möbius energy, are approximated by polygonal knot energies often naively derived from their smooth counterparts. In this respect, one would be interested in the behaviour of the polygonal knot energy as the number of edges in the polygonal knot is multiplied so that it approaches a smooth knot. As a minimum, one would expect a polygonal knot energy not to diverge as the number of edges is multiplied without bound. Secondly, the sequence of representative knots of type $\mathcal{K}$ at the energy minimum of an energy function should approach a "smooth" knot as the number of edges is increased.

Some apparently well defined knot energies of polygonal knots fail these tests, and it is surprisingly hard to find energies on polygonal knots that behave well (in the sense of being semi-ideal). First, these ideas are made precise by the following two definitions:

Definition 15: (Asymptotically Finite Polygonal Knot Energies) Let f be an energy function of equilateral polygonal knots and let $f_n(\mathcal{K})$ be the infimum of the energy of polygons of length n and knot type $\mathcal{K}$, as in Definition 13. Then f is an asymptotically finite knot energy function if

$$\limsup_{n\to\infty} f_n(\mathcal{K}) < \infty, \tag{31}$$

for any knot type $\mathcal{K}$. $\diamondsuit$

Secondly, as $n \to \infty$, one would expect the a representative polygon of knot type $\mathcal{K}$ and minimal knot energy to approach a "smooth" limit. Energy functions with this property will be called asymptotically smooth.

Definition 16: (Asymptotically Smooth Polygonal Knot Energies) Let f be an energy function of equilateral polygonal knots and let $f(P_n)$ be the energy of a polygon P_n of knot type $\mathcal{K}$. Let the edges in P_n be labeled $\{e_1, e_2, \ldots, e_n\}$ and suppose that θ_i is the excluded angle between edges e_i and e_{i+1} while θ_n is the excluded angle between e_n and e_1. Then f is asymptotically smooth if there exists a constant $M > 0$ such that for all polygons and all $n > 0$,

$$\theta = \max_i\{\theta_i\} \leq \left[\frac{M}{n}\right] f(P_n). \tag{32}$$

In particular, M is independent of n. $\diamondsuit$

The equilateral stick energy in Example 3 is basic and strong. It is not charge, nor asymptotically finite or smooth. One may expand Definition 14 to knot energies on the wider class of polygons. In that case it follows that the stick energy in Example 2 is basic and strong, but not charge nor asymptotically finite or smooth. It is not clear how to expand the definition to include lattice edge energies (Example 4), but one may argue that the lattice curvature energy in Example 5 is basic and strong, but not charge.

Example 17: [Polygonal Curvature Energy] Let P_n be a equilateral polygonal knot and define $\theta = \max_i \theta_i$. Consider the energy function $\mathcal{E}_\theta(P_n) = n\theta$. This energy function is basic, since its global minimum is on regular (planar) polygons R_n, in which case $\mathcal{E}_\theta(R_n) = 2\pi$. However, it is not strong, since all knots with braid index k will the same minimum energy less than or equal to $2k\pi$ in the $n \to \infty$ limit. Nor it is charge, since self-intersecting polygons will have finite energy. However, it can be checked that it is both asymptotically smooth and finite.

Define instead $\mathcal{E}_\theta^2 = n^2\theta$. This energy is again basic and not strong nor charge. It is also asymptotically smooth. To see this, observe that $\theta = \mathcal{E}_\theta^2/n^2 \leq \mathcal{E}_\theta^2/n$. On the other hand, $\mathcal{E}_\theta^2$ is not asymptotically finite, since $\theta > 2\pi/n$, and so $\mathcal{E}_\theta^2 \geq 2\pi n$. △

Example 18: [Squared Curvature Energy] The energy function

$$E_\theta(P_n) = \sum_{i=1}^{n} \theta_i^2 \tag{33}$$

is basic. It is not charge, nor strong. It is asymptotically finite but not smooth. To see this, consider a regular polygon R_n, and subdivide the edges k times to obtain Q_{nk}. Then $E_\theta(R_n) = E_\theta(Q_{nk})$, but $n\theta \to \infty$. △

Example 19: [The Physical Knot Energy] This energy was first studied in reference.[36] There are three terms in this energy. The first is an electrostatic repulsion between vertices, the second an elastic energy, and the third a curvature energy between adjacent edges. If the energy is restricted to equilateral polygons, then the elastic energy can be ignored. The curvature energy between adjacent edges e_i and e_{i+1} is given by

$$E^c(P_n) = \sum_{i=1}^{n} cot^2((\pi - \theta_i)/2) \sim \sum_{i=1}^{n} \theta_i^2/4, \tag{34}$$

and so is similar to the energy in Example 18.

This leaves the electrostatic energy. This type of knot energy (usually implemented as a repulsion between vertices or edges in the polygon) is a popular one, versions of which was presented in numerous studies.[16,21,23,36,61,62] The electrostatic energy[36] is a repulsion between vertices, defined by

$$E^q(P_n) = \sum_{i \neq j} \left(\frac{q_i q_j}{\|v_i - v_j\|} \right) \tag{35}$$

where q_i is the charge of the i-th vertex v_i in the polygon P_n where the edges have unit length. This energy is basic. It is not known to be strong, although this is claimed.[16] It is not charge, since edges may intersect. One may check that

$$E^q(P_n) \geq n \sum_{k=1}^{\lfloor (n-1)/2 \rfloor} \frac{1}{k-1} \tag{36}$$

and so E^q is not asymptotically finite. It is not known whether this energy is asymptotically smooth. △

Example 20: [The MD-Energy] The MD-energy[60,61] is similar to the above, and is defined by

$$E_{MD}(K) = \sum_{e_1, e_2} \frac{\ell(e_1)\ell(e_2)}{(M_d(e_1, e_2))^2} \tag{37}$$

where $\ell(e)$ is the length of the edge e, $M_d(e_1, e_2)$ is the minimum distance between e_1 and e_2, and the summation is over all pairs of non-adjacent edges in the knot K. It is obvious that this energy is charge, and it is known to be strong.[60] It is also basic, but not known to be asymptotically smooth. Since $E_{MD}(K_n) \geq n$ if K_n is a knot composed of n edges, this energy is not asymptotically finite.[16] Hence, like the physical knot energy, this energy is asymptotically weak. △

Computer simulations[36] demonstrates the inherent weakness in a knot energy that is asymptotically weak. For large values n, the minimal knot conformation appears to approach a regular polygon with a small part that is locally a knotted ball-pair. This can be explained as follows: If a small number of edges in a large regular polygon is replaced by a sequence of edges in a small 3-ball that changes the polygon into a almost regular polygon with a (small) knotted arc (so that the entire polygon is a knot), then almost all the vertices in the polygon remains unchanged, and the numbers of pairs of vertices that remain unchanged is of order $O(n^2)$. Since the changed vertices is confined to a small 3-ball, the contribution to the electrostatic energy by the addition of vertices is of order $O(n)$. In other words, the knot has energy equal to the energy of a regular polygon plus a small contribution from the new vertices. Since the electrostatic energy is basic, this shows as the number of vertices is increased in any knot-type, that the minimal energy conformation will occur on a knot conformation that expels a regular polygon in the limit. All knot types will have the same limiting knot energy, and while it is not known, this energy is presumable not asymptotically smooth either. This argument also works if the number of new vertices is of order $O(n^\epsilon)$, $0 \leq \epsilon < 1$.

If the argument above is made rigorous, then it will also prove that the physical knot energy is not strong, since any knot type can be tied "locally" in this manner. It is generally not a simple task to modify energy functions to find semi-ideal and asymptotically ideal energy functions. One may attempt to substract the energy of a regular n-gon from the Physical Knot Energy. This may turn it into an energy function with better properties, but this has not been examined. The MD-energy approximates the Mobius energy, and one may expect them to satisfy the same properties, perhaps with some readily understood differences.

Example 21: [The DEJ-Energy] Let P_{2n} be an equilateral polygon of length $2n$ with edges $\langle e_1, e_2, \ldots, e_{2n}\rangle$. Define $d_{i,k}$ to be the shortest (Euclidean) distance between e_i and e_{i+k}, where $i+k \equiv i+k-2n$ whenever $i+k > 2n$. Similarly, let $s_{i,k}$ be the shortest chemical distance (thus, measure along the polygon) between edges e_i and e_{i+k}. Define the energy

$$E_d(P_{2n}) = \max_{\substack{1\leq i\leq 2n \\ 2\leq k\leq n}} L^2 \left(\frac{1}{d_{i,k}^2} - \frac{1}{s_{i,k}^2} \right). \tag{38}$$

the following is known about the DEJ-energy:

Theorem 22: The DEJ-energy is an asymptotically ideal energy.[16] △

A thickness of polygonal knots have also defined.[52] This energy is known to be basic and charged, and is conjectured to be strong. It is also known to be asymptotically finite and smooth.

More is known about asymptotically ideal polygonal knot energy functions. In particular, it is possible to say something about the limiting process as one considers the sequence of polygonal knots that minimize such an energy function.

Define

$$C^k(\cdot) = \{f \mid S^1 \to I\!R^3,\ \ f^{(k)} \text{ is continuous}, \|f'(t)\| = 1\}, \tag{39}$$

the space of all k-times continuously differentiable embeddings of the circle $S^1 \subset I\!R^3$ with arclength parameterization. If $\mathcal{K}$ is a knot type, then

$C^k(\mathcal{K}) \subseteq C^k(\cdot)$ is the space of all embeddings in $C^k(\cdot)$ that are knots of type $\mathcal{K}$.

In this event a strong result is known:

Theorem 23: (Dai and Diao[11]) If E is an asymptotically smooth and finite energy function of equilateral polygons, then any sequence $\langle P_n \rangle$ of minimizers of E of length n within a knot type $\mathcal{K}$ has a subsequence that converges to a curve K in the space $C^1(\cdot)$. If in addition, E is a charge energy function, then $K \in C^1(\mathcal{K})$. $\diamondsuit$

This theorem, for example, applies to the energy function in Example 21.

4. Thickness Energies

Thickness energies are defined on simple closed curves in 3-space. The definition is on curves in C^k, the set of simple closed curves that are k times differentiable.

Definition 24: The function $f : C^k \to I\!R^+$ is scale invariant if for any curve $\alpha \in C^k$ and $\lambda > 0$ it is the case that $f(\lambda\alpha) = f(\alpha)$. f is an energy function on C^k if it is scale invariant.

If $f(\alpha)$ is the energy of a curve in C^k, and $C^k(\mathcal{K})$ are all the curves in C^k of knot type $\mathcal{K}$, then the minimum energy of the knot type $\mathcal{K}$ is

$$f(\mathcal{K}) = \inf_{\alpha \in C^k(\mathcal{K})} f(\alpha). \tag{40}$$

$\diamondsuit$

Normally, one would expect that f should also be continuous on C^k in the definition above, but this often proves hard to demonstrate for any proposed energy function. The properties of C^k energy functions are similar to those of polygonal energy functions:

Definition 25: Let $f : \alpha \to I\!R^+$ be an energy function on C^k, where $k > 0$. Then

(1) f is a basic energy function if $f(\alpha)$ is the absolute minimum within C^k if an only if α is a circle.
(2) f is a strong energy function if for any given positive number $a > 0$ there are only finitely many knot types $\mathcal{K}$ such that $f(\mathcal{K}) < a$.
(3) f is a charge energy function if $f(\alpha) \to \infty$ whenever α approaches a curve with self-intersections.

(4) f is a tight energy if $f(\alpha_i)$ goes to infinity for any sequence of embeddings α_i of unit length and the same knot type $\mathcal{K}$, where α_i contains a knotted arc of knot type $\mathcal{K}$ in a ball B_i, and the diameter of B_i approaches zero.
(5) f is a semi-ideal energy function if it satisfies conditions (1), (2) and (3).
(6) f is ideal if it is semi-ideal and tight.

◇

The crossover energy in Example 1 is strong, but is neither basic, charge nor tight. The curvature energies $C_p(\mathcal{K})$ and $C_m(\mathcal{K})$ in Example 6 can be seen to be basic, but they are generally not strong or charge. One can show that the curvature energies are tight. The Gromov distortion in Example 7 is basic and charge, but not strong[15,23,44] nor tight.[15]

The Möbius energy E_M in Example 23 is basic,[22] strong and charge, but is known to be not tight.[44] Hence, it is semi-ideal. The energy E_j^p is known[22] to be basic in the case that $p = 1$ and $j = 2$ and in many other cases.[1] E_j^p is strong only if $0 < j \leq 2$ and $jp > 2$.[44] It is also known to be charge and tight if $0 < j \leq 2$ and $jp > 2$,[44] but in the case that $jp \leq 2$, then it is not tight.

In the case of the open rope energy in Example 9 one may consider a straight rope between the bounding planes as the unknot. The minimum of the open rope energy is on the unknot and this energy could be considered basic. It is also known that E^o is strong, charge and tight, and is an ideal knot energy.[17] The ideas underlying the open thickness has been expanded to links as well.[19]

Example 26: [The Disk Thickness Energy] The Disk Thickness of C^2 knots was defined in Definition 11. The thickness of a knot type is the supremum over all unit length C^2 curves of knot type $\mathcal{K}$:

$$t_D(\mathcal{K}) = \sup_{K \in C^2(\mathcal{K})} \{t_D(K) \mid K \text{ is a unit length curve}\}. \tag{41}$$

I have noted that the length of the shortest double critical chord is the double critical self distance $D(K)$ of the knot K, and that the disk thickness of a curve is closely related to this in Theorem 12: for a C^2 knot K, the disk thickness is related to the minimum of double critical chords $D(K)$ and its maximal curvature κ by

$$t_D(K) = \min\{D(K)/2, 1/\kappa\}. \tag{42}$$

The disk thickness can be used to define an ideal energy function on knots:

Definition 27: Let $E_D(K) = L/t_D(K)$. The disk thickness energy of a knot type $\mathcal{K}$ is defined by

$$E_D(\mathcal{K}) = \inf_{K \in \mathcal{K}} E_D(K). \tag{43}$$

$\diamondsuit$

Theorem 28: [15] Let K be a C^2 knot of length L. Then $E_D(\mathcal{K})$ is an ideal energy function. $\triangle$

There are also other definitions of knot thicknesses.[9,19,25] In particular, one would like to define a thickness for C^1-curves; an example of such a thickness is the *maximal thickness.*

Example 29: [Maximal Thickness] Let $c > 0$ and define the c-neighbourhood of a simple closed C^1-curve K of knot type $\mathcal{K}$ by

$$K_c = \{x \in I\!R \,\big|\, \text{there is a point } z \in K \text{ such that } \|x - z\| \le c\}, \tag{44}$$

where $\|\cdot\|$ is the usual Euclidean metric. Consider a plane Σ_x normal to K at x. The intersection $\Sigma_x \cap K_c$ has a component C_x containing x. This component is a topological (meridian) disk that only cuts K in x for c small enough. For small $c > 0$ it is the case that K_c is homotopic to K by strong deformation retract along the meridian disks.[18] The supremum of c for which this is true defines the maximal thickness:

$$t_M(K) = \sup\{c > 0 \,\big|\, K \sim K_t \text{ for all } 0 < t < c\} \tag{45}$$

where $K \sim K_t$ via strong deformation retract.

It is known that this thickness is not continuous with respect to small changes in K.[18] A knot energy may be defined using $t_M(K)$: This energy is basic, strong and charge, and is also tight, and so ideal.

Theorem 30: (Maximal Thickness Energy) Let K be a closed C^1-curve of length L. Define $E_M(K) = L/t_M(K)$. The maximal thickness energy of a knot type $\mathcal{K}$ is the infimum of $E_M(K)$ for all $K \in \mathcal{K}$:

$$E_M(\mathcal{K}) = \inf_{K \in \mathcal{K}} E_M(K). \tag{46}$$

$E_M(\mathcal{K})$ is an ideal energy of C^1 knots. $\triangle$

It is believed that the disk thickness and maximal thickness of knot types coincide on the energy minima of knot types:

Conjecture 1: [19] *Let $\mathcal{K}$ be any knot type. Then there exists a C^1 knot $K \in \mathcal{K}$ such that*

$$E_D(K) = E_M(K) = E_D(\mathcal{K}) = E_M(\mathcal{K}). \tag{47}$$

$\diamondsuit$

This conjecture remains unproven. In the case of links, one can demonstrate that maximally thick links are obtained when some components are stadium curves, and so are in C^1 and not in C^2.[9] It it generally thought that some knots will have the same property, but an example is lacking. $\triangle$

Example 31: [CKS-thickness] Let $r(x, y, z)$ be the radius of the unique circle through the points x, y and z in $\mathbb{R}^3$. If K is a knot, then the *local* thickness of K at $x \in K$ is

$$\tau_x(K) = \inf_{\substack{y,z \in K \\ x \neq y \neq z \neq x}} r(x, y, x). \tag{48}$$

The CKS-thickness of the curve is defined[9] as

Definition 32: The CKS-thickness of a C^1 knot K is defined by

$$\tau_K = \inf_{x \in K} \tau_x(K). \tag{49}$$

If $\mathcal{K}$ is the class of all C^1 knots of length L, then the CKS-energy of the knot type $\mathcal{K}$ is

$$E_\tau(\mathcal{K}) = \inf_{K \in \mathcal{K}} \frac{L}{\tau_K}. \tag{50}$$

It is known that there exists C^1-energy minimizers with bounded curvature for $E_\tau(\mathcal{K})$ for any knot (or link) type.[9] It is also shown there that

$$E_\tau(\mathcal{K}) \geq 2\pi \left(2 + \sqrt{2}\right) \tag{51}$$

for any non-trivial knot type $\mathcal{K}$. Naturally, $E_\tau(\emptyset) = 2\pi$. $\triangle$

Other definitions of thickness can be found in the literature.[18] Notably, Gonzalez and Maddocks defined the *global radius of curvature* of a closed space curve.[25]

5. Conclusions

In this paper I have attempted to give a broad overview about knot energies. I have focused on what one could consider their ideal properties, such as being basic, charge, strong or tight. There remains some unanswered questions, even in the case of the well-studied examples provided in this paper: For example, could one find simple techniques to extend the known stick numbers and lattice edge numbers to higher values? In the case of lattice edge numbers, the proof that $m(3_1) = 24$ is a combinatorial case analysis, and therefore lengthy and involved. One could imagine that this method could be extended to prove that $m(4_1) = 30$, but after that the proofs becomes too long to handle with ease. It may be the case that the method of proof could be computerized by turning it into an algorithm. Then these numbers could be proven by using faster and faster computers. This is a fascinating possibility that deserves further attention.

There are also significant computer simulations to estimate the thicknesses of knots[18,45,54,62] in the literature, and in at least one case was a thickness determined experimentally.[17] Thicknesses for polygonal knots have also been determined,[19,46] see also the website of E. Rawdon.[53] These calculations can be used to find upper bounds on the rope lengths of given knot types. For example, the minimal rope length of the trefoil is less than or equal to 32.744 (since it is known that there exists a $C^{1,1}$ knot whose length is at most 32.744). This value is probably a good estimate of minimum rope length, since numerical computations give values close to it (an estimate[19] gave 32.79). There are a variety of lower bounds in the literature. Notably, Y. Diao proved that the minimal rope length of the trefoil is at least 24,[14] and that was later improved to the very sharp lower bound of 31.32.[12] More experimental work involving DNA or charged polymers are also ongoing.[2,10,21,64,65] Conjecture 1 was first stated in the work of Diao et. al.[18] and remains unproven.

References

1. A. Abrams, J. Cantarella, J.H.G. Fu, M. Ghomi, R. Howard, 2003. *Circles Minimize Most Knot Energies.* Topology **42**, 381-394.
2. J. Arsuaga, M. Vázquez, S. Trigueros, D.W. Sumners and J. Roca, 2002. *Knotting Probability of DNA Molecules Confined in Restricted Volumes: DNA Knotting in Phage Capsids.* Proc. Natl. Acad. Sci. USA **99**(8), 5373-5377.
3. K. Brakke, 1989. *Evolver.* Geometry Center, University of Minnesota. Program for visualizing the minimizing of energy functions of knots.

4. G. Buck and J. Orloff, 1993. *Computing canonical conformations for knots.* Topo. and its Appl. **5**, 247-253.
5. J.A. Calvo, 2001. *The Embedding Space of Hexagonal Knots.* Topo. and its Appl. **10**, 137-174.
6. J.A. Calvo, 2001. *Geometric Knot Spaces and Polygonal Isotopy.* J. Knot Theory and its Ramifications **10**, 245-267.
7. J.A. Calvo, 2002. *Characterizing Polygons in $\mathcal{R}^3$.* in *Physical Knots*, eds. J. A. Calvo, K. C. Millett, and E. J. Rawdon. Contemporary Math. **304**, 37-53.
8. J.A. Calvo and K.C. Millett, 1998. *Minimal Edge Piecewise Linear Knots.* In *Ideal Knots*, eds. A. Stasiak, V. Katritch and L.H. Kauffman. Series on Knots and Everything **19**. World Scientific, 107-128.
9. J. Cantarella, R. Kusner and J. Sullivan, 2002. *On the Minimum Ropelength of Knots and Links.* Inventiones Math. **150**, 257-286.
10. N.J. Crisona, R. Kanaar, T.N. Gonzalez, E.L. Zechiedrich, A. Klippel and N.R. Cozzarelli, 1994. *Progressive Recombination by Wild-type Gin and an Enhancer-independent Mutant. Insight into the Mechanisms of Recombination Selectivity and Strand Exchange.* J. Mol. Biol. **234**, 437-457.
11. X. Dai and Y.Diao, 2001. *The Minimum of Knot Energy Functions.* J. Knot Theory and its Ramifications **9**, 713-724.
12. E. Denne, X. Diao and J.M. Sullivan, 2004. *Quadrisecants Give New Lower Bounds for the Ropelength of a Knot.* Preprint.
13. Y. Diao, 1993. *Minimal Knotted Polygons on the Cubic Lattice.* J. Knot Theory and its Ramifications **2**, 413-425.
14. Y. Diao, 2003. *The Lower Bounds of the Length of Thick Knots.* J. Knot Theory and its Ramifications **12**, 1-16.
15. Y. Diao, C. Ernst and E.J. Janse van Rensburg, 1996. *Properties of Knot Energies.* In *Topology and Geometry in Polymer Science*, eds. S.G. Whittington, D.W. Sumners and T. Lodge. Proceedings of the 1995-1996 IMA program on Mathematical Methods in Material Science, Workshop 8, 10-16 June 1996.
16. Y. Diao, C. Ernst and E.J. Janse van Rensburg, 1997. *In Search of a Good Polygonal Knot Energy.* J. Knot Theory and its Ram. **6**, 633-657.
17. Y. Diao, C. Ernst and E.J. Janse van Rensburg, 1997. *Knot Energies by Ropes.* J. Knot Theory and its Ram. **6**, 799-807.
18. Y. Diao, C. Ernst and E.J. Janse van Rensburg, 1999. *Thicknesses of Knots.* Math. Proc. Camb. Phil. Soc. **126**, 293-310.
19. Y. Diao, C. Ernst and E.J. Janse van Rensburg, 1999. *Knots with Minimal Energies*, contributed to *Ideal Knots.* Series on Knots and Everything, Vol 19. Eds. A. Stasiak V. Katritch and L.H. Kauffman. World Scientific, Singapore.
20. Y. Diao, C. Ernst and X. Yu, 2004. *Hamiltonian Knot Projections and Lengths of Thick Knots.* Top. and its Appl. **136**, 7-36.
21. P.G. Dommersnes, Y. Kantor and M. Kardar, 2002. *Knots in Charged Polymers.* Phys. Rev. E **66**, 031802-1-8.
22. M.H. Freedman, Z.-X. He and Z. Wang, 1994. *Möbius Energy of Knots and*

Unknots. Ann. Math. **139**, 1-50.

23. S. Fukuhara. *Energy of a Knot*, 1988. In *A Fete of Topology*, eds. Y. Matsumoto, T. Mizutani and S. Morita. Academic Press, Inc. 443-451.
24. E. Furstenberg, J. Li and J. Schneider, 1998. *Stick Knots*. Chaos, Solitons and Fractals **9**(4/5), 561-568.
25. O. Gonzalez and J.H. Maddocks, 1999. *Global Curvature, Thickness, and the Ideal Shape of Knots*. Proc. Natl. Acad. Sci. USA **96**, 4769-4773.
26. M. Gromov, 1978. *Homotopical Effects of Dilation*. J. Diff. Geom. **13**, 303-310.
27. M. Gromov, 1983. *Filling Riemannian Manifolds*. J. Diff. Geom. **18**, 1-147.
28. M. Gromov, J. Lafontaine, and P. Pansu, 1981. *Structures Métriques pour les Variétés Riemannienses*. Cedic/Fernand Nathan, Paris (1981).
29. E.J. Janse van Rensburg, 1998. *Minimal Lattice Knots*. In *Ideal Knots*, eds. A. Stasiak, V. Katritch and L.H. Kauffman. Series on Knots and Everything**19**, 88-106. World Scientific (1998).
30. E.J. Janse van Rensburg and S.D. Promislow, 1995. *Minimal Knots in the Cubic Lattice*. J. Knot Theory and its Ram. **4**, 115-130.
31. E.J. Janse van Rensburg and S.D. Promislow, 1999. *The Curvature of Lattice Knots*. J. Knot Theory and its Ram. **8**, 463-490.
32. G.T. Jin and H.S. Kim, 1993. *Polygonal Knots*. J. Korean Math. Soc. **30**, 371-383.
33. R.B. Kusner and J.M. Sullivan, 1996. *Möbius Energies for Knots and Links, Surfaces and Submanifolds*. In *Geometric Topology*, 570-604. International Press (1996).
34. R.B. Kusner and J.M. Sullivan, 1996. *On the Distorsion and Thickness of Knots*. In *Topology and Geometry in Polymer Science*, eds. S.G. Whittington, D.W. Sumners and T. Lodge. Proceedings of the 1995-1996 IMA program on Mathematical Methods in Material Science, Workshop 8, 10-16 June 1996.
35. H.A. Lim, M.T. Carroll and E.J. Janse van Rensburg,1992. *Electrophoresis of Knotted DNA in a Regular and Random Electrophoretic Medium*. In *Biomedical Modelling and Simulation*, eds. J. Eisenfeld *et. al.* Elsevier Science Publications (1992) 213-223.
36. T.J. Ligocki and J.A. Sethian, 1994. *Recognizing Knots Using Simulated Annealing*. J. Knot Theo, and its Ram. **3** (1994) 477-496.
37. R.A. Litherland, J. Simon, O. Durumeric and E. Rawdon, 1999. *Thickness of Knots*. Topo. and its Appl. **91**, 233-244.
38. M. Meissen, 1997. *Lower and Upper Bounds on Edge Numbers and Crossing Numbers of Knots*. Ph.D. thesis, University of Iowa (1997).
39. K.C. Millett, 1994. *Knotting of Regular Polygons in* 3*-space*. J. Knot Theory and its Ramifications **3**, 263-278.
40. J. Milnor, 1953. *On the Total Curvature of Closed Space Curves*. Math. Scand. **1**, 289-296.
41. H.K. Moffatt, 1990. *The Energy Spectrum of Knots and Links*. Nature **347**, 367-369.
42. S. Negami, 1991. *Ramsey Theorems for Knots, Links and Spatial Graphs*.

Trans. Amer. Math. Soc. **324**(2), 527-541.
43. J. O'Hara 1991. *Energy of a Knot*, Topology **30**, 241-247.
44. J. O'Hara, 1992. *Family of Energy Functionals of Knots.* Top. Appl. **48**, 147-161. - 1994, *Energy Functionals of Knots II.* Top. Appl. **56**, 45-61.
45. P. Pieranski, 1998. *In Search of Ideal Knots.* In *Ideal Knots*, eds. A. Stasiak, V. Katritch and L.H. Kauffman. Series on Knots and Everything**19**, 20-41. World Scientific (1998).
46. P. Pieranski, J. Baranska and E. Rawdon, 2004. *Ropelength of Tight Polygonal Knots.* Preprint on http:/mathcs.duq.edu/ rawdon/index.html.
47. P. Pieranski, S. Przybyl and A. Stasiak, 2001. *Tight Open Knots.* Eur. Phys. J. E **6**, 123-128.
48. B. Podlésny, 2003. *Minimum Stick Numbers for Knots and Links.* Preprint.
49. R. Randell, 1988. *A Molecular Conformation Space.* In *Studies in Physical and Theoretical Chemistry, Math/Chem/Comp 1987* **54**, 125-140. Eds. R.C. Lacher (1988).
50. R. Randell. *A Molecular Conformation Space.* In *Studies in Physical and Theoretical Chemistry, Math/Chem/Comp 1987* **54**, eds. R.C. Lacher (1988) 141-156.
51. R. Randell, 1994. *An Elementary Invariant for Knots.* J. Knot. Theory and its Ramifications **3**, 279-286.
52. E. Rawdon, 1998. *Approximating the Thickness of a Knot.* In *Ideal Knots*, eds. A. Stasiak, V. Katritch and L.H. Kauffman. Series on Knots and Everything **19**, 143-155. World Scientific (1998).
53. E. Rawdon, 2003. *Can Computers Discover Ideal Knots?* Experimental Math. **12**, 287-303.
54. K. C. Millett and E.J. Rawdon, 2003. *Energy, Ropelength, and other Physical aspects of Equilateral Knots.* J. Comp. Phys. **186**(2), 426-456.
55. E.J. Rawdon and R.G. Scharein, 2002. *Upper Bounds for Equilateral Stick Numbers.* Contemporary Math. **304**, 55-75.
56. R.L. Ricca, 1998. *New Developments in Topological Fluid Mechanics: From Kelvin's Vortex Knots to Magnetic Knots.* In *Ideal Knots*, eds. A. Stasiak, V. Katritch and L.H. Kauffman. Series on Knots and Everything**19**, 255-273. World Scientific (1998).
57. D. Rolfson, 1990. *Knots and Links.* Publish or Perish Inc.
58. R.G. Scharein, 1998. *Interactive Topological Drawing.* Ph.D. thesis, University of British Columbia.
59. R.G. Scharein, 1998. *KnotPlot.* Program for drawing, visualizing and minimizing energy functions on knots.
60. J. Simon, 1994. *Energy Functions for Polygonal Knots.* In *Random Knotting and Linking*, eds. K.C. Millet and D.W. Sumners. Series on Knots and Everything **7**, 67-88. World Scientific (1994).
61. J. Simon, 1998. *Energy Functions for Knots: Beginning to Predict Physical Behaviour.* In *Ideal Knots*, eds. A. Stasiak, V. Katritch and L.H. Kauffman. Series on Knots and Everything **19**, 151-182. World Scientific (1998).
62. A. Stasiak, J. Dubochet, V. Katritch and P. Pieranski, 1998. *Ideal Knots and their Relation to the Physics of Real Knots.* In *Ideal Knots*, eds. A.

Stasiak, V. Katritch and L.H. Kauffman. Series on Knots and Everything **19**, 1-19. World Scientific (1998).

63. J. Sullivan, 2001. *Approximating Rope Length by Energy Functions.* In *Physical Knots: Knotting, Linking, and Folding Geometric Objects in* $\mathcal{R}^3$. Eds. J.A. Calvo, K.C. Millett and E. Rawdon. Contemporary Math. **304**.
64. A.V. Volodogski, N. Crisona, B. Laurie, P. Pieranski, V. Katritch, J. Duboshet and A. Stasiak, 1998. *Sedimentation and Electrophoretic Migration of DNA Knots and Catenanes.* J. Mol. Biol. **278**, 1-3.
65. S. Wasserman and N.R. Cozzarelli, 1991. *Supercoiled DNA-directed Knotting by T4 Topoisomerase.* J. Biol. Chem. *266*, 20567-20573.

CHAPTER 3

UNIVERSAL ENERGY SPECTRUM OF TIGHT KNOTS AND LINKS IN PHYSICS

Roman V. Buniy* and Thomas W. Kephart†

Department of Physics and Astronomy
Vanderbilt University
Nashville, TN 37235
**roman.buniy@vanderbilt.edu,*
†thomas.w.kephart@vanderbilt.edu

We argue that a systems of tightly knotted, linked, or braided flux tubes will have a universal mass-energy spectrum, since the length of fixed radius flux tubes depend only on the topology of the configuration. We motivate the discussion with plasma physics examples, then concentrate on the model of glueballs as knotted QCD flux tubes. Other applications will also be discussed.

1. Introduction

It is known from plasma physics that linked magnetic flux tubes are much more stable than an unknotted single loop [1]. Linked flux tubes carry topological charge, and this can be thought of as a conserved (at least to lowest order) physical quantum number. Similarly, knotted flux tubes carry topological quantum numbers, and one can think of a knot as a self-linked loop. The topological charges are described by knot polynomials that are related to projections of knots or links into a plane where the crossings of the loops are assigned various attributes. Following each line around its loop generates the polynomials. Several types of polynomials have been studied in the literature (see e.g. Refs. [2, 3]): Alexander, Conway, HOMFLY, Jones, Kauffman, etc., with increasing levels of precision for distinguishing knots. For example, the simplest knot, the trefoil, has a chiral partner (mirror image) that is not detected by the simpler polynomials, but is by the more sophisticated ones. Hence, a pair of knots with different polynomials are different, but the converse is not necessarily true. It is still an unsolved problem to find a set of polynomials that distinguishes all non-homeomorphic knots/links. Similar results hold for braids, and we will also discuss these objects below.

2. Review of previous physical results on tight knots and links

If the loops have fixed uniform thickness and circular cross-section (we will eventually discuss how one can relax this condition), then each knot and link has a completely specified length if the configuration is tight, i.e., is of the shortest length with the tubes undistorted and non-overlapping. If tubes have uniform cross sections, as can be approximately the case with magnetic or electric flux tubes carrying quantized flux, or for a polymer or even a piece of spaghetti, then the length of the tight knot is proportional to the mass (or energy) of the knot. This, we claim, generates a universal mass (energy) spectrum for knotted/linked configurations of objects of this type. The lengths of tight knots were not studied until the mid-1990s [4], and only recently have accurate calculations of large numbers of tight knots [5] and links [6] become available. These results now make it possible to examine physical systems and compare them with the knot spectrum. The first physical example studied was tightly knotted DNA [4]. More recently, we have examined the glueball spectrum of QCD [7]. These particles [8] are likely to be solitonic states [9] that are solutions to the QCD field equations. While QCD will be our main focus in this chapter, there are many more cases where tight knots may play a role. We first proceed with an analysis of flux tubes in plasma physics. The lack of controllable quantum flux renders this case somewhat less interesting than its generalization to QCD. We will not go into any experimental details here, but we hope the experts in the areas discussed will take our general perspective into account when analyzing their data.

In order to decide if a system of flux tubes falls into the universal class of having a tight knot energy spectrum, we must first investigate the time scales involved. These are the lifetime of the soliton τ_s and the relaxation time τ_r necessary to reach the ground state of a tight knot configuration. The soliton lifetime (or the corresponding decay width $\Gamma_s = 1/\tau_s$) can depend on several factors. These include the effects of flux tube breaking, rearrangement, and reconnection. The partial width for flux tube breaking is non-zero if the production of particle/anti-particle pairs is energetically possible, for example monopole/anti-monopole ($M\bar{M}$) pairs or color monopole/anti-monopole ($M_C\bar{M}_C$) pairs for magnetic flux (or color magnetic flux) or quark/anti-quark ($q\bar{q}$) pairs for color electric flux tubes. The partial widths can vary widely depending on the particle masses (e.g., $m_q \ll M$, so we expect $q\bar{q}$ pairs to be easier to produce than $M\bar{M}$ pairs), interaction strengths (this, for instance, enhances $M\bar{M}$ pair production versus $q\bar{q}$ pair production), and boundary conditions (tube shape and length). Rearrangement is a quantum effect where, for example, in a double donut arrangement, the loops can tunnel free of each other. Finally, reconnection is another effect where tubes break and re-attach in a different configuration. Such behavior has been seen in plasma physics, and is of major importance in understanding a variety of astrophysical systems. For a review see the last chapter of the text by Bellan [10]. All these processes change topological charge, and their partial widths compete more or less favorably with each other depending on the parameters that describe the system.

3. Exact calculations

While no knot lengths have been calculated exactly, it is possible to calculate the exact lengths of an infinite number of links and many braids [6]. For links these calculations are possible in the case where each individual elements of the link lies in a plane. For braids, exact calculations are possible when the elements of the braid are either straight sections or where their centerlines follow helical paths. The shortest of all links, the double donut, is exactly calculable. The two elements lie in perpendicular planes and are tori of equal length. The shortest non–trivial braid is a helically twisted pair. "Weyl's tube formula" is the ideal tool for calculating the volume of a flux tube [11, 12]. The formula states that for a tube of constant cross-section σ normal to a path of length l, the volume of the tube is just $V_T = l\sigma$. This is a remarkable result that holds in flat 3D. In higher dimensions or curved space, the result is more complicated, but here we need only the simple 3D case. This means that if we have an analytic form for the path and a circular cross-section we can find V_T. This leads to the fact that there is a class of exactly calculable links and braids. Since there are no known analytic forms for the path of tight knots, their volumes can only be calculated numerically, but once we have an estimate of the length, we can then also estimate the volume and therefore the energy for a corresponding physical system. Braids are also of physical interest. For the simplest example, a tightly twisted pair, the path of the center lines of such tubes are helices, and so the lengths $l(\theta_P, h)$ depend on the pitch angle θ_P, the radius of the tubes a, and the height of the braid h. Hence the volume is $V_T = 2\sigma l(\theta_P, h)$ where $\sigma = \pi a^2$. The volume of triple, etc., helical tight braids can also be found exactly; however, as with knots, the volumes of topologically non-trivial tight braids (those where the elements are woven together) can only be found approximately. (The helical twisted pair can also become non-trivially knotted/linked by identifying top and bottom boundaries, but we will not pursue this possibility here.) While the simple helically twisted braid has a volume that depends on the pitch angle which can potentially be adjusted by experimental conditions (see below), the tight knots and links derived from braids have no such adjustable parameter.

4. Plasma physics

Before going on to our main example of QCD, let us stop here to discuss tight links of flux in electromagnetic plasma. This example is conceptually somewhat simpler and provides motivation for what is to come.

Movement of fluids often exhibits topological properties (for a mathematical review see e.g. Ref. [13]). For conductive fluids, interrelation between hydro- and magnetic dynamics may cause magnetic fields, in their turn, to exhibit topological properties as well. For example, for a perfectly conducting fluid, the (abelian) magnetic helicity $\int d^3x \, \epsilon^{ijk} A_i \partial_j A_k$ is an invariant of the motion [14], and this quantity can be interpreted in terms of knottedness of magnetic flux lines [15]. Let us discuss this in some detail since its implications are central to our more general results.

4.1. *Magnetic relaxation*

A perfectly conducting, viscous and incompressible fluid relaxes to a state of magnetic equilibrium without a change in topology [1]. The system approaches a state of magnetic equilibrium by decreasing its magnetic energy, which is achieved by contraction of the magnetic field lines. In the case of trivial topology, where field lines are unknotted and unlinked closed curves that can be contracted to a point without crossing each other, such magnetic relaxation proceeds uninterrupted. For example, the initial toroidal field configuration upon contraction deforms into a configuration of poloidal fields confined to a tube perpendicular to the original torus. Such configuration is still unstable since small disturbances augmented by the magnetic pressure lead to the increase of length of the tube and decrease of its cross-section. The relaxation eventually leads to a state with zero fields (vacuum).

If, however, the topology of the initial magnetic fields is non-trivial (for example, when flux tubes are knotted or linked), the relaxation stops when flux tubes are tightly knotted or linked. This happens because the "freeze-in" condition (see below) forces topological restrictions on possible changes in field configurations and so any initial knots and links of field lines remain topologically unchanged during relaxation. The energy of a final (equilibrium) state is determined by topology and is bounded from below. One such bound is proportional to $|\mathcal{H}|$, where $\mathcal{H}$ is the magnetic helicity (see Ref. [1] for details).

4.2. *Abelian helicity*

Consider an abelian gauge potential 1-form $A = A_i dx^i$ and the corresponding field-strength 2-form $F = \frac{1}{2} F_{ij} dx^i dx^j$. The helicity for the field inside volume V is defined by

$$\mathcal{H} = \int_V AF. \tag{1}$$

Under a gauge transformation, $A \rightarrow A + d\psi$ and $F \rightarrow F$, and so using the Bianchi identity $dF = 0$ and the Stokes theorem we find

$$\mathcal{H} \rightarrow \mathcal{H} + \int_{\partial V} \psi F. \tag{2}$$

The helicity is thus gauge invariant if the normal component of the field F vanishes on the surface ∂V.

It is easy to calculate the helicity for two linked flux tubes with fluxes Φ_1 and Φ_2. Considering first infinitely thin tubes (centered around the curves C_1 and C_2) and integrating over their cross sections we find

$$\mathcal{H} = \Phi_1 \int_{C_1} A + \Phi_2 \int_{C_2} A. \tag{3}$$

The Stokes theorem now leads to

$$\mathcal{H} = 2n\Phi_1\Phi_2, \tag{4}$$

where n is the Gauss linking number of the two tubes (the algebraic number of times that one tube crosses the surface spanned by the other tube). It is straigtforward to generalize this to the case of linked and/or self-linked thick flux tubes.

4.3. *Non-abelian helicity*

Now we begin to step toward QCD by considering a non-abelian plasma. Since both A and F are conserved during the plasma motion, any combination of A and F is a candidate for a conserved quantum number $\mathcal{H}$. The choice of expression is narrowed by requiring that $\mathcal{H}$ is a topological quantity. In particular, this requires that $\mathcal{H}$ be a surface integral. By analogy with the abelian case, for a conserved non-abelian helicity [16], we choose the corresponding expression with topological properties,

$$\mathcal{H} = \int_V \left(AdA + \tfrac{2}{3}A^3\right). \tag{5}$$

A (time-independent) gauge transformation $A \to g^{-1}Ag + g^{-1}dg$ leads to

$$\mathcal{H} \to \mathcal{H} - \tfrac{1}{3}\mathrm{tr}\int_V \left(g^{-1}dg\right)^3 + \mathrm{tr}\int_{\partial V} gdg^{-1}A. \tag{6}$$

We consider only a limited set of gauge transformations such that do not change the condition $v^\mu A_\mu = 0$; this implies the constraint $v^i\partial_i g = 0$. Because of this constraint only two of the components $g^{-1}\partial_i g$ are independent and so $(g^{-1}dg)^3 = 0$. The helicity is invariant if the normal component of A vanishes on the surface ∂V. This condition is more restrictive than the one needed in the abelian case and implies the latter.

4.4. *"Freeze-in" condition*

In a perfectly conductive relativistic non-abelian plasma, the electric field vanishes in the local frame moving with the plasma. The Lorentz transformation to the rest frame gives $v^\nu F_{\mu\nu} = 0$, where $v^\mu = (1, v^i)$ is the local plasma velocity and

$$F_{\mu\nu} = \partial_\mu A_\nu - \partial_\nu A_\mu + [A_\mu, A_\nu] \tag{7}$$

is the field strength in terms of the gauge potential A_μ. (The equation $v^\nu F_{\mu\nu} = 0$ is the appropriate generalization of its familiar abelian non-relativistic counterpart [17] $E_i = \epsilon_{ijk}v^j B^k$.) In the "hydrodynamic gauge" $v^\mu A_\mu = 0$, from the definition of the field strength $F_{\mu\nu}$ we obtain

$$\partial_0 A_i + (\partial_k A_i)v^k + A_k(\partial_i v^k) = 0 \tag{8}$$

and

$$\partial_0 F_{ij} + (\partial_k F_{ij})v^k + F_{ik}(\partial_j v^k) + F_{kj}(\partial_i v^k) = 0. \tag{9}$$

As a fluid particle moves in space, the rate of change of a local quantity is given by the Lagrange derivative $d/dt = v^\mu\partial_\mu$. Applying this operator to differential forms $A = A_i dx^i$ and $F = \frac{1}{2}F_{ij}dx^i dx^j$, and using Eqs. (8), (9) and

$$(d/dt)dx^i = (\partial_j v^i)dx^j, \tag{10}$$

we find that A and F are constants of motion:

$$dA/dt = 0, \tag{11}$$

$$dF/dt = 0. \tag{12}$$

In other words, for a curve C and a surface S moving with the fluid, integrals $\int_C A$ and $\int_S F$ do not change in time: the magnetic lines are "frozen" into the plasma. Note the integral $\int_C A$ is usually called a Wilson loop in the high-energy physics literature and $\int_S F$ is related to the Wilson loop via Stokes' theorem.

The freeze-in condition derived in this subsection generalizes the well-known (see e.g. Ref. [17]) "freeze-in" condition in MHD to non-abelian case. The non-relativistic analog of this result was obtained in Ref. [18].

5. QCD and Glueballs

5.1. *QCD*

What is the ideal physical system in which to discover and study tight knots and links? We claim it is Quantum Chromodynamics (QCD), so in order to explain our reasoning we first pause to briefly summarize QCD [8].

To lowest order, the standard model of particle physics can be broken into two major sectors, one describing the electroweak processes for leptons, quarks, photons and intermediate vector bosons, like electron-neutrino scattering, and the other describing the strong interactions. The electroweak processes are easier to deal with because the coupling constants of this sector are small (e.g., the fine structure constant $\alpha = e^2/4\pi$ is approximately $1/137$) so that perturbative calculations can be preformed to a high precision and are in beautiful agreement with experiment. On the other hand, the strong interactions which describes the interactions of quarks and gluons, while in principle completely described by QCD, are much more difficult to deal with because at low energy the coupling constant $\alpha_S = {g_S}^2/4\pi$ is $O(1)$. This, for instance, makes the quark-antiquark ($q\bar{q}$) bound state problem analytically intractable.

In somewhat more detail, the QCD is a gauge theory with the lagrangian density

$$\mathcal{L} = -\tfrac{1}{4}F^a_{\mu\nu}F^{a\mu\nu} + \bar{\psi}^i_q \gamma^\mu (D_\mu)_{ij}\psi^j_q - m_q \bar{\psi}^i_q \psi^i_q, \tag{13}$$

where the field strength is

$$F^a_{\mu\nu} = \partial_\mu A^a_\nu + \partial_\nu A^a_\mu - g_S f_{abc} A^b_\mu A^c_\nu \tag{14}$$

and the covariant derivative is

$$(D_\mu)_{ij} = \delta_{ij}\partial_\mu + \tfrac{1}{2}\lambda^a_{ij} A^a_\mu. \tag{15}$$

Here the ψ^j_qs are the Dirac spinor quark fields with color (i) and flavor (q) indices, the Yang-Mills fields A^a_μ (connection) describe the gluons, the f_{abc} are the structure constants for the $SU(3)$ gauge group, and the λ^a_{ij}'s are the triplet representation matrix elements for the quarks. Quarks come in six flavors with six associated quantum numbers conserved by the strong interactions but all except electric charge can be violated by the electroweak interactions. Color charges are confined due to the fact that $SU(3)$ is an unbroken symmetry with asymptotic freedom. I.e., the strong coupling α_S is energy dependent and becomes small at high energy but is large at low energies where it becomes $O(1)$ at a few hundred MeV. Then the theory becomes non-perturbative and confining (color charges cannot be isolated) with all observables color singlet states. These singlets are described by the quark model and are either bosonic $q\bar{q}$ bound state called mesons (pions, kaons, etc.) or fermionic qqq states called baryons (protons, neutrons, etc.). Besides the mesons and baryons there are a number states that do not fit neatly into the quark model. These go by the names of hybrid states, exotic states and glueballs. Hybrid states and exotic states are thought to be unusual combinations of quarks e.g., $qq\bar{q}\bar{q}$ bosonic states, or $qqqq\bar{q}$ fermionic states, most of which have net flavor charges. The glueballs on the other hand are thought to be made of gluons with at most some virtual quark content, hence no flavor charges.

We are now in a position to support our claim that QCD is the ideal physical system in which to discover and study tight knots and links. Here are our reasons:

(1) QCD is a solidly based part of the standard model of particle physics, and much about color confinement and the quark model is already well understood in this context, making much previous work transferable to the problem of tightly knotted flux tubes in this theory.
(2) Unlike plasmas, fluids or other condensed matter systems where flux tubes are excitations of some media with many parameters that could hide universal behavior, flux knots in QCD can exist in the vacuum. Thus continuum states are absent and there are no media parameters

to vary and obscure the universality. Hence, the results in QCD can be far less ambiguous.

(3) The hadronic energy spectrum has been measured over a large range of energies (140 MeV to 10 GeV) and already many hundreds of states are known. We expect that among these, about 15 can be classified as tightly knotted/linked flux tubes states, since as we will see, this is roughly the number of states with the appropriate quantum numbers to qualify as ground state glueballs. These states must also have no valence quarks (i.e., no flavor quantum numbers such as strangeness, charm etc.) in order to be classified as glueballs.

(4) Knotted solitons in QFT are already known to exist [19], [20].

(5) One can efficiently search for new glueball states at accelerators. (Also, data from older experiments still exist and can be reanalyzed to check the predictions of new states described below.)

5.2. *Knot energies*

Consider a hadronic collision that produces some number of baryons and mesons plus a gluonic state in the form of a closed QCD flux tube (or a set of tubes). From an initial state, the fields in the flux tubes quickly relax to an equilibrium configuration, which is topologically equivalent to the initial state. (We assume topological quantum numbers are conserved during this rapid process.) The relaxation proceeds through minimization of the field energy. Flux conservation and energy minimization force the fields to be homogeneous across the tube cross sections. This process occurs via shrinking the tube length, and halts to form a "tight" knot or link. The radial scale will be set by $\Lambda_{\mathrm{QCD}}^{-1}$. The energy of the final state depends only on the topology of the initial state and can be estimated as follows. An arbitrarily knotted tube of radius a and length l has the volume $\pi a^2 l$. Using conservation of flux Φ_E, the energy becomes $\propto l(\mathrm{tr}\Phi_E^2)/(\pi a^2)$. Fixing the radius of the tube (to be proportional to $\Lambda_{\mathrm{QCD}}^{-1}$), we find that the energy is proportional to the length l. The dimensionless ratio $\varepsilon(K) = l/(2a)$ is a topological invariant and the simplest definition of the "knot energy" [21].

Many knot energies have been calculated by Monte Carlo methods [4] and certain types can be calculated exactly (see below), while for other cases simple estimates can be made (see Table 1). For example, the knot energy of the connected product of two knots K_1 and K_2 satisfies

$$\varepsilon(K_1 \# K_2) < \varepsilon(K_1) + \varepsilon(K_2). \tag{16}$$

A rule of thumb is

$$\varepsilon(K_1\#K_2) \approx \varepsilon(K_1) + \varepsilon(K_2) - (2\pi - 4). \tag{17}$$

The order of magnitude of the error in this process is seen to be approximately 5% or less. For example, the knot energies of a granny knot, $3_1\#3_1^*$ (the connected sums of two trefoil knots of opposite chirality), and of a square knot, $3_1\#3_1$ (equal chirality), satisfy Eq. (17) with errors of 4.9% and 5.6% respectively, as compared with the Monte Carlo results (see Table 1). The connected sum or larger knots will give relatively smaller errors since a smaller fraction of the knots are involved in the sum. These results are reliable as long as neither K_1 nor K_2 change their local configurations significantly after the connected summing.

Most of the knot energies in Table 1 have been taken from Ref. [4], but we have independently calculated the energy of 2_1^2, 4_1^3 and 6_1^4 exactly and the energy for several other knots and links approximately. We find $\varepsilon(2_1^2) = 4\pi \approx 12.57$, to be compared with the Monte Carlo value 12.6. We also find $\varepsilon(4_1^3) = 6\pi + 2$ and $\varepsilon(6_1^4) = 8\pi + 3$, where there are no Monte Carlo comparisons available, or needed. Other exactly calculable links can be found in Ref. [6] and an example of a link with energy $10\pi + 5$ is shown in Fig. 1.

5.3. *Model*

In our model, the chromoelectric fields F_{0i} are confined to knotted/linked tubes. After an initial time evolution, the system reaches a static equilibrium state which is described by the energy density

$$\mathcal{E}_E = \tfrac{1}{2}\mathrm{tr}F_{0i}F^{0i} - V. \tag{18}$$

Similar to the MIT bag model [22], we have included a constant potential energy V needed to keep the tubes at a fixed cross-section. The chromoelectric flux Φ_E is conserved and we assume flux tubes carry one flux quantum. To account for conservation of the flux, we add to Eq. (18) the term

$$\mathrm{tr}\lambda\{\Phi_E/(\pi a^2) - n^i F_{0i}\}, \tag{19}$$

where n^i is the normal vector to a section of the tube and λ is a Lagrange multiplier. The energy density should be constant under variations of the degrees of freedom, the gauge potentials A_μ. This leads to the system of equations

$$D^0(F_{0i} - \lambda n_i) = 0, \tag{20}$$

$$D^i(F_{0i} - \lambda n_i) = 0, \tag{21}$$

Fig. 1. An example of an exactly calculable link.

which has the constant field

$$F_{0i} = (\Phi_E/\pi a^2)n_i \tag{22}$$

as its solution. With this solution, the energy is positive and proportional to l and thus the minimum of the energy is achieved by shortening l, i.e., tightening the knot.

The case of chromomagnetic flux tubes can be similarly considered. This requires the confinement of color magnetic flux tubes which is possible if there are light quarks in the spectrum of the theory [23].

Lattice calculations, QCD sum rules, electric flux tube models, and constituent glue models agree that the lightest non–$q\bar{q}$ states are glueballs with quantum numbers $J^{++} = 0^{++}$ and 2^{++} (see Ref. [24]). We will model all J^{++} states (i.e., all f_J and f'_J states listed by the PDG [8]), some of which will be identified with rotational excitations, as knotted/linked chromoelectric QCD flux tubes. We proceed to identify knotted and linked QCD flux tubes with glueballs, where we include all f_J and f'_J states. The lightest candidate is the $f_0(600)$, which we identify with the shortest knot/link, i.e., the 2_1^2 link; the $f_0(980)$ is identified with the next shortest knot, the 3_1

trefoil knot, and so forth. All knot and link energies have been calculated for states with energies less then 1680 MeV. Above 1680 MeV the number of knots and links grows rapidly, and few of their energies have been calculated. However, we do find knot energies corresponding to all known f_J and f'_J states, and so can make preliminary identifications in this region. (We focus on f_J and f'_J states from the PDG summary tables. The experimental errors are also quoted from the PDG. There are a number of additional states reported in the extended tables, but some of this data is either conflicting or inconclusive.)

Our detailed results are collected in Table 1, where we list f_J and f'_J masses, our identifications of these states with knots and the corresponding knot energies.

In Fig. 2 we compare the mass spectrum of f_J states with the identified knot and link energies. Since errors for the knot energies in Ref. [4] were not reported, we conservatively assumed the error to be 1%. A least squares fit to the most reliable data (below 1680 MeV) gives

$$E(G) = (23.4 \pm 46.1) + (59.1 \pm 2.1)\varepsilon(K) \quad [\text{MeV}], \tag{23}$$

with $\chi^2 = 9.1$. The data used in this fit is the first seven f_J states (filled circles in Fig. 2) in the PDG summary tables. Inclusion of the remaining seven (non-excitation) states (unfilled circles in Fig. 2) in Table 1, where either the glueball or knot energies are less reliable, does not significantly alter the fit and leads to

$$E(G) = (26.9 \pm 24.9) + (58.9 \pm 1.0)\varepsilon(K) \quad [\text{MeV}], \tag{24}$$

with $\chi^2 = 10.1$. The fit (23) is in good agreement with our model, where $E(G)$ is proportional to $\varepsilon(K)$. Better HEP data and the calculation of more knot energies will provide further tests of the model and improve the high mass identification.

For comparison, we have fitted the same data in Fig. 3 except that the first glueball is missed out; this results in $\chi^2 = 231$. Similarly, we have fitted the same data in Fig. 4 except that this time the first link is missed out; this results in $\chi^2 = 355$. This is strong evidence that our identification is appropriate.

In terms of the bag model [22], the interior of tight knots correspond to the interior of the bag. The flux through the knot is supported by current sheets on the bag boundary (surface of the tube). Knot complexity can be reduced (or increased) by unknotting (knotting) operations [2, 3]. In terms of flux tubes, these moves are equivalent to reconnection events [25].

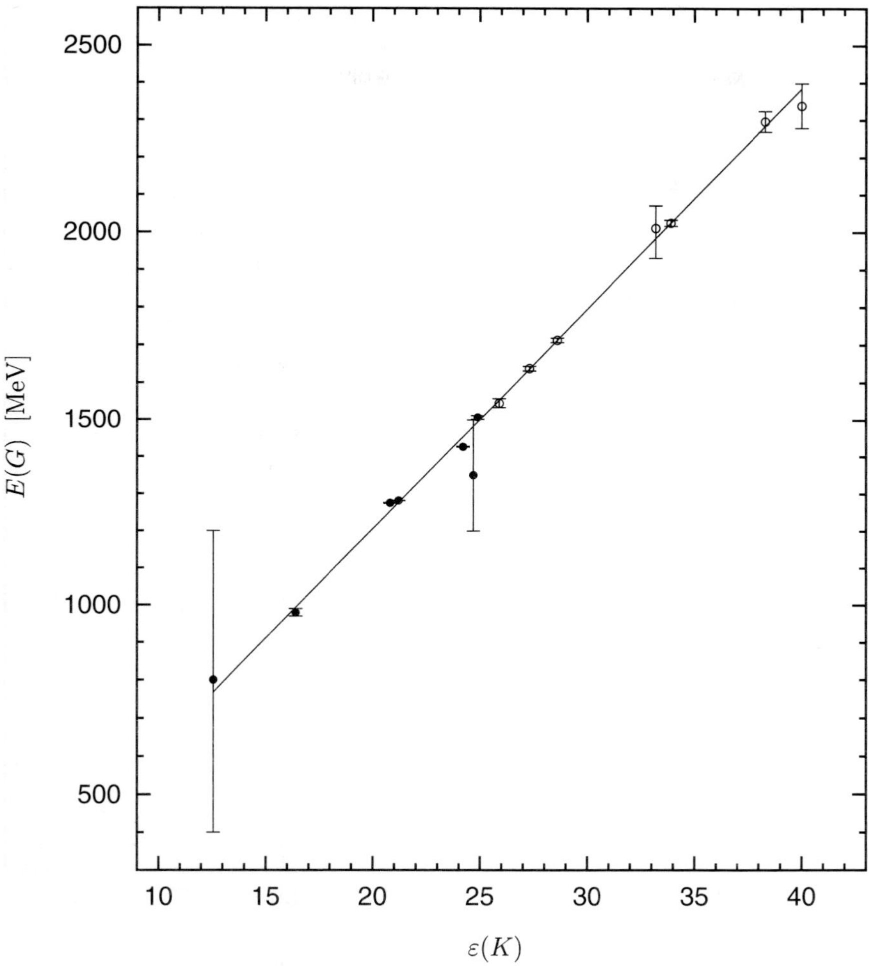

Fig. 2. Relationship between the glueball spectrum $E(G)$ and knot energies $\varepsilon(K)$. Each point in this figure represents a glueball identified with a knot or link. The straight line is our model and is drawn for the fit (23).

Hence, a metastable glueball may decay via reconnection. Once all topological charge is lost, metastability is lost, and the decay proceeds to completion. Two other glueball decay processes are: flux tube (string) breaking; this favors large decay widths for configurations with long flux tube components; and quantum fluctuations that unlink flux tubes; this would tend to broaden states with short flux tube components. As yet we are not able to go beyond providing a phenomenological fit to these qualitative obser-

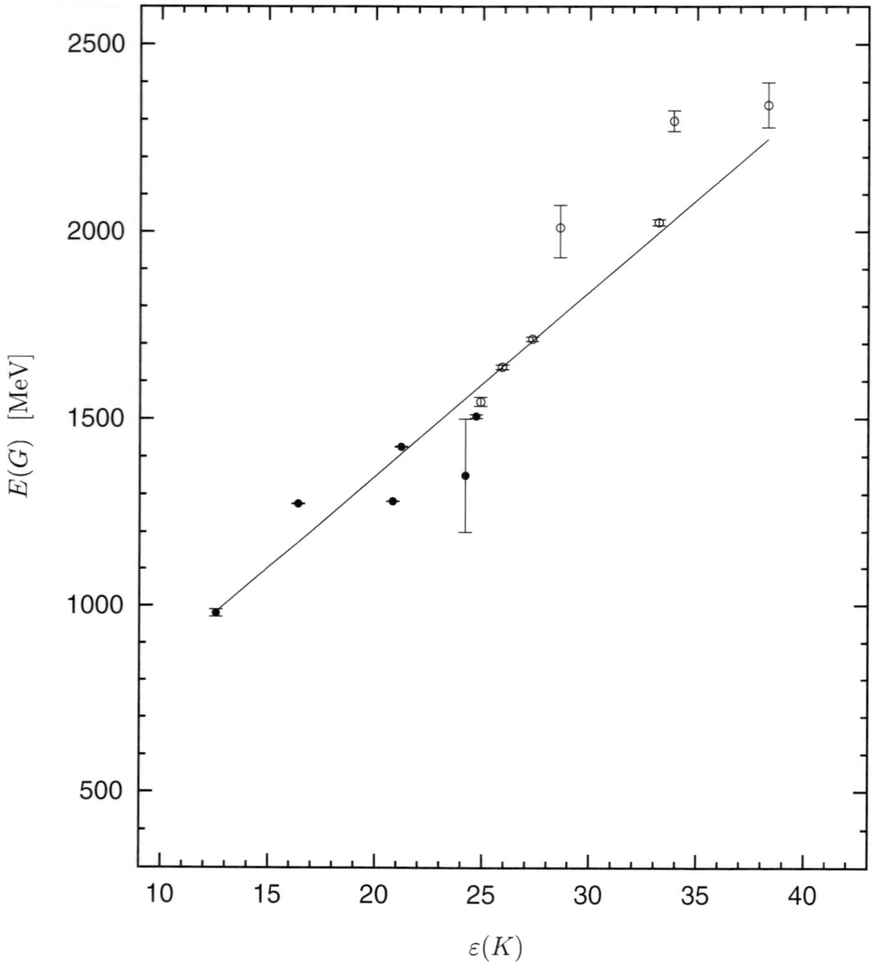

Fig. 3. The same as in Fig. 2, but the first glueball candidate [the $f_J(600)$] eliminated.

vations [7], but hope to be able to do so in the future.

We have assumed one fluxoid per tube. (At the quantum level, flux is not a continuous variable. The quantum unit of flux–a fluxoid–is proportional to $\hbar$, just like angular momentum in quantum mechanics is quantized in units of $\hbar$.) There may be states with more than one fluxoid, but these would presumably have somewhat fatter flux tubes with higher flux densities and higher energies. For example, the two fluxoid trefoil knot 3_1 would certainly have $\varepsilon(K) > 2\,\varepsilon(3_1)$ and a fairly reliable estimate gives $\varepsilon(K) \approx 2\sqrt{2}\,\varepsilon(3_1)$.

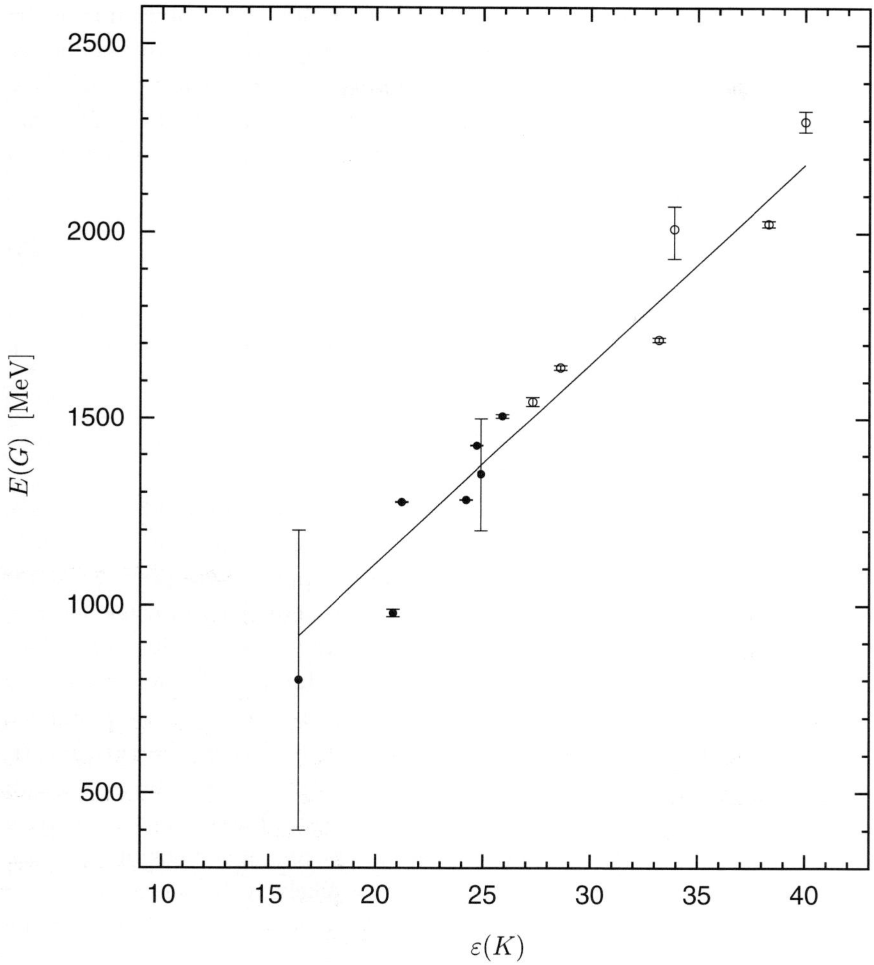

Fig. 4. The same as in Fig. 2, but the first knot/link (the 2_1^2) eliminated.

Hence most multifluxoid states would be above the mass range of known glueball candidates.

6. Discussion and conclusions

We expect all tight knots and links to be described by smooth curves with minimum radius of curvature $2a$ where a is the tube radius. Consider how we would approximate such a curve on a square lattice. The simplest nontrivial example is the double donut. We consider one of its loops (a circle) which we call Γ to be centered on a particular lattice point P. The radius of the circle R is some number of lattice spacings away from P (we assume we are in a lattice plane), along the direction of one of P's nearest neighbors. Assume all other points on Γ are at a distance $d \geq R$ away from P. If we minimize the length L of Γ, then the ratio of L to the circumference of a circle of radius R is

$$\frac{L_{\min}}{2\pi R} = 4/\pi, \tag{25}$$

which is the best approximation (about a 27% error) we can have on a square lattice. Similar arguments show we can do somewhat better on a hexagonal lattice where we can achieve

$$\frac{L_{\min}}{2\pi R} = 2\sqrt{3}/\pi, \tag{26}$$

about a 10% error. However, neither of these approximations are stunningly successful, nor does decreasing the lattice spacing improve the approximations. We assume other knots and links will be approximated on the lattice with similar accuracy. It is interesting to note here that both lattice estimates are too high and that lattice QCD typically predicts glueball masses above 1 GeV, while Monte Carlo predictions for tight knots leads to a lighter glueball spectrum with lightest state just below 800 MeV. Lattice QCD are certainly more sophisticated and involved than the simple estimates we have just made, but we do expect requiring knots/links to lie on a lattice to give glueball masses on the high side. Furthermore, the tube cross section on a lattice is not circular, inhibiting tight packing, and the amount of curvature energy on the lattice is likely to be more than for a smooth curve. We expect these effect to also contribute to increasing the lattice glueball mass estimates.

Let us return to continuum physics and consider a slab of material that can support flux tubes. We have in mind a superfluid or superconductor, but are not limited to these possibilities. Assume further that the flux tubes carry one and only one unit of flux. Next consider manipulating these flux tubes. For instance, consider a hypothetical superconductor where the flux tubes are pinned at the bottom of the slab, say by being attracted to the poles of some magnetic material, and at the top of the slab they are each associated with the pole of a movable permanent magnet, perhaps a magnetic whisker, or fine solenoid. If the north poles of a flux tube is on the top surface, we call it a (+) flux tube (and (−) for the south pole at the top). Since the flux tubes are pinned on the bottom of the slab we can maneuver the top ends into a braid. For example, if there are two flux tubes we can rotate them around each other to make a helically twisted pair. If there are three or more flux tubes we can twist them or braid them. If we measure the forces on the magnets when the flux tubes are being manipulated and use these to find the energy needed to form a braided configuration, then we gain information on the tension of the flux tubes and on the energy stored in a tight braid. If we have both (+) and (−) flux tubes we can bring (+) – (−) pairs together, annihilate the ends so the tube can be pulled into the interior of the superconductor. If we do this at both the top and bottom surfaces for a braided pair then it should be possible to form a bulk knot (similar for links). Assuming they have time to relax to tight configurations, the energy released by the eventual decay of these structures in the bulk should correspond to the universal energy spectrum described above. Perhaps there is no system with all the properties of our ideal superconductor; however, there are systems that contain many if not most of the desired features.

Another collection of physical systems of potential interest for its ability to support vortices, and so knots and braids, are the atomic Bose-Einstein condensates. For example, dilute ^{87}Rb atoms at 80 nK. Recently results have been presented demonstrating laser stirring of these condensates to produce vortices [26]. As many as seven vortices have been seen in a 0.1 mm^2 region. One could hope that advancement in these techniques could lead to more complicated solitonic structures, in particular knots, links, and braids. The advantage of these systems is the experimental accessibility of these structures, and the large percentage of the volume of the system taken up by the solitons. This could potentially dramatically improve the signal-to-noise ratio in the study of tight knots in condensed matter systems.

We have argued that knotted/linked solitonic physical systems have universal mass-energy spectra and have demonstrated this with a detailed example from QCD. We have provided other examples that are good candidates but where the universality has yet to be seen. A system is a candidate if it contains line solitons that can (somewhat paradoxically) relax to their tight knot/link ground state in a time shorter than their decay time. Future work along these lines could tie knot theory to many subfields of physics or to other sciences.

References

1. H. K. Moffatt, J. Fluid Mech. **159**, 359 (1985).
2. D. Rolfsen, *Knots and Links*, Publish or Perish, 1990.

3. L. H. Kauffman, *Knots and Physics*, World Scientific, 2001.
4. V. Katritch, et al., Nature **384**, 142 (1996); V. Katritch, et al., Nature **388**, 148 (1997).
5. K. Millett and E. Rawdon, J. of Comp. Phys. **186**, 426 (2003).
6. R. V. Buniy, T. W. Kephart, M. Piatek and E. Rawdon, in preparation.
7. R. V. Buniy and T. W. Kephart, Phys. Lett. B **576**, 127 (2003).
8. K. Hagiwara et al., Phys. Rev. D **66**, 010001 (2002).
9. H. B. Nielsen and P. Olesen, Nucl. Phys. B **61**, 45 (1973).
10. P. M. Bellan, *Spheromaks : a practical application of magnetohydrodynamic dynamos and plasma self-organization*, Imperial College Press, London 2000.
11. H. Weyl, Am. J. Math. **61**, 461 (1939).
12. A. Gray, *Tubes*, Addison-Wesley, 1990.
13. V. I. Arnold and B. A. Khesin, *Topological Methods in Hydrodynamics*, Springer, 1998.
14. L. Woltier, Proc. Nat. Acad. Sci. **44**, 489 (1958).
15. H. K. Moffatt, J. Fluid Mech. **35**, 117 (1969).
16. R. Jackiw, V. P. Nair, and So-Young Pi, Phys. Rev. D **62**, 085018 (2000).
17. L. D. Landau and E. M. Lifshitz, *Electrodynamics of Continuous Media*, Pergamon Press, 1984.
18. D. D. Holm and B. A. Kupershmidt, Phys. Rev. D **30**, 2557 (1984).
19. A. J. Niemi, *Given at APCTP - ICTP Joint International Conference (AIJIC 97) on Recent Developments in Nonperturbative Quantum Field Theory, Seoul, Korea, 26-30 May 1997*
20. Y. M. Cho, arXiv:hep-th/0409247.
21. H. K. Moffatt, Nature **347**, 367 (1990); S. G. Whittington, D. W. Sumners, and T. Lodge, editors, *Topology and Geometry in Polymer Science*, Springer, 1998; R. A. Litherland, J. Simon, O. Durumeric, and E. Rawdon, Topology Appl. **91**, 233 (1999); G. Buck and J. Simon, Topology Appl. **91**, 245 (1999).
22. A. Chodos, R. L. Jaffe, K. Johnson, Charles B. Thorn, and V. F. Weisskopf, Phys. Rev. D **9**, 3471 (1974); T. DeGrand, R. L. Jaffe, K. Johnson, and J. E. Kiskis, Phys. Rev. D **12**, 2060 (1975).
23. A. S. Goldhaber, Phys. Rept. **315**, 83 (1999).
24. For further discussion of glueballs and their quantum numbers see, for instance G. B. West, Nucl. Phys. Proc. Suppl. **54**A, 353 (1997); N. A. Tornqvist, arXiv:hep-ph/9510256; J. Terning, arXiv:hep-ph/0204012; R. C. Brower, Int. J. Mod. Phys. A16S1C, 1005 (2001); M. Suzuki, Phys. Rev. D **65**, 097507 (2002); M. Teper, Nucl. Phys. Proc. Suppl. **109**, 134 (2002).
25. E. Priest and T. Forbes, *Magnetic Reconnection: MHD Theory and Applications*, Cambridge University Press, 2000.
26. M. Matthews *at al.*, Phys. Rev. Lett. **83**, 2498 (1999); K. Madison *at al.*, Phys. Rev. Lett. **84**, 806 (2000).

Table 1. Comparison between the glueball mass spectrum and knot energies.

State	Mass	K [a]	$\varepsilon(K)$ [b]	$E(G)$ [c]
$f_0(600)$	$400 - 1200$	2^2_1	$12.6\ [4\pi]$	768 [766]
$f_0(980)$	980 ± 10	3_1	16.4	993
$f_2(1270)$	1275.4 ± 1.2	$2^2_1 * 0_1$	$[6\pi + 2]$	[1256]
$f_1(1285)$	1281.9 ± 0.6	4_1	21.2	1277
		4^2_1	(21.4)	(1289)
$f_1(1420)$	1426.3 ± 1.1	5_1	24.2	1454
$\{f_2(1430)$	$\approx 1430\}$ [d]	5_1	24.2	$1454 + \delta'$
$f_0(1370)$	$1200 - 1500$	$3_1 * 0_1$	(24.7)	(1484)
$f_0(1500)$	1507 ± 5	5_2	24.9	1496
$\{f_1(1510)$	$1518 \pm 5\}$	5_2	24.9	$1496 + \delta$
$f'_2(1525)$	1525 ± 5	5_2	24.9	$1496 + 3\delta$
$\{f_2(1565)$	$1546 \pm 12\}$	5^2_1	(25.9)	(1555)
$\{f_2(1640)$	$1638 \pm 6\}$	6^3_3	((27.3))	((1638))
		$(2^2_1 * 0_1) * 0_1$ [e]	$[8\pi + 3]$	[1686] [f]
$f_0(1710)$	1713 ± 6	6^3_2	((28.6))	((1714))
		$3_1\#3^*_1$	28.9 (30.5)	1732 (1827)
		$3_1\#3_1$	29.1 (30.5)	1744 (1827)
		$2^2_1 * 2^2_1$	$[8\pi + 4]$	[1745]
		6_2	29.2	1750
		6_1	29.3	1756
		6_3	30.5	1827
		7_1	30.9	1850
		8_{19}	31.0	1856
		8_{20}	32.7	1957
$f_2(2010)$	2011^{+60}_{-80}	7_2	33.2	1986
$f_4(2050)$	2025 ± 8	8_{21}	33.9	2028
		8_1	37.0	2211
		$10_{161,162}$	37.6	2247
$f_2(2300)$	2297 ± 28	8_{18}, 9_1	38.3	2288
$f_2(2340)$	2339 ± 60	9_2	40.0	2389
		10_1	44.8	2672
		11_1	47.0	2802

[a] Notation n^l_k means a link of l components with n crossings, and occurring in the standard table of links (see e.g. Ref. [2]) on the kth place. $K\#K'$ stands for the knot product (connected sum) of knots K and K' and $K * K'$ is the link of the knots K and K'.

[b] Values are from Ref. [4] except for our exact calculations of 2^2_1, $2^2_1 * 0_1$, and $(2^2_1 * 0_1) * 0_1$ in square brackets, our analytic estimates given in parentheses, and our rough estimates given in double parentheses.

[c] $E(G)$ is obtained from $\varepsilon(K)$ using the fit (23).

[d] States in braces are not in the Particle Data Group (PDG) summary tables.

[e] This is the link product that is not $2^2_1 * 2^2_1$.

[f] Resonances have been seen in this region, but are unconfirmed [8].

CHAPTER 4

KNOT DYNAMICS IN A DRIVEN HANGING CHAIN: EXPERIMENTAL RESULTS

Andrew Belmonte

The W. G. Pritchard Laboratories, Department of Mathematics, Pennsylvania State University, University Park, PA 16802, USA
E-mail: belmonte@math.psu.edu

Knotted segments are a natural occurence in linear chains subjected to fluctuations. We study experimentally the size and dynamics of knots in a shaken metal chain. For a given set of conditions, we find that the time it takes a knot to slip off the end of the chain depends sensitively on its topology.

1. Introduction

The topological study of knots has in recent years received many new ideas from the *physical* study of knots, or from physical considerations of ideal knots, such as thickness or crossings in projection; this volume contains several chapters on progress made in these directions (for overviews see e.g.[1,2]). In this chapter, we will discuss the results of a physical experiment with a shaken hanging chain. At certain frequencies, spontaneous knots arise. The study of the dynamics of knots with our chain experiment may lead to new considerations in the theory of knots.

Practically all strings or chains, subjected to complicated motions, will end up knotted - from dangling ropes on sailboats to hair to extension cords to DNA[3,4] and other polymers[5]. The competing effects which lead to spontaneous knots are the random exploration of space performed by a fluctuating chain versus the excluded-volume effect - the chain cannot pass through itself. In describing the physics of a real knot, including its ability to bind, one must consider the detailed distribution of tension in the knot, relative to the self-contact points[6,7]. Thus both tension properties and knot topology contribute to strength, corresponding to the well-known fact that some knots hold much better than others. We do not consider here the role

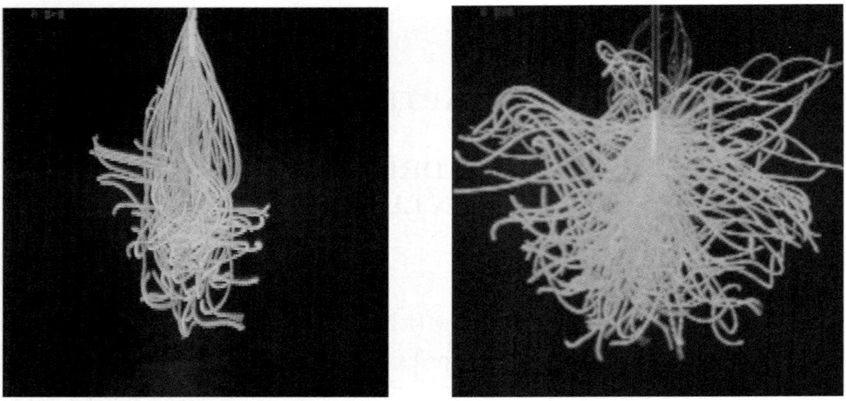

Fig. 1. Experimental collage of the position of a shaken hanging chain at successive times, for $L = 22.7$ cm at two different frequencies (side view). About 8 seconds of data are shown, with $\Delta t = 1/30$ s.

of fluctuations[8], or of charge (long range self-repulsion)[9], in tightening or loosening knots.

2. Our Experimental System

Our experimental study of knots was the accidental result of another study, on the dynamics and patterns of a periodically shaken hanging chain[10]. We used a thin stainless steel chain made of beads of diameter $D = 2.4$ mm, maximum inter-bead spacing 3.4 mm, and linear mass density $\rho = 0.095$ g/cm. The setup is a simple wall-mounted bracket which holds a stepper motor and linear translation stage. The chain is driven sinusoidally by vertically oscillating its upper end via a wheel attached to the translation stage and to the motor, at frequencies f from 1 to 4 Hz and amplitudes $A = 1.3$ cm, 2.0 cm, and 2.7 cm. For more details see Belmonte *et al* [10].

In that experimental study we found several transitions between distinct dynamic states, varying the driving frequency f at fixed amplitude A, with chain lengths L ranging from 7 cm to 80 cm. At certain frequencies we observe a wildly energetic motion (Fig. 1), during which the chain explores its surrounding space, occasionally colliding with itself. For different L or A these same transitions are seen at different frequencies. By introducing the nondimensional amplitude $\varepsilon \equiv A/L$ and frequency $\omega \equiv 2\pi f\sqrt{L/g}$, the transition frequencies collapse onto continuous curves in (ε, ω) parameter space, as shown in Fig. 2. These boundaries were shown by the linear stabil-

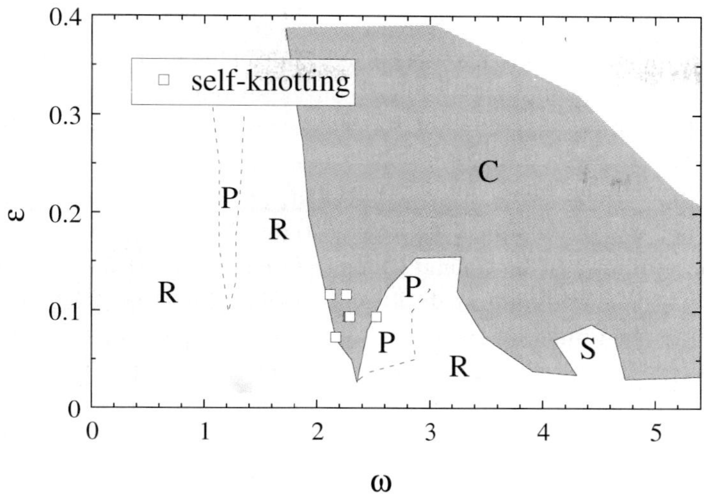

Fig. 2. Points in the (ϵ, ω) parameter space of the hanging chain for which self-knotting (spontaneous knots) have been observed. The letters label the dynamic states of the chain[10]: rod-like (R), pendular (P), stellate (S), and complex (C).

ity analysis of a vertical hanging chain to correspond to Arnol'd tongues[10].

Note that the same metal chain was used independently at Los Alamos to study the dynamics of knots under slightly different conditions: the chain was bounced on a vibrating plate[11,12].

2.1. *Spontaneous Knots*

At certain frequencies in the region of parameter space labelled C for 'complex' in Fig. 2, we noticed previously that the seemingly random motion of the chain could lead to a knot[10]. This is apparently well-known to people who perform certain activities involving long ropes or cords (sailors, rock concert roadees, climbers). The knots vary in complexity, though an early puzzle was that the most frequently observed knot was the figure-eight, and not the simpler trefoil[10]; this is surprising because a simple assumption of random chain motion predicts that knots with fewer crossings should occur more often. The apparent rarity of spontaneous trefoils was explained by the observation that, if a trefoil was tied by hand into the chain, it would rapidly slip off the free end[10].

3. How Long is a Knot?

Although physical knots have been used by humans for centuries, from a mathematical viewpoint the earliest considerations of knots were topological. In this setting the knot is comprised of a knotted curve embedded in $\mathbb{R}^3$; the curve is zero-dimensional (it has no width). Topologically, all representations of a given knot are equivalent, i.e. all are equally valid. Thus it does not matter whether the knot is tight or loose. However, with the beginnings of interest in distinguishing different configurations or representations of the same knot, such as symmetric, relaxed, or with respect to some minimized function such as total curvature[2], more physical considerations were introduced. One could say that the study of knots has expanded from pure topology to physics[1].

A particularly physical concept is the "length" or "ropelength" of a knot; overviews of the history of this idea can be found in[1,2,13]. The basic question is: how much length ΔL of the curve is taken up in a knot? This is a meaningless question without some width or thickness to the curve in which the knot has been tied. Much work has already been done on closed knots (that is, standard topological knots), both numerical and analytic. Recently, these studies have been extended to open knots, which are more often seen in the physical world. Our study involves the dynamics of open knots, including both tying and untying, as well as what we focus on in this chapter: the motion of the knot along the chain.

3.1. *Previous Work*

Pierański *et al.* used the numerical SONO ('shrink on no overlap') algorithm[14] to find 'minimal energy' configurations of several open knots[15]. These knots can be thought of as being in some specifically defined 'tightest' configuration, which results in a certain amount of total length ΔL in the knot itself - the 'ideal length', or knot displacement length. In Fig. 3 we compare previously tabulated values of this ideal length for open and closed knots, both computational[15] and experimental (tied with climbing rope)[16]. Note that we have adjusted the normalization factor for the data in Diao *et al.* to be the diameter and not the radius[16], which gives an extra factor of 2.

3.2. *Open Knot Length in Our Chain*

Using the metal chain from our experiments described above[10], we measured the displacement length ΔL for several open knots. An illustration of

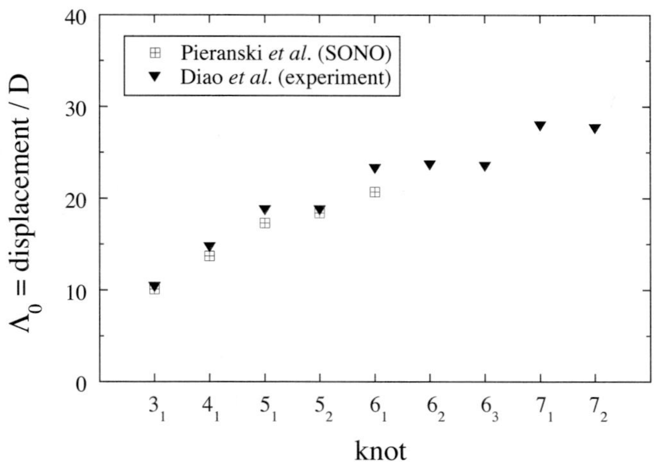

Fig. 3. Comparison of the knot displacement length for open knots seen numerically (with the SONO algorithm)[15] and experimentally[16], as a function of knot type.

this displacement was given in Pierański *et al.*[15] In our system, the knots were tied by hand, and jostled lightly until they were compact; no attempt was made to tighten them by pulling. The reason was that we wanted to obtain a knot size which was appropriate to the dynamic processes experienced by a knot in the shaking chain, in its 'native configuration'. The experimental values of ΔL for several knots in our experimental chain are listed in Table 1.

The open circles in Figure 4 represent the normalization $\Lambda_0 = \Delta L/D$, using $D = 2.4$ mm for the diameter of our chain. However, it is clear from examining these knots that the diameter is not playing the same role as it does in the SONO simulations of ideal knots. This is because our knots are *stiff*; the bending stiffness of the chain provides a hard lower limit to the radius of curvature allowed in the knot configuration. In other words, there is a second lengthscale, the minimum bending radius R of the chain itself, which becomes important because $R > D/2$. Here the diameter plays the same limiting role for the SONO knots, because volume exclusion effectively

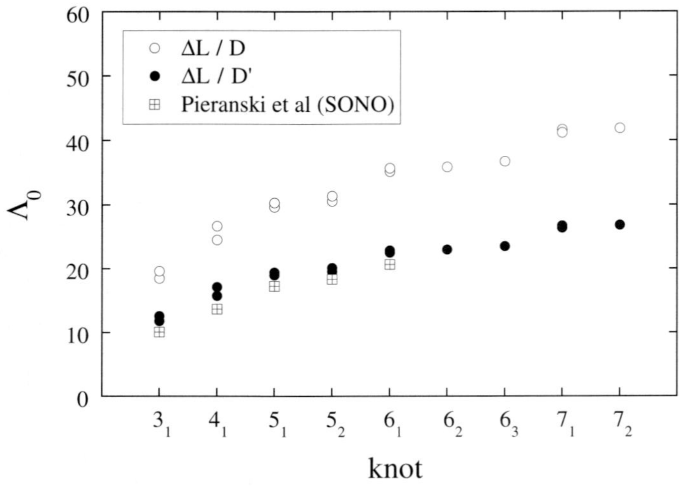

Fig. 4. The normalized displacement length Λ_0 of open knots as measured in our chain, compared to numerical results using the SONO algorithm[15].

defines a bending stiffness for the knot. We will therefore define a knot to be a 'stiff knot' if it is tied in a rope for which $R > D/2$. In this case, the minimum distance across a loop in the knot is either the diameter D or the radius of curvature to centerline of chain: $D' = R - (D/2)$. The difference is determined by whether or not the point of closest approach is in the local plane of curvature. Roughly speaking, the increased stiffness acts as a larger diameter in the plane of curvature (for some of the crossings). For

Table 1. Data on knots in our chain presented in this chapter.

Open knot	ΔL ± 1.5 [mm]	T_u [s]
3_1	44.6	14.0 ± 6.6
4_1	60.0	66.5 ± 1.3
5_1	70.3	12.6 ± 1.2
5_2	72.5	55.2 ± 21.4
6_1	82.9	241.5 ± 89.0
6_2	84.0	119.8 ± 30.0
6_3	86.0	-
7_1	97.1	22.1 ± 10.2
7_2	98.0	-

T_u: for knot tied 10 cm from bottom, with chain $L = 120$ cm, shaken at $f = 3.6$ Hz.

our chain we have $D' = 3.65$ mm, and using this lengthscale we redefine the knot displacement length as $\Lambda_0' = \Delta L/D'$. We plot our experimentally measured Λ_0' in Figure 4, and find that it is in much better agreement with the displacement length for open SONO knots[15]. This suggests that the knots we observe in our chain are close to an 'ideal' configuration as defined for *open stiff knots.*

The consideration of stiff knots, like the consideration of the role of fluctuations, brings interesting physical effects into the theory of knots. In each case more room is introduced into the knot configuration, which should affect quantities such as the ropelength. Note that for our chain, the "flexibility" f as defined elsewhere[17] is approximately

$$f = \frac{\kappa}{2/D} \simeq 0.3;$$

by definition $f = 1$ for a perfectly flexible rope[17], thus our knots are indeed rather stiff.

4. Untying Dynamics: Dependence on Knot Type

It was observed initially that a trefoil knot (3_1) would slip off in our chain experiment, whereas a 4_1 knot did not appear to move: when tied 10 cm from the free end of a 28 cm chain, a 3_1 takes about 10 seconds[10]. This observation helped to explain the apparent rarity of spontaneous trefoils in the experiment. It was only later, when different frequencies were tried more systematically, that we observed that a 4_1 would *also* slip off the chain, albeit not nearly as quickly as a 3_1. This opened up the study of the dependence of the slipping motion on knot type.

To test the dependence of slipping motion on knot type, we fixed all other experimental parameters, such as the driving frequency (3.6 Hz), amplitude (2.7 cm), and chain length (120 cm), and tied various knots at a fixed distance from the end of the chain. The time T_u for the knot to untie was measured for knots tied 10 cm from the free end of the chain. As expected from our previous study[10], the 4_1 knot took longer to untie than the 3_1. However, we were surprised to find that there is a strong and complicated dependence of T_u on knot type, ranging from an average of 14 s for the 3_1 to greater than 5 minutes for the 6_3 - in fact, we were not able to measure T_u for this knot.

The average values for T_u as a function of knot type are listed in Table 1, and plotted in Fig. 5. The dependence of T_u on knot type appears quite complicated; note additionally that $T_u(9_1) \simeq 140$ s. For each knot at least

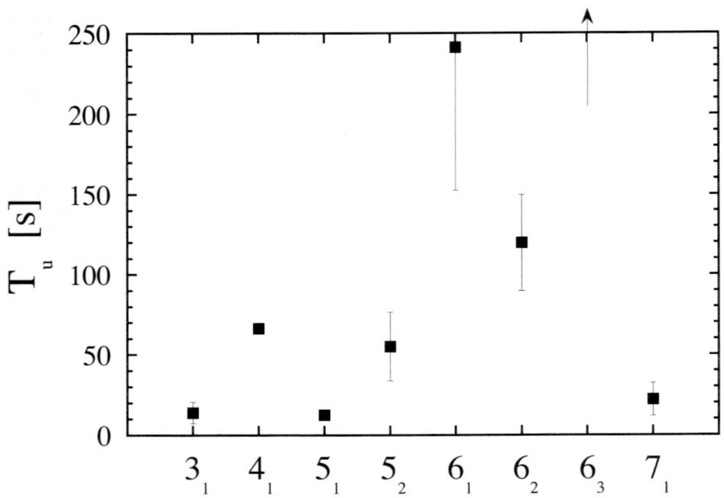

Fig. 5. Dependence of the untying time on knot type, shown for an initial starting length of 10 cm from the bottom (chain with L = 120 cm, f = 3.6 Hz); the error bars indicate the standard deviation in the distribution of times for the given knot. The arrow represents the fact that the 6_3 knot did not untie during the course of the experiment.

four trials were made, and the individual times measured varied widely. The standard deviations are also reported in Table 1 (as $\pm$) and are plotted as error bars in Fig. 5. This wide variation in untying time suggests a stochastic process, which is currently being modeled. It is not clear to us why some knots display more variance in their untying times than others.

Overall, we do not currently understand the role of topology in determining the apparent speed of a knot on the shaking chain. However, it is interesting to compare the order of knot type from fastest to slowest: $3_1, 5_1, 7_1, 5_2, 4_1, 6_2, 6_1$, with the ordering of 'knot strength' measured by Pieranski *et al* [18], running from easiest to break to hardest, based on the tension required: $3_1, 5_1, 5_2, 7_1, 6_1, 6_2, 4_1$.

Following a suggestion made after a seminar at the University of Pennsylvania in October 2002 (A. Ulyanov, private communication), we found experimentally that the torus knots fall relatively faster than the other

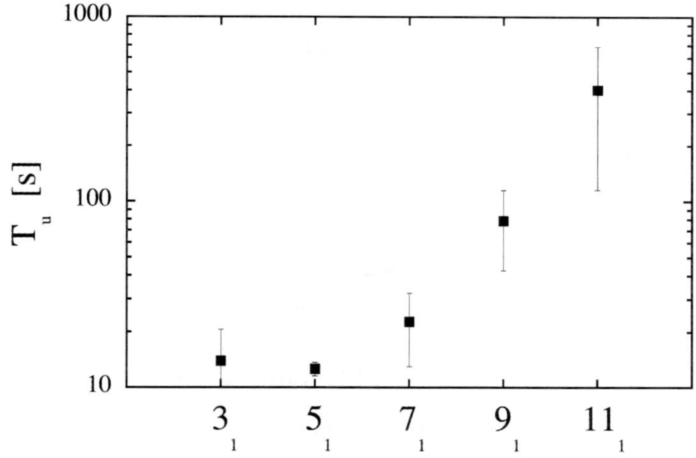

Fig. 6. Dependence of the untying time on knot type for torus knots, under the same conditions as in Fig. 5.

knots. By restricting the plot to only torus knots, as shown in Fig. 6, we at least find an approximately increasing function. A study focusing only on the dynamics of the motion of torus knots and other knot families (such as the twist knots) is currently in progress.

5. Conclusions

We have presented here a new experimental system in which certain physical aspects of knots can be studied, specifically the dynamics of tying and untying, and the slipping motion of a knot along a fluctuating chain. Surprisingly, the slipping motion is very sensitive to topological aspects of the knots, with torus knots being relatively the most efficient at slipping on the chain. However the exact topological quantity which couples to the knot motion is not yet known. This experimental study raises interesting new questions on the coupling of physical properties and topology in knots, for which much remains to be understood.

Acknowledgments

This work was supported by the Alfred P. Sloan Foundation and the National Science Foundation (CAREER grant DMR-0094167). I would like to thank M. J. Shelley and A. Ulyanov for helpful discussions, J. Banyas,

R. Geist, and M. C. Sostarecz for experimental assistance, and G. Buck and A. Stasiak for comments on the manuscript.

References

1. A. Stasiak, V. Katritch, L. H. Kauffman, Eds., *Ideal Knots* (World Scientific, 1998).
2. J. O'Hara, *Energy of Knots and Conformal Geometry* (World Scientific, 2003).
3. S. A. Wasserman and N. R. Cozzarelli, *Science* **232**, 951 (1986).
4. D. Sumners, *Math. Intelligencer* **12**, 71 (1990).
5. P. G. de Gennes, *Scaling Concepts in Polymer Physics*, (Cornell Press, Ithaca, NY, 1979).
6. B. F. Bayman, *Am. J. Phys.* **45**, 185 (1977).
7. J. Walker, *Scientific American* (August 1983), p.120.
8. R. Metzler, A. Hanke, P. Dommersnes, Y. Kantor, M. Kardar, *Phys. Rev. Lett.* **88**, 188101 (2002).
9. P. Dommersnes, Y. Kantor, M. Kardar, *Phys. Rev. E* **66**, 031802 (2002).
10. A. Belmonte, M. J. Shelley, S. T. Eldakar, and C. H. Wiggins, *Phys. Rev. Lett.* **87**, 114301 (2001).
11. E. Ben-Naim, Z. A. Daya, P. Vorobieff, and R. Ecke, *Phys. Rev. Lett.* **86**, 1414 (2001).
12. M. B. Hastings, Z. Daya, E. Ben-Naim, R. Ecke, *Phys. Rev. E* **66**, 25102 (2002).
13. J. Cantarella, R. Kusner, and J. M. Sullivan, *Invent. Math.* **150**, 257 (2002).
14. P. Pierański, in Ref. 1, p. 20.
15. P. Pierański, S. Przybył, A. Stasiak, *Eur. Phys. J. E* **6**, 123 (2001).
16. Y. Diao, C. Ernst, & E. J. Janse van Rensburg, in Ref. 1, p. 52.
17. G. Buck, E. J. Rawdon, *Phys. Rev. E* **70**, 11803 (2004).
18. P. Pierański, S. Kasas, G. Dietler, J. Dubochet, A. Stasiak, *New J. Phys.* **3**, 10.1 (2001).
19. A. Stasiak, V. Katritch, J. Bednar, D. Michoud, and J. Dubochet, *Nature* **384**, 122 (1996).

CHAPTER 5

BIARCS, GLOBAL RADIUS OF CURVATURE, AND THE COMPUTATION OF IDEAL KNOT SHAPES

M. Carlen,* B. Laurie,** J.H. Maddocks,* J. Smutny*

**Institute of Mathematics B*
Swiss Federal Institute of Technology
Lausanne, CH-1015, Switzerland
E-mail: mathias.carlen, john.maddocks, jana.smutny@epfl.ch

***17 Perryn Road*
London, W3 7LR, England
Email: ben@algroup.co.uk

We combine the global radius of curvature characterisation of knot thickness, the biarc discretisation of space curves, and simulated annealing code to compute approximations to the ideal shapes of the trefoil and figure-eight knots. The computations contain no discretisation error, and give rigorous lower bounds on thickness of the true ideal shapes. The introduction of a precise definition of a contact set for an approximately ideal shape allows us to resolve previously unobserved features. For example, in our approximations of both the ideal trefoil and figure-eight knots, local curvature is within a rather small tolerance of being active, i.e. achieving thickness, at several points along the knot.

1. Introduction

An ideal shape $\bar{\mathcal{K}}$ of a knot is the particular geometrical configuration of the prescribed knot type that maximises the minimal distance from self-intersection in a non-dimensional, scale invariant way involving the *rope length* $(\lambda/\Delta)[\mathcal{K}]$:

Definition 1: An *ideal shape*[3,15,35] of the knot isotopy class $[\mathcal{K}^*]$ is a knot shape $\bar{\mathcal{K}}$ that minimises the functional length/thickness within the isotopy class, i.e. an ideal knot shape is a solution of

$$\frac{\lambda(\mathcal{K})}{\Delta[\mathcal{K}]} \to \min$$

subject to $\mathcal{K} \in C(S^1, \mathbb{R}^3)$, $\mathcal{K} \simeq \mathcal{K}^*$.

Here $\lambda(\mathcal{K})$ is the arc length of the knot shape $\mathcal{K}$, while $\Delta[\mathcal{K}]$ is the thickness or distance from self-intersection, whose definition is discussed below.

It is now known[5,11,13] that ideal knot shapes exist in the space of $C^{1,1}$-curves for each knot isotopy class, so that curvature exists almost everywhere. But a sharper classification of regularity of ideal shapes is not well understood. For example there is no known example for either knots or links that excludes the possibility that ideal shapes could be C^∞-curves except at a finite number of points. At least one ideal shape, namely the circle which is the ideal shape for the unknot, is C^∞, while for links it is known that there are various ideal shapes assembled from straight line segments and arcs of circles which fail to be C^2 at a finite number of points.[5]

It has been shown[17] that the thickness Δ of a curve $\mathbf{q} \in C^2$ (with parametrisation $\mathbf{q}(s)$ for $s \in I$) is governed by two numbers: the minimal radius of curvature and half of the minimal distance between *double critical points* or *points of stationary approach*, i.e. pairs of distinct points $\mathbf{q}(s)$ and $\mathbf{q}(\sigma)$ and associated curve tangents $\mathbf{q}'(s)$ and $\mathbf{q}'(\sigma)$ satisfying

$$\mathbf{q}'(s) \cdot (\mathbf{q}(s) - \mathbf{q}(\sigma)) = \mathbf{q}'(\sigma) \cdot (\mathbf{q}(s) - \mathbf{q}(\sigma)) = 0. \tag{1}$$

More precisely when we denote by *dc* the set of arguments $(s, \sigma) \in I \times I$ with $s \neq \sigma$ that satisfy (1), then[17]

$$\Delta[\mathbf{q}] = \min\left\{\min_{s \in I} \rho(s), \tfrac{1}{2} \min_{(s,t)\in dc} |\mathbf{q}(s) - \mathbf{q}(t)|\right\}, \tag{2}$$

where $\rho(s)$ is the classic radius of curvature.

Another approach to characterisation of thickness is in terms of *global radius of curvature*.[12] The global radius of curvature of a continuous curve $\mathbf{q}$ at $s \in I$ is defined by

$$\rho_g(s) := \inf_{\substack{\sigma, t \in I \\ s \neq \sigma \neq t \neq s}} \mathsf{ppp}(s, \sigma, t), \qquad s \in I, \tag{3}$$

where $\mathsf{ppp}(s, \sigma, t)$ denotes the radius of the circle passing through the three points $\mathbf{q}(s)$, $\mathbf{q}(\sigma)$, and $\mathbf{q}(t)$. When the curve $\mathbf{q}$ is simple and smooth, the function ρ_g is continuous and $0 \leq \rho_g(s) \leq \rho(s)$ for all $s \in I$. Moreover, the global radius of curvature function $\rho_g(s)$ at s is achieved either by the local radius of curvature $\rho(s)$, or by a circle passing through $\mathbf{q}(s)$ and some distinct point $\mathbf{q}(\sigma)$ at which the circle is tangent, so that to evaluate $\rho_g(s)$ it suffices[12] to consider the minimisation (3) but now with the restriction $\sigma = t$, i.e.

$$\rho_{\mathsf{pt}}(s) := \inf_{\substack{\sigma \in I \\ s \neq \sigma}} \mathsf{pt}(s, \sigma), \qquad s \in I, \tag{4}$$

where the two argument function $\mathsf{pt}(s, \sigma)$ denotes the radius of the circle that passes through the two points $\mathbf{q}(s)$ and $\mathbf{q}(\sigma)$, and that is also tangent to the curve at $\mathbf{q}(\sigma)$. For smooth, closed curves the two functions $\rho_{\mathsf{pt}}(s)$ and $\rho_g(s)$ coincide.[12,14,33]

The thickness functional Δ_{pt} can be defined by[12,14,33]

$$\Delta_{\mathsf{pt}}[\mathbf{q}] := \inf_{s \in I} \rho_{\mathsf{pt}}(s) = \inf_{\substack{s, \sigma \in I \\ s \neq \sigma}} \mathsf{pt}(s, \sigma), \tag{5}$$

and for a closed, smooth $\mathbf{q}$

$$\Delta_{\mathsf{pt}}[\mathbf{q}] = \Delta[\mathbf{q}]. \tag{6}$$

The characterisation of $\Delta[\mathbf{q}]$ using the global radius of curvature function ρ_{pt} is analytically attractive because it is explicit and because it simultaneously captures both possibilities manifested in (2), namely that the thickness is achieved by a local radius of curvature, or that it is achieved by half of a critical self-distance. The global radius of curvature function $\rho_g := \rho_{\mathsf{pt}}$ also provided the first known necessary condition for ideality, namely on an ideal shape $\rho_g(s)$ is constant on curved segments.[12,30]

The properties of thickness and various global radii of curvature of curves (and other manifolds) under weakened regularity assumptions continues to be an active area of research.[4,8,9,13,14,29,30,33] In all these treatments a recurring issue in the analysis is to know whether or not the case of local curvature is active in either (2) or the realisation of $\rho_{\mathsf{pt}}(s)$ in (4).

Although the ideal knot problem is easy to state, the only known ideal shape is that for the unknot, where it is a circle. (As mentioned earlier, other simple cases arise for links, where various shapes assembled from straight line segments and arcs of circles have been shown to be ideal.[5]) Accordingly, it is of interest to have numerical computations of ideal knot shapes that allow quantities such as local and global curvatures to be approximated accurately.

The purpose of this article is to outline a computational approach to finding approximate ideal knot shapes via simulated annealing applied to the biarc discretisation of a space curve, and to present our first results in the particular cases of the trefoil and figure-eight knots. The structure of the presentation is as follows. In section 2 we describe properties of the two argument $\mathsf{pt}(s, \sigma)$ function, its relation to the definition of exact and approximate contact sets, and criteria for assessing closeness to ideality. Then in section 3 we introduce the biarc discretisation and contrast its properties with those of prior approaches. The key features of the computational scheme using biarcs are outlined in section 4, while our results for, respectively, the trefoil and figure-eight knots are presented in sections 5 and 6. Finally in section 7 we discuss the strengths and weaknesses of our approach. A much more detailed treatment of much of the material described here can be found in the thesis of Smutny.[33]

2. Criteria for the assessment of closeness to ideality

In order to be able to assess the quality of different computational approximations to ideal shapes it is desirable to establish criteria with which to estimate closeness to ideality of different shapes. Because Definition 1 involves minimisation of rope length, curves of the prescribed knot class $\mathcal{K}$ for which the arc length $\lambda(\mathcal{K})$ and thickness $\Delta[\mathcal{K}]$ can be computed to a known accuracy provide rigorous upper bounds for the ideal rope length, and the value of the upper bound provides a relative ranking of approximations. However in the absence of good lower bounds,[a] closeness of a given shape to ideality must be assessed by other means, and ideally these means should be robust for numerics. We now describe various criteria of this type, and later apply them to the data for trefoil and figure-eight knot shapes presented in sections 5 and 6.

The first such test arises because, as already mentioned in section 1, the global radius of curvature function ρ_{pt} is constant and equal to the thickness on curved segments of ideal configurations. Thus the variation $[\max \rho_{\mathsf{pt}} - \min \rho_{\mathsf{pt}}]$ along curved segments of an approximation gives a measure of closeness to ideality. It is our experience that this test alone is not a stringent discriminator of closeness to ideality. Our second, and much more demanding, test is based upon the *contact* and *μ-contact* sets of a configu-

[a]Recently[6] it has been shown analytically that any nontrivial knot has rope length greater than 31.22. The very generality of this result means that it cannot be a tight bound for any knot other than the trefoil, and even for the trefoil this rigorous bound is probably unduly pessimistic. Thus the result does not seem to be sufficiently sharp for our specific purpose of judging quality of computations.

ration. These contact sets are in turn based on properties of the pt and ρ_{pt} functions.

We first remark that it is the two-argument function $\mathsf{pt}(s,\sigma)$ that determines both the global radius of curvature function ρ_{pt} via (4) and the curve thickness via (5), so that it is of interest to understand its properties. For any closed curve $\mathbf{q}(s) \in C^{1,1}$ with positive thickness, the function $\mathsf{pt}(s,\sigma)$ is doubly periodic and positive in $I \times I$, and is either infinite or continuous away from the diagonal. It is not, in general, symmetric in its arguments on $I \times I$, but it can be shown that its global minimisers are symmetrically located, i.e. they arise either on the diagonal at points of the form (s,s), or at pairs of double critical points (s,σ) and (σ,s) with $s \neq \sigma$ at which (1) is satisfied. Informally, the *contact set* of a curve is the set of points (s,σ) (with $s = \sigma$ possible) that realise the global minimum of pt, or equivalently, the thickness in (5). For a general curve $\mathbf{q}$—and in particular even for a curve that is arbitrarily close to, but not precisely ideal—the contact set is likely to be a single pair of points.

We note that pt is bounded below by half of the two-point Euclidean distance function, i.e. by $\mathsf{pp}(s,\sigma) := \frac{1}{2}|\mathbf{q}(s) - \mathbf{q}(\sigma)|$. At double critical points, and therefore at any off-diagonal global minimiser of pt, $\mathsf{pp}(s,\sigma) = \mathsf{pt}(s,\sigma)$, but the global minimum of pp is always zero, and is achieved along the diagonal for any curve. Thus the usual Euclidean distance is not well-suited to the computation of contact sets.

Plots of the pt function on approximations to the ideal trefoil and figure-eight knots are provided in sections 5 and 6. Another function that will be plotted is the multi-valued extension $\hat{\rho}_{\mathsf{pt}}(s)$ of the global radius of curvature function ρ_{pt} defined in (4). The function $\hat{\rho}_{\mathsf{pt}}(s)$ is defined by replacing the global infimum over σ appearing in (4) by local minimisation in σ, so that each branch of $\hat{\rho}_{\mathsf{pt}}$ is of the form $\hat{\rho}_{\mathsf{pt}}(s) = \mathsf{pt}(s,\hat{\sigma}(s))$ where the function $\hat{\sigma}(s)$ is locally defined by the property that for each s, $\mathsf{pt}(s,\cdot)$ is locally minimised at $\sigma = \hat{\sigma}(s)$. It can be shown[33] that local minima of $\mathsf{pt}(s,\sigma)$ in σ can arise either a) on the diagonal, b) at local maxima or minima of $\mathsf{pp}(s,\sigma)$ in σ, or c) for σ such that the centre of the circle realising $\mathsf{pt}(s,\sigma)$ lies on the polar axis[37] to the curve $\mathbf{q}$ at $\mathbf{q}(\sigma)$.

Various contact sets can now be defined rigorously as follows:

Definition 2: For a closed, non-intersecting curve $\mathbf{q} \in C^{1,1}(I,\mathbb{R}^3)$ we denote by $\mathcal{C}(s,\sigma,\sigma)$ the circle through $\mathbf{q}(s)$ and tangent to the curve at $\mathbf{q}(\sigma)$. Then we define

(1) the *contact set* χ to be the set

$$\chi := \{(s,\sigma) \in I \times I;\ \mathsf{pt}(s,\sigma) = \Delta[\mathbf{q}]\},$$

and the *set of contact points* in three dimensional space to be the set

$$\mathbb{C} := \{\mathsf{c} \in \mathbb{R}^3;\ \mathsf{c} \text{ is the centre of } \mathcal{C}(s,\sigma,\sigma) \text{ and } (s,\sigma) \in \chi\},$$

(2) and for $\mu > 0$ the *μ-contact set* χ_μ to be the set

$$\chi_\mu := \{(s,\sigma) \in I \times I;\quad \mathsf{pt}(s,\sigma) \leq \Delta[\mathbf{q}](1+\mu) \text{ and } \mathsf{pt}(s,\cdot) \text{ has a local minimum in } \sigma\},$$

and the *set of μ-contact points* in three dimensional space to be the set

$$\mathbb{C}_\mu := \{\mathsf{c} \in \mathbb{R}^3;\ \mathsf{c} \text{ is the centre of } \mathcal{C}(s,\sigma,\sigma) \text{ and } (s,\sigma) \in \chi_\mu\}.$$

Note that both the exact contact set χ and any approximate contact set χ_μ can contain contact of the type (s,s), i.e., local curvature may be active.

The ideas motivating Definition 2 are the following. For a generic curve Γ the contact set χ will typically be a single pair of a point (s^*,σ^*) and its reflexion (σ^*,s^*), with χ_μ, for μ sufficiently small, being two short curve segments $(s,\hat{\sigma}(s))$ containing (s^*,σ^*), and $(\sigma,\hat{s}(\sigma))$ containing (σ^*,s^*). For an exactly ideal shape, constancy of ρ_{pt} implies that the exact contact set χ should be much larger, and will contain points of the form $(s,\hat{\sigma}(s))$ for each s in any curved segment, with, presumably, the contact set being made up, at least locally, of one or more curves in the (s,σ) plane generated by the local minimiser of $\mathsf{pt}(s,\cdot)$ moving with the parameter s. For an approximately ideal shape the exact contact set χ should again be a small number of isolated points, but for μ small the approximate contact set χ_μ should explode to a much larger set, containing one or more branches of curves of the form $(s,\hat{\sigma}(s))$ for all s in curved segments. It is also reasonable to anticipate ranges of μ for which χ_μ is invariant or changes very little. These expectations are borne out by the computations described in sections 5 and 6.

The sets of contact points $\mathbb{C}$ and $\mathbb{C}_\mu$ provide information in three dimensional configuration space that is different from the contact sets χ and χ_μ in two dimensional parameter space. For example the set of contact points $\mathbb{C}$ of the circle (namely the ideal shape of the trivial knot) contains one single point, namely the centre of the circle, but $\chi = I \times I$ because the function $\mathsf{pt}(s,\sigma)$ is constant on any circular arc.

We remark that it is also of interest[14,33] to repeat the development described in this section, but with local or global minimisation of $\mathsf{pt}(s,\sigma)$

over its second argument σ being replaced by minimisation over its first argument s. This leads to another global radius of curvature function ρ_{tp}, which is closely related to, but different from ρ_{pt}, and different approximate contact sets.

3. Why compute with Biarcs?

Several groups have performed numerical computations using either a sequence of points or piece-wise linear segments as the curve discretisation, divers definitions of thickness and various minimisation methods, both for closed knots and links[12,15,16,20,24,27,34] and open knots,[7,23] as well as in related optimal packing problems.[18,36] These computations produce approximations of rope length which appear to be satisfactory to three or four significant digits. However the computed approximations do not provide upper bounds for the corresponding ideal rope length because no error estimate between the discrete and the underlying continuous problem is typically given, and the point or piece-wise linear discretisations do not match the known minimal $C^{1,1}$ regularity of the actual solution. Another disadvantage of point or piece-wise linear discretisations is that tangents are discontinuous, and higher-order quantities such as curvature can only be interpreted in a finite difference sense.

An apparently related issue is that it is problematic to compute accurate contact sets with a point discretisation. One of the first attempts in this direction[12] considered the three-dimensional set of contact points for the ideal trefoil based on visualisation of the diameters of all circles through three of the discretisation points that realise, to a rather small tolerance, the thickness of the set of points. That computation already indicated that contact sets are very sensitive to error in shape. Meanwhile, a series of three articles[20,21,22] concluded that a version (based on the Euclidean distance function pp rather than a global radius of curvature) of what we call the two-dimensional parameter contact set of the ideal trefoil could not be resolved using a point discretisation. We note in particular that it seems impossible to address the issue of whether or not local curvature is active in realising thickness using a piece-wise linear discretisation and a Euclidean distance evaluation of thickness, both because of the inexact interpretation of local curvature, and the fact that in evaluating thickness using the usual distance function some (usually imprecisely characterised) number of neighbouring points or line segments must be ignored due to the fact that the pp function vanishes along the diagonal.

A notable exception to the above critique of computations based on a point or piece-wise linear discretisation is provided by the work of Rawdon.[24,25,26] Given an equilateral (i.e. constant edge length) piece-wise linear approximation to an ideal shape Rawdon constructs an associated $C^{1,1}$ curve by dint of inscribing arcs of circles into the corners of the piece-wise linear shapes. He also obtains estimates of the thickness of such curves, and thereby does achieve rigorous upper bounds on rope length.

The computations that we present are based upon another discretisation exploiting arcs of circles, namely the *biarc* discretisation of a space curve. Geometrically, biarcs can be described as a special sequence of arcs of circles that are joined continuously and with continuous tangents. Computationally, biarcs can be described as a special non-uniform rational quadratic spline that achieves Hermite interpolation of a sequence of point-tangent data. Biarcs have been discussed in the Computer-Aided Design (CAD) literature,[2,19,28,31,32] but we are unaware of a comprehensive mathematical treatment prior to the thesis of Smutny,[33] which can be consulted for a full discussion of the results that are only sketched here. We remark that the precise connexions between our biarc discretisation and Rawdon's inscribed circle construction remain to be fully worked out. One difference is that our biarc computations are based throughout on a point-tangent data format:

Definition 3: *Point-tangent data* is of the form $(\mathbf{q}, \mathbf{t}) \in \mathcal{J} := \mathbb{R}^3 \times S^2$, where $S^2 \subset \mathbb{R}^3$ denotes the unit 2-sphere. A *point-tangent data pair* is of the form $((\mathbf{q}_0, \mathbf{t}_0), (\mathbf{q}_1, \mathbf{t}_1)) \in \mathcal{J} \times \mathcal{J}$ with $\mathbf{q}_0 \neq \mathbf{q}_1$.

Definition 4: The point-tangent data pair $((\mathbf{q}_0, \mathbf{t}_0), (\mathbf{q}_1, \mathbf{t}_1)) \in \mathcal{J} \times \mathcal{J}$ will be called *proper* if

$$\begin{aligned} &(\mathbf{q}_1 - \mathbf{q}_0) \cdot \mathbf{t}_0 > 0, \\ \text{and} \quad & \\ &(\mathbf{q}_1 - \mathbf{q}_0) \cdot \mathbf{t}_1 > 0. \end{aligned} \tag{7}$$

In other words the point-tangent data pair is proper if the chord $(\mathbf{q}_1 - \mathbf{q}_0)$ and the two tangents $\mathbf{t}_0$ and $\mathbf{t}_1$ are compatibly oriented. Point-tangent data sampled from an underlying $C^{1,1}$ curve will be proper if the sampling is sufficiently fine.

Definition 5: A *proper biarc* $(\mathbf{a}, \bar{\mathbf{a}})$ is a pair of circular arcs in $\mathbb{R}^3$, joined continuously and with continuous tangents, that interpolate a proper point-tangent data pair. The common end point $\mathbf{m}$ of the two arcs $\mathbf{a}$ and $\bar{\mathbf{a}}$ is the *matching point* of the biarc.

Two examples of proper biarcs are illustrated in Figure 1.

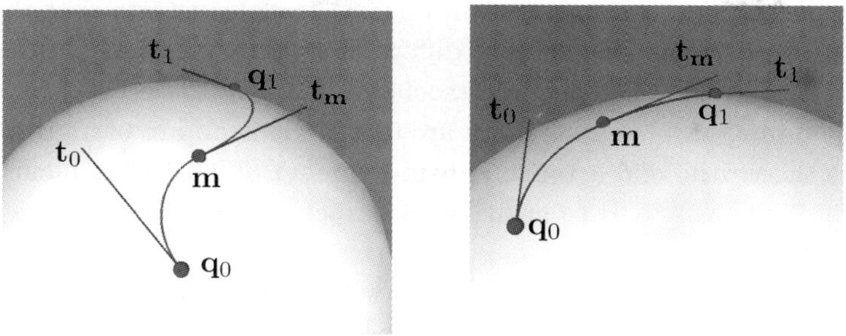

Fig. 1. A *biarc* is a pair of circular arcs, assembled with a common tangent $\mathbf{t}_m$ at a matching point $\mathbf{m}$, that interpolate a given pair of point-tangent data $((\mathbf{q}_0, \mathbf{t}_0), (\mathbf{q}_1, \mathbf{t}_1))$. The biarc lies entirely on the (generically unique) sphere $\mathcal{S}$ defined by the data. In the first case the data forces an inflexion point and the construction is clearer. However the second case is more typical.

It happens that there is a one-parameter family of proper biarcs interpolating a given proper point-tangent data pair, with possible matching points $\mathbf{m}$ lying on a certain circular arc passing through the two data points. We always compute with the *mid-point* proper biarcs that are uniquely defined via the choice of a specific, simple matching rule, namely the matching point is taken half-way along the circular arc of possible matching points.[33]

Definition 6: A *proper, mid-point, biarc curve* $\boldsymbol{\beta}$ is a space curve assembled from proper, mid-point biarcs in a C^1 fashion, where the biarcs interpolate a sequence $\{(\mathbf{q}_i, \mathbf{t}_i)\}$ of proper point-tangent data.

As biarc curves are assembled from arcs of circles they are both a geometrically natural, higher-order extension of piece-wise linear Lagrange interpolation of points, and a geometrically elementary alternative to cubic splines for Hermite interpolation of point-tangent data. In particular there is a unique proper, mid-point, biarc curve that interpolates a given sequence $\{(\mathbf{q}_i, \mathbf{t}_i)\}$ of proper point-tangent data, and its construction is both explicit and simple.

Biarc curves have several attractive properties. As to the standard quantities of differential geometry it is simple to compute arc length, the tangent vector field is defined everywhere and is continuous, curvature is defined almost everywhere and is piece-wise constant, and, wherever the curvature is positive, the principal normal and the torsion angle between successive circular arcs are also well-defined. If the point-tangent data defining the biarc curve is assumed to be drawn from an underlying base curve of known regularity, then Taylor expansions can be used to demonstrate that the biarc curve also provides simple finite difference approximations to quantities, such as curvature and torsion, associated with the base curve itself. Finally, the tangent indicatrix of an arc of a circle is an arc of a great circle, so that the writhe and average crossing number of closed biarc curves can be computed easily and to arbitrary accuracy by use of the spherical area formulas of Fuller.[10]

Biarc curves are especially attractive for the more specialised purposes of computing global radius of curvature functions, thickness and contact sets. The definition (4) of ρ_{pt} and the characterisations (2), (5), and (6) of curve thickness, remain valid for biarc curves. Moreover an iterative subdivision algorithm based on (2) can be constructed to evaluate thickness on a given, closed biarc curve to a prescribed, but arbitrarily small tolerance. Thus rope length of a biarc curve can be evaluated to an arbitrary tolerance. As biarc curves are themselves $C^{1,1}$-curves we can therefore obtain rigorous upper bounds for rope length of ideal knot shapes via construction of biarc curves of the appropriate knot type. Finally, on biarc curves it can be shown that local minima of $\mathsf{pt}(s, \sigma)$ in σ can arise only a) close to the diagonal (more specifically on the same circular arc as s), b) at local maxima or minima of $\mathsf{pp}(s, \sigma)$ in σ, or c) for σ at a junction between two arcs. A consequence of these facts is that on a biarc curve the functions $\rho_{\mathsf{pt}}(s)$ and $\hat{\rho}_{\mathsf{pt}}(s)$ can be evaluated exactly for any given s, and the contact sets $\boldsymbol{\chi}$ and $\boldsymbol{\chi}_\mu$ introduced in Definition 2 can be computed to an arbitrary precision.

4. Simulated annealing with biarcs

The approximately ideal shapes that will be presented here were obtained from simulated annealing computations using an upgraded version of an existing code that was originally based on a piece-wise linear discretisation.[16] The key ingredients in a simulated annealing approach are a) fast and accurate evaluation of rope length, and b) a set of random moves to search configuration space. The properties of biarc curves outlined in the previous section allow rope length to be evaluated straightforwardly because both arc length and thickness can be evaluated efficiently and accurately to a prescribed tolerance. (In fact in our computations the thickness was evaluated up to a relative error of 10^{-12}, that is the maximal error divided by the lower bound of thickness is smaller than 10^{-12}, and to compute contact sets accurately a tolerance of this order of magnitude seems appropriate.) The basic data format in the code was updated from a list of points to a list of point-tangent data. Then the allowed moves were taken to be random and independent changes in each point and each tangent, but with different, and adaptive, scales for point and tangent moves.

The starting biarc configurations were obtained by fitting a biarc curve to previously computed piece-wise linear approximations of ideal shapes. The initial configurations for the trefoil and figure-eight were generated from respectively data of Laurie[16] and Pieranski.[b] [20,34] In the first instance our objective was to obtain smoother and more accurate approximations of ideal shapes close to known approximations. In particular we did not attempt high-temperature Monte Carlo, which could be expected to explore all of configuration space. Rather we performed low-temperature simulated annealing, which remained close (in the C^0 sense) to the initial configurations. During the computations, and as a function of observations of curvature plots, various local and manual mesh refinement were performed via the simple mechanism of splitting some biarcs into pairs of biarcs.

Once the simulated annealing computation was stopped, plots of global radius of curvature functions and contact sets were obtained from an independent post-processing on the computed biarc curves using Matlab scripts. In particular the all important length and thickness of the computed biarc approximation of ideal configurations were each evaluated twice, using two independently written codes, with complete agreement to the specified tolerance.

5. Results for the 3.1-knot

Our final computed trefoil shape involved 264 biarcs, rather non-uniformly spaced with additional biarcs having been added in three regions of rapid change in the radius of curvature function. We denote each of the associated 528 arcs by $\mathbf{a}_i$, their radii by r_i, the mid-point of each arc by $\mathbf{M}_i$, $1 \leq i \leq 528$, and the overall arc curve by $\boldsymbol{\alpha}$. The total time for the simulated annealing computation was several months on a single CPU.

[b] whom we thank for providing his data

5.1. *The numbers*

The final bounds of rope length, the length $\lambda(\boldsymbol{\alpha})$ of the arc curve, the minimal radius $\min_i r_i$, and upper and lower bounds for thickness $\Delta[\boldsymbol{\alpha}]$ are

$$\begin{aligned} \lambda(\boldsymbol{\alpha}) &= 0.99999999997863, \\ \min_i r_i &= 0.03054053096312, \\ 0.03053951779966 \leq \Delta[\boldsymbol{\alpha}] &\leq 0.03053951779968, \\ 32.74445937679887 \leq \frac{\lambda(\boldsymbol{\alpha})}{\Delta[\boldsymbol{\alpha}]} &\leq 32.74445937682155. \end{aligned}$$

Note that the minimum arc radius $\min_i r_i$ is only very slightly higher than the thickness $\Delta[\boldsymbol{\alpha}]$ of the biarc curve. We repeat that the only numerical errors in these data are associated with the prescribed relative tolerance in thickness evaluation, namely 10^{-12}, and double precision arithmetic error, which is usually assumed to be 12 significant digits or more. In particular we have very tight bounds both from above and below on the thickness and rope length of the biarc curve $\boldsymbol{\alpha}$. As the configuration $\boldsymbol{\alpha}$ is of the appropriate knot type and is therefore a true competitor, the upper bound on rope length for $\boldsymbol{\alpha}$ is also an upper bound for rope length of the true ideal knot shape.

5.2. *The shape*

To display the shape of the knot, the centre of mass and principal axes $\mathbf{v}_1$, $\mathbf{v}_2$, and $\mathbf{v}_3$ of inertia of the set of end points of all arcs were computed (and ordered with associated eigenvalues $\lambda_1 = 3.1161 \cdot 10^{-3} \leq \lambda_2 = 3.1347 \cdot 10^{-3} \leq \lambda_3 = 5.2057 \cdot 10^{-3}$). Figure 2 (a) shows the projection of the end points of all arcs onto the $\mathbf{v}_1$-$\mathbf{v}_2$ plane, whereas (b) is the projection of the tangents at the end points of all arcs onto the $\mathbf{v}_1$-$\mathbf{v}_2$ plane. In Figure 2 the polar coordinates of (c) the end points and (d) end-point tangents are plotted against arc length, with the lower continuous curve being the $\mathbf{v}_3$ coordinate, the upper continuous curve being the radial coordinate $r = \sqrt{v_1^2 + v_2^2}$, and the discontinuous curve being the angle ϕ in radians (modulo 2π and scaled by a factor to make variations visible on the same plot). As the curve $\boldsymbol{\alpha}$ is C^1, these plots could be sampled as finely as desired. Both the curve and the tangent curve appear to be very smooth, but the derivative of the tangent, i.e. the curvature, is large in three small regions, as can be observed in the Figures 2 (b) and (d). Figures 2 (a)-(d) suggest that the ideal shape of the trefoil is close to having a period three, rotational symmetry about the $\mathbf{v}_3$ axis. Rotational symmetry would imply a double eigenvalue of the inertia matrix. Therefore one measure of asymmetry is $\frac{\lambda_2-\lambda_1}{\lambda_1+\lambda_2+\lambda_3} = 1.6233 \cdot 10^{-3}$.

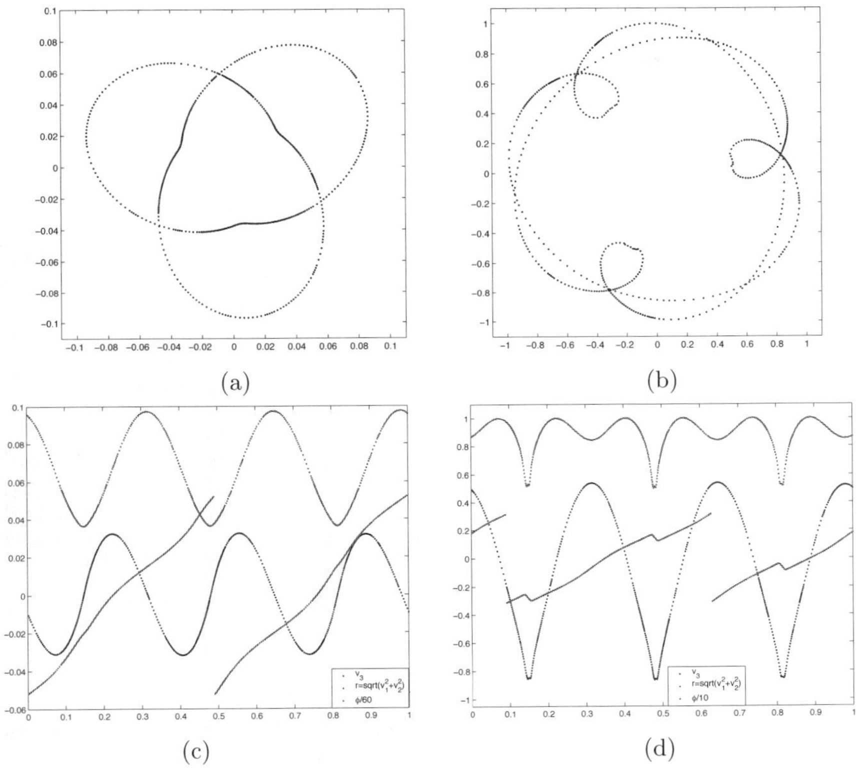

Fig. 2. Trefoil knot: Projection onto the plane of the $\mathbf{v}_1$-$\mathbf{v}_2$ inertia axes of (a) the end points of all arcs, and (b) the unit tangents at the end points of all arcs. Parts (c) and (d) are plots of the corresponding cylindrical polar coordinates v_3, $r = \sqrt{v_1^2 + v_2^2}$ and (scaled) angle ϕ.

Figure 3 depicts the radii of the 528 arcs scaled by the thickness Δ. There are three regions with radii approximately equal to 2Δ and three regions with a high variation, running from approximately 2Δ to Δ to 2.8Δ and back to Δ to end again at 2Δ. The dips in local radii come within a factor 10^{-5} of the thickness. The regions with approximately constant curvature belong to the large exterior loops visible in Figure 2 (a) and Figure 8. As can be seen from torsion plots and the two dimensional contact map, cf. Figure 7, these parts are close to being, but are not precisely, circular arcs.

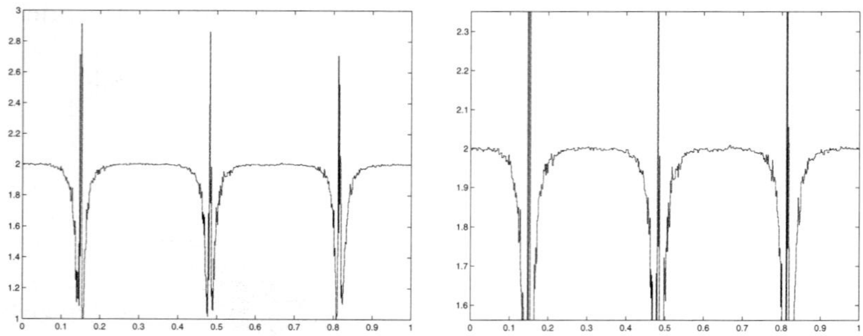

(a) (b)

Fig. 3. Trefoil knot: (a) Histogram of the radii of the arcs scaled by the thickness (non-dimensional), (b) a magnification of (a).

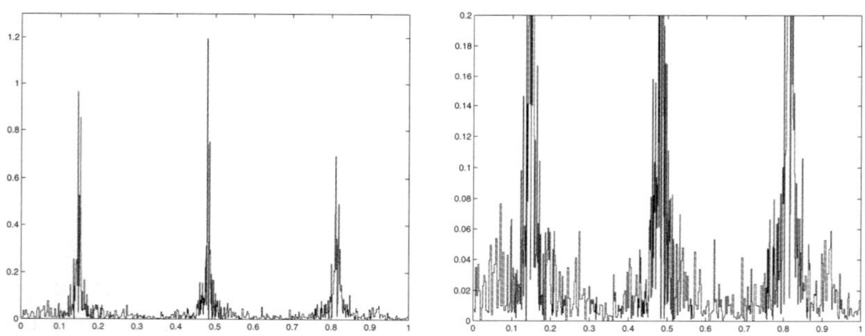

(a) (b)

Fig. 4. Trefoil knot: Plots of (a) torsion angle, i.e. the (unsigned) angle between the planes of successive arcs in radians and (b) magnification of (a).

The (unsigned) angle between the planes of adjacent arcs is plotted in Figure 4. There are three regions with large angles that correspond to the regions with a high variation of the radii, that is to the parts of the curve "inside the knot." Note that the angle is given in radians, that is, the maximal value of around 1.2 between adjacent arcs corresponds to an angle of around 70 degrees. These extremely large values could lead to the speculation that the Frenet frame of the underlying ideal curve may be discontinuous, and that the associated ideal centreline curve may not be C^2 at nearby points.

One necessary condition for a configuration to be ideal is constancy of ρ_{pt} on curved segments. The computed shape is curved everywhere, so for this shape to be truly ideal the function ρ_{pt} should be constant everywhere.

We can evaluate ρ_{pt} with an error within that of double precision arithmetic at a discrete number of points, specifically on the mid-points of all the arcs, to obtain a sampling $\rho_{\mathsf{pt}}(\mathbf{M}_i)$, for which $\max_i \rho_{\mathsf{pt}}(\mathbf{M}_i) - \min_i \rho_{\mathsf{pt}}(\mathbf{M}_i) = 1.6992 \cdot 10^{-8}$, i.e., rather close to constant.

5.3. *The contact sets*

Figure 5 and Plate 1 are visualisations of the two-dimensional function $\mathsf{pt}(s, \sigma)$ evaluated on our computed biarc trefoil using a colour map with blue for low values and red for high. More precisely the figure is a mesh of values for all pairs $1 \leq i, j \leq 528$, where for each i (vertical) and j the plotted function is $\mathsf{pt}(i, j) := \min_{\sigma \in \mathbf{a}_j} \mathsf{pt}(\mathbf{M}_i, \sigma)$. Because the mesh is quite non-uniform there is a noticeable distortion from a uniform sampling in arc length, but this serves to accentuate regions of rapid change where the biarcs were subdivided. Note that the global minima (blue) lie in two rather flat valleys, and that at three places along the diagonal there are localised double dips reaching down to close to the global minimum.

The dark blue dots in Figure 5 are superposed on the colour map at indices (i, j) for which $\mathsf{pt}(\mathbf{M}_i, \sigma)$ has a local minimum in σ on the arc $\mathbf{a}_j$. In other words as the index i varies, the dark blue dots provide a sampling of the multi-valued, global radius of curvature functions $\hat{\rho}_{\mathsf{pt}}(s)$. Any such point is an element of the μ-contact set $\boldsymbol{\chi}_\mu$ introduced in Definition 2 if $\mathsf{pt}(i, j)$ is smaller than $\Delta(1 + \mu)$. (With this definition a dark blue dot is placed at each diagonal pair (i, i) because pt is constant on circular arcs, and therefore realises a local minimum in σ, albeit not a strict local minimum, along the diagonal of a biarc curve.)

We remark that in the three (taking account of double periodicity of the plot), approximately square, turquoise regions, there are many vertical dark blue line segments of what is seemingly numerical noise. These plateaus, at an intermediate height of $\mathsf{pt}(i, j) \approx 2\Delta$, correspond to the large, approximately circular, exterior arcs of the trefoil on which the $\mathsf{pt}(s, \sigma)$ function is nearly constant. Note that in each of the deep valleys there are three 'curves' of local minima in the index j at fixed index i.

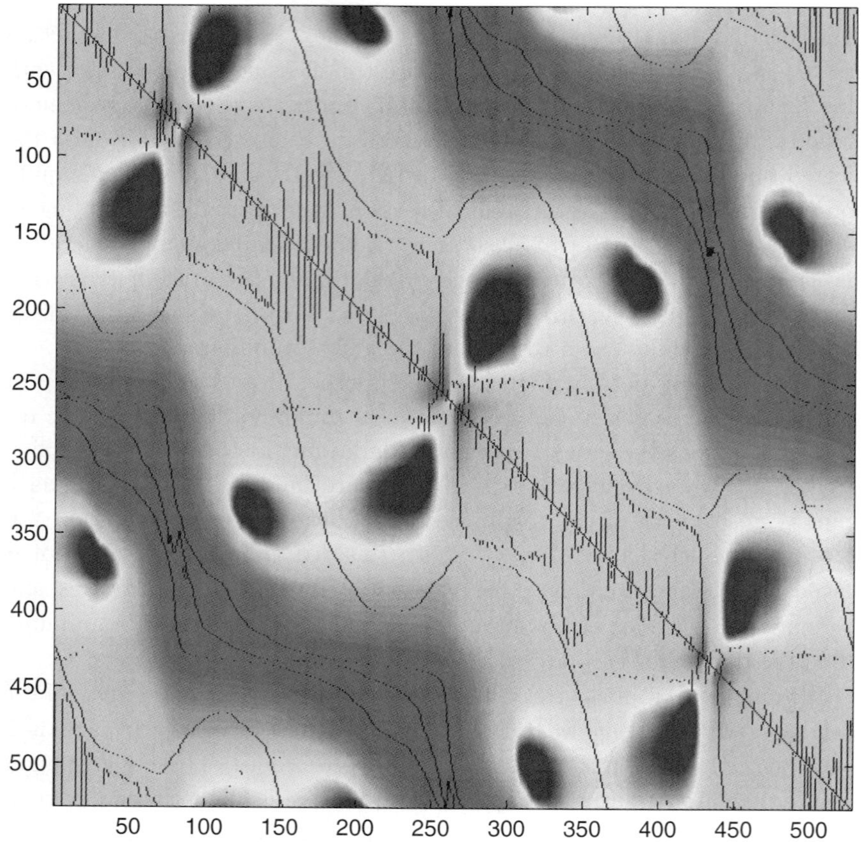

Fig. 5 and Plate 1. Trefoil knot: A colour plot of the function $\mathsf{pt}(i,j) := \min_{\sigma \in \mathbf{a}_j} \mathsf{pt}(\mathbf{M}_i, \sigma)$ evaluated on each arc of the approximately ideal 264-biarc trefoil knot, with blue for low and red for high values. In addition there is a dark blue dot superposed at indices j with local minima of $\mathsf{pt}(s,\cdot)$ realised in the arc $\mathbf{a}_j$. Note that the plot is a uniformly spaced function of indices corresponding to a non-uniform mesh in arc length.

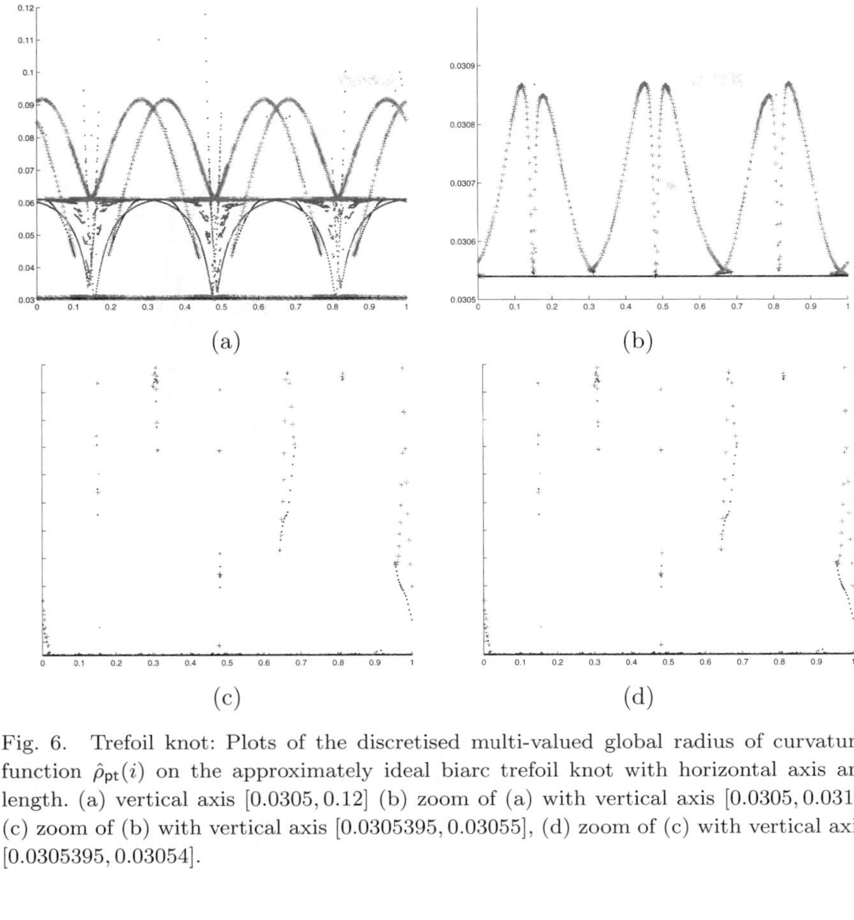

Fig. 6. Trefoil knot: Plots of the discretised multi-valued global radius of curvature function $\hat{\rho}_{\mathsf{pt}}(i)$ on the approximately ideal biarc trefoil knot with horizontal axis arc length. (a) vertical axis $[0.0305, 0.12]$ (b) zoom of (a) with vertical axis $[0.0305, 0.031]$, (c) zoom of (b) with vertical axis $[0.0305395, 0.03055]$, (d) zoom of (c) with vertical axis $[0.0305395, 0.03054]$.

Figure 6 (a)-(d) plots the multi-valued, global radius of curvature functions $\hat{\rho}_{\mathsf{pt}}(s)$ along arc length in various magnifications. The points in Figure 6 (a) correspond to all the dark blue points in Figure 5, but with the values of σ projected out. The points in Figure 6 (b) correspond to the lowest line in (a) and also to the three 'curves' in the deep valleys of Figure 5. The dots in Figure 6 (c) correspond to the lowest line of part (b) along with the lowest parts of the higher structure, and finally, (d) is an ever stronger magnification of the lowest line in (c).

It is now possible to select an appropriate $\mu > 0$ to fix the μ-contact set $\boldsymbol{\chi}_{\mu}$, i.e., the subset of the dark blue points in Figure 5 for which the minima of $\mathsf{pt}(i, j)$ is smaller or equal to $\Delta(1 + \mu)$. And similarly for $\boldsymbol{\chi}_{2\mu}$. In Figure 6 (d) there is a gap above a layer of candidates. We therefore

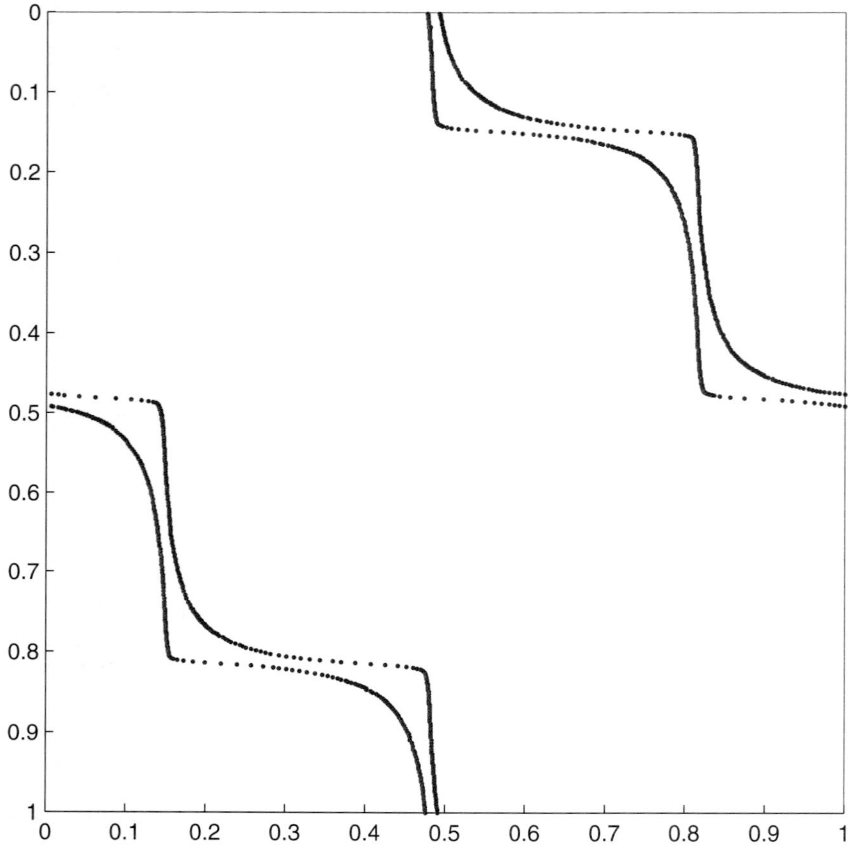

Fig. 7. Trefoil knot: The μ-contact set χ_{μ^*}, with $\mu^* = 8.1861 \cdot 10^{-6}$, of the approximately ideal 264-biarc trefoil knot plotted as a function of arc lengths s vertical and σ horizontal.

set $\mu^* := 8.1861 \cdot 10^{-6}$, which is approximately half of the height of Figure 6(d).

Figure 7 displays the μ-contact set χ_{μ^*} now as a function of arc length, i.e., the distortion arising from the non-uniform mesh has been eliminated. The contact set appears to be two smooth lines. With our biarc discretisation we could sample this curve in the variable s as finely as desired, but this plot is only for arc lengths s_i corresponding to the mid-points of each arc. As a consequence nearly horizontal segments appear to be sampled less frequently. The contact set appears to be two smooth lines. In fact, χ_{μ^*} has 1058 elements so that the average number of μ-contacts per arc is $\frac{1058}{2 \cdot 264} \approx 2.00$. We expect the set χ_{μ^*} to be relatively invariant for small increases in μ^*. In fact, the number of elements of $\chi_{2\mu^*}$ is 1062, that is $\chi_{2\mu^*} \setminus \chi_{\mu^*}$ contains only four additional contacts, and the change of elements relative to the number of arcs is $\frac{4}{2 \cdot 264} = 0.0076$.

We remark that local radii are not active in the contact set for $\mu^* = 8.1861 \cdot 10^{-6}$, but are remarkably close to being active. For example local curvature does form part of the contact set $\chi_{5\mu^*}$. Thus it is quite possible that on the true ideal shape local curvature does achieve thickness at six distinct points.

The next goal is to understand the three dimensional set of μ-contact points. Figure 8 and Plate 2 provide two visualisations in three dimensional space of our biarc knot. In part (a) the actual knot is drawn with a green sphere at each mid-point $\mathbf{M}_i$ of the arcs $\mathbf{a}_i$, along with a red sphere at each μ-contact point in $\mathbb{C}_{\mu^*}$, and blue lines for each associated contact chord. A *contact chord* associated with an element (s, σ) of the contact set χ_μ is the straight line segment with end points $\boldsymbol{\alpha}(s)$ and $\boldsymbol{\alpha}(\sigma)$. The mid-points of all contact chords are exactly the set of μ-contact points $\mathbb{C}_\mu$ in three dimensional space because the contact chords are exactly perpendicular to the curve tangents at one end point, cf. Definition 2. At this resolution and ball radius, the union of the red, μ-contact-point spheres form a tube. And the centreline of this tube is another (non-intersecting, but far from ideal) trefoil knot. The visualisation in part (a) suggests that the contact chords form a surface. This conjecture is further supported by the visualisation in part (b), which is a shading of a triangulation of all the contact chords. The resulting surface is an oriented, closed, single-sheeted surface with an interior. The actual trefoil knot is the sharp edge of the surface, which is emphasised by shading the surface in a periodic fashion such that the two portions of the surface meeting along an edge are contrasted. With the biarc discretisation these figures could be recomputed with an arbitrarily fine sampling. We emphasise that while in the data used to generate Figure 8 one end of each contact chord is by construction at an arc midpoint $\mathbf{M}_i$, the other end point has to be calculated as some general point along the arc curve $\boldsymbol{\alpha}$. The computation of the contact set is sufficiently delicate that restricting to the finite dimensional problem of chords with both end points at an arc midpoint does not yield a smooth surface.

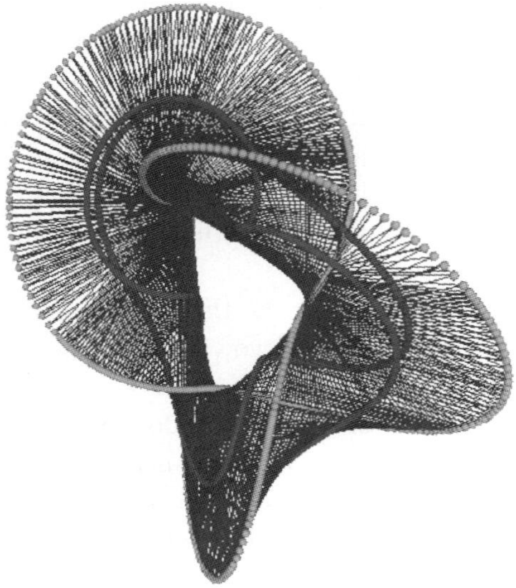

(a)

(b)

Fig. 8 and Plate 2. Trefoil knot: Two visualisations of the approximately ideal, biarc trefoil knot in three dimensional space. In part (a) the green spheres are centred at the mid-points $\mathbf{M}_i$ of the arcs $\mathbf{a}_i$, the red balls are the set of contact points $\mathbb{C}_{\mu^*}$ with $\mu^* = 8.1861 \cdot 10^{-6}$, and the blue line segments are the associated contact chords. At this resolution the red spheres overlap to form a tube, whose centreline is another trefoil. The contact chords appear to form a surface, which is confirmed in the visualisation of part (b), which is a shaded triangulation of the blue contact chords.

6. Results for the 4.1-knot

The shape we obtained after severals weeks of simulated annealing involved a 416 biarc, or 832 arc, configuration. As before we denote the 832 arcs by $\mathbf{a}_i$, their radii by r_i, the mid points of the arcs $\mathbf{a}_i$ by $\mathbf{M}_i$, $1 \leq i \leq 832$, and the arc curve by $\boldsymbol{\alpha}$. In this case the arcs were close to uniform in length. The results and figures to be presented in sections 6.1-6.3 are, up to different scales, the precise analogues of those described in sections 5.1-5.3 for the trefoil knot. We therefore do not repeat the detailed explanations, but instead concentrate on describing the conclusions to be drawn. In general it will become clear that the presented computations for the 4.1-knot are noticeably less converged than those for the 3.1-knot.

6.1. *The numbers*

The bounds we obtain are

$$\begin{aligned} \lambda(\boldsymbol{\alpha}) &= 0.99999999927731, \\ \min_i r_i &= 0.02374402362244, \\ 0.02374401039630 \leq \Delta[\boldsymbol{\alpha}] &\leq 0.02374401039633, \\ 42.11588449404105 \leq \frac{\lambda(\boldsymbol{\alpha})}{\Delta[\boldsymbol{\alpha}]} &\leq 42.11588449409459. \end{aligned}$$

Thickness was again evaluated up to a relative error of 10^{-12}, and we can again observe that $\min_i r_i$ is only slightly higher than the thickness $\Delta[\boldsymbol{\alpha}]$; in fact in this case $\min_i r_i$ is much closer to the thickness $\Delta[\boldsymbol{\alpha}]$ than the respective values for the trefoil knot, with the difference now being only of the order of 10^{-8}.

6.2. *The shape*

To display the shape of the knot, the centre of mass and principal axes $\mathbf{v}_1$, $\mathbf{v}_2$, and $\mathbf{v}_3$ of inertia of the set of end points of all arcs were computed (ordered with corresponding eigenvalues $\lambda_1 = 2.3526 \cdot 10^{-3} \leq \lambda_2 = 2.7632 \cdot 10^{-3} \leq \lambda_3 = 2.8144 \cdot 10^{-3}$). Figure 9 (a) shows the projection of the end points of all arcs onto the $\mathbf{v}_2$-$\mathbf{v}_3$ plane, whereas (b) is the projection of the unit tangents at the end points of all arcs. Parts (c) and (d) are the corresponding polar coordinates of end points, and end-point tangents plotted as functions of arc length. The projections of both the curve and the tangent curve onto the $\mathbf{v}_2$-$\mathbf{v}_3$ plane have an approximate period four symmetry. Here $\frac{\lambda_3-\lambda_2}{\lambda_1+\lambda_2+\lambda_3} = 6.4567 \cdot 10^{-3}$, and in this measure seem further from rotational symmetry than the trefoil computations described in section 5.2. Note that for the 4.1 knot the $\mathbf{v}_1$ component of both the curve and the tangent curve are approximately period two symmetric, but do not exhibit period four symmetry.

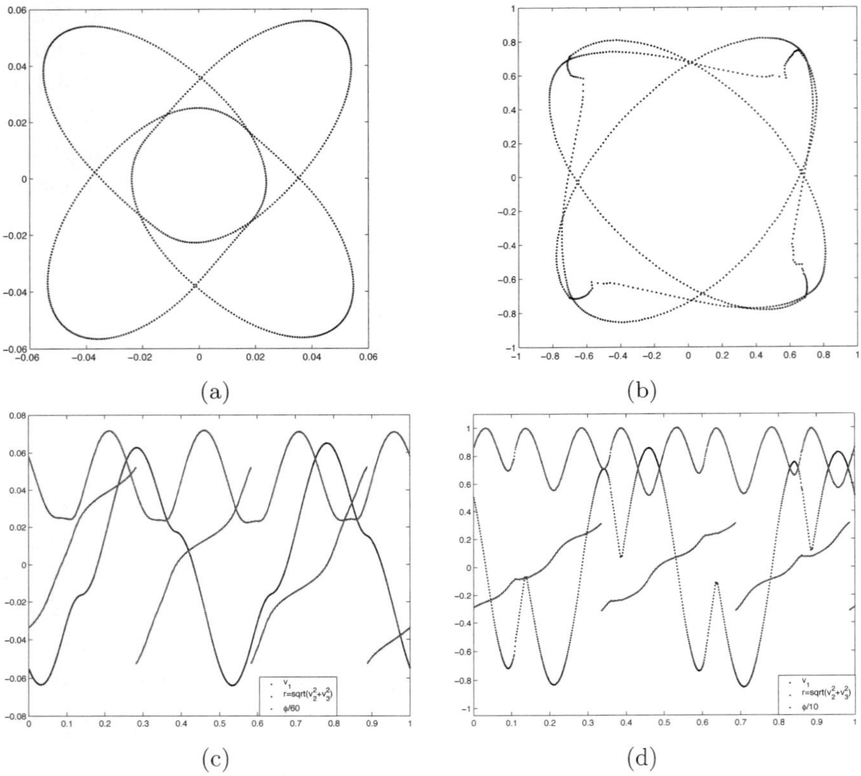

Fig. 9. Figure-eight knot: Projection onto the plane of the $\mathbf{v}_2$-$\mathbf{v}_3$ inertia axes of (a) the end points of all arcs, and (b) the unit tangents at the end points of all arcs. Parts (c) and (d), plots of the corresponding polar coordinates v_1, $r = \sqrt{v_2^2 + v_3^2}$ and (scaled) angle ϕ. While the $\mathbf{v}_2$-$\mathbf{v}_3$ projections are approximately period four symmetric, the v_1 components have an approximate period two symmetry.

Figure 10 depicts the radii of the 832 arcs scaled by the thickness Δ. As for the 3.1 knot there are regions (now four) with almost constant radii equal 2Δ corresponding to the large loops visible in Figure 9 (a). But the radii of our 4.1 arc curve take values from Δ to 55Δ. More precisely, there are four very short parts (approximately two to four arcs) which have very large radii. On the other hand, there are four even shorter parts (approximately two arcs) situated "inside the knot" where the radii are close to achieving the thickness Δ.

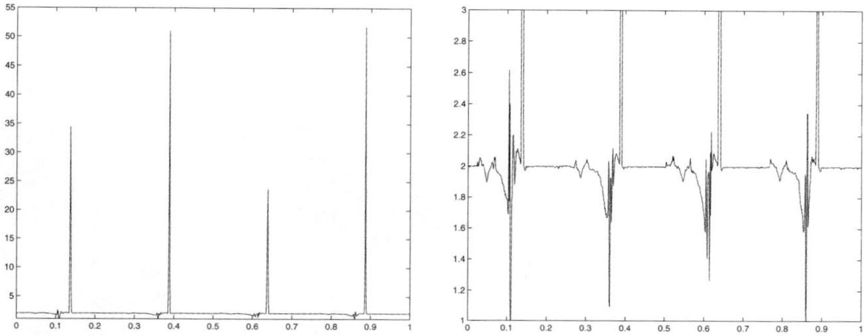

(a) (b)

Fig. 10. Figure-eight knot: (a) Histogram of the radii of the arcs scaled by thickness (non-dimensional), (b) a magnification of (a).

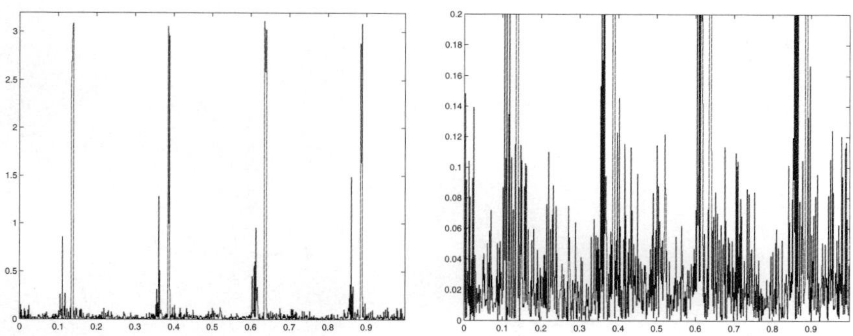

(a) (b)

Fig. 11. Figure-eight knot: Plots of (a) torsion angle, i.e. the (unsigned) angle between the planes of successive arcs in radians, and (b) magnification of (a). The spikes in torsion angle occur at the same arc lengths as the spikes in radii shown in Figure 10.

The (unsigned) angle between the planes of adjacent arcs is plotted in Figure 11. There are four regions with large angles that correspond to the regions with a high variation of the radii. In fact in the region where the radii are small, i.e. close to Δ, the torsion angles have a local maximum, and in the region where the radii are very large the torsion angles have another higher maximum. Note that the angle is given in radians, that is,

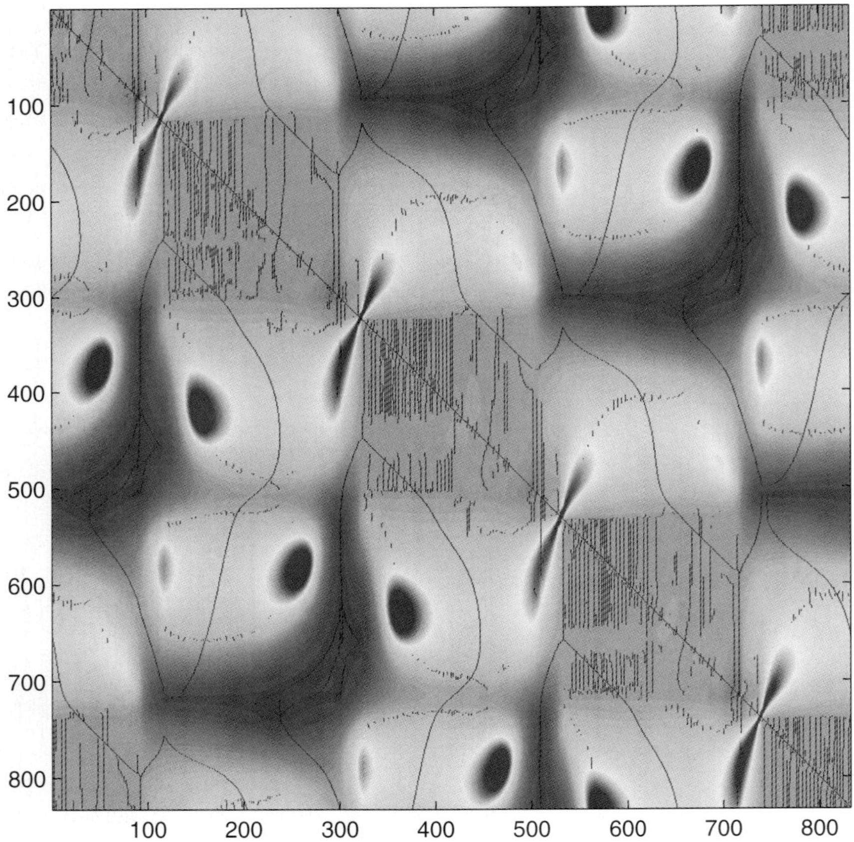

Fig. 12 and Plate 3. Figure-eight knot: A colour plot of the function $\mathsf{pt}(i,j) := \min_{\sigma \in \mathbf{a}_j} \mathsf{pt}(\mathbf{M}_i, \sigma)$ evaluated on each arc of the approximately ideal biarc figure-eight knot, with blue for low and red for high values. In addition there is a dark blue dot superposed at each index j associated with a local minimum.

the maximal value of about 3 corresponds to an angle of almost 180 degrees. In other words, at this maximum the arcs are almost straight, but the planes flip by approximately 180 degrees. It seems plausible that the underlying ideal shape could have a discontinuous Frenet frame at these points.

Just as for the trefoil we can evaluate ρ_{pt} with an error within that of double precision arithmetic at a discrete number of points. For sampling on the mid-points of all the arcs of our 4.1-knot, we obtain $\max_i \rho_{\mathsf{pt}}(\mathbf{M}_i) - \min_i \rho_{\mathsf{pt}}(\mathbf{M}_i) = 3.5178 \cdot 10^{-5}$, i.e., rather close to constant, but a larger variation by three orders of magnitude than on our 3.1-knot configuration. On the other hand, restricting the computation to the parts of the knot where the radii are not high, the variation that is obtained is of the order $10^{-7} - 10^{-6}$. This result could be interpreted in two ways: Either the biarcs approach straight line segments, where the global radius of curvature is not expected to be constant, or the computation is not sufficiently converged. Probably both effects are making a contribution.

6.3. *The contact sets*

Figure 12 and Plate 3 are visualisations of the two-dimensional function $\mathsf{pt}(s, \sigma)$ analogous to Figure 5 and Plate 1, but now evaluated on our computed biarc figure-eight knot. Now the global minima (blue) form four disconnected valleys, the dips along the diagonal are much less pronounced, but the four intermediate-height plateau regions associated with the exterior loops are better defined. In each of the four minimising valleys the dark blue dots (which, as before, indicate local minima in the index j of the discretely sampled function $\mathsf{pt}(i, j)$) form topologically nontrivial curves with corners.

Figure 13 (a)-(c) plots the multi-valued, global radius of curvature function $\hat{\rho}_{\mathsf{pt}}(s)$ along arc length in various magnifications. The points in Figure 13 (a) correspond to all the dark blue points in Figure 12, but with the values of σ projected out. The points in Figure 13 (b) all lie on the lowest visible line in (a), while Figure 13 (c) corresponds to the lowest line of part (b) along with the lowest parts of the higher structure.

We now have to select an appropriate value of μ for the approximate contact sets. In contrast to the plots for the trefoil, cf. Figure 6 (d), the gap above a layer of candidates in Figure 13 (c) is less clear cut. Nevertheless we set $\mu^* := 4.2115 \cdot 10^{-6}$, which is about half of the height of Figure 13 (c). Figure 14 displays the corresponding μ-contact set χ_{μ^*}. It seems that the set χ_{μ^*} has four distinct connected components, but the computation is

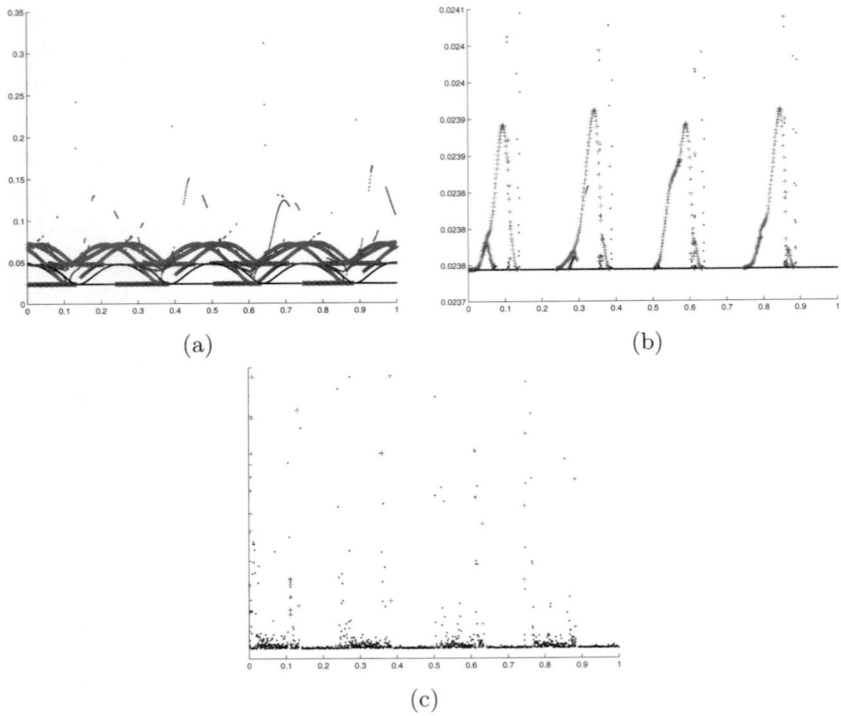

Fig. 13. Figure-eight knot: Plots of the discretised multi-valued global radius of curvature function $\hat{\rho}_{\text{pt}}(i)$ on the approximately ideal biarc figure-eight knot with horizontal axis arc length. (a) vertical axis $[0.00, 0.35]$ (b) zoom of (a) where the vertical axis is $[0.0237, 0.0241]$, (c) zoom of (a) where the vertical axis is $[0.0237440, 0.0237443]$.

sufficiently far from convergence that this conclusion can only be tentative. The apparent undersampling of nearly horizontal segments of the contact set is even more pronounced here than in the case of the trefoil, but as the exact contact set is known to be symmetric about the trailing diagonal this effect is not a serious difficulty. Here $\boldsymbol{\chi}_{\mu^*}$ has 1213 elements, so that the average number of μ-contacts per arc is $\frac{1213}{2 \cdot 416} \approx 1.46$, which is much less than for the trefoil knot. We expect $\boldsymbol{\chi}_{\mu^*}$ to be relatively invariant for small increases in μ, and in fact, the number of elements of the μ-contact set $\boldsymbol{\chi}_{2\mu^*}$ is 1238, that is $\boldsymbol{\chi}_{2\mu^*} \backslash \boldsymbol{\chi}_{\mu^*}$ contains only 25 additional contacts. The change of elements relative to the number of arcs is $\frac{25}{2 \cdot 416} = 0.03$. We remark that in contrast to the trefoil, the radii of two arcs are active in $\boldsymbol{\chi}_{\mu^*}$, so that at this tolerance local radius of curvature is already active for the figure-eight knot.

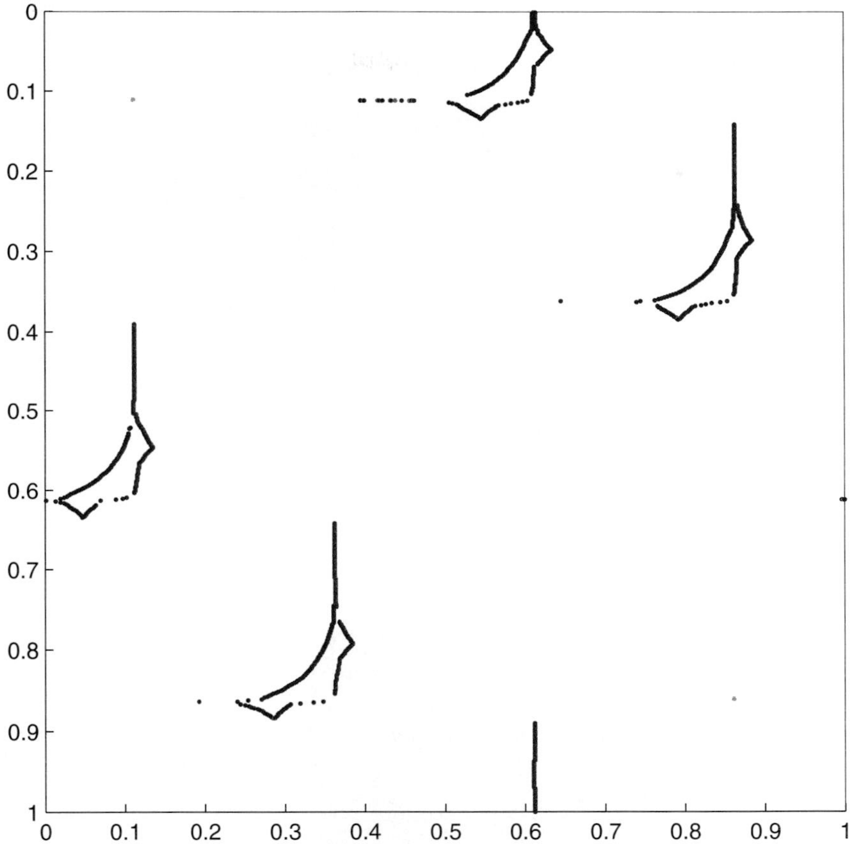

Fig. 14. Figure-eight knot: The μ-contact set χ_{μ^*} with $\mu^* = 4.2115 \cdot 10^{-6}$ for the approximately ideal biarc figure-eight knot. The two dots on the diagonal indicate that two contacts of the type (s, s), i.e. local curvature being active, are contained in χ_{μ^*}.

The next goal is to understand the three dimensional structure of our figure-eight knot. Figure 15 and Plate 4 show two projections of the knot, with the same visualisation conventions for the knot, contact points, and contact chords as in Figure 8(a). For the figure-eight knot the set of μ-contact points $\mathbb{C}_{\mu^*}$ is radically different from the one for the trefoil. Most of the (mid)points have either one or two associated contact points, but there is also a small number of points having no contact. The arcs of such points have very large radii. The visualisations in Figure 15 are an essentially exact representation of both the approximate computation of the knot shape, and the associated contact points with the prescribed tolerance μ^*. Of course it

Fig. 15 and Plate 4. Figure-eight knot: Two images of the approximately ideal, figure-eight knot (the green balls are centred at the mid-points $\mathbf{M}_i$ of the arcs $\mathbf{a}_i$), the set of contact points $\mathbb{C}_{\mu^*}$ in three dimensional space (red balls), and associated contact chords (blue line segments), all for $\mu^* = 4.2115 \cdot 10^{-6}$.

is interesting to extrapolate from this precise visualisation of approximate data, to obtain imprecise conjectures as to the contact set for the truly ideal knot. Given the state of convergence of the computations for the figure-eight, such an extrapolation can only be rather tentative. For example the red spheres centred at the contact points now do not all overlap nearest neighbours to form locally a contact line—rather it seems that there are 'missing' segments in some regions. Nevertheless the image suggests that for the ideal figure-eight knot the set of contact points $\mathbb{C}$ should be composed of two disjoint components, which are similar to each other, and topologically non-trivial, having a closed loop with two tails, with, moreover, corners.

7. Discussion

We believe the shapes of the 3.1 and 4.1 knots described in sections 5 and 6 to be the best known approximations to the corresponding ideal knot configurations. Our methods for the evaluation of thickness and the ρ_{pt} function on a given biarc curve are rigorous and accurate to a prescribed tolerance. In this way we achieved the upper bounds on rope length for the trefoil (with 528 arcs) of 32.74446, and for the figure-eight knot (with 832 arcs) of 42.11588. The only bounds close to these, known to us, were achieved by Rawdon,[26] using the method of inscribing arcs of circles onto a piecewise linear shape, of 32.90 and 32.77 for the trefoil with respectively 160 and 1332 linear segments, and 42.38 for the figure-eight knot with 208 linear segments. (In this regard we reiterate that purely piece-wise linear, or point computations do not provide rigorous upper bounds for rope length because of the presence of discretisation error and corners at which the thickness vanishes.) Thus the values we find for thickness are the best published, but involve only improvement in the third or fourth digit.[c]

In point of fact we believe that it is of more interest to understand the properties of the curve realising the optimal rope thickness rather than the value of thickness itself, and it seems that rather significant changes in, for example, local curvature of the configuration can be necessary to obtain extremely small improvements in rope length. The adoption of the biarc discretisation in our computations allows a close and detailed inspection of the approximately ideal shapes, including rather accurate tangent and curvature information. In particular the exploitation of biarcs has allowed

[c]And of course best computations are a shifting target in time. For example in another Chapter of this volume[1] Rawdon's method of inscribed circles is used to obtain the upper bound of 32.74338 for a piece-wise linear trefoil with 2544 edges.

us to start to address the issue of whether or not local curvature is active on ideal knot shapes, and, when combined with the introduction of the rigorous Definition 2 of approximate contact sets, to start to discuss the detailed geometry of which points in an ideal shape realise thickness. Such accurate computations also permit the examination of known necessary conditions for optimality, such as constancy of ρ_{pt}, and offer insights which may lead to the discovery of further necessary, and perhaps sufficient, analytic conditions for optimality.

A striking and previously unobserved feature in our simulations of both the 3.1 and 4.1 knots is the rapid and large scale variations in local curvature in, respectively, three and four narrow regions corresponding to the curve passing through the 'centre' of the knot. This phenomenon was first observed in a 192-biarc, 3.1 computation, whereupon the number of biarcs was doubled in the three regions of interest to obtain the 264-biarc shapes described here. Upon continued simulated annealing the features persisted, and indeed sharpened. For this reason, and the fact that the features have the appropriate, approximately period three, respectively four, behaviour we believe them to be real, and not numerical artifacts. In both the 3.1 and 4.1 knots the same features are associated with both a) extremely large torsion angles, suggesting a singularity in the second derivative, and b) small radii of curvature that approach the lower bound provided by the thickness. It has been generally assumed that for simple knots like the trefoil, local radius of curvature was never close to achieving thickness. We now suspect that this belief has been based upon limitations of previously adopted numerical schemes, either due to the general exclusion of nearest neighbour effects as done by many authors using point or piecewise linear discretisations, or due to the curve shortening algorithm that was used.[12]

One additional piece of evidence concerning the achievement of thickness locally is provided by a third simulation involving a 568-biarc discretisation of the composite +3.1#-3.1-knot. This simulation, with rope length 58.27448, is much less converged than either the 3.1 or 4.1 computations described above, as can be seen both from the large variations in the local curvature plot shown in Figure 16, and the fact that the variation $\max \rho_{\mathsf{pt}} - \min \rho_{\mathsf{pt}} \approx 2 \cdot 10^{-4}$ in the global radius of curvature function on curved segments is much higher than the respective values for the 4.1 and 3.1 simulations. Although the local radius of curvature function ρ is not yet at all smooth, the global radius of curvature function ρ_{pt} is in comparison already quite close to constant on curved segments. We include the preliminary results from this simulation because of two features. First there are

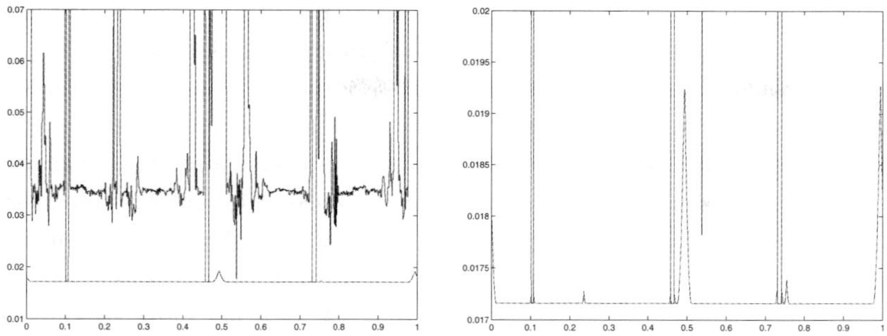

(a) (b)

Fig. 16. Plot of ρ (upper curve) and ρ_{pt} on a 568-biarc simulation of a composite +3.1#-3.1-knot. Part (b) is a magnification of part (a).

upward spikes in ρ_{pt} on segments of the curve that are close to straight, which confirms for our biarc simulations a phenomenon first observed in point discretisation simulations.[12] Second, as indicated in Figure 16 (b), there are multiple adjacent arcs on which ρ_{pt} is achieved by local curvature, and which achieve thickness to within a relative error of 10^{-7}. This strongly suggests that local curvature is active in realising thickness for the ideal shape of the +3.1#-3.1-knot.

Computation with biarcs has yielded an improved understanding of the rather complicated, approximate contact sets, which give a measure of closeness to ideality. The resolution of our simulations seems sufficient to conclude that, as shown in Figure 8, the contact chords of the 3.1 knot span a surface, and that the contact points themselves form a trefoil knot. In contrast, as shown in Figure 15 the contact chords associated with $\boldsymbol{\chi}_{\mu^*}$ of the figure-eight knot appear to be split into two disjoint components, with the contact points $\mathbb{C}_{\mu^*}$ in each component forming a loop with two tails.

In summary we claim that biarc curves are a rather efficient choice of space discretisation for computations involving self avoiding curves, particularly so for optimal packing problems such as the simulation of ideal knot shapes. Nevertheless the computations presented here perhaps raise as many questions as they answer. For example, simulated annealing could be carried out for both 3.1 and 4.1 knots, but with period three, respectively two, rotational symmetry forced *a priori*. The results of such computations should clarify whether or not the current simulations, which predict shapes that are close to, but not exactly symmetric, are picking up real, but small deviations from symmetry, or whether the simulations are merely reflecting a very slow elimination of asymmetry, perhaps due to the entirely local nature of the moves used in the simulated annealing.

Our stochastic approach of simulated annealing, which requires no derivative information, is largely dictated by the fact that little is currently known analytically to be able to formulate and justify a constrained gradient flow on biarcs that would be guaranteed to converge, essentially because it is far from simple to be able to write down the appropriate constrained derivatives. It seems likely that an appropriate numerical gradient flow would lead to more efficient computation, at least close to the ideal shape. Similarly, while our algorithm for computing thickness is robust and accurate to an arbitrary precision, it is likely that still more efficient, and faster algorithms could be constructed. For example one might compute thickness by deriving estimates of the error between the minimum of $\rho_{\mathsf{pt}}(s_i)$ at a discrete set of points s_i and the continuous minimum over arc length, combined with a bisection iteration.

Even within the specific approach of simulated annealing we do not believe the current code implementation to be particularly efficient. Rather our primary objective was to construct a robust code to test the efficacy of the biarc space discretisation. With that established, many improvements could be made. For example, we do not understand the optimal values for parameters such as temperature and cooling rate that need to be set during the simulation. In another direction we observe that in this first implementation only the simplest possible moves were used, namely random, small, independent, changes to an individual node point or tangent. It seems likely that significant gains in efficiency could be obtained by the introduction of a set of more sophisticated moves. For example, because we now know how to compute the contact set accurately, one could bias the moves of non-active points to be arc length shortening. It also seems likely that the addition of some nonlocal, cooperative moves could speed convergence. One potentially interesting choice for global moves seems to be inversion in a sphere, which maps a biarc curve to another biarc curve. Moreover arbitrary pairs of inversions leave the knot type unaltered, and pairs can be constructed either to give very large deformations, or to be close to the identity. In addition to a more suitable choice of moves, a more suitable choice for the biarc matching rule, in replacement of the simple mid-point rule adopted here, or indeed using the matching parameters as additional free variables in the simulated annealing, might also improve efficiency of computation.

References

1. J. Baranska, P. Pieranski and E. Rawdon, Ropelength of tight polygonal knots, *this volume.*
2. K. M. Bolton, Biarc curves, *Computer-Aided Design* **7** (1975) 89-92.
3. G. Buck and J. Simon, *Energy and length of knots*, in Lectures at Knots96, ed. Suzuki, S., World Scientific Publishing, Singapore (1997) 219–234.
4. J. A. Calvo, K. C. Millett and E. J. Rawdon (Eds), *Physical Knots: Knotting, Linking, and Folding. Geometric Objects in* $\mathbb{R}^3$, Contemporary Mathematics **304** (2002).
5. J. Cantarella, R. B. Kusner and J. M. Sullivan, On the minimum ropelength of knots and links, *Inventiones mathematicae* **150**(2) (2002) 257-286.
6. E. Denne, Y. Diao and J. M. Sullivan, Quadrisecants give new lower bound for the ropelength of a knot, arXiv:math GT/0408026 (2004).
7. G. Dietler, P. Pieranski, S. Kasas and A. Stasiak, The rupture of knotted strings under tension, *Contemporary Mathematics* **304** (2002) 217-222.
8. O. C. Durumeric, Thickness Formula and C^1-Compactness for $C^{1,1}$ Riemannian Submanifolds, arXiv: math.DG/0204050 (2002).
9. O. C. Durumeric, Local structure of ideal shapes of knots, arXiv:math GT/0204063 (2002).
10. F. B. Fuller, Decomposition of the Linking Number of a Closed Ribbon: A Problem from Molecular Biology, *Proc. Natl. Acad. Sci. USA* **75**(8) (1978) 3557-3561.
11. O. Gonzalez and R. de la Llave, Existence of ideal knots, *J. of Knot Theory and its Ramifications* **12**(1) (2003) 123-133.
12. O. Gonzalez and J. H. Maddocks, Global curvature, thickness, and the ideal shapes of knots, *Proc. Natl. Acad. Sci. USA* **96**(9) (1999) 4769-4773.
13. O. Gonzalez, J. H. Maddocks, F. Schuricht and H. von der Mosel, Global curvature and self-contact of nonlinearly elastic curves and rods, *Calc. Var. Partial Differential Equations* **14**(1) (2002) 29-68.
14. O. Gonzalez, J. H. Maddocks and J. Smutny, Curves, circles, and spheres, *Contemporary Mathematics* **304** (2002) 195-215.
15. V. Katritch, J. Bednar, D. Michoud, R. G. Scharein, J. Dubochet and A. Stasiak, *Geometry and physics of knots*, Nature **384** (1996) 142–145.
16. B. Laurie, *Annealing Ideal Knots and Links: Methods and Pitfalls*, Chpt. 3 of [35].
17. R. A. Litherland, J. Simon, O. Durumeric and E. Rawdon, *Thickness of knots*, Topology and its Applications **91** (1999) 233–244.
18. A. Maritan, C. Micheletti, A. Trovato and J. R. Banavar, Optimal shapes of compact strings, *Nature* **406**(6793) (2000) 287-290.
19. A. W. Nutbourne and R. R. Martin, *Differential geometry applied to curve and surface design*, Vol. 1: Foundations, Ellis Horwood, Chichester, 1988.
20. P. Pieranski, *In Search of Ideal Knots*, Chpt. 2 of [35].
21. P. Pieranski and S. Przybyl, Ideal trefoil knot, *Physical Review E*, **64** (2001).

22. P. Pieranski and S. Przybyl, In Search of the Ideal Trefoil Knot, *Contemporary Mathematics* **304** (2002) 153-162.
23. P. Pieranski, S. Przybyl and A. Stasiak, Tight open knots, *European Physical Journal E*, **6**(2) (2001) 123–128.
24. E. J. Rawdon, *Approximating the Thickness of a Knot*, Chpt. 9 of [35].
25. E. J. Rawdon, Approximating smooth thickness, *J. of Knot Theory and its Ramifications* **9** (1) (2000) 113–145.
26. E. J. Rawdon, Can computers discover ideal knots?, *Experimental Mathematics*, **12**#3 (2003), 287-302.
27. E. J. Rawdon, Program TOROS (Thickness Or Ropelength Optimizing System), http://www.mathcs.duq.edu/˜rawdon/manual.html
28. J. Schonherr, Smooth biarc curves, *Computer-Aided Design* **25** (1993) 365-370.
29. F. Schuricht and H. von der Mosel, Global curvature for rectifiable loops, *Mathematische Zeitschrift*, **243** (2003) 37-77.
30. F. Schuricht and H. von der Mosel, Characterization of ideal knots, *Calc. Var.*, **19** (2004) 281–305.
31. T. J. Sharrock, *Surface Design with Cyclide Patches*, PhD Thesis, University of Cambridge, Engineering Department, 1985.
32. T. J. Sharrock, Biarcs in three dimensions, in *The Mathematics of Surfaces II*, Ed. R. R. Martin, Oxford Univ. Press, New York, (1987) 395-411.
33. J. Smutny, *Global radii of curvature, and the biarc approximation of space curves: In pursuit of ideal knot shapes*, PhD Thesis No. 2981, EPFL, 2004.
34. A. Stasiak, J. Dubochet, V. Katritch and P. Pieranski, *Ideal Knots and Their Relation to the Physics of Real Knots*, Chpt. 1 of [35].
35. A. Stasiak, V. Katritch and L. H. Kauffman (Eds), *Ideal knots*, Ser. Knots Everything **19**, World Sci. Publishing, River Edge, NJ, (1998).
36. A. Stasiak and J. H. Maddocks, *Best packing in proteins and DNA*, Nature **406** (2000) 251–253.
37. D. J. Struik, *Lectures on Classical Differential Geometry*, Second Edition, Dover, New York (1988).

CHAPTER 6

KNOTTED UMBILICAL CORDS

Alain Goriely
Department of Mathematics and Program in Applied Mathematics,
University of Arizona, AZ85721
E-mail: gориely@math.arizona.edu
Homepage: http://www.math.arizona.edu/~goriely

"These vessels, whose origin is already known, wind around each other, like the twigs which form the handel of a basket. Sometime the arteries creep round the vein, like ivy round a tree; and sometimes the vein does the same round the arteries. The vein often folds itself into a kind of loops of different lengths, or forms itself into a species of knots subject to become varicious... The cord from its great length, sometimes get tied into one or more knots. These do not prevent the ordinary development of the foetus nor occasion its death as imagined by some" Baudelocque (1789).[1]

1. Introduction

To the delight of topologists, there are many examples of knots in Nature starting at the level of DNA, proteins, polymers, fish, cables and magnetic braids in the solar corona.[2,3,4,5,6,7,8,9] The purpose of this Note is to discuss and document the formation of knots in umbilical cords, a biological system which exhibits knots and that has seemingly eluded the scrutiny of mathematicians and physicists.

2. Description

An umbilical cord is a cordlike structure in the pregnant female mammal connecting the placenta to the abdominal wall of the fetus. Its primary function is to carry oxygen and blood to the fetus and return waste products. In pregnant human females, it is normally composed of two arteries and one vein.[10] The umbilical vein carries oxygenated blood from the placenta to the fetus while the umbilical arteries carry deoxygenated blood and wastes from the fetus back to the placenta. The arterial vessels are embedded in a sturdy gel-like structure known as the Wharton's jelly. The vein is twisted around the arteries and the arteries are twisted within the cords to form a triple helix. Typical diameters of an umbilical cord at delivery are around 1.7cm.[a] The length of the umbilical cord at birth (38 weeks) averages[12] 57.4cm with standard deviation 12.6cm but on rare occasions cords as long as 300cm and as short as 6.7cm have been observed.[b] Cords below 40cm occur only in 6% of the cases and are known as short cords. Short cords may cause traction, placental abruption and even cord rupture during vaginal delivery and are associated with many different problems at birth (such as low Apgar scores and hypotonia) and neurological abnormalities later in life. Cords must be sufficiently long (at least 32cm) as to allow the fetus to exit safely the womb. However, if a cord becomes too long it can loop around foetal parts and produce distress through cord compression.

3. Knots

The feature of interest for this Note is the possibility of knot formation in umbilical cords. Physicians use the word knot for two different structures. The so-called "false knots" correspond to the looping of the umbilical vessels due to high vascular torsion inside the Wharton's Jelly of the umbilical cord (See Figure1 (Color Plate 1A)). Their bulky appearance is sometime mistaken for true knots. A common superstition is that the number of false knots determines the number of children the mother is to have. However, they have little clinical significance, no interesting topology, and will not be discussed further. The "true knots" are knots in the common sense (see Figure 2 (Color Plate 1B))[14] but not in the restrictive mathematical sense since, obviously, the cord is not closed. However, there is no ambiguity in the definition of a mathematical knot for an umbilical cord since it is usually localized and pulled tight. One can virtually close the knot by first pulling the end of the cord so that it is contained inside a sphere with its ends on the sphere and imagining a tube connecting the two ends of the cord without

[a]Unless specified otherwise, the data used in this paper is taken from Benirschke and Kaufmann (2000).[11]

[b]Leonardo Da Vinci remarked that the typical length of the umbilical cord at birth is equal to the length of the newborn infant, a rather good estimate.[13]

any intersection with the sphere itself. To avoid pointless arguments, when referring to knots, we assume here that the central curve of the cord has been closed away from the knot in such a manner.

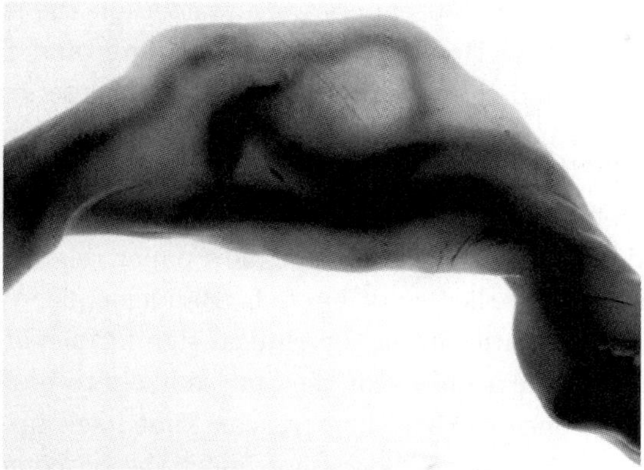

Fig. 1. A false knot is a enlargement of the umbilical cord due to excessive vascular growth. Picture Courtesy of the University of Utah Placental Bank (Reproductive Genetics Research Lab.)

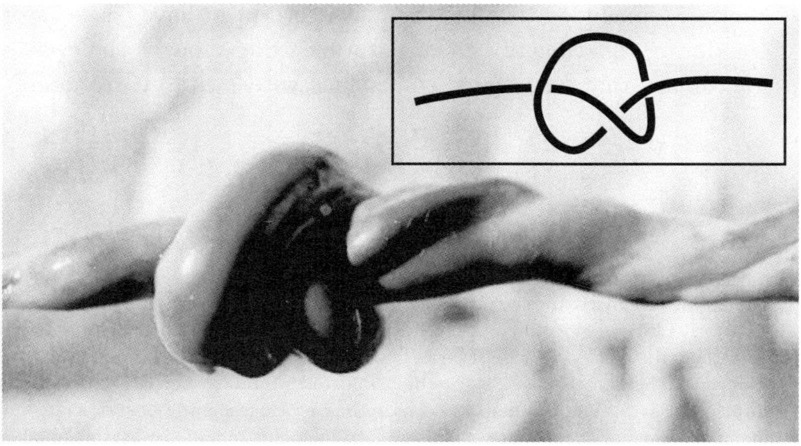

Fig. 2. A true knot occurs when the fetus passes through a loop and forms a knot. Copyright ©1996 Massachusetts Medical Society. All rights reserved.

4. History

Umbilical knots are a rather peculiar and puzzling feature of the umbilical cord. They have attracted the attention of physicians and midwives for centuries and have been discussed in the medical literature extensively. Prior to 1948, more than 150 papers were written on the subject as cited in the review papers by Browne[15] (1923), Lundgren and Boice[16] (1939), Hennessy[17] (1944), and Spivack[18] (1946). Most of these articles are published cases, that is, anecdotal accounts of observed knots in umbilical cords. The earliest paper addressing exclusively the problem of umbilical knots is due to M. Baudelocque who published in 1842 an article[c] entitled "Sur les noeuds du cordon ombilical"[19] Baudelocque's paper relates his own observations and discusses earlier work by J. L. Baudelocque, William Smellie (See Figure 3) and Mauriceau on the clinical significance of knotted cords and their formation. He argues that the knot is likely to be formed at birth ("It is difficult to conceive that the cord can knot itself up to three times at the same place") and that it does not affect the outcome of pregnancy ("If a dead infant has a knotted umbilical cord, one should look for another cause of death"). Unbeknown to the modern medical literature, it seems that the earliest report of knots in umbilical cord is to be found in the work of Louise Bourgeois dating back to 1609.[20] In there, she described how she delivered to a baby with a knotted umbilical cord ("le nombril noué a droit noeud"), the woman had complained of belly pains days before giving birth and Bourgeois concluded that the umbilical cord must have passed around the fetus during this time of "great agitation". She thought the event was so extraordinary that she offered to produce witnesses to corroborate her account.[d]

[c]This paper is cited in Hennessy[17] but wrongly attributed to Jean-Louis Baudelocque (1745-1810). J.-L. Baudelocque was considered the leading French obstetrician of his time. He attended Napoleon's wife, Marie Thérèse and published the important treatise "L'art des accouchements" in 1789 in which knotted umbilical cords is also discussed (see quotation above). The Baudelocque family has produced many obstetricians and is still remembered in Paris' renowned Clinique Baudelocque, one of the first public health clinics ever established (1890).

[d]Louise Bourgeois (1563-1636) was a midwife and attended queen Marie de Medicis for six deliveries. Her Observations is the first book of obstetrics written in French by a woman. Her remarkable life and work has been rediscovered recently and is the object of many interesting publications.[21,22,23]

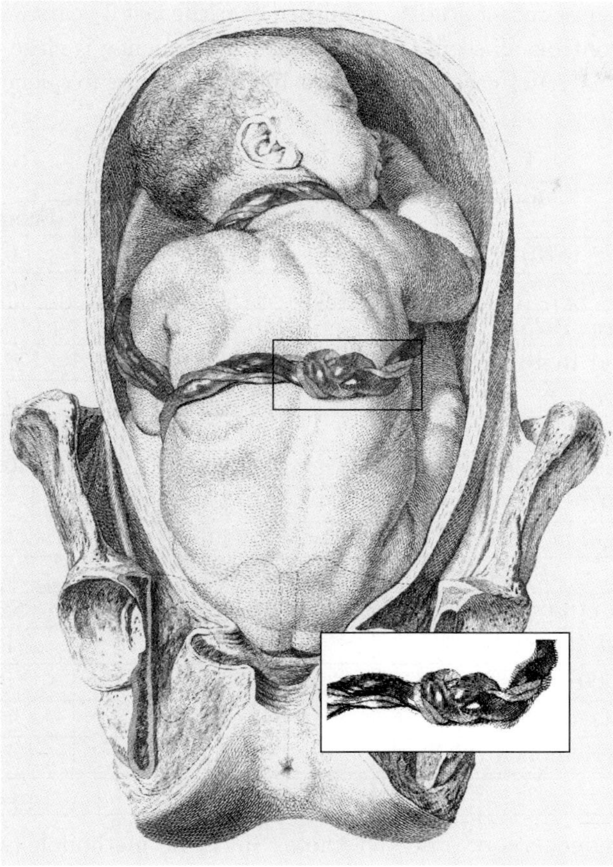

Fig. 3. Table XXIX from Smellie's "An abridgement of the practice of midwifery: and a set of anatomical tables with explanations" (1786). Note also the nuchal coil (the loop around the neck) typical of long umbilical cords. Smellie found the problem sufficiently interesting to discuss it in four different places in his book and include an engraving of a knot. William Smellie (1698-1763) was the most famous and best known teacher of man-midwives in London. He published three treatises on midwifery and was virulently critiqued by English writers opposed to the practice of midwifery by men.

5. Knotting frequency

How often are knotted umbilical cords observed? True knots are an uncommon but not rare occurrence. Most papers cite the frequency of knot formation as being between 0.3% and 2.1% of all pregnancies. The earlier studies date back to Chantreuil's thesis in 1875. Over the last 125 years, more than 15 studies on knotting frequencies have been published (see Table 1). The most recent studies published over the last 3 years with rigorous statistical controls all seem to agree that the frequency is close to 1% (the grand total is 2,104 cases for 214,124 birth, that is a frequency of about 1.0%).

Authors	**(# cases)/(# birth)**	**Frequency**
Chantreuil (1875) [16]	6/1,000	0.6%
Mundé (1879) [16]	2/1,000	0.2%
Von Hecker (1925) [15]	115/31,590	0.4%
Terlizzi and Rossi (1959) [24]	48/15,416	0.3%
Dippel (1964) [25]	21/1,009	2.1%
Spellacy *et al.* (1966) [26]	180/17,190	1.0%
Recasens *et al.* (1968) [24]	5/800	0.6%
Hartge (1979) [24]	33/3,400	1%
Blickstein *et al* (1987) [27]	57/4,650	1.2%
McLennan *et al* (1988) [28]	6/1,115	0.5%
Sepulveda *et al* (1995) [29]	18/5,575	0.3%
Joura *et al* (1998) [30]	286/22,531	1.3%
Sornes (2000) [31]	216/22,012	1.0%
Hershkovitz *et al*(2001) [32]	841/69,139	1.2%
Airas and Heinonen (2002)[33]	288/23,0272	1.2%
TOTAL	**2,104/214,124**	**1.0%**

Table 1. Frequencies of observed knots. Since the methodology varies between all these studies, the total is only given as an indication of the most likely frequency of knotting.

6. Contributing factors

Almost all studies conclude that the mean cord length is higher in knotted cords than in normal cords (*e.g.* 84cm versus 59cm from Sornes[31] (2000)). Airas and Heinonen (2002) found that the risk factors associated with true knot formation are multiparity (2 or more pregnancies), previous miscarriages, and obesity.[33] Similarly, Hershkovitz *et al.* identified the following groups at risk: grandmultiparous women (5 or more pregnancies), pregnancies complicated with hydramnios (excess of amniotic fluid), patient who underwent genetic amniocentesis (a prenatal test in which a small amount of amniotic fluid is removed and examined) and those with chronic hypertension.[32] Male gender was also found to be a significant contributor to knot formation.

These risk factors point to three basic simple chief causes to knot formation: (1) the cord length: Increased fetal activity is associated with long umbilical cords (this is related to the interesting assumption that one of the stimuli for growth in cords is the tension generated through fetal activity[34]), also males have slightly longer cords than females and knots are commonly found in Chimpanzees who have a relatively much longer cord[35]; (2) large uterine volume: Factors such as multiparity, hydramnios , obesity, and diabetes are all associated with large uterine volume or lax uterine wall; (3) high fetal activity: amniocentesis is known to produce large fetal activity. Therefore, these studies suggest that the two relevant geometric features are: long cords and large uterine volume and that the important dynamical feature is fetal activity. These effects can be combined into the following simple model.

7. A simple model

To simplify the purely geometrical problem of knot formation we consider a free floating ball of radius R attached to a string of length L of negligible thickness and anchored at one point. The ball is free to move randomly in a 3 dimensional space with the constraint that the length of the string remains constant (See Figure 4.A). Will the ball eventually pass through a loop and knot itself? Clearly, the string must be long enough since if $L/R < \lambda_{\text{crit}} = 2\pi$, the string cannot slide around the sphere to create a trefoil knot. Since the thickness of the string is considered negligible, the only relevant parameter in the problem is the ratio $\lambda = L/R$. Intuitively, if λ is large enough, we expect that knotting will occur frequently and, conversely as $\lambda \to \lambda_{\text{crit}}$, no knotting will be observed. The precise manner in which the knotting frequency depends on λ is not known. However, similar problems have been extensively studied in Brownian motion.[36,37,38,39,40,41] These studies reveal that the probability of forming a knot in self-avoiding polygonal walks increases with the length n of the polygon as

$$P(n) = 1 - e^{-\alpha n + o(n)} \tag{1}$$

for some positive constant α. Therefore, it is reasonable to expect that the situation in our model will be similar and that the likelihood of knot formation increases with the parameter λ. Depending on the geometry, the critical value of λ may be different. For instance, if a string of length L is attached on a cylinder of height $2h$ and radius r, the minimal length to pass the cylinder around the loop is $l + 2\pi r$ where $h \leq l \leq 2h$.

Fetuses are not perfect spheres and different choices can be made for an effective radius R. Here we give two different choices roughly corresponding to an upper and a lower bound for the problem. First, as an upper bound, we use the weight of the fetus and compute the radius (referred to as the "ball radius") of the sphere with equal weight (assuming that the density of the fetus is close to that of water). Second as a lower bound, we use the crown-rump length as the diameter of our ball. In Figure 5 we show the evolution of the parameter λ along the pregnancy for these two choices of radii. The dotted line corresponds to $\lambda = 2\pi$, that is, if the fetus was a perfect sphere with diameter equal to the crown-rump length, the only geometric possibility of knot formation for average umbilical cords and fetuses would occur between 10 and 15 weeks. The most important features of Figure 5 is the peak appearing at 12-13 week of gestation and the increase of λ between 9 and 15 weeks.

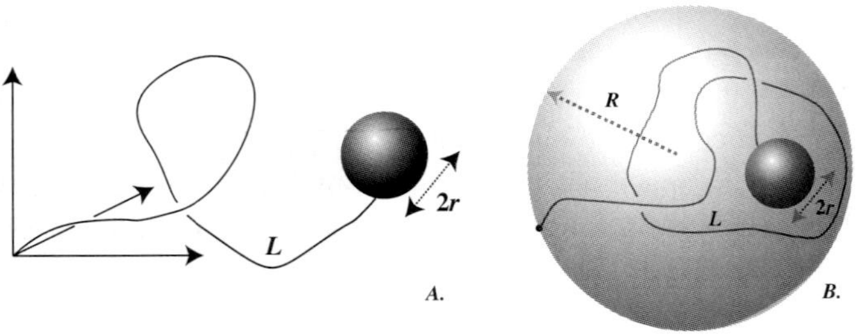

Fig. 4. In the first model (A), a ball attached to a string is free to move in a three-dimensional space. The string is fixed at one point and the ball performs random motion, dragging the string behind itself. The goal of this model is to understand the interplay between the motion of the ball and the constraint due to the string on the formation of knots. In the second model (B), the same ball is attached to a string itself attached to the inside surface of a sphere. This model introduces a new length in the problem (the radius of the outer sphere) and is aimed at understanding the effect of the amniotic sac volume on the formation of knots.

A fetus is not allowed to move freely in space, its motion is restricted by the amniotic sac. The smaller the sac with respect to the size of the fetus, the less likely the fetus will be able to move around and form loops. Eventually, close to gestation (typically after 20 weeks), very few large scale

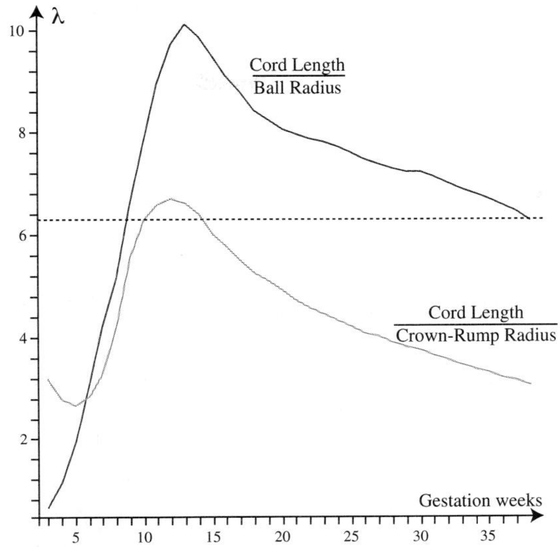

Fig. 5. Evolution of the ratios cord length to ball radius and crown-rump length as a function of gestation time (counted in weeks). The ball radius corresponds to the equivalent radius of a sphere of density one of equal fetus weight. The dotted line corresponds to the critical value $\lambda_{\text{crit}} = 2\pi$. That is, if the fetus was a perfect sphere with radius given by half the crown-rump length, knotting could only occur in a very small window of time around week 13.

motion is possible. Ideally, we can look at it as restricting the motion of our ball into a larger sphere (See Figure 4.B). To quantify this effect, we show on Figure 6, the ratio of the placenta radius to the ball radius. Here again, we clearly see a peak at week 13 followed by a sharp decay and a plateau around 20 weeks where fetus motion will be impeded by the sac. This simple analysis suggests that the time where knot formation is most likely to occur is between 9 and 18 weeks with a peak around 13-14 week. Furthermore, if we assume that the increase in probability for knot formation follows a similar law as the one given by Equation (1), we see that small increases in the length of the umbilical cord will result in a large increase in the probability of knot formation.

How does this geometric prediction compare to the accepted knowledge in the medical literature? It is usually believed that knot formation occurs between 9 and 12 weeks' gestation.[32] Blickstein *et al.*[27] argue that knot formation should occur between week 9 and week 28 (when cord length does not increase significantly). Our analysis is also consistent with the onset of large scale motions of the fetus: "By 8 to 10 weeks, the fetus

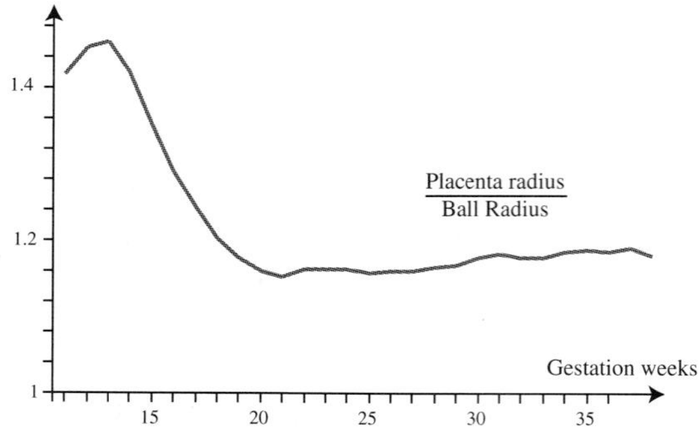

Fig. 6. Evolution of the placenta radius to the ball radius as a function of gestation time (in week). After 20 weeks, the relative room for fetal motion does not change.

typically exhibits a series of abrupt movements characterized by sudden arching of the back and extension of the limbs. These movements cause the fetus to rise up in the amniotic fluid pool and then to sink slowly at completion of the movement"[42] Finally, the correlation with patients who underwent amniocentesis is also consistent with these results since this test is usually performed between week 15 and 18.

This analysis does not exclude the possibility of knot formation during labor when the fetus, during ejection, is passed through a loop. However, Browne[15] argues that these knots could easily be detected from knots formed earlier as they should not retain their shape when untight. If a true knot forms itself in the second trimester, the shape of the knot is completely set during further growth even when the cord is not pulled.

8. Clinical significance

A central question for physicians is the relevance of knots for the final outcome of the pregnancy: "defining a high risk group of patients for true knot of cord is of utmost importance in order to decrease perinatal morbidity and morbidity".[32] This question has been hotly debated since the first paper by Baudelocque (see quotes above). The risks could be multiple. First, if the cord is pulled too tight it could compress itself and impairs perfusion so that oxygen and nutrient could not reach the fetus, possibly triggering asphyxia and an increase in cardiac load[43] potentially leading to fetal distress. To test this idea, Browne[15] conducted experiments on the perfusion of cords. In his experiment, he collected a fresh umbilical cord and made a slack knot, then he attached a weight to the cord to simulate tension and perfused it with increasing pressure until circulation was established. His *in vitro* experiment showed that "even a slack knot may be sufficient to interfere with, if not completely to obstruct, the cord circulation". Modern statistical analyses by Airas and Heinonen[33] have shown that fetuses with umbilical knots have a four-fold increased risk of stillbirth during pregnancy, showing a clear association between fetal demise and cord knots, a finding supported by other studies.[30,31] However, it is not clear in these cases that the umbilical knot was the main reason for fetal death since asphyxia and increased cardiac load can be caused by nuchal coils which are themselves correlated with long cords and henceforth umbilical knots. Sornes[31] argues that: "the mere presence of of a knot on the cord cannot in my opinion be a 'cause of death'. When these cases occur, a more thorough investigation into the whole of the placenta, and the umbilical cord in its entire length, is called for in order to elucidate the real cause of death." Moreover, these findings may not be clinically significant since: (1) umbilical knots are extremely difficult to detect by ultrasonography;[29] (2) if a knot is detected, it is not clear what medical care could be given to prevent the tightening of the cord and fetal death associated with asphyxia.

The second risk associated with umbilical knots is the possibility of cord tightening during delivery when the cord is stretched. However, cord knots does not seem to affect birth since the Apgar scores and the level of obstetrical intervention are equal.[28,30,31]

Whereas knot formation is relatively begin in singletons, it becomes significant for the so-called MoMo twins (monoamniotic, monochorionic twins).[44] These MoMo twins are usually monozygotic and share the same uterine sac without any membrane to keep them apart. This is by itself a rare occurrence in twins (about 1 to 2%) but when it presents itself there is a high mortality rate as high as 50-62% and congenital anomalies in 15-20% of cases due to entanglement of the two cords (see Figure 7 (Color Plate 2A)).

9. Complex and multiple knots

Most of the umbilical knots observed are not properly described or classified. Since trefoil knot are the simplest and easiest to make (passing once through as single loop), it is likely that the vast majority of umbilical knots are trefoil. Nevertheless, more complex and multiple knots have also observed. The only comprehensive statistical analysis of multiple knots was performed by Sornes[31] who observed 11 cases of double knots, 201 cases of

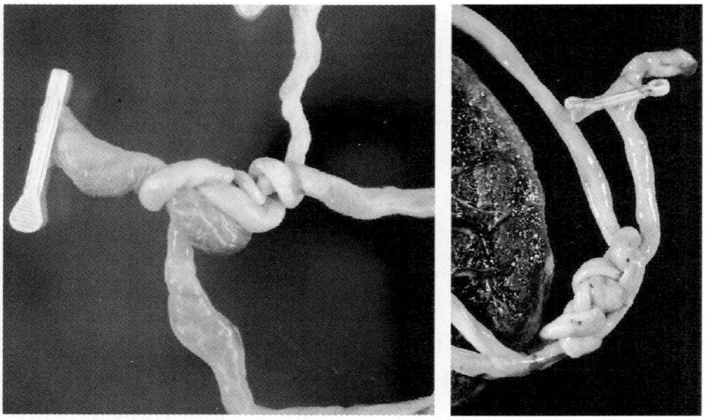

Fig. 7. Cord entanglement in MoMo twins. This is a very serious condition with high mortality rate. Picture Courtesy of the University of Utah Placental Bank (Reproductive Genetics Research Lab.)

simple knots in 22012 births. Therefore, the probability of forming a double knot is about 68/10000 much larger than the probability of finding two different cords with a single knot (that is about $(1\%)^2 = 1/10000$). This is not surprising since a long cord in a large intrauterine volume has a higher probability of forming a single and hence a second knot. Different authors report multiple or complex knots found in cords. Beside the usual trefoil knot (3_1 in the standard notation), the figure-eight knot[11] (4_1) and the 5_2 knot have also been observed (remarkably, in this last case, the knot could be observed by ultrasonography[45]). Multiple knots also come with various topologies such as a double trefoil knot[28], a trefoil knot and a figure-eight knot[46] (see Figure 8 (Color Plate 2B)), and the most elaborate knot reported: a double figure-eight knot[47] (in this case the authors also tried to find a mechanistic explanation of the knot formation and realized that it could be obtained if the fetus passed through a double twisted loop–see also Hartge[24] for similar mechanisms for simpler knots).

10. Handedness and perversion

Umbilical cords have also a very interesting fine geometric structure. The arteries are longer than the vein which is itself longer than the jelly and they are wrapped around each other so that the umbilical cord forms a triple helix. The handedness of this helical structure is another puzzling feature of the umbilical cord that has been the object of many studies[48,49] and was first discussed by Berengerius in 1521. The umbilical cord has up to 40 he-

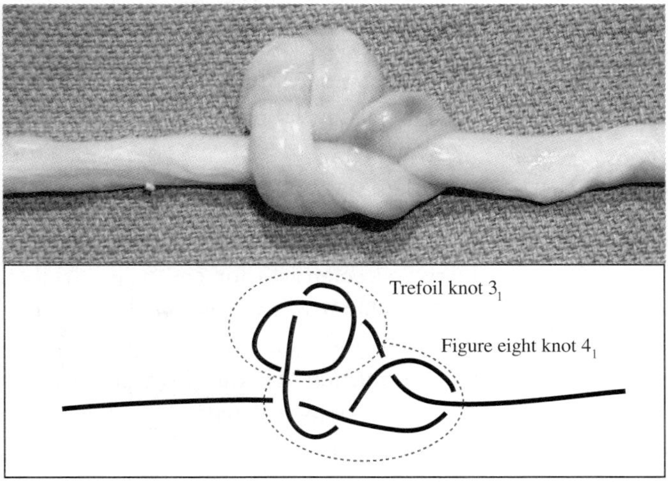

Fig. 8. An example of complex knot in an umbilical cord, a composition of an eight-knot with a trefoil knot. The infant was in good condition, without any clinical evidence of intrapartum asphyxia. Copyright ©1996 Massachusetts Medical Society. All rights reserved.

lical turns and handedness can be observed as early as 42 days' gestation.[12] Umbilical cords can be either left-handed, right-handed, straight or with mixed helicity.[50] The ratio of left-handed to right-handed cords is about 7 to 1 similar to the average ratio of right-handed to left-handed adults. However, there is no statistical correlation between the handedness of the cord and hand preferences. [51] An interesting feature that can shed light on the mechanism that selects handedness in cords is the existence of cords with both left and right handed structures. These particular cords with mixed handedness account for 2% to 26% of all cases.[49,52] The transition from left to right handedness is a common occurrence in filamentary structures and is known as *perversion*. It is found in the formation of tendrils in vines,[53,54] in bacterial flagella,[55,56] in the shape of certain bacteria such as spirochetes,[57] in some mutant forms of *B. subtilis*,[58] and in the microscopic structure of cotton fibers.[59] The mechanics and mathematics of perversion has been discussed in length by Goriely, Tabor and McMillen[60,61] who showed that the inversion of helicity if caused by both differential growth and twist blockage of the filamentary structure constraining the filament to writhe and shape itself as helices with zero total twist (the twist of each left and right helices canceling each other). However, to date, there is no model for the growth of umbilical cords that would explain the difference or handedness, its inversion or even the occurrence of helicity. Both genetic and mechanical factors seem important as indicated by the correlation between umbilical cord helicity in monozygotic twins[48] and experiments on the effect of tension on fetal activity in laboratory rats.[13,62]

11. Conclusions

The formation of knots in umbilical cords is an uncommon feature of the umbilical cord that has intrigued scientists for centuries. As early as the 18th century, it was suggested that excessively long cords are the most likely to form knots. Modern data and simple mathematical ideas supports this view and suggest that the formation of a knot is mostly a geometric and dynamical event rather than a physiological pathology. Therefore knotted umbilical cords provide us with a simple and beautiful system to motivate and illustrate the theory of physical knots in long chains.

Mathematicians love knots, they are a simple and elegant construction and knot theory has tentacular connections to various branches of mathematics. However, despite the fact that the microscopic world is mostly filamentary, physical knots in nature are scarce but involved in important processes (such as the ones found in DNA molecules). Most typical ratios of length scales or physical blockage prevent the formation of knots which can, as in the case of umbilical knots, create serious problems and malfunctions. The understanding of mechanisms and geometrical features that prevent knots from forming in many filaments but allow them in particular sytems could be a fascinating new chapter in the theory of physical knots.

Acknowledgments It is my pleasure to thank Dr. William Madden (University Medical Center in Tucson) for his help. I am also indebted to Christoph Lüthy for some references in early medical history and to John Maddocks and the Bernoulli center in Lausanne for their hospitilaty in the Summer 2003 when this paper took shape. This work is supported by the NSF grant DMS-0307427.

References

1. J. L. Baudelocque. *L'art des accouchements v.1. Translation from An abdrigment of Mr Heath's translation of Baudelocque's Midwifery, 1807 Philadelphia.* , Paris, 1789, v. 1.
2. M. A. Krasnow, A. Stasiak, S. J. Spengler, F. Dean, Th. Koller, and N. R. Cozzarelli. Determination of the absolute handedness of knots and catenanes of DNA. *Nature* 304: 559–560, 1983.

3. J. M. Sogo, A. Stasiak, M. L. Martnez-Robles, D. B. Krimer, P. Hernndez, and J. B. Schvartzman. Formation of knots in partially replicated DNA molecules. *J. Mol. Biol.*, 286:637–643, 1999.
4. J. J. Tyson and S. H. Strogatz. The differential geometry of scroll waves. *Intl. J. of Bifur. and Chaos*, 1:723–744, 1991.
5. L. H. Kauffman. *Knots and physics.* World Scientific, Singapore, 1993.
6. I. Tabor and I. Klapper. The dynamics of knots and curves I. *Nonlinear Science Today*, 4:7–13, 1994.
7. C. C. Adams. *The Knot Book: An Elementary Introduction to the Mathematical Theory of Knots.* W H Freeman , 1994.
8. P. G. Dommersnes, Y. Kantor, and M. Kardar. Knots in charged polymers. *Phys. Rev. E*, 66:# 031802, 2002.
9. W. R. Taylor, B. Xiao, S. J. Gamblin, and K. Lin. A knot or not a knot? SETting the record 'straight' on proteins. *Comput. Biol. Cham.*, 27:11–15, 2003.
10. P. Malpas and A. M. Symonds. Observations on the structure of the human umbilical cord. *Surgery Gynec. Obstet*, 123:746–750, 1966.
11. K. Benirschke and P. Kaufmann. *Pathology of the human placenta. Fourth edition.* Springer, New York, 2000.
12. G. Ente and P. H. Penzer. The umbilical cord: normal parameters. *J. Roy. Soc. Helath*, August:138–140, 1991.
13. A. C. Moessinger, W. A. Blanc, P. A. Marone, and D. C. Polsen. Umbilical cord length as an index of fetal activity: experimental study and clinical implications. *Pediatr. Res.*, 16:109–112, 1982.
14. R. B. Altman and J. E. Merino. Knotted umbilical cord. *New England J. Med.*, 334:573–573, 1996.
15. F. J. Browne. On the abnormalities of the umbilical cord which may cause antenatal death. *J. Osbt. Gynec. Brit. Emp.*, 32:17–48, 1923.
16. A. T. Lundgren and W. A. Boice. True knotting of the umbilical cord. *Illinois Medical Journal*, 76:451–458, 1939.
17. J. P. Hennessy. True knots of the umbilical cords. *Am. J. Obstet. & Gynec.*, 48:528–536, 1944.
18. M. Spivack. The anatomic peculiarities of the human umbilical cord and their clinical significance. *Amer. J. Obst. Gynec.*, 52:387–401, 1946.
19. M. Baudelocque. Sur les noeuds du cordon ombilical. *Revue Médicale Francaise et Etrangére*, 3:355–360, 1842.
20. L. Bourgeois. *Observations diverses sur la stérilité, perte de fruits, fécondité, accouchements et maladies des femmes et enfants nouveau-nés suivi de instructions à ma fille.* Jean Dehoury, Paris, 1609.
21. Pelcé P. (Ed.). *Perspective in Physics: Dynamics of curved fronts.* Academic Press Inc., San Diego, 1988.
22. B. Sheridan. At birth : the modern state, modern medicine, and the royal midwife Louise Bourgeois in seventeenth century france. *Dynamis*, 19:145–166, 1999.
23. W. Perkins. *Midwifery and medicine in early modern France Louise Bourgeois.* University of Exeter Press, Exeter, 1996.

24. R. Hartge. Uber das vorkommen von Nabelschnurknoten. *Geburtsh. u. Frauenheik.*, 39:976–980, 1979.
25. A. L. Dippel. Maligned umbilical cord entanglements. *Am. J. Obstet. & Gynec.*, 88:1012–1021, 1964.
26. W. N. Spellacy, H. Gravem, and R. O. Fisch. The umbilical cord complications of true knots, nuchal coils and cords around the body. *Ber. ges. Gynak. Geburtsh*, 94:296, 1966.
27. I. Blickstein, Z. Shoham-Schwartz, and M. Lancet. Predisposing factors in the formation of true knots of the umbilical cord-analysis of morphometric and perinatal data. *Int. J. Gynaecol. Obstet.*, 25:395–398, 1987.
28. H. McLennan, E. Price, M. Urbanska, N. Craig, and M. Fraser. Umbilical cord knots and encirclements. *Aust. NZ J. Obstet. Gynaecol.*, 28:116–119, 1988.
29. W. Sepulveda, A. H. Shennan, S. Bower, P. Nicolaidis, and N. M. Fisk. True knot of the umbilical cord: a difficult prenatal ultrasonographic diagnosis. *Ultrasound Obstet. Gynecol.*, 5:106–108, 1995.
30. E. A. Joura, H. Zeiler, and M. O. Sator. Epidemiologie und klinische wertigkeit von echten nabelschnurknotten. *Wien. Klin. Wochenschr.*, 110:232–235, 1998.
31. T. Sornes. Umbilcal cord knots. *Acta Obstet. Gynecol, Scand.*, 79:157–159, 2000.
32. R. Hershkovitz, T. Silberstein, E. Sheiner, I. Shoham-Vardi, G. Holcberg, M. Katz, and M. Mazor. Risk factors associated with true knots of the umbilical cord. *Eur. J. Obstet. Gynecol. Reprod. Biol.*, 98:36–39, 2001.
33. U. Airas and S. Heinonen. Clinical significance of true umbilical knots: A population-based analysis. *American J. Perinatology*, 19:127–132, 2002.
34. A. E. Miller, M. C. Jones, and D. W. Smith. Tension: the basis of umbilical cord growth. *The journal of Pediatrics*, 101:844, 1982.
35. C. Naaktgeboren and A. M. van Wagtendonk. Wahre knoten un der nabelschnur nebst bemerkungen uber plazentophagie bei menschenaffen. *Z. Saugetierk*, 31:376–382, 1966.
36. W. S. Kendall. Knotting of Brownian motion in 3-space. *J. Lond. Math. Soc*, 19:378–384, 1979.
37. D. W. Summers and S. G. Whittington. Knots in self-avoiding wlaks. *J. Phys. A*, 21:1689–1964, 1988.
38. D. W. Summers and S. G. Whittington. Detecting knots in self-avoiding walks. *J. Phys. A.*, 23:1471–1472, 1990.
39. E. J. J. van Rensburg, D. W. Summers, E. Wasserman, and S. G. Whittington. Entanglement complexity of self-avoiding walks. *J. Phys. A.*, 25:6557–6566, 1992.
40. E. J. J. van Rensburg, D. W. Summers, E. Wasserman, and S. G. Whittington. The writhe of a self-avoiding polygon. *J. Phys. A.*, 26:L981–L986, 1993.
41. M. C. Tesi, E. J. J. Vanrensburg, E. Orlandini, and S. G. Whittington. Knot probability for lattice polygons in confined geometries. *J. Phys. A*, 27:347–260, 1994.

42. F. A. Manning. *Fetal Medicine.* Appleton & Lange, Norwalk, Connecticut, 1995.
43. J. Aranyosi, T. Major, B. Fülesdi, and J. Zatik. Fetal arterial redistribution indicating true umbilical cord knot. *European J. Obstert. Gynecol. Reprod Biol.*, 106:225–226, 2003.
44. J. S. Krüsel, S. v. Eckardstein, and T. Schenzer. Doppelter nabelschnurknoten bei monoamniotischer gemini-gravidität als ursache des intrauterinen fruchttods beider zwillinge. *Zentralbl. Gynakol.*, 116:497–499, 1994.
45. J. C. Collins, R. J. Muller, and Ch L. Collins. Prenatal observation of umbilical cord abnormalities: A triple knot and torsion of the umbilical cord. *Am. J. Obstet. & Gynec.*, July:102–104, 1993.
46. W. Camann and J. Marquardt. Complex umbilical-cord knot. *New England J. Med.*, 349:159–159, 2003.
47. J. R. Robins. A complex true knot of the umbilical cord. *Br. J. Clin. Pract.*, 49:164–165, 1995.
48. H. W. Edmonds. The spiral twistof the normal umbilical cord in twins and singletons. *Am. J. Obst. & Gynec.*, 67:102–120, 1954.
49. B. D. Chaurasia and B. M. Agarwal. Helical structure of the human umbilical cord. *Acta anat.*, 103:226–230, 1979.
50. S. Fletcher. Chirality in the umbilical cord. *Brit. J. Obstet. Gynec.*, 100:234–236, 1993.
51. R. V. Lacro, K. L. Jones, and K. Benirschke. The umbilical cord twsit: origin, direction, and relevance. *Am. J. Obstet. Gynecol.*, 157:833–838, 1987.
52. W. Blackburn, N. R. Cooley, and E. A. Manci. Correlations between umbilical cord structure–composition and normal and abnormal fetal devlopment. In R. A. Saul, editor, *Proceedings of the Grennwood Genetics Conference*, pages 180–181. Jacobs Press, Clinton, SC, 1988.
53. A. Gray. Note on the coiling of tendrils. *Proceedings of the American Academy of Arts and Science*, 4:98–100, 1858.
54. Ch Darwin. *The Movements and Habits of Climbing Plants.* Appleton, New York, 1888.
55. I. Yamashita, K. Hasegawa, H. Suzuki, F. Vonderiszt, Y. Mimori-Kiyosue, and K. Namba. Structure and switching of bacterial flagellar filaments studied by X-ray diffraction. *Nature structural biology*, 5:125–132, 1998.
56. R. E. Goldstein, A. Goriely, G. Hubber, and C. Wolgemuth. Bistable helices. *Phys. Rev. Lett.*, 84:1631–1634, 2000.
57. S. F. Goldstein and N. W. Charon. Motility of the spirochete *Leptospira. Cell Motility and the Cytoskeleton*, 9:101–110, 1988.
58. M. J. Tilby. Helical shape and wall synthesis in a bacterium. *Nature*, 266:450–552, 1977.
59. L. Waterkeyn. Light microscopy of the cotton fibre. In *Cotton fibres: Their development and properties.* International Institute for Cotton, Manchester, U.K., 1985.
60. A. Goriely and M. Tabor. Spontaneous helix-hand reversal and tendril perversion in climbing plants. *Phys. Rev. Lett.*, 80:1564–1567, 1998.
61. T. McMillen and A. Goriely. Tendril perversion in intrinsically curved rods.

J. Nonlinear Science, 12:241–281, 2002.
62. S. Barron, J. A. Foss, and E. P. Riley. The effect of prenatal cocaine exposure on umbilical cord length in fetal rats. *Neurotoxicol. Teratol.*, 13:503–506, 1991.

CHAPTER 7

MODELLING DNA AS A FLEXIBLE THICK POLYMER: DNA ELASTICITY AND PACKAGING THERMODYNAMICS

Cristian Micheletti

International School for Advanced Studies (S.I.S.S.A.) and INFM, Via Beirut 2-4, 34014 Trieste, Italy

Davide Marenduzzo

Department of Physics, Oxford University, 1 Keble Road, Oxford OX1 3NP, United Kingdom

We consider a thick polymer model with the purpose of providing a coarse-grained description for double-stranded DNA (dsDNA). The model is used to gain insight into how the intrinsic thickness of DNA affects the behaviour of the biomolecule subject to compaction. Our reference system is provided by the process of DNA packaging inside a viral capsid that has been recently characterized through delicate single-molecule experiments. The simplest of the two models considered here is able to provide a satisfactory quantitative accord with experimental measurement by using, as only inputs, the known dsDNA diameter and the base-pair spacing. In a second model we further incorporate a standard bending rigidity term obtaining a better agreement with experiments.

1. Introduction

Recent developments of experimental techniques, such as atomic force microscopes, optical tweezers, etc., have allowed to characherise the physical properties of individual biomolecules such as proteins, DNA, RNA etc. The wealth of detailed experimental results on the mechanical properties of such molecules have further stimulated the development of stuitable models for the elasticity of biomolecules.

A recent set of such experiments was carried out on the process of packaging of DNA inside the capsid of a bacteriophage.[1] The bacteriophage used in these studies was the $\phi29$ virus. Under carefully-controlled conditions it was possible to monitor the loading process of the 6.6 μm long double-stranded DNA (dsDNA) inside the viral head (a prolate 54 nm by 42 nm icosahedron[1,2,3]). By using suitable force-feedback measurements it has been possible to carefully measure the force exerted by the motor at various stages of the loading process, and hence the total energy required for the whole packaging process. It has been found[1] that the internal force exerted by the packaged DNA against the ϕ 29 capsid walls (which has to be overcome by the portal motor), displays a dramatic increase during the final loading stages. In particular, the maximum value attained by this internal force is $\sim$ 50 pN. This is a strikingly large number when compared to typical forces encountered at the molecular and cellular scale. Such high packaging forces are also not expected on the basis of standard models which describe DNA elasticity by means of bending-rigidity terms. In addition, a recent molecular dynamics simulation used to characterize the *dynamical* packaging of DNA in a spherical cavity has shown that at high packing densities, the resistance to packaging was not dominated by the DNA self-interaction (whether attractive or repulsive). These two facts seem to indicate that the DNA resistance to packaging is dominated by entropic effects which are not fully captured by conventional bending-rigidity effects in traditional beads-and-strings models. Other interesting theoretical works on this subject have recently appeared.[4,5]

These observations stimulated the investigation of an alternative model for DNA, where the traditional coarse-grained description of DNA as a succession of beads connected by strings is replaced by one as a self-avoiding thick tube.[6] It is expected that the proper treatment of the DNA intrinsic thickness may severely reduce the configurational entropy with respect to beads-and-strings models, and thus account for the unexpectedly high packaging force found in experiments.

In order to elucidate in a clear way how the DNA thickness impacts on the thermodynamics of the packaging process we have intentionally kept the thick polymer model at the simplest level and thus avoided the introduction of any explicit treatment of the (electrostatic) self-interaction energy between base pairs (for a discussion on the accuracy of this approximation see also Ref. 7).

As we shall show below, the appropriate treatment of the finite thickness of the biopolymer *does* allow to account for a variety of elastic properties of DNA in a natural way, even in the absence of additional phenomenological parameters, such as the bending rigidity. For example, it will be shown that the mere difference in thickness of single-stranded and double-stranded

DNA can account satisfactorily for the large change in the respective persistence lengths. More importantly, by considering the force necessary to compactify thick chains of an increasing number of beads, we estimate the packaging force (and work) which turns out to be in remarkable quantitative agreement with experimental measures for $\phi 29$ under the same density conditions. The present analysis involves finite-size scaling techniques to extrapolate results obtained in equilibrium stochastic simulations of chains with up to 200 beads. The most compact conformations obtained for such lengths reveal the clear tendency of thick polymers to occupy the spherical cavity by adopting a spool-like conformation, consistently with previous expectations[3,8,9] and with the recent numerical studies of Kindt *et al.* and of Arsuaga *et al.* with different models.[10,4]

2. The model

Due to the large number of atomic constituents of both inorganic and biological polymers, it is necessary to resort to suitably simplified models in order to characterize their physical behaviour by means of analytical or numerical studies. Customarily the polymeric chain is coarse-grained into a succession of discrete beads. Besides capturing the effect of chain connectivity through the introduction of suitable strings or springs between consecutive beads, it is usually necessary to introduce appropriate interactions between the beads in order to capture the salient physical features of the given system. For the case of double-stranded DNA typical models include effects such as strands self-interaction (either attractive or repulsive) and bending rigidity. Thus, the effective Hamiltonian of a discretised chain of equispaced beads, $(\vec{r}_1, \vec{r}_2, ...\vec{r}_N)$, reads:

$$\mathcal{H} = \sum_{ij} V_2(r_{ij}) - \frac{K}{a} \sum_i cos(\theta_i), \tag{1}$$

where r_{ij} is the distance between beads i and j, V_2 is a pairwise interaction between the beads (aimed at capturing the polymer self-interaction), a is the separation of consecutive beads, K/a is the bending rigidity, and θ_i is the angle formed by the bonds t_i and t_{i+1} ($t_i = \vec{r}_i - \vec{r}_{i-1}$).

The bending rigidity term in Eq. (1) represents the lowest order term in the expansion of the system in terms of the angle deviations, $\{\theta_i\}$. Clearly, as the bending deformation becomes more and more pronounced, the validity of this approximation will be reduced. One extreme example is provided by cases where the polymer centerline (represented by the chain of beads) is locally bent beyond the radius curvature imposed by the polymer intrinsic thickness. This physically unrealizable situation receives a finite energy penalty per bead, $\bar{e}$, according to the model Hamiltonian of Eq. (1). If the system thermal energy, $K_B T$ is comparable to $\bar{e}$, some of the configurations attained by the centerline under thermal equilibrium will be unphysical (in the sense that they do not have a counterpart for the thick polymer).

By using concepts derived in the context of ideal shapes of knots,[11,12,13,14,15,16,17] it is possible to generalise the model Hamiltonian so as to incorporate, in the simplest possible way, the finite thickness of the polymer and explore the consequences of the elastic properties of DNA (particularly in the compact phase).

To this purpose we shall consider DNA as a discretised tube of uniform thickness, Δ. The finite thickness impacts on two distinct features of polymer conformations. On one hand it will constrain the local radius of curvature to be not less than Δ to avoid singularities.

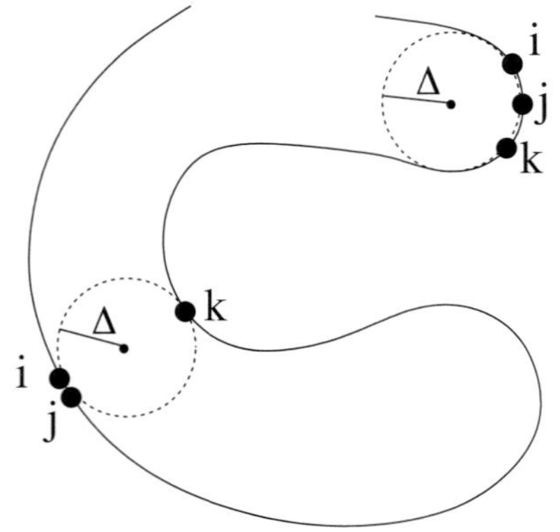

Fig. 1. Sketch of a curve which is a viable centerline for a tube of thickness Δ. The radii of the circles going through any triplet of points, r_{ijk} are not smaller than Δ.

On the other hand there is also a non-local effect due to the fact that any two portions of the tube, at a finite arclength separation, cannot interpenetrate. Equivalently, the centerlines of the two portions need to be at a distance greater than 2Δ (see Fig. 1). In traditional beads-and-springs models it is only this second effect that is taken into account through a hard-core repulsion enforced by the pairwise potential in Eq. 1. Interestingly, one needs to go beyond pairwise interactions to account for the above mentioned effects in discretised polymer chains. In fact, the requirement on the local radius of curvature can be enforced by finding the radii of the circles going through any consecutive triplet of points and ensuring that each of them is greater than Δ. The non-local effect can be addressed within the same framework by considering the minimum radius among circles going through any non-consecutive triplet of points. This radius is precisely the distance of minimum approach and, again, has to be greater than Δ. In summary, a practical implementation for the finite thickness, Δ, of the tube associated to the discretized centerline is reflected in the fact that the radii of the circles going through any triplet of distinct points has to be greater than Δ.[15] Other implementations of the thickness constraint are also possible; for example the algorithm discussed by Rawdon[16] allows to approximate a thick tube as a succession of suitably joined cylinders. For the same level of discretisation this approach produces smoother curves than the algorithm discussed previously albeit at the expense of computational time (and hence has not been adopted in the present context).

One is thus led to consider the following Hamiltonian for the unconstrained, non self-interacting, double-stranded DNA:

$$\mathcal{H}(\Gamma) = \sum_{ijk} V_3(r_{ijk}) \tag{2}$$

where V_3 is the three-body potential used to enforce the thickness Δ of the chain[13,15,17,18,19,20] and Γ is the configuration of the dsDNA. The argument of V_3 is the radius of the circle going through the triplet of distinct points i, j, k and has the form

$$V_3(r) = \begin{cases} 0 & \text{if } r > \Delta, \\ +\infty & \text{otherwise.} \end{cases} \tag{3}$$

In addition to this term, the tube ends were modelled as hemispheres by introducing two "phantom" beads essentially coinciding with the terminal beads of the tube.

For the case of dsDNA, the natural spacing of the effective beads, is provided by the base-pair separation, $a = 0.34$ nm, while Δ is determined as the dsDNA hydration radius: $\Delta \approx 1.25$ nm. Therefore, since the maximum allowed bending angle per base-pair, θ_{max}, must satisfy $\cos\theta_{max} > 1 - a^2/(2\Delta^2)$ one has $\theta_{max} \approx 0.27$. On the other hand, the standard angle deviation per nucleotide obtained from the traditional bending-rigidity model[21] is $\langle\theta\rangle \approx \sqrt{2a/\xi}$, where $\xi = 50$ nm is the observed persistence length.[22] This yields $\langle\theta\rangle = 0.12$, which is comparable to θ_{max}. As a consequence, it may be anticipated that the inclusion of an appropriate thickness treatment will lead to important corrections to the Hamiltonian of Eq. 1, especially in situations when DNA is tightly packed.

In the following we shall consider first how a finite polymer thickness impacts on its effective persistence length and its resistance to mechanical stretching. Finally, in section IV we shall consider how the severe reduction in configuration entropy operated by the thickness term gives an important contribution to DNA resistance to packaging.

In all cases, we will consider two models for the biopolymer chain. In the first instance, we can fully neglect the bending rigidity K/a and consider a self-avoiding *flexible* tube model as given by Hamiltonian (2). The semi-flexible tube model represents instead a refinement to the previous one obtained by including the bending rigidity term so that the energy function becomes

$$\mathcal{H}(\Gamma) = \sum_{ijk} V_3(r_{ijk}) - \frac{K}{a} \sum_i cos(\theta_i) \ . \tag{4}$$

3. Persistence length & Stretching curves

In this section, we discuss how the elastic properties, namely the persistence length and the elongation versus applied force stretching curves, of a flexible and semi-flexible tube depend on its thickness and bending rigidity.

In order to make analytical progress possible, we neglect the DNA self-interaction, V_2, and only consider the thickness constraint on local triplets (i.e. those involving triplets of neighbouring points, $\vec{r}_{i-1,i,i+1}$). In this case the persistence length ξ of the semiflexible (discrete) tube can be calculated exactly since ξ is related to to the canonical-ensemble average $\langle \cos(\theta) \rangle$ through:

$$a\xi^{-1} = -\frac{1}{\log \langle \cos(\theta) \rangle}. \tag{5}$$

One thus obtains the inverse persistence length of a semiflexible tube as

$$a\xi^{-1} = -\log\left(1 - \frac{a}{\beta K} + \frac{a^2}{2\Delta^2} \frac{1}{\exp\frac{\beta k a}{2\Delta^2} - 1}\right) \tag{6}$$

Two noteworthy limiting behaviours of the persistence length are obtained when either the bending rigidity term or the thickness constraint are absent. In the latter situation, when $K >> a$ (continuum limit) one obtains the traditional result $\xi = \beta K$. The other limit, on the other hand, corresponds to the persistence length of a discretised thick tube with no bending rigidity and yields:

$$a\xi^{-1} = -\log\left(1 - \frac{a^2}{4\Delta^2}\right) \tag{7}$$

Therefore, even in the absence of an explicit bending penalty, the restrictions imposed by Δ on the local radius of curvature are sufficient to produce an effective persistence length. It is instructive to calculate this thickness-induced persistence length for dsDNA and compare it to the experimental value. By using the previously mentioned values for a and Δ one obtains, $\xi = 19$ nm which is about one half of the correct experimental value $\xi_{exp} \approx 50$ nm. This theoretical estimate, which takes into account the mere effect of steric hindrance of the chain does not include the contribution of electrostatic repulsion of the phosphate groups to the apparent persistence length[22] but appears to capture the correct order of magnitude of the observed persistence length.

For the case of single-stranded DNA the natural spacing of the effective beads is provided by the nucleotide separation, $a = 0.5$ nm while the experimental value for the thickness $\Delta = 0.4$ nm. This yields a value of ξ equal to about 1 nm which agrees with the experimental data for ssDNA in solutions of high ionic strength (which screen the phosphate charges) for which values for ξ of the order of 1 nm are observed.[23] Another interesting case of biopolymers is provided by naturally-occurring proteins. Taking a as the typical spacing of C_α atoms of consecutive residues, $a = 3.8$ Å, and setting the typical peptide thickness as $\Delta \approx 2.5$ Å, as suggested by other studies, one obtains a thickness induced persistence length equal to $\xi = 0.55$ nm which is in reasonable agreement with the experimental persistence length of polypeptides $\xi = 0.4$ nm.[24,25]

These facts highlight how for DNA and proteins, the effects of a finite thickness can be hardly neglected since they alone appear to account satisfactorily well for the observed persistence lenghts. This is not expected to be true for all polymers. In fact, it is known from elasticity theory, that the bending rigidity coefficient for an elastic rod with uniform density increases with the fourth power of its radius. This law appears to be well-obeyed for organic polymers ranging from alkanes to microtubules.[21] Due to this fact, when the rod cross-section is sufficiently large, thermal excitations will not be able to produce configurations with e.g. local curvature exceeding the limit imposed by the rod thickness. In such situations the incorporation of an explicit thickness treatment is not expected to alter the equilibrium properties of the rod (except under extreme packing conditions).

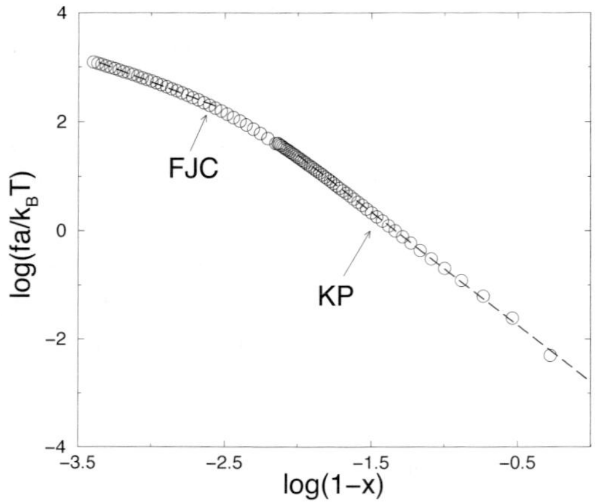

Fig. 2. Relationship between the relative elongation, x, at given applied force, f, for the flexible tube model ($\Delta = 1$ and $K = 0$). The plot highlights the crossover between two regimes where the chain behaves as the Kratky-Porod model (worm-like chain) and as a freely-jointed chain, denoted with KP and FJC, respectively. In both regimes the force versus extension relationship is well described by the law $f\ (1-x)^{-}\alpha$. A linear regression analysis of the plotted data shows that the stretching exponent α is approximately 2.1 and 1.02 for the KP- and FJC-like regimes, respectively.

A better characterization of the elasticity properties of the semiflexible tube requires going beyond the mere computation of its persistence length. Here we do not aim at doing this in a systematic way, but we confine ourselves to the quantities which are most easily available in experiments, namely the detailed shape of the stretching curves recorded when a polymer or biopolymer is subjected to an external force (i.e. by means of a soft cantilever or of laser tweezers trapping its extremities). This study also allows us to compare our model with existing ones, since the stretching curves are among the best characterized quantities in single molecule physics.

In Fig. 2 we show the stretching curve for a flexible tube with $\Delta = 1$ obtained with a transfer matrix approach. Qualitatively similar results are obtained with the $K \neq 0$ and different thickness. As expected, the (semi-)flexible tube model yields extension versus force data reminiscent of those obtained with the Kratky-Porod model.[26] In particular, we observe a low-force regime in which the extension is linear with the applied force (due to the lack of self-avoidance in the local triplet treatment), followed by a regime, typical of the continuum worm-like chain, in which $f \sim \frac{1}{(1-x)^2}$. For very large forces, the freely jointed chain behaviour $f \sim \frac{1}{(1-x)}$ is found again, as recently reported by a few researchers working with the discretized worm-like chain.[27,28,29,30] The picture arising from our calculation is thus that, in the swollen and stretched phase, the semiflexible tube behaves analogously to the Kratky-Porod model . The difference between the two models is, on the other hand remarkable in the collapsed or confined phase, as discussed later.

An important theoretical aspect associated to the configurations of semi-flexible thick polymer models discussed so far is that no torsional penalty has been introduced. This simplification may be legitimate when the bead-spacing a is dictated by the intrinsic "granularity" of the polymer in terms of units that may be relatively free in orientation with respect to each other. This requirement is clearly violated when taking the continuum limit ($a \to 0$) of the model Hamiltonian. As a result of the absence of torsional constraints at infinitely many points, when $a \to 0$ the persistence length of Eq. (7) diverges. A meaningful continuum limit can be recovered by introducing a term which disfavours changes of the torsional axis, $\vec{\tau}_i$ along the chain.

For example, if this term is chosen as

$$\mathcal{H}_t = -\frac{K_t}{a} \sum_i \vec{\tau}_i \cdot \vec{\tau}_{i+1}, \tag{8}$$

and neglecting polymer self-interactions and non-local thickness constraints, one can devise a soluble recursive scheme to find the decay of the tangent-tangent correlation. In particular, if the torsional rigidity, K_t/a, is small compared to Δ/a then the presistence length is found to depend on Δ and K_t through:

$$\xi \propto \Delta^2/K_t \ . \tag{9}$$

Interestingly, if K_t is not small compared to Δ then the tangent-tangent correlations as a function of the arclength separation exhibits an oscillatory behaviour instead of a monotonic decay.

It is to be stressed that these delicate aspects of the continuum limit appear when the polymer is in the swollen or extended phase. In a compact phase, e.g. in the presence of a confining pressure (or of attractive self-interactions), the persistence length of a thick polymer instead scales with Δ so that the continuum limit is well defined in this regime.

4. Loading curves & DNA packaging

As stated earlier, our main goal is to describe the equilibrium thermodynamics of a self-avoiding flexible or semiflexible thick chain of N beads confined in a sphere of radius R. In particular, we aim at calculating the free energy, $F_N(R)$, of a chain of N beads confined inside a sphere of radius R. If the process of packaging is carried out adiabatically then the work required to confine the unconstrained chain into the spherical cavity is given by:

$$\Delta G_N(R) = F_N(R) - F_N(\infty) \ . \tag{10}$$

During the DNA loading process inside a phage capsid, a portion of the DNA tail that protrudes out of the phage head is fed inside the cavity by exerting a certain force. Whithin our approach this force is obtained by differentiating $\Delta G_N(R)$ with respect to N:

$$f \propto \frac{\partial \Delta G_N(R)}{\partial N} \ . \tag{11}$$

It is worth pointing out that this force estimate does not treat the free energy coming from the DNA tail outside the cavity in a realistic way. Incorporating this effect would certainly lead to a more realistic description of the loading process. However, the contribution of the floating tail to the free energy is expected to become less and less important compared to that of the confined portion as the cavity is filled.

Before coming to the presentation of the details of the numerical studies of DNA packaging, it is useful to consider a mean-field model (inspired by the treatment in Refs. 31, 32) which, while incorporating the basic features relevant for polymer packing, is still amenable to analytic treatment. This model consists of a polymer chain of N beads embedded in a discrete three-dimensional space and subject to a contact interaction, ϵ, and a bending rigidity (corner penalty), h. Within a mean-field picture[31,32] it is possible to obtain explicit expressions for the free energy loss, F_N, of a chain with N beads confined in a cube of side L (the lattice spacing is taken as the length unit). This is

$$\frac{\beta F_N(x)}{N} = \frac{1-x}{x}\log(1-x) + \beta\epsilon(1-3x) - \log\left[\frac{2+4\exp(-\beta h)}{e}\right], \quad (12)$$

where x is the density $x = N/L^3$ and $\beta = 1/(K_B T)$. It is important to notice that the free-energy per bead, F_N/N depends on N and L only through the density x; this fact will be exploited to analyze the results for our confined off-lattice chains with finite thickness.
From this free energy one can estimate the loading force as in Eq. (11):

$$f(x) = \frac{\partial}{\partial N}[F_N(x) - F_N(0)] = \frac{-1}{\beta}[\ln(1-x) + 6\beta\epsilon x] \quad (13)$$

Despite the simplified nature of the mean field model, the approximate force of expression (13) provides a useful starting point for both understanding recent packaging studies as well as suggesting improvements to traditional models. The crucial observation is that in the tight-packing limit, $x \to 1$, the divergence of $f(x)$ is entirely controlled by the logarithmic term which is independent of both ϵ and h. This result provides a theoretical framework for explaining the numerical findings of Ref. 10 who ascertained that the force loading curve was not too sensitive to the potential strength and sign at high packing densities. In other words, from this simple mean-field treatment we have an indication that a resistance to dense confinement comes from entropic effects rather then polymer self-interactions or flexibility. This is also in agreement with the picture proposed in Ref. 7. It is reasonable to expect that in the tightly packed regime different portions of the dsDNA will be at close contact with one another, so that in order to have a good estimate of the entropic loss due to the confinement of the polymer into the capsid steric constraints need to be accurately modelled.

Note that for low packing, instead, the mean field loading curve is sensitive to the details of the self-interactions between the beads, and predicts a linear increase with the force with x. More detailed arguments, based on the calculation of the osmotic pressure in a semidilute polymeric system[33] show that the actual limiting law in the discretized model for $x \to 0$ would be $f \sim x^{1/(3\nu-1)}$ in $d = 3$, where ν is the critical exponent connected with the growth with chain length of the gyration radius of the polymer in the unconfined phase. If at $x = 0$ the polymer is in the swollen phase, then $f \sim x^{1/(3\nu-1)}$ for $x \to 0$.

It is also interesting to note that this theory allows treating the part of the chain outside the sphere as well. If instead of being free, the chain outside the cube is under the action of a (large) stretching force, as occurs in some experiments, one can find that within the mean field treatment proposed above the difference in the loading curve amounts to only a constant term.

While the soluble models, leading to e.g. the results of Eq. (13) and Eq. (6), are useful for gaining insight in the basic physics of thick polymers, they cannot be used to fully characterize more complicated instances such as the packaging process. To address this issue it is, therefore, necessary to resort to stochastic numerical simulations.

In principle, the free energy, $F_N(R)$ could be obtained through a Monte Carlo simulation where the conformation space is restricted to only those structures that can be confined in a sphere of radius not larger than R. This is, however, impractical due to the fact that in situations of high density (small R) it will be very difficult to move through distinct compact structures while respecting the confining constraints.

A useful alternative is to work in the conjugated ensemble where, instead of keeping the radius of the confining sphere, R, fixed, one applies a uniform pressure, P that compactifies the chain. In this situation, the Boltzmann weight of a configuration, Γ is given by:

$$e^{-(E+PV)/K_B T} \tag{14}$$

where the temperature T is taken as the room temperature, $T = 300\,K$, E is the structure energy calculated via Eq. (2) for the flexible tube model (or Eq. (4) for the semiflexible one) and V is the volume of the spherical hull associated to Γ. To find the radius of the spherical hull enclosing Γ we first determine the maximum distance, $\bar{r}$ from the centre of mass of any of the beads in Γ. Since these beads lie on the centerline of a tube of thickness Δ, the required radius is $R = \bar{r} + \Delta$. It should be noted that, in principle, the smallest confining sphere is not necessarily centred on the centre of mass of Γ. However, we have verified, by an explicit numerical search of the location of the smallest sphere centre, that this approximation is very good when dealing with structures with good overall compactness, which are the focus of this investigation.

The equilibrium properties of the thick polymer were characterised by a multiple Markov chain simulation[34] where several copies of the system each at a different pressure were evolved simultaneously at a given temperature. The evolution was controlled by the Metropolis acceptance of elementary chain distortions involving crankshaft, pivot and slithering moves.[35] A preliminary measure of correlation times allowed for the collection of statistically independent chain configurations. From the distribution (histogram) of the various hull volumes encountered at various pressures one can use a multiple histogram technique (a generalization of the one treated in Ref. 36) to recover the density of states, $W(R, E)$ corresponding to the number of viable structures (i.e. compatible with the preassigned thickness Δ), with hull radius R and energy equal to E.

The required free energy, $F(R)$ is then obtained from

$$e^{-\beta F(R)} \propto \sum_{E} \sum_{R'<R} W(R', E) . \tag{15}$$

The reconstructed free energy, $F(R)$, is defined up to an additive constant, which reflects the fact that the density of states can be obtained only up to a multiplicative constant. This additive constant has to be set appropriately in order to simplify the scaling analysis of $F_N(R)$ as a function of N at fixed β. For this reason we have taken the unconstrained situation, $R = \infty$, as the reference case associated with zero free energy: $F_N(R = \infty) = 0$.

The full equilibrium thermodynamics was thus obtained for tubes with a discrete number of beads, N. Values of N up to 200 were considered here. As was argued before, in the limit of large N the free energy should be extensive in N at fixed filling fraction, N/R^3. Therefore we expect $F_N(R)$ to obey the following scaling relations:

$$\lim_{N\to\infty} F_N(R) \sim N\, g\left(\frac{N}{R^3}\right) , \tag{16}$$

From this relation one can obtain the force required to feed additional beads inside the cavity:

$$a\, f_N(R) = \frac{\partial F}{\partial N} \sim g\left(\frac{N}{R^3}\right) + \frac{N}{R^3} g'(x)\,|_{x=N/R^3} . \tag{17}$$

Note that the same scaling theory applies also to expressions (12) and (13). These relations will be obeyed only approximately for any finite chain, though such corrections decay for increasing values of N, as shown later. We first consider the case in which the dsDNA being packaged is modelled by a flexible tube. In this case, due to the absence of the bending rigidity penalty, the energy E can take only two values, 0, and ∞ for configurations that are respectively compatible or incompatible with the thickness requirement.

In Fig. 3 we have plotted the reduced free energy, $g_N = F_N(R)/N$ as a function of N/R^3 for different values of N. It is apparent that, as the

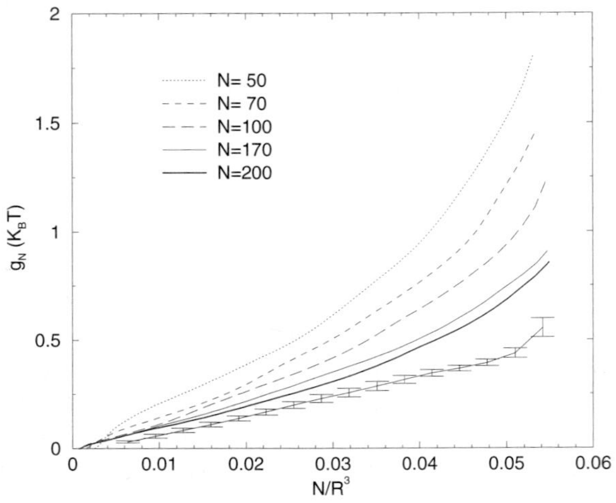

Fig. 3. The reduced free energy as a function of the packing density, N/R^3 for the indicated values of N. The bottom curve is obtained by extrapolating the finite-size curves for $N \to \infty$.

size of the system, N, increases, the curves for g_N deviate less from each other, indicating the decrease of finite-size corrections from the limit curve of expression (16). Indeed, the systematic decay of such corrections for increasing N is visible in Fig. 4 where we have reported the trend of g_N as a function of N^{-1} at a fixed density. As is visible in the figure, the data are well interpolated by a linear regression with correlation coefficient $r \sim 0.98$. Through such regression one can estimate the error on the extrapolated value for g as the error on the intercept. The observed correlation coefficient is statistically significant, since it pertains to 16 data points and is suggestive that finite-size corrections decay approximately as N^{-1}. However, from the systematic decrease of the linear regression slope observed when fitting data for the larger tube lengths only, it can be argued that the convergence is faster than $1/N$ and may also be compatible with decays such as $N^{-1.5}$.

The same rate of convergence of finite-size corrections is inherited by the force obtained from (17) by using g_N instead of g. By considering several values of N one obtains a succession of curves for the loading force that, again, appear to approach a limiting curve from above, as is visible in Fig. 5. This is in agreement with the physical intuition that a larger force is required to pack a shorter (but equally thick) chain to the same target

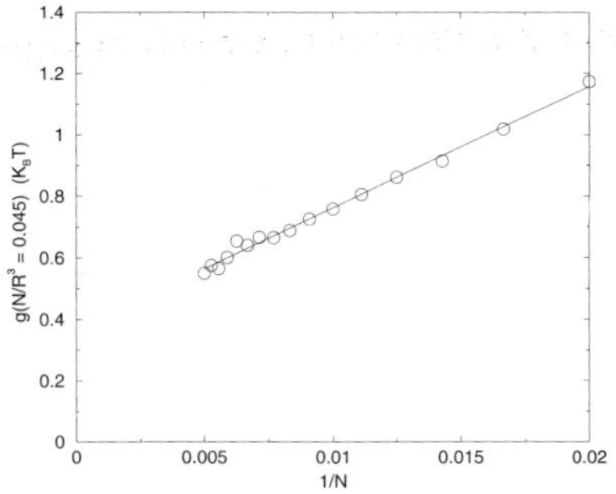

Fig. 4. Curves for the reduced free energy at the fixed density $N/R^3 = 0.045$ for values of N in the range $50 \leq N \leq 200$.

density.

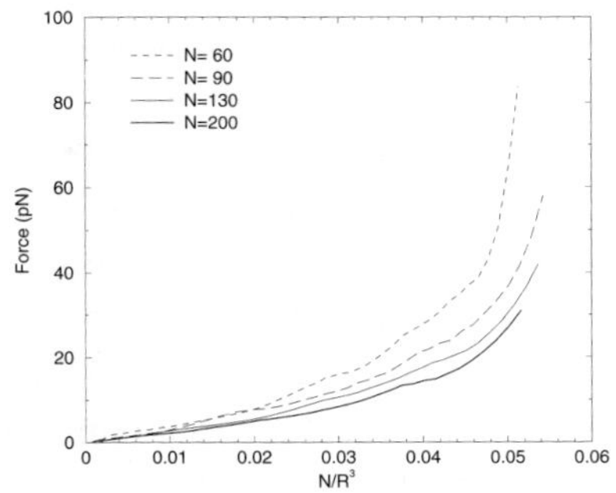

Fig. 5. Curves for the force necessary to apply to package the thick tube to the required density, N/R^3. The curves for the force are obtained from simulations with the indicated number of beads, N.

As for the corrections to g, the force curves also appear to approach the limit curve of interest at a rate somewhat faster than $1/N$, as visible in Fig. 6a. Nevertheless, since we cannot give a precise account of the correction exponent we calculated the limit force by using the conservative $1/N$ decay estimate and by fitting only the data of the largest tube lengths, $100 < N \leq 200$. The results are provided in Fig. 6b and can be compared with the experimental curve of Ref. 1 as discussed below.

If one considers the prolate $\phi 29$ capsid as an ellipsoid, the reference density at full genome packaging as $x = N/R^3 = 0.063$. The largest densities

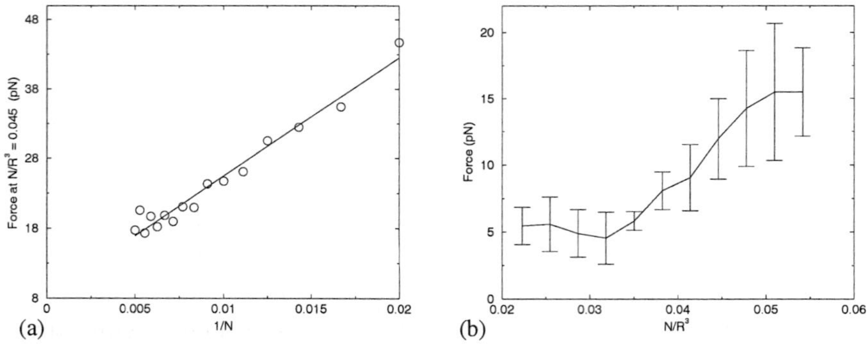

Fig. 6. (a) Curves for the packaging force at the fixed density $N/R^3 = 0.045$ for values of N in the range $50 \leq N \leq 200$. (b) Plot of the extrapolated force required to package an infinitely long tube to the preassigned density, N/R^3. The force is expressed in picoNewtons. The estimated packing density of the fully-loaded $\phi 29$ capsid corresponds to 0.063.

reached in our simulation correspond therefore to about 85% of this value. Neglecting the contribution of the protruding tail in the $\phi 29$ experiment one can then compare the experimental force at 85% genome packaging, 34 pN, with our peak force of 16 pN, which appears to be correct within only a factor of two. We stress that this finding exploits only the severe restrictions in configuration space operated by the finite thickness of DNA itself; it is striking that this entropic effect appears to account for half the force observed in the experiment. Furthermore, the limit curve for the force has been obtained with the conservative $1/N$ estimate of the decay of finite-size corrections; this arguably leads to an underestimation of the force required to package the DNA chain.

Even if these results are quite significant and directly point to the relevance of including the hydration thickness steric constraint (as a three body term in the Hamiltonian) in the theoretical treatment, they give only a semi-quantitative agreement. In order to improve on this, we have considered the more refined semiflexible tube model.

As a first step we provide an upper bound for the confining force increment due to the added bending rigidity term. To this purpose let us consider a thick chain that has been compactified so to nearly fill a given cavity. It is assumed that packing conditions are such that there is almost no structural freedom left for how to add an additional bead inside the hull. In particular, the bead has to be added so to achieve the tightest local curvature of the chain. Since the lower bound for the curvature is the thickness itself, Δ, the additional packaging force is estimated as:

$$\Delta f_{b.r.} = \frac{K_B T \xi}{2\Delta^2} \, . \tag{18}$$

Using $\xi = 50$ nm as in Ref. 10 and $T = 300\,K$ one obtains that an upper bound for the force increment due to bending rigidity is $\Delta f_{b.r.} \approx 66$ pN. To obtain a more accurate estimate of Δf we have undertaken additional studies where the numerical scheme previously discussed for the confined thick chain was generalised to include explicitly the bending rigidity term of Eq. (4) so that $\beta K = \xi$.

The augmented difficulty of packaging the chain in this situation resulted in the fact that the maximum density achieved was $N/R^3 = 0.050$, corresponding to 80% of the estimated packing fraction of ϕ29. The confining force for a chain of 200 beads was 39 pN. This corresponds to an extra force of around 12 pN, (the pure thickness-induced force at the same filling fraction is ~ 27 pN, see Fig. 7). This value overestimates the bending rigidity contribution in the thermodynamic limit due to the systematic decrease of the force as a function of chain length. However, even allowing for the overemphasized contribution of bending rigidity due to finite-size effects one can already conclude that the finite-thickness effects are comparable with those associated with bending rigidity.

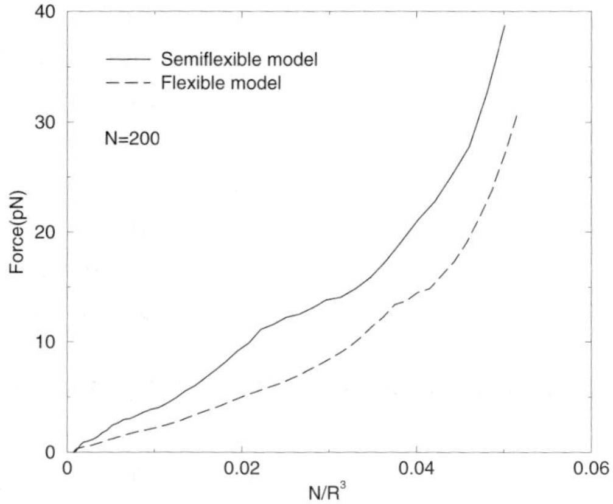

Fig. 7. Curves for the packaging force as a function of density for $N = 200$ beads. Results for the flexible and semiflexible models are shown with a dashed and continous lines, respectively.

This conclusion remains true when finite-size extrapolation is carried out on this second model using chains of 40, 80, 100, 120, 160 and 200 beads. From the thermodynamic data a total force of 24±7 pN can be extrapolated at the highest densities $N/R^3 = 0.050$. Despite the still simplistic nature of the second model considered here, the extrapolated force appears to be fully consistent with the experimental results for ϕ29 at the corresponding filling fraction of 80%, where a total force of 27 pN was recorded. Thus, the different calculations presented here all indicate that the contribution to the resistance of packaging arising from the finite DNA thickness is comparable to that of other terms usually considered in coarse-grained models of DNA. This indicates that various phenomenological properties of DNA can be accounted for in a more accurate and complete way by including an explicit modelling of DNA thickness besides the traditional terms in the effective Hamiltonian.

5. Perspectives and Conclusions

In conclusion, we have presented a novel model to study the problem ds-DNA packaging. In this model, dsDNA is described as a discrete flexible or semi-flexible thick polymer. The model parameters are the thickness of the hydrated dsDNA, the bending rigidity, and the discretization spacing (taken as the base-pair spacing).

We have first characterized the elastic properties of flexible thick polymers by showing how the polymer thickness impacts on the persistence length and the relative elongation under mechanical stretching. The behaviour shown by the thick polymer model in this latter case strongly resembles that of a Kratky-Porod chain or a freely-rotating chain. On the other hand, the behaviour of the thick polymer under compaction is dramatically different from that of Kratky-Porod chain due to the severe configurational limitations imposed by the thickness on the local and global radii of curvature. Despite the minimal ingredients introduced in the flexible and semi-flexible models, both of them yield a very satisfactory agreement with the packaging force measured in experiments.

A desirable improvement of the model would be constituted by the inclusion of DNA self-interactions (accounting also for the presence of monovalent or polyvalent counterions in solution). It has already been observed that suitable quantities of monovalent counterions can induce an 'apparent DNA thickness' which affects strongly the observed persistence length and knotting probabilities.[22,37] The model proposed here would allow to interpret these previous findings within a framework where the apparent thickness can be incorporated naturally.

Acknowledgments

We are indebted to S. Harvey, A. Maritan, T. Odjik, E. Rawdon, A. Rosa, R. Podgornik and A. Stasiak for illuminating discussions. The material discussed here has been presented at the conference *Knots, random walks and biomolecules*, co-organized by J. H. Maddocks and A. Stasiak and sponsored by the Bernoulli Centre of Swiss Federal Institute of Technology in Lausanne. We acknowledge support from INFM, MIUR COFIN 2003 and EPSRC.

References

1. Smith, D. E., Tans, S. J., Smith, S. B., Grimes, S., Anderson, D. L. & Bustamante, C. (2001). The bacteriophage ϕ29 portal motor can package DNA against a large internal force. *Nature,* **413**, 748–752.
2. Cerritelli, M. E., Cheng, N. Q., Rosenberg, A. H., McPherson, C. E., Booy, F. P. & Steven, A. C. (1997). Encapsidated conformation of bacteriophage T7 DNA. *Cell,* **91**, 271–280.
3. Riemer, S. C. & Bloomfield, V. A. (1978). Packing of DNA in bacteriophage heads: some considerations on energetics. *Biololymers,* **17**, 785–794.
4. Arsuaga, J., Tan, R. K. Z., Vazquez, M., Sumners, D. W. & Harvey, S. C. (2002). Investigation of viral DNA packaging using molecular mechanics model. *Biophys. Chem.* **101**, 475–484.
5. Purohit, P. K., Kondev, J. & Phillips, R. (2003). Mechanics of DNA packaging in viruses. *Proc. Natl. Acad. Sci. USA,* **100**, 3173–3178.
6. Marenduzzo, D. & Micheletti, C. (2003). Thermodynamics of DNA packaging inside a viral capsid: the role of DNA intrinsic thickness. *J. Mol. Biol.* **330**, 485–492.
7. Strey, H. H., Podgornik, R., Rau, D. C. & Parsegian, V. A. (1998). DNA-DNA interactions. *Curr. Opin. Struc. Biol.* **8**, 309–313.
8. Hansen, P. L., Podgornik, R. & Parsegian, V. A. (2001). Osmotic properties of DNA: critical evaluation of counterion condensation theory. *Phys. Rev. E,* **64**, 021907.
9. Odijk, T. (1998). Hexagonally packed DNA within bacteriophage T7 stabilized by curvature stress. *Biophysical Journal,* **75**, 1223–1227.
10. Kindt, J., Tzlil, S., Ben-Shaul, A. & Gelbart, W. M. (2001). DNA packaging and ejection forces in bacteriophages. *Proc. Natl. Acad. Sci. USA,* **98**, 13671–13674.
11. Buck, G. & Orloff, J. (1995). A simple energy function for knots. *Topol. Appl.* **61**, 205–214.
12. Buck, G. (1998). Four-thirds power law for knots and links. *Nature,* **392**, 238–239.

13. Katrich, V., Olson, W. K., Pieranski, P., Dubochet, J. & Stasiak, A. (1997). Properties of ideal composite knots. *Nature,* **388**, 148–151.
14. Litherland, R., Simon, J., Durumeric, O. & Rawdon, E. (1999). Thickness of knots. *Topol. Appl.* **91**, 233–244.
15. Gonzales, O. & Maddocks, J. H. (1999). Global curvature, thickness and the ideal shapes of knots. *Proc. Natl. Acad. Sci. USA,* **96**, 4769–4773.
16. Rawdon, E. (2000). Approximating smooth thickness. *J. Knot Th. Ram.* **9**, 113–145.
17. Stasiak, A. & Maddocks, J. H. (2000). Best packing in proteins and DNA. *Nature,* **406**, 251–253.
18. Banavar, J. R., Gonzales, O., Maddocks, J. H. & Maritan, A. (2003). Self intersection of strands and sheets. *J. Stat. Phys,* **110**, 35–50.
19. Maritan, A., Micheletti, C., Trovato, A. & Banavar, J. R. (2000). Optimal shapes of compact strings. *Nature,* **406**, 287–290.
20. Banavar, J., Maritan, A., Micheletti, C. & Trovato, A. (2002). Geometry and physics of proteins. *Proteins: Structure Function and Genetics,* **47**, 315–322.
21. Boal, D. (2002). *Mechanics of the cell.* Cambridge University Press, Cambridge.
22. Rybenkov, V. V., Cozzarelli, N. R. & Vologodskii, A. V. S. (1993). Probability of DNA knotting and the effective diameter of the double helix. *Proc. Natl. Acad. Sci. USA,* **90**, 5307–5311.
23. B. Tinland, A. Pluen, J. S. & Weill, G. (1997). Persistence length of single-stranded DNA. *Macromolecules,* **30**, 5763–5765.
24. Rief, M., Gautel, M., Oesterhelt, F., Fernandez, J. & Gaub, H. E. (1997). Reversible unfolding of individual titin immunoglobulin domains by AFM. *Science,* **276**, 1109–1112.
25. Rief, M., Pascual, J., Saraste, M. & Gaub, H. E. (1999). Single molecule force spectroscopy of spectrin repeats: low unfolding forces in helix bundles. *J. Mol. Biol.* **286**, 553–561.
26. Daune, M., (1999). *Molecular biophysics.* Oxford University Press, Oxford.
27. Rosa, A., Hoang, T. X., Marenduzzo, D. & Maritan, A. (2003). Elasticity of semiflexible polymers with and without self interactions. *Macromolecules,* **36**, 10095-10102.
28. Livadaru, L., Netz, R. R. & Kreuzer, H. J. (2003). Stretching response of discrete semiflexible polymers. *Macromolecules,* **36**, 3732.
29. Lamura, G., Burkhardt, T. W. & Gompper, G. (2001). Semiflexible polymer in a uniform force field in two dimensions. *Phys. Rev. E,* **58**, 061801.
30. Storm, C. & Nelson, P. C. (2003). Theory of high-force DNA stretching and overstretching. *Phys. Rev. E,* **67**, 051906.
31. Doniach, S., Garel, T. & Orland, H. (1996). Phase diagram of a semiflexible polymer chain in a theta solvent: application to protein folding. *J. Chem. Phys.* **105**, 1601.
32. Lise, S., Maritan, A. & Pelizzola, A. (1998). Bethe approximation for a semi-flexible polymer chain. *Phys. Rev. E,* **58**, R5241.
33. Doi, M. & Edwards, S. F. (1989). *The Theory of polymer dynamics.* Clarendon Press, Oxford.

34. Tesi, M., van Rensburg, E. J., Orlandini, E. & Whittington, S. (1996). Monte-Carlo study of the interacting self-avoiding walk in three dimensions. *J. Stat. Phys.* **82**, 155–181.
35. Sokal, A. D. (1997). Monte Carlo methods for the self-avoiding walk. *Nuclear Physics B,* **Suppl. 47**, 172–179.
36. Ferrenberg, A. M. & Swendsen, R. H. (1989). Optimized Monte-Carlo data-analysis. *Phys. Rev. Lett.* **63**, 1195–1198.
37. Baumann, C. G., Smith, S. B., Bloomfield, V. A. & Bustamante, C. (1996). Ionic effects on the elasticity of single DNA molecules. *Proc. Natl. Acad. Sci. USA,* **94**, 6185–6190.

CHAPTER 8

MONTE-CARLO SIMULATIONS OF GEL-ELECTROPHORESIS OF DNA KNOTS

C. Weber[1], M. Fleurant[2], P. De Los Rios[2] and G. Dietler[3,*]

[1]Institut Romand de Recherche sur les Matériaux - IRRMA,
[2]Laboratoire de Biophysique Statistique - ITP,
[3]Laboratoire de Physique de la Matière Vivante - IPMC,
Faculté des Sciences de Base,
Ecole Polytechnique Fédérale de Lausanne,
CH-1015 Lausanne, Switzerland.
E-mail: giovanni.dietler@epfl.ch

Gel electrophoresis allows to separate knotted DNA (nicked circular) of equal length according to the knot type. We present a computer simulation of knotted charged chains moving in an external electric field through grids of obstacles. Using a simple Monte-Carlo algorithm, the dependence of the electrophoretic migration of the DNA molecules on the type of knot was investigated at a low electric field regime. The results are in qualitative agreement with electrophoretic experiments done under conditions of low electric fields: in particular the electrophoretic mobility increases quasi linearly with the mean average crossing number calculated for variously knotted chains under modeled conditions of gel electrophoresis.

1. Introduction

Gel electrophoresis of linear and circular DNA has been since long time a topic on which numerical simulations and analytical models have been applied [1,2,3,4,5,6,7,8]. Most experimental and theoretical studies of the electrophoresis process deal with linear or circular DNA [9,10,11,12]. But DNA comes also in knotted form. Various classes of enzymes (topoisomerases and site-specific recombination enzymes) produce different types of knots or catenanes by acting on circular DNA molecules [13,14]. Being able to study which knots are produced by a given enzyme in prescribed conditions implies being able to perform some sort of "knot spectroscopy", which can be

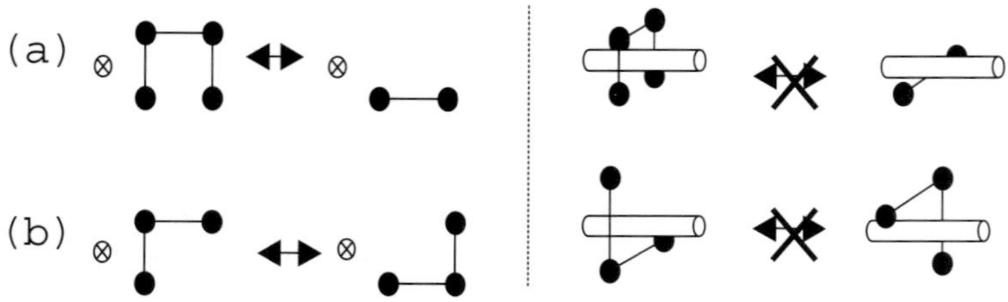

Fig. 1. Monte-Carlo moves in the BFACF : (a) creation / destruction of a handle and (b) flip of a handle. On the left side, the movement is not crossing a gel rod and is allowed, on the right side the movement is forbidden.

done for example by electron microscopy or by Atomic Force microscopy, where knots are observed one by one. Yet, if large numbers of knots need to be classified, then some high throughput technique is needed. Such a technique is gel electrophoresis. Indeed, experimental work has shown a linear relationship between the distance of electrophoretic migration on agarose gel of different types of DNA knots (all with the same number of base pairs) and the average crossing number of the ideal geometrical representations of the corresponding knots (closely related to the complexity of the knot) [18]. As a consequence, the type of a knot can be simply identified by measuring its position on the gel, without the need of microscopy experiments as required until recently. At low electric field the usual observation is that the more complex the knot is, the higher is its mobility. A simple intuitive explanation for this behavior is that the compactness of a knot increases with it complexity (for a constant string length) and the friction coefficient (with the hydrodynamic radius of the knot and the viscosity of the solvent) is smaller, resulting in higher mobilities. A more refined calculation of the friction coefficient relies on Kirkwood-Riseman formula [19]:

$$\gamma = \left(\sum_{i=1}^{N} \zeta_i\right) \left[1 + \left(6\pi\eta_0 \sum_{i=1}^{N} \zeta_i\right)^{-1} \sum_{i,j=1(i\neq j)}^{N} \zeta_i \zeta_j R_{ij}^{-1}\right]^{-1} \tag{1}$$

where the chain is modelled by N beads of radius σ_i and friction coefficent $\zeta_i = 6\pi\eta_o\sigma_i$, and R_{ij} is the distance between beads i and j. The term $\frac{1}{R_{ij}}$, due to hydrodynamic interactions between beads, in the second factor of equation (1) explains the observed behavior: more compact molecules have smaller distances R_{ij}, and thus a smaller friction coefficient.

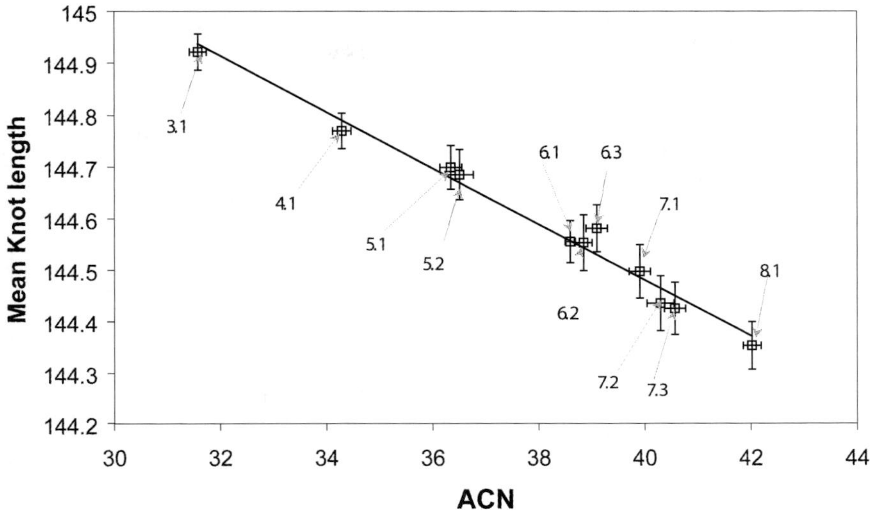

Fig. 2. Mean knot length versus ACN in a medium without gel for $C = 0.2$. We have checked that the length variation is not affecting significantly the drift speed.

Recently, it was observed by two-dimensional agarose gel electrophoresis that when the strength of the electric field is increased, the electrophoretic mobility of DNA knots changes behavior [17,21] and the opposite behavior is observed: more complex knots cover smaller distances than simple ones.

The presence of two regimes of weak and strong electric field can be captured with a simple argument. If the transverse diffusion explores distances d much larger than the typical size of the knot, ξ, then, on average, the knot will be able to avoid a collision with the gel and the knots will drift as if they were in a pure solution, with just a slight renormalization of the friction coefficient. If instead $d \ll \xi$, then whenever a knot is on a collision course with a gel strand, it can not avoid it. As a consequence, after impinging over the obstacle, the knot needs to crawl around it in order to free itself. Crawling around an obstacle is much more difficult for more complex knots than for simple ones, due to the self-avoidance constraint. Following this argument, the two electric field regimes are separated by a critical field. Here we want to study more in details the low electric field regime using lattice Monte-Carlo simulations.

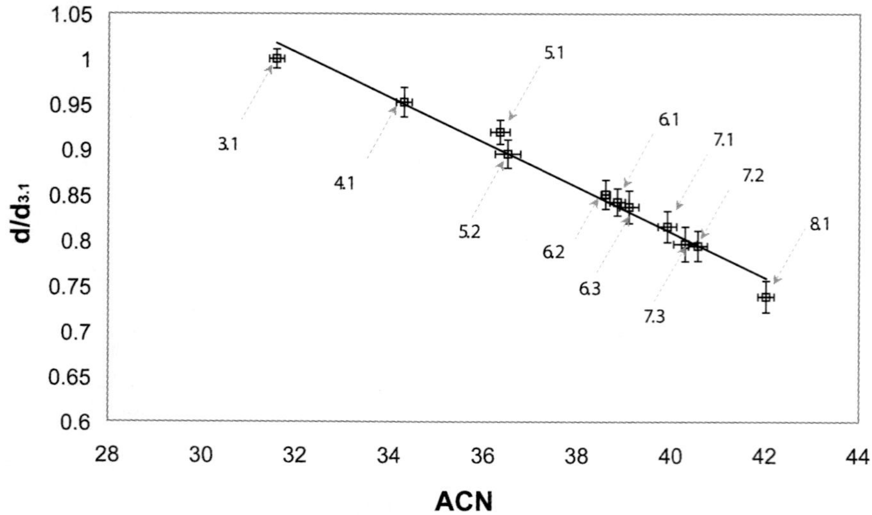

Fig. 3. Distance of migration (normalized to knot 3.1) in a medium without gel versus ACN for $C = 0.2$.

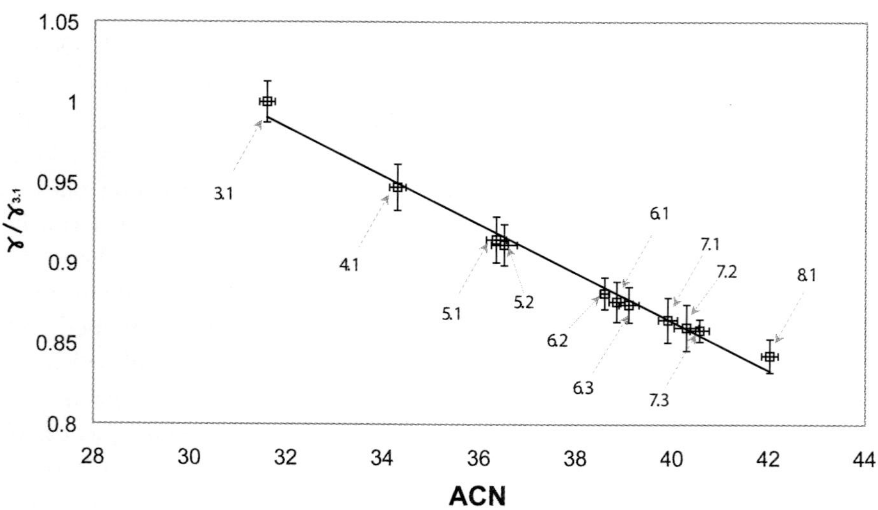

Fig. 4. Kirkwood coefficient (normalized to knot 3.1) measured in a medium without gel versus ACN for $C = 0.2$.

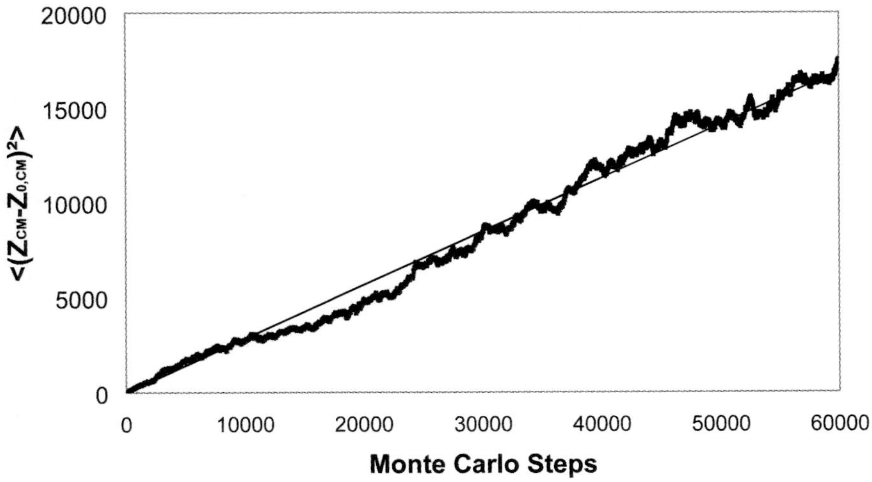

Fig. 5. Center of Mass longitudinal component versus Monte-Carlo steps in a medium with gel for $C = 0$ and $b = 20$ (3.1 knot). The average was done on 100 different knot trajectories.

2. Methods

DNA knots are modelled by closed self-avoiding walks (SAWs) composed of N segments of length a on a three-dimensional cubic lattice (the lattice constant a is comparable to the persistence length of the DNA molecules). The gel is a set of stacked two dimensional grids forming a sublattice with a mesh size b (= gel parameter, b is a commensurate value of the lattice constant a) and perpendicular to the applied electric field which is chosen to be along the $\hat{z}$ axis. The gel lattice is shifted by the quantity $(\frac{a}{2}, \frac{a}{2}, \frac{a}{2})$ compared to the knot lattice, so that no points of the knot lie on the gel (see also Fig. 7). Knots are not allowed to cross the gel network. The coordinates of the N monomers in the configuration at time t are written as:

$$\bar{r}(t) = (\vec{r}_1(t), \vec{r}_2(t), ..., \vec{r}_N(t)) \tag{2}$$

with constraints $\|\vec{r}_j(t) - \vec{r}_{j+1}(t)\| = a$.

The dynamics is followed using the BFACF algorithm [23]. Two types of moves are allowed: (a) the creation/destruction of a handle and (b) the flip of a corner into the mirror position (see Fig. 1). The first move clearly does not preserve the knot length, which can vary by ± 2 at every step. However, due to an elasticity functional (4) that puts a penalty on chains departing from the original length the chains only slightly change their average length (see Fig. 2 and its discussion).

The BFACF algorithm preserves knot classes, within which it is ergodic [24]. Self-avoidance is imposed by disallowing monomers to visit any site which is already occupied by other monomers. Furthermore, knots are not allowed to cross gel rods, so that corner flips and handle creation/destruction are forbidden when the surface drawn by moving segments intersects with a grid representing the matrix of a gel (see Fig. 1).

Under an external uniform electric field $\vec{\mathcal{E}} = (0, 0, \mathcal{E})$, the electrostatic energy at time t is given by:

$$E_q(t) = -\frac{q}{N(t)} \sum_{j=1}^{N} \vec{r}_j(t) \cdot \vec{\mathcal{E}} = -\frac{q}{N(t)} \sum_{j=1}^{N} r_j^z(t) \cdot \mathcal{E} \tag{3}$$

where $N(t)$ is the length of the knot at time t and q is the total charge carried by the knot. We consider in the simulation that each knot vertex carries a net charge $q/N(t)$. The knot length $(N(t))$ is also associated with an elastic energy

$$E_{el}(t) = \frac{1}{2} K \left[N(t) - N_0\right]^2 \tag{4}$$

where K is the spring constant and N_0 the rest length of the knot. In the simulation a value $K/k_B T = 0.1$ and different values of $C = q \cdot a \cdot \mathcal{E}/k_B T$ were considered. The rest length was set to 150. The knot energy is then $E(t) = E_q(t) + E_{el}(t)$.
At each timestep, we choose a point at random on the chain and propose alternatively one of the two moves. If it satisfies the self-avoiding and gel-avoiding constraints, it is accepted with a probability given by the Metropolis algorithm: if the energy of the new trial configuration, E_{trial}, is lower than that of the previous configuration, $E_{old} = E(t)$, the move is accepted and $\bar{r}(t+1) = \bar{r}_{trial}$; otherwise, the probability of acceptance of the trial configuration is equal to $\exp\{-[E_{trial} - E(t)]\}/k_B T$. If the move is rejected, then $\bar{r}(t+1) = \bar{r}(t)$. One Monte-Carlo step consist of N_0 single iterations attempts, where N_0 is the rest length of the knot.

After a knotted configuration is randomly generated, the knot type is obtained by calculating its Alexander polynomial [25]. Then, we let the system freely relax to thermodynamic equilibrium in the absence of an external field ($\mathcal{E} = 0$) until correlations from the initial configuration have disappeared. Then the electric field is switched on, and we let the knots

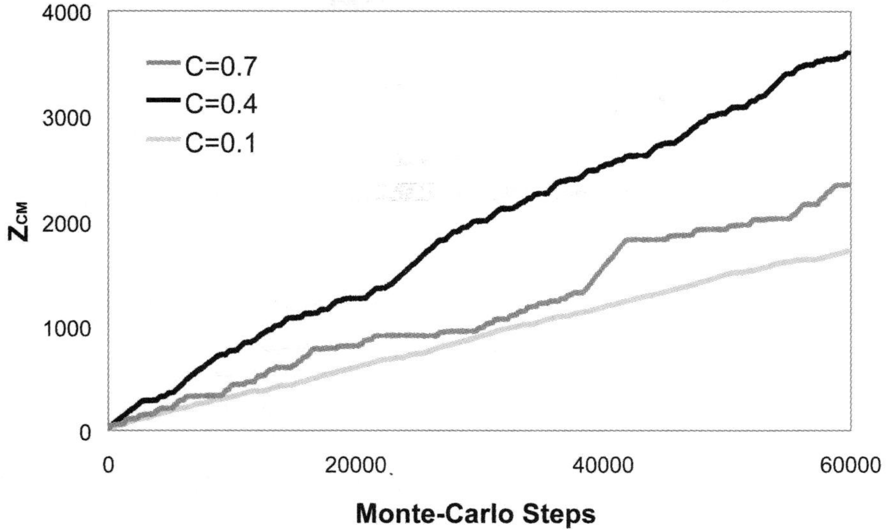

Fig. 6. Center of Mass longitudinal component versus Monte-Carlo steps in a medium with gel ($b = 20$) for $C = 0.1$, $C = 0.4$ and $C = 0.7$ for a 3.1 knot.

migrate on the lattice. The quantities we compute are the position of the center-of-mass and the average crossing number (ACN) of the knot along a trajectory. The ACN is measured by projecting the knots coordinates on a randomly chosen plane and counting the number of intersections for this particular projection. The procedure is then repeated and the calculated ACN is obtained by the average of crossing numbers for 100 random projections. The error on the ACN measurement is obtained by the dispersion of these 100 values.

Time is measured in Monte-Carlo iterations, length in lattice spacing. The initial length N_0 of our polymers was set to 150, and the mean length of the knot depends generally on the electric field and on the gel parameter. It also depends slightly on the mean ACN (see Fig. 2). However, we have checked that this length variation does not affect significantly the drift speed (less than 0.5%). The average length is slightly shorter that N_0, since the probability of shortening the polymer is a slightly larger than the probability of lengthening it due to the self-avoiding condition, however due to elastic energy functional taken into account during Monte-Carlo moves the average length stays close to the initial value. The gel parameter was set to $b = 5, 10, 20$ (in units of a), corresponding to a relatively small pores

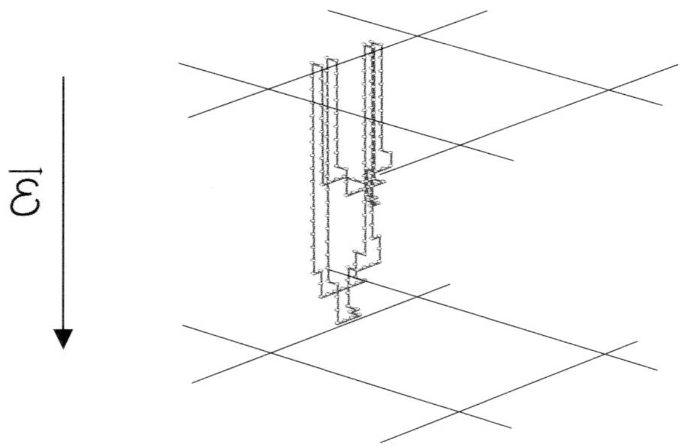

Fig. 7. Snapshot of a 7.1 knot moving in a medium with gel at $C = 0.7$ and $b = 20$. The knot is stucked around a gel rod in a U-Shape configuration.

gel for $b = 5$ and to a big pores gel for $b = 20$. For each initial knot, $2 \cdot 10^5$ iterations were performed. The center-of-mass position and the calculated ACN have been calculated every 20 Monte-Carlo steps, and they were then averaged over the trajectories obtained by the migration of 100-200 different initial knots (to obtain an accuracy of about 10%).

One problem with the Monte-Carlo algorithm is that more complex knots have a smaller drift velocity than less complex ones even in absence of the gel, when time is measured in Monte-Carlo steps (see Fig. 3).

This is due to the fact that already in the absence of the gel, the moves are hindered by the complexity of the knot. In order to correct this problem, we used Kirkwood-Riseman formula (1) to compute the friction coefficient γ of every knot (see Fig. 4): since then $v = q\mathcal{E}/\gamma$, we can find the specific time rescaling necessary to go from Monte-Carlo time (t_{MC}) to real time (t_R): $t_R = (\gamma v_{MC}/q\mathcal{E})t_{MC}$, where v_{MC} is the velocity measured using t_{MC}. Once we find the time conversion (one for every knot class) in the absence of the gel, then we apply it throughout our simulations in the gel.

3. Results

Let us begin with the study of the dependence of the center of mass position with respect to the number of Monte-Carlo steps. As a first test, we have done a simulation without electric field ($C = 0$). We observe that the knot

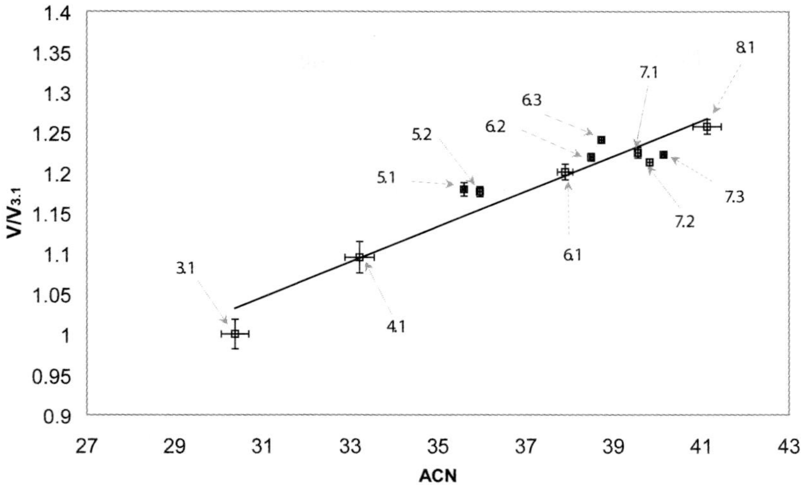

Fig. 8. Velocity of DNA knots (normalized to knot 3.1) in a medium with gel versus ACN for $C = 0.1$ and $b = 20$.

is following a diffusive random walk, i.e. we observe that the mean square distance covered by the center of mass grows linearly with the number of Monte-Carlo iterations (see Fig. 5). When we set the electric field on, we observe two different behaviors for the $\hat{z}$ component of the center of mass position (see Fig. 6) : at high temperature or weak electric field, $C = 0.1$, the velocities of the knots are well defined and the center of mass is moving with constant speed.

On the other hand, at high electric field, $C = 0.4, 0.7$, we can observe plateaus in the center of mass position versus time. Indeed, if the electric field is strong, or temperature is low, knots tend to hang over obstacles and take a U-shape configuration (see Fig. 7). When a knot hits a gel rod, it can easily remain trapped there, because the probability of a backward step is very small, and it is a growing function of the temperature T.

Trapping in U-shape conformations introduces plateaus (see Fig. 6) in the migration distance as a function of time for individual knots, hence reducing the average migration velocity. To check that the plateaus are not an artifact of the Monte-Carlo algorithm, we have checked that the center of mass distance versus time for the chains doesnt show any large plateaus in the range $C = 0 - 1.5$.

Now we study the dependence of the velocity with respect to the average crossing number (see Fig. 8). We restricted as a first step the simulation to low electric field, $C = 0.1$, and we find, due to the simulation time rescaling, a fairly linear relationship between the velocity and the average crossing number : at low electric field, knots with higher complexity are moving faster than simpler knots. These results are in agreement with experiments.

4. Conclusion

The intuitive view of a knot making its way through the pores of the gel would have as a consequence that more complex (thus more compact) knots migrate faster than simple knots, because they are less disturbed by the gel. This is indeed what happens in weak electric fields. We presented some results of a Monte-Carlo simulation of DNA knots in a gel in three dimensions. In summary, our model allows to explain the low electric field behavior observed in experimental DNA knots gel electrophoresis. Finally, varying the length of the knots and carrying simulations at high electric field would probably bring some further insights into the mechanisms of the migration in a gel.

Acknowledgments

We thank A. Baldereschi, G. D'Anna, J. Roca, J. Schwartzman, A. Stasiak, and X. Zotos for help and fruitful discussions. A preliminary version of this paper was presented at the conference "Knots, random walks and biomolecules" (co-organized by J.H. Maddocks and A. Stasiak and sponsored by the Bernoulli Centre of the Swiss Federal Institute of Technology in Lausanne) held in Les Diablerets, Switzerland during 14-17 July 2003. This work was partially supported by the Swiss National Science Foundation (Grant Nr. 21-50805.97).

References

*. Corresponding author: Giovanni Dietler, Laboratoire de Physique de la Matière Vivante, Faculté des Sciences de Base, Ecole Polytechnique Fédérale de Lausanne, CH-1015 Lausanne. Tel: (+41) 21 693 0446, FAX (+41) 21 693 0422, email: giovanni.dietler@epfl.ch

1. Noolandi, J., Rousseau, J., Slater, G.W., Turmel, C., & Lalande, M., (1987) *Phys. Rev. Lett.* **58**, 2428-2431.
2. Deutsch, J.M., (1988) *Science* **240**, 922-924 .
3. Viovy, J.L., (1988) *Phys. Rev. Lett.* **60**, 855-858.
4. Doi, M., Kobayashi, T., Makino, Y., Ogawa, M., Slater , G.W., & Noolandi, J., (1988) *Phys. Rev. Lett.* **61**, 1893-1896.

5. Noolandi, J., Slater, G.W., Lim, H.A., & Joanny, J.-F., (1989) *Science* **243**, 1456-1458.
6. Duke , T.A.J. & Viovy, J.L., (1992) *Phys. Rev. Lett.* **68**, 542-545.
7. Aalberts, D.P., (1995) *Phys. Rev. Lett.* **75**, 4544-4547.
8. Semenov, A.N. & Joanny, J.-F., (1997) *Phys Rev. E* **55**, 789-799.
9. Schwartz, D.C. & Koval, M., (1989) *Nature* **338**, 520-522.
10. Smith, S.B., Aldridge, P.K., & Callis, J.B., (1989) *Science* **243**, 203-206.
11. Sturm, J. & Weill, G., (1989) *Phys. Rev. Lett.* **62**, 1484-1487.
12. Wang Z.L., & Chu, B., (1989) *Phys. Rev. Lett.* **63**, 2528-2531.
13. Dean, F.B., Stasiak, A., Koller, T., & Cozzarelli, N.R., (1985) *J. Biol. Chem.* **260**, 4975-4983.
14. Spengler, S.J., Stasiak, A., & Cozzarelli, N.R., (1985) *Cell* **42**, 325-334.
15. Duplantier, B., Jannink, G., & Sikorav, J.L., (1995) *Biophys. J.* **69**, 1596-1605.
16. Arsuaga, J., Vazquez, M., Trigueros, S., Sumners, D., & Roca, J., (2002) *PNAS* **99**, 5373-5377.
17. Trigueros, S., Arsuaga, J., Vazquez, M.E., Sumners, D.W., & Roca, J., (2001) *Nucl. Acid. Res.* **29**, e67.
18. Stasiak, A., Katritch, V., Bednar, J., Michoud, M., & Dubochet, J., (1996) *Nature* **384**, 122-122.
19. De la Torre, J.G., & Bloomfield, V.A., (1981) *Quarterly Reviews of Biophysics* **14**, 81-139.
20. Vologodskii, A.V., Crisona, N.J., Laurie, B., Pieranski, P., Katritch, V., Dubochet, J., & Stasiak, A., (1998) *J. Mol. Biol.* **278**, 1-3.
21. Sogo, J., Stasiak, A., Martinez-Robles, M.L., Krimer, D.B., Hernandez, P., & Schvartzman, J.B., (1999) *J. Mol. Biol.* **286**, 637-643.
22. This value is estimated from Atomic Force Images of complex knots. The knots were kindly provided by Dr. J. Roca and the AFM images were taken by Dr. F. Valle.
23. Berg, B., & Foerster, D., (1981) *Phys. Lett. B* **106**, 323-326; Aragão de Carvalho, C., Caracciolo, S., & Fröhlich, J., (1983) *Nucl. Phys. B* **215**, 209-248; see also Lim, H.A., (1996) *Intl. J. Mod. Phys. C* **7**, 217-271 and references therein.
24. van Rensburg, E.J.J., & Whittington, S.G., (1991) *J. Phys. A* **24**, 5553-5567.
25. Alexander, J.W., (1923) *Trans. Amer. Math. Soc.* **20**, 275-306.

CHAPTER 9

ATOMIC FORCE MICROSCOPY OF COMPLEX DNA KNOTS

F. Valle[1,*], M. Favre[1], J. Roca[2] and G. Dietler[1]

[1]Laboratoire de Physique de la Matière Vivante - IPMC,
Faculté des Sciences de Base,
Ecole Polytechnique Fédérale de Lausanne,
CH-1015 Lausanne, Switzerland.
[2]Instituto de Biologia Molecular de Barcelona, CID-CSIC,
Jordi Girona 18-26,
08034 Bercelona, Spain.

E-mail: francesco.valle@unibo.it

Complex knots on circular closed DNA molecules extracted from virus capsids have been observed by Atomic Force Microscopy after being adsorbed onto a surface. A characterization of the knot conformation according to the adsorption procedure is presented. The adsorption procedure has been previously characterized by measuring the scaling properties of a linear DNA molecule belonging to a different virus. The high resolution of the AFM images has allowed us to identify the chirality of some DNA crossings without any specific sample treatment. The high complexity of the knot population prevents us from identifying the knot types. The present results concern the determination of the three dimensional knot conformation from the two dimensional images.

1. Introduction

Knotted DNA appears frequently both in nature and in laboratory [1,2,3].DNA recombinases [4,5] and topoisomerases [6,7] produce distinct type of knots when acting on circular DNA. Ligation of DNA free-ends also results in different degrees of knotting when carried under specific conditions[8,9]. The characterization and analysis of the knots produced in all these processes is a tool to infer physical properties of the duplex DNA, and to reconstruct the architecture of the complexes that enzymes or subcellular structures form with DNA. So far, two main techniques were used to identify and quantify the types of knots: electron microscopy (EM) and gel electrophoresis. EM was the first technique able to visualize DNA knots and allowed to identify their chirality but upon a laborious sample preparation[10]. The procedure consists in coating the double stranded DNA with RecA protein to make it thicker but conserving the knot's topology. Electrophoretic techniques have been recently improved allowing to separate individual populations of knots within complex mixtures of knotted, unknotted and linear molecules[11]. The highest separation is achieved by exploiting the different relative mobility of knots according to the voltage applied to the gel[12,13]. Atomic Force Microscope (AFM), first introduced by Binnig, Quate and Gerber[14] in 1986, is a scanning probe microscopy technique. A probe, consisting in a very soft cantilever with a tip at its end, is scanned over a surface. The interaction forces between the tip and the sample lead to the deflection of the cantilever. Changes of the deflection parameters allow to reconstruct the topography of the scanned surface. More recently, due to the softness of most biological samples, the imaging technique has been improved by the so called dynamic mode of operation. In this case the tip is not in contact with the sample but is oscillated over it, touching the surface only at the end of its oscillation. In this way, the load applied by the tip to the sample is strongly reduced[15]. In this work we used AFM to get images of DNA knots adsorbed onto a surface to see their conformations. As a source of knots we used DNA isolated from bacteriophage P4 capsids[16]. Most of these DNA molecules are knotted closed circles that result from cohesive end-joining of the 11.6 kb phage genome. The P4 phage DNA is highly suitable for this study for two reasons. First, these DNA knots are complex and heterogeneous[16]. Secondly, these knotted molecules are closed in a non-covalent way and therefore contain no confounding supercoiling, so they can be analyzed directly by gel electrophoresis without requiring enzymatic nicking. All the samples used for these experiments were naked DNA molecules not subject to any preliminary treatment.

2. Methods

Knotted DNA was obtained by infecting the E.coli strain C-8001 with P4 vir1 del22 phages, following the procedure described by Isaksen et al.[17] with minor modifications. After bacterial lysis the released capsids were precipitated with PEG 8000, and then solubilized in a buffer containing 10 mM MgCl2, 10 mM Tris-HCl pH 7.2 and 130 mM ammonium acetate. P4 tailless capsids were banded by cesium chloride centrifugation in an

NVT65 rotor for 14 h at 45 000 r.p.m. The banded particles were dialyzed against P buffer. DNA was extracted twice with phenol, once with phenol/chloroform, precipitated with ethanol, and resuspended in TE buffer (10 mM Tris-HCl pH 8, 1 mM EDTA) to a concentration of 1 mg/ml. Linear Lambda-Mix DNA was purchased from Fermentas Gmbh (Nunningen, Switzerland) and it has been supplied as a stock solution with 0.5 mg/ml nominal concentration; it contains 14 fragments of different lengths (from 1503 up to 48502 base pairs) obtained by restriction enzymes cleavage of the whole DNA of the bacteriophage Lambda. For AFM imaging, the sample was diluted in the same buffer to a final DNA concentration of 0.5 μg/ml. According to the deposition protocol the solution could be adjusted with different ion concentrations to achieve the desired ionic strength. A small droplet (10 μl) of sample was then placed onto the appropriate surface and left to adsorb for approximately 10 minutes at room temperature; the sample was then rinsed with nanopure (Ultra High Quality) water (USF Elga, High Wycombe, England) and blown dry with clean air. The surface used for DNA imaging is AP-mica. This surface consists in freshly cleaved mica which has been exposed to the vapor of 3-Aminopropyl-triethoxysilane. This amino terminating molecule binds to the mica surface, exposing the amino groups which are positively charged at physiological pH. The surface is thus able to bind DNA by its highly negatively charged phosphate backbone. To achieve a good silanization of the mica surface a protocol developed in the Lyubchenko laboratory was used[18]. Images were collected using a Nanoscope IIIa (Veeco Inc, Woodbury, NY, USA) operated in tapping mode in air. The ultrasharp non-contact silicon cantilevers (NT-MDT Co., Zelenograd, Moskow, Russia) used had a nominal tip radius of <10 nm and were driven at oscillation frequencies in the range of 150 to 300 kHz. During imaging, the mica surface was scanned at a rate of one line per second. Images were simply flattened using the Nanoscope III software and no further processing was done.

3. Characterization of the conformation of a DNA molecule bound to a surface

As described in sections 1 and 2 the AFM is a surface technique, thus to obtain images of macromolecules one has to put them onto a surface. Consequently, the first task to accomplish in order to understand AFM images of a DNA molecule is to examine the effects of the surface-binding procedure on the DNA conformation. To this end we performed an analysis of linear

DNA molecules adsorbed on to mica surface in terms of Flory exponents. Linear DNA molecules were chosen because the end-to-end distance can be used to evaluate the scaling behavior and this is a quantity easily accessible from AFM images. We proceeded as follow: 1) long linear and heterogeneous DNA molecules (Lambda Mix DNA) were bound onto AP-mica, 2) we digitized the contour of the DNA molecules from images obtained by AFM (Fig. 1, left panel), 3) for each contour length the corresponding end to end distance was determined, 4) we analyzed the scaling behavior of the end-to-end distance vs the DNA contour length by averaging the values obtained over many molecules. Then the averaged end-to-end distance vs contour length was plotted on a log-log plot. We obtained a double power law. A first regime, for short distances is followed by a second one with a smaller exponent after a certain cross-over length[19]. Figure 1 depicts the graph fitted using the equation

$$f(x) = c\left(\frac{x}{\ell_o}\right)^{\nu_1}\left(1+\frac{x}{\ell_o}\right)^{\nu_2-\nu_1} \tag{1}$$

corresponding to a double power law (with exponents ν_1 and ν_2) with a cross over length (l); c is a normalization constant. The values obtained from the fit are: ν_1=1.03±0.02, ν_2=0.589±0.006 , l=44±6 nm and c= 49±5 nm. This relation means that for contour lengths shorter than l, the DNA behaves as a rigid molecule (exponent ν_1=1.03±0.02), for longer values the molecule is described by a Self Avoiding Random Walk (SAW) with the three-dimensional exponent (ν_2=0.589±0.006)[20]. From the point of view of our experiments this means that DNA molecules, upon binding onto AP-mica, do not equilibrate on the surface but conserve a conformation corresponding to the two-dimensional "projection" of the three-dimensional DNA.

The value of the cross-over length is in good agreement with the persistence length measured in other studies, under the same ionic conditions, using microscopy techniques[21,22]. The measured three-dimensional exponent for a molecule bound on a surface can be explained by the conservation of the fractal dimension of an object upon its projection into a space with a larger space dimension[23]. This kind of analysis allows us to characterize the conformation on the surface and, in this case, to draw some conclusion about the conformation that the same molecules assume in the bulk solution before adsorption. In fact it has been shown that, under conditions of strong adsorption used in this work, the DNA molecules do not equilibrate. Therefore the adsorbed molecules trace a path resulting from

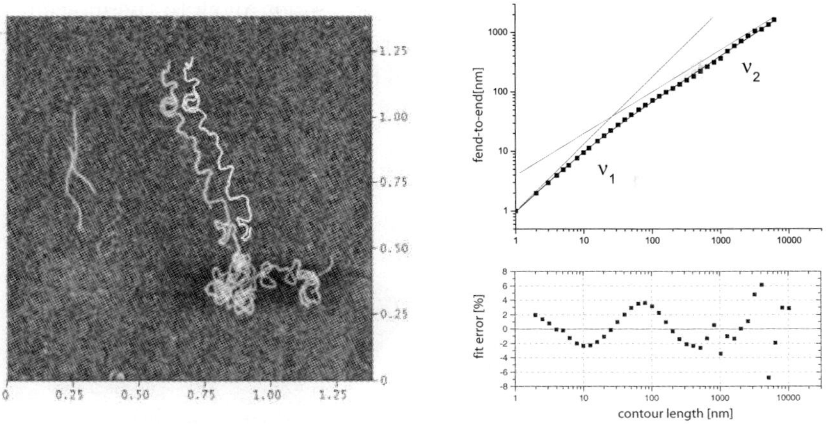

Fig. 1. On the left an AFM image of a long Lambda DNA molecule bound onto an AP-mica surface. On the same image, near the molecule has been reported a digitization of a part of the DNA chain; this is the kind of objects which are then analyzed. On the right a plot of the end to end distance vs the contour length averaged over many molecules. The fit is reported with the two exponents measured.

a "projection" of the three-dimensional conformation. This rules out the possibility that the observed conformations are the consequence of a rearrangement on the surface. The detailed characterization performed using linear DNA molecules allows us to interpret the conformation of the closed circular knotted DNA molecules adsorbed under the same conditions.

4. Images of knotted DNA

DNA molecules extracted from the bacteriophage P4 are knotted with high probability. Differences in knot mobility by gel electrophoresis reflected a broad and very high knot complexity. So far, little direct visualization of these molecules was available[24]. We deposed on AP-mica different aliquots of the sample containing the whole DNA extracted from the virus capsids. Based on the electrophoresis analysis this sample should contain a mixture of linear molecules (less than 10%) and circular closed ones, both unknotted and knotted, the last representing the large majority of the sample. When DNA molecules are adsorbed on the surface under conditions of strong binding, as it has been previously shown, their conformation reflects the one assumed in the bulk solution. In the AFM images under these conditions the large majority of the molecules found on the surface were highly knotted, as seen in figure 2. This is in agreement with previous estimations[25]. Because of the deposition procedure, which does not allow to the molecule to equilibrate on the surface, not all the crossings present are due to the knot topology. Some crossings may originate from the projection. In figure 2 are

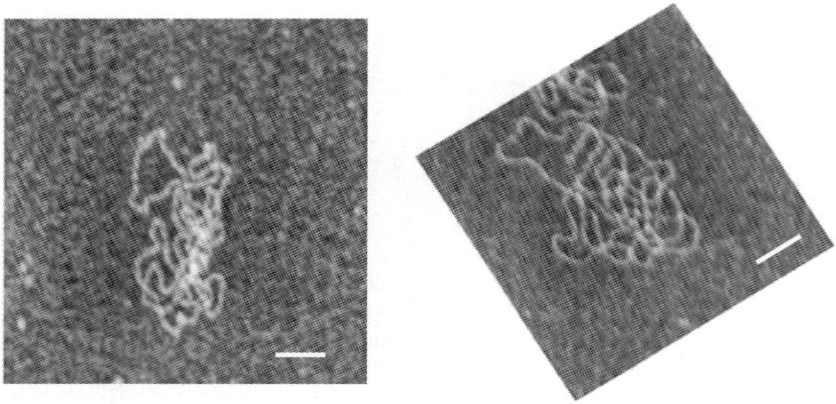

Fig. 2. Images of DNA knots adsorbed onto AP-mica. Under strong binding conditions, the conformation of the knots in two-dimensions is the projection of the three-dimensional conformation (The scale bars are 100 nm). The number of crossings can be estimated for this kind of molecules to be 24 and 25 for the left and the right molecule respectively.

shown AFM images of the knots. Some important features must be stressed here. First, at most DNA crossings, AFM allows to distinguish which double strand is passing below or above (see figure 3). Then the chirality of each imaged molecule could be fully determined. However, the determination of the sign of all crossing, that is necessary for the determination of the chirality of the whole knot, is hampered by their density and the complexity of the knotted molecules. In fact, in the region where the density of crossings is too high it is sometimes even hard to identify a single crossing. Second important aspect of the imaged molecules is that the crossings are distributed over all the contour of the molecules. This distribution indicates a weak knot localization in three-dimensions, as it has been theoretically shown[26] for knots of high complexity. Hence most crossings visible in the previous images (Fig. 2) are due to the knots (unknotting experiments have been performed on the same molecules to confirm this result; data not shown). The number of crossings can be estimated for molecules like those reported in figure 2 and it is in good agreement with the number determined by gel electrophoresis (average crossings number = 25) and reported in a previous work[25]. The analysis of the Flory exponents performed on few linear

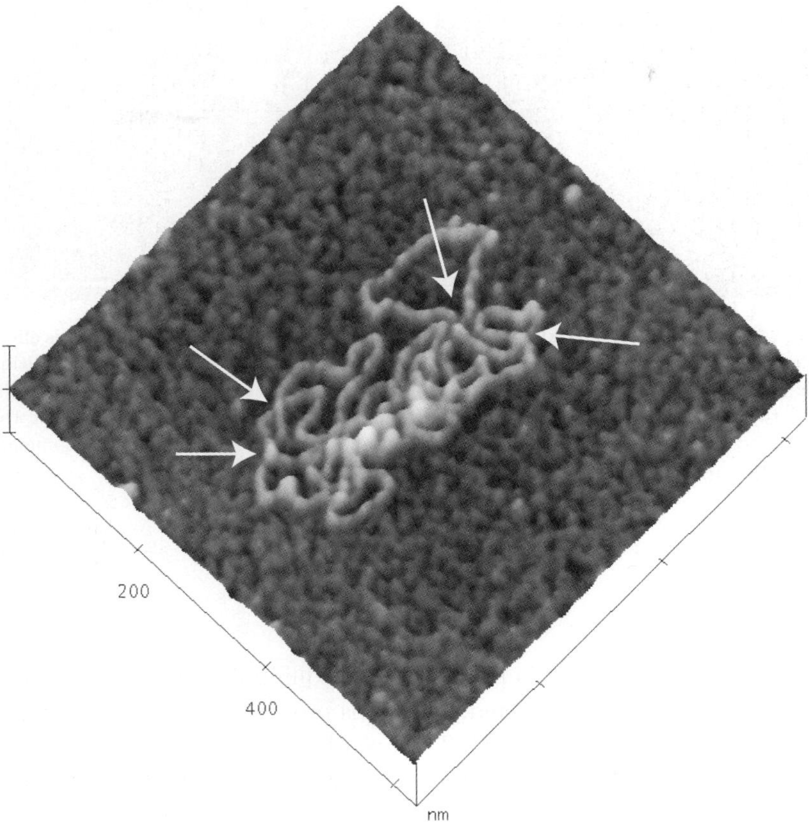

Fig. 3. Three dimensional rendering of the AFM image of a knotted DNA molecule adsorbed onto AP-mica. In the figure it is possible to see how, in AFM images, one can distinguish the orientation of some crossing, by identifying which double strand is passing above and which one below. This kind of identification is possible only for the regions where the density of crossings is not too high.

molecules found in this sample gave the same results as presented in section 2. It is then interesting to see how the knotted DNA molecules behave when adsorbed under conditions that allow them to partially equilibrate on the surface. We observed that under such conditions knotted molecules looked like having less crossings and the remaining crossings confined in a relatively small portion of the DNA chain. In figure 4 are shown two ex-

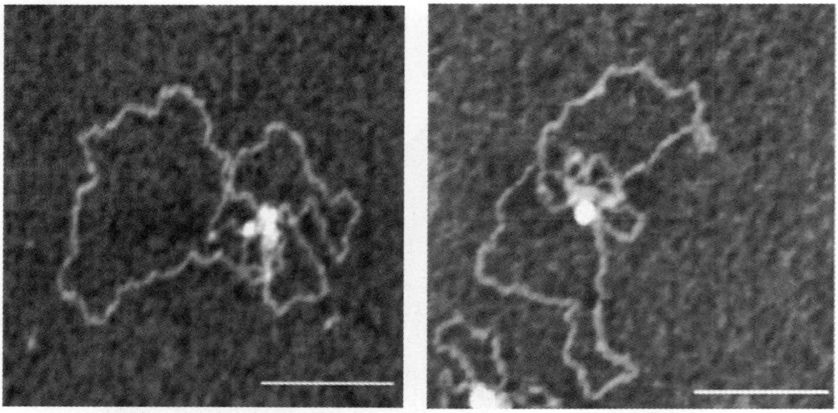

Fig. 4. AFM images of DNA knots which were weakly bound to the AP-mica. This effect was achieved by increasing the ionic strength of the solution (following the addition of increasing concentrations of NaCl up to concentrations of 300 mM) to screen the electrostatic interaction between the DNA and the surface. It can be seen that the crossing are usually concentrated in a relatively small region of the chain, a phenomenon called "knot localization". (Scale bar 100 nm)

amples of this behavior. It is likely that the equilibration process removes the non-topological crossings and, while this process takes place, the topological crossings are concentrated. The resulting molecules are almost fully relaxed with a small portion where the knot is relatively tight. One can observe that the sizes of these molecules, although smaller than the unknotted ones, are larger than those strongly bound onto AP-mica without equilibration (Fig. 2). The concentration of the crossings along the DNA chain is in good agreement with the theoretical studies of Metzler et al.[27].

The use of naked molecules is critical for this kind of studies, because the main parameters of the DNA double helix are not altered by any preparation technique. Thus, the effects on the conformation can be ascribed only to the different level of equilibration on the surface achieved by different adsorption protocols. The ability to see the details of each crossing will allow us to correlate these studies with the chirality of each knot (figure 3); a step that has not yet been done because, due to the high complexity of the available knot population, as explained before in this section, it is not possible to distinguish the orientation of all the crossings in these molecules.

Acknowledgments

The authors wish to thank Dr. Andrzej Stasiak (University of Lausanne) for the useful discussions and suggestions. A preliminary version of this paper was presented at the conference Knots, random walks and biomolecules (co-organized by J.H. Maddocks and A. Stasiak and sponsored by the Bernoulli Centre of the Swiss Federal Institute of Technology in Lausanne) held in Les Diablerets, Switzerland during 14-17 July 2003. This work was supported by the Swiss Natural Science Foundation (project 2100-063746.00/1) and partly by the Roche Research Foundation.

References

*. Corresponding author: Francesco Valle, presently at Department of Biochemistry, University of Bologna, Via Irnerio 48, 40126 Bologna, Italy. Tel: (+39)0512094388 Fax: (+39)0512094387, email: francesco.valle@unibo.it
1. Liu L. F., Perkocha L., Calendar R. & Wang J .C., (1981) *Proc. Natl. Acad. Sci. USA* **78**, 5498-5502 .
2. Menissier J., de Murcia G., Lebeurier G. & Hirth L., (1983) *Embo J.* **2**, 1067-71
3. Ryan K. A., Shapiro T. A., Rauch C. A., Griffith J. D. & Englund P. T., (1988) *Proc. Natl. Acad. Sci. USA* **85**, 5844-8.
4. Wasserman S. A., Dungan J. M. & Cozzarelli N. R., (1985) *Science* **229**, 171-4.
5. Spengler S. J., Stasiak A. & Cozzarelli N. R., (1985) *Cell* **42**, 325-34.
6. Wasserman S. A. & Cozzarelli N. R., (1991) *J. Biol. Chem.* **266**, 20567-73
7. Roca J., Berger J. M. & Wang J. C., (1993) *J. Biol. Chem.* **268**, 14250-5.
8. Rybenkov V. V., Cozzarelli N. R. & Vologodskii A. V., (1993) *Proc. Natl. Acad Sci USA* **90**, 5307-11.
9. Shaw S. Y. & Wang J. C., (1993) it Science **260**, 533-6.
10. Krasnow M. A., Stasiak A., Spengler S. J., Dean F., Koller T. & Cozzarelli N. R., (1983) *Nature* **304**, 559-60.
11. Vologodskii A. V., Crisona N. J., Laurie B., Pieranski P., Katrich V., Dubochet J. & Stasiak A., (1998) *J. Mol. Biol.* **278**, 1-3.
12. Trigueros S., Arsuaga J., Vazquez M. E., Sumners D.W. & Roca J., (2001) *Nucl. Acids Res* **29**, e67.
13. Sogo J. M., Stasiak A., Martinez-Robles M. L., Krimer D. B., Hernandez P. & Schvartzman J. B., (1999) *J. Mol. Biol.* **286**, 637-43.
14. Binnig G., Quate C. F. & Gerber C., (1986) *Phys. Rev. Lett.* **56**, 930-933.
15. Garcia R. & Prez R., (2002) *Surf. Sci. Rep.* **47**, 197-.301
16. Liu L. F., Davis J. L. & Calendar R., (1981) *Nucl. Acids Res.* **9**, 3979-89.
17. Isaksen M., Julien B., Calendar R. & Lindqvist B. H., in *Methods in Molecular Biology:DNA Topoisomerase Protocol* Vol. 1 (Humana Press, Totowa, 1999), p. 69.
18. Lyubchenko Y. L. & Shlyakhtenko L. S., (1997) *Proc. Natl. Acad Sci USA*

94, 496-501.
19. Favre M., Valle F., De Los Rios P., Rosa A. & Dietler G. (submitted)
20. Le Guillou J. C. & Zinn-Justin J., (1977) *Phys. Rev. Lett.* **39**, 95-98.
21. Bednar J., Furrer P., Katrich V., Stasiak A. Z., Dubochet J. & Stasiak A., (1995) *J. Mol. Biol.* **254**, 579-94.
22. Berge T., Jenkins N. S., Hopkirk R. B., Waring M. J., Edwardson J. M. & Henderson R. M., (2002) *Nucleic Acids Res.* **30**, 2980-6.
23. Mandelbrot B. B., in *The Fractal Geometry of Nature* (W. H. Freeman and Company, Oxford, 1982), p. 91
24. Wolfson J. S., McHugh G. L., Hooper D. C. & Swartz M. N., (1985) *Nucl. Acids Res.* **13**, 6695-702.
25. Arsuaga, J., Vazquez, M., Trigueros, S., Sumners, D., & Roca, J., (2002) *PNAS* **99**, 5373-5377.
26. Metzler R., (2002) *New J. Phys.* **4**, 91.1-91.10.
27. Metzler R., Hanke A., Dommersnes P.G., Kantor Y. & Kardar M., (2002) *Phys. Rev. Lett.* **88**, 188101.

CHAPTER 10

PROTEIN FOLDS, KNOTS AND TANGLES

William R. Taylor
Division of Mathematical Biology,
The National Institute for Medical Research,
The Ridgeway, Mill Hill, London NW7 1AA, U.K.
E-mail: wtaylor@nimr.mrc.ac.uk

Protein folds are considered as simple directed chains. Local smoothing of their coordinate sets (without chain passage) leads to a simple method to detect knots in open chains. A trefoil and figure-of-eight protein knot are described. Extension of the algorithm to incorporate hydrogen-bond cross-links allows protein pseudo-knots to be considered. The general connectivity of a protein chain is considered as a graph and a novel concept of topological accessibility described. These ideas are then applied to semi-random folds.

1. Introduction

Proteins are linear polymer chains, typically 100—1000 units (residues) in length. Most proteins fold into a compact globular shape and the path of the chain through this structure is called the fold of the protein. The fold of a protein is specific and is determined by the sequence of residues along the chain. Ever since they were first seen, protein folds have remained a source of fascination, and even wonder. The different folds adopted by proteins become more enigmatic when it is realised that they bear little relationship to the function of the protein. Proteins with different folds can perform the same function (for example; catalyse the same reaction) while proteins with the same fold can perform different functions and, beyond clear evolutionary relationships (homology), it is not possible to predict a function from a fold with any degree of certainty. One of the greatest mysteries in molecular biology is why proteins adopt the folds that are observed: are they simply the result of (pre)historical accident or do they represent a partial or complete manifestation of the folds allowed by 'laws' that we do not yet know?

As an approach towards answering some of these questions, attempts have been made to establish a system to describe protein folds, thereby allowing protein structures to be compared and classified in the hope of establishing some correlations between fold and function. However, the fold of a protein is a difficult thing to quantify and many descriptive terms have been used, often derived from the imagination of those working in the field — resulting in names such as "jelly-roll" or "Greek-key" or after the crystallographers who determined the first example of the fold, such as the "Rossmann fold". More commonly, the name comes from a particular function with which the fold was once exclusively associated but has later been found in more diverse proteins: as has been seen with the TIM-barrel, the immunoglobulin (Ig) fold and the oligosaccharide-binding (OB) fold. While these names provide useful mnemonics for different aspects of the structure or function, an automated analysis requires a more general and systematic scheme.

In a search for a more systematic approach, attention is often directed towards the field of topology. However, when viewed from a topological perspective, protein chains are very uninteresting as they are just one-dimensional strings and are all the same. This is true when only the backbone of the chain is considered but proteins can form covalent cross-links between the side-chains of some or their residues. This adds some topological excitement as the cross-links can form linked rings and close the chain into a knot. Much has been written about protein knots formed by covalent cross-linking but I will not say much about it in this chapter as almost all protein cross-links are formed after the protein fold has been formed and as such are not fundamental to the nature of the fold. They act more to stabilise a fold in the way that some pieces of tape might stabilise an origami.

Rather than abandon the application of topological ideas to proteins; two courses can be followed: 1) extend the concept of a knot to apply to open strings and, 2) extend the definition of a cross-link to encompass, not just covalent links, but all the types of interaction found within a protein structure. Both these approaches will be elaborated below.

2. The 'Topology' of open chains

Unlike DNA, protein chains are very short (relative to their bulk) and the range of features cannot be expected to be very great. Rather than finding complex linked chains or different knot topologies (as in DNA), it is rare to find a protein chain that can even be considered as a knot. Until recently, the few folds that were reported to be knotted (without considering post-translational cross-links) have one end of the chain barely extending through a loop by a few residues. Attempts have been made to identify these 'knots' by extending the termini of the chain to 'infinity' where they can be considered linked, so forming a proper knot or link[1]. However, if the termini do not clearly protrude through the knotting loops then whether a knot is detected will depend on the direction in which the termini are extended. To avoid this problem, the termini can be extended in all directions giving a probabilistic measure of how close the chain is to being knotted[2].

2.1. *Shrinking chains*

An approach that is related, but different, from extending the termini is to keep the termini where they lie and shrink the rest of the chain[a]. The approach has been applied both to proteins [3] and to ideal open knots [7]. Neither terminus linking nor chain shrinking can produce an absolute solution to the problem of defining a knot in an open chain: an extended terminus may pass arbitrailly either side of a segment of chain but equally, a chain segment may shrink either side of a terminus. Both approaches are helped, however, by the preferred occurrence of chain termini on the surface of the protein [4] and often close together [26].

2.1.1. *Chain smoothing*

Chain shrinkage can be done gradually through repeated local averaging: in a chain of length N consisting of a set of coordinate vectors a $(a_1, a_2, \ldots a_N)$ representing the α-carbon of each residue, each position a_i can be replaced by the average of itself and its two neighbours;

$$a_i^{t+1} = (a_{i-1}^t + a_i^t + a_{i+1}^t)/3, \quad \forall i, 1 > i < N, \tag{1}$$

where t marks the time step in the iteration.

[a]This approach is conceptually equivalent to, but was developed independently from, the SONO algorithm [6]

In Figure 1, the result of the application of this algorithm over five iterations is shown for a small protein structure (Protein databank code: 3chy) that contains both α-helices and β-strands. After three cycles, both these secondary structure elements are reduced to almost straight lines and with further iteration, the chain becomes increasingly smooth, eventually ending as a straight line connecting the two termini (which remain in the same position).

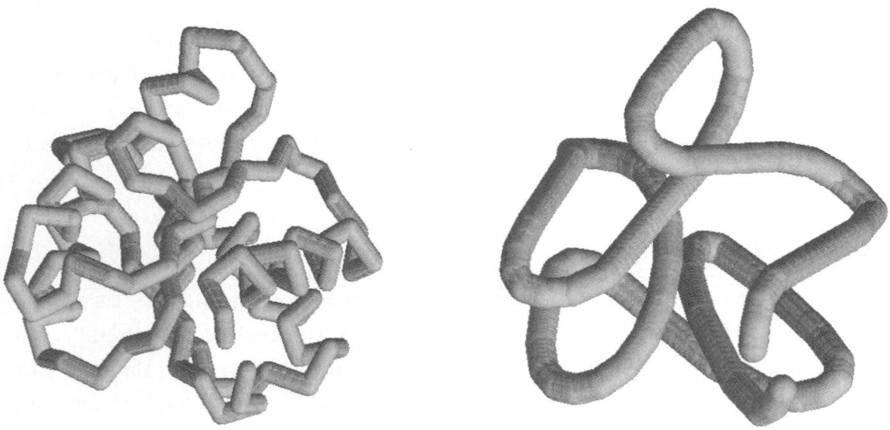

Fig. 1. The native structure of a small protein, before (left) and after smoothing (right). The structures are represented by connecting their α-carbon atoms (one per residue) to show the fold of the chain. The smoothed structure resulted from five iterations of the algorithm described in the text in which each position is replaced by the average of itself and its two neighbours. (The structures have the same orientation).

2.1.2. *A self-avoiding chain*

As it stands, this simple algorithm will allow the chain to pass through itself and if we are interested in topological-like properties, this is a very undesirable property. To avoid this, each move ($a_i^t \rightarrow a_i^{t+1}$) was checked to ensure that the two triangles formed by the points $\{a_{i-1}^t, a_i^t, a_i^{t+1}\}$ and $\{a_{i+1}^t, a_i^t, a_i^{t+1}\}$ were not intersected by any other line segment in the chain. If they were, then the new position (a_i^{t+1}) was not accepted. (Figure 2).

One practical problem with this algorithm was that if two line segments are blocked from moving, their attached points might still be able to take small steps, bringing the lines increasingly closer together. In practice, their approach can become so close that the numerical accuracy of the computer is insufficient to discriminate whether the move is allowed or forbidden and, in a numerical equivalent of quantum tunnelling, the lines can pass through each other. This was avoided by representing each line between residues by a tube 0.5Å in radius and each point by a sphere of the same radius.

The adoption of a thick chain allowed the introduction of a further practical device. Occasionally, configurations would become stuck — as in a log-jam. And although these would eventually work themselves free, it was found that the process could be greatly speeded by gradually reducing the tube radius by a very small amount throughout the course of the iterations. (Typically, by 1% per iteration). This acted as an algorithmic oil to ease the passage of bumping tubes.

Repeated iteration of the smoothing algorithm will now still reduce most chains to a straight line, but slightly more slowly because of bumping. However, more importantly, if the chain contains a knot, then it will never become a straight line and this can be used as a test for knots.

2.1.3. *Removing redundant points*

The adoption of a thick chain allowed a further, more fundamental, addition to the algorithm. If a point (i) lies within the tube formed by linking its adjacent points ($i-1$ and $i+1$), then the middle point is redundant to the calculation and can be removed. Through smoothing, it can never increase its deviation from linearity and although it is possible that large movements of the flanking points may shift the position of the $i-1$,$i+1$ tube, this is an unlikely event and even in this situation, the removal of point i will not affect the topology of the chain: that is, no line segments can have been allowed to erroneously pass through each other. The same argument also applies to the middle point also when its neighbours approach each other closely (that is the chain segment deviates most from being linear). If points $i-1$ and $i+1$ come within a tube diameter of each other, then the acute triangle $i-1$,i,$i+1$ cannot be penetrated by any other line segment and so the middle point can be removed without fear of allowing any forbidden chain passage.

These features not only improved execution time but led to an even simpler test for knots as any chain that can be reduced to just its two termini is not knotted. Chains with more than two residues remaining are

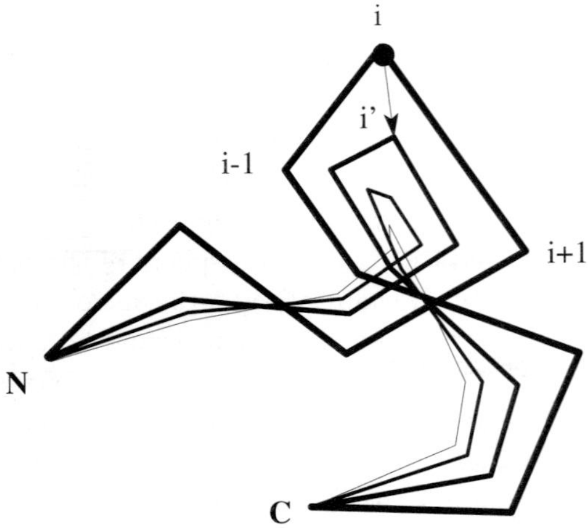

Fig. 2. The basic chain smoothing algorithm. Protein chains are drawn schematically as lines connecting the central carbon atom in the backbone of each residue unit running from the amino (N) terminus to the carboxy (C) terminus. Beginning at the second residue, for each residue point (i) in the starting conformation, the average coordinate of i, $i-1$ and $i+1$ was taken as the new position (i') for the residue. This procedure was then repeated, and the results of this are progressively smoother chains, shown as a series of fainter lines. Note that the termini do not move. With each move, it was checked that the chains did not pass through each other. This was implemented by checking that the triangles $\{i'-1,\ i,\ i'\}$ and $\{i,\ i', i+1\}$ (dashed lines in the Figure) did not intersect any line segment $\{j'-1, j'\}$ ($j < i$) before the move point or any line $\{j, j+1\}$ ($j > i$) following.

either knots or tangles in which a group of moves have become 'grid-locked' and have not been allowed enough time (or enough oil) to become untangled. Most chains of a few hundred residues are reduced to their termini in around 50 iterations. If by 500 iterations a chain was still not reduced to two points, then the resulting configuration was analysed in more detail. In Figure 3, the algorithm is applied to two protein chains — one of which has a particularly simple fold (called a TIM-barrel) and the other is knotted. The latter will be considered in more detail below.

2.1.4. *A simple knot nomenclature*

As the termini are now well separated from the knot-site, they can be unambiguously joined and analysed as a 'proper' circular knot. This might be done using one of the knot-invariant polynomials (See Ref. 8). However, the few knots encountered in proteins are so simple that they do not require

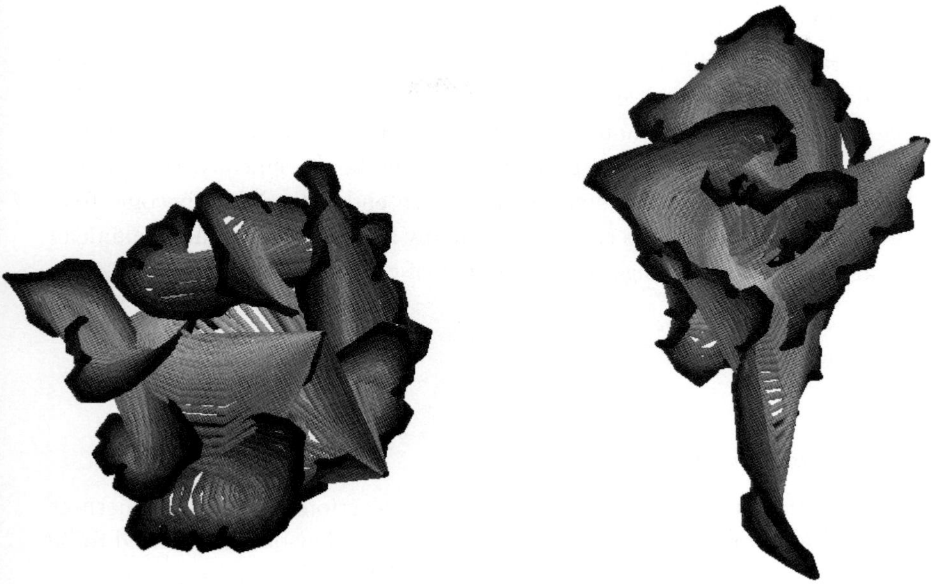

Fig. 3. Applying the smoothing algorithm described in the text to protein structures produces a series of increasingly smoothed chains, (shown superposed in the figures). (left) applied to a protein that has no knots (triosephosphate isomerase, [1tph1]) results in a straight line joining the termini. To reach this stage took 52 smoothing iterations. (right) applied to the knotted protein (the carboxy-terminal domain of acetohydroxy acid isomeroreductase, [1yveI]), a straight line is never attained and a small knot remains deep in the core part of the protein.

any sophisticated analysis. From a theoretical perspective, protein knots are different from conventional knots. Not only are protein knots directional but they also they have a unique break-point (between the termini) which is not taken into consideration by any of the polynomial forms.

As a working tool, a simpler method was adopted to characterise these open knots based on the Dowker knot notation[8]. In this, each crossover in a two-dimensional projection of the knot is characterised by its handedness. Beginning at the amino terminus, recording the handedness of successive crossovers as 1 or 0 generates a binary number which can be used as a reasonably unique descriptor for simple knots. To minimise the effects of projection, each knot was rotated around the axis defined by the two termini and the smallest numeric descriptor recorded.

2.1.5. *Finding the minimal knot segment*

An advantage of the algorithm described above is that it is not sensitive to the direction of projection of the termini and can therefore be used to define the exact region of the chain that gives rise to the knot. This allows knots in proteins to be characterised by how deep they lie: specifically, the number of residues that must be removed from each end before they become free. In addition, if the knot is complex, then any simpler knots that remain can be monitored by a change in the number of crossovers. (Figure 4).

2.2. *Knots in proteins*

Applying this method to a non-redundant selection of protein structures revealed a surprisingly large number of knots [9]. A few of these proved to be unresolved tangles, including slip-knots. Of the eight remaining structures, five were right-handed trefoils, one was a left-handed trefoil and two were figure-of-eight (or Flemish) knots. These latter two are of particular interest as they include an additional crossover above the trefoil and are therefore less likely to be formed by random chain wandering during folding.

The location of the knots were determined by a series of deletions from both termini of the protein chain. This revealed that one of the the figure-of-eight knots was barely tied and was held by just the last eight residues. By contrast, the knot in acetohydroxy acid isoisomerase, which is in the carboxy terminal domain of the protein, remained until 70 residues were deleted from the carboxy terminus and 245 residues (including a complete domain) were removed from the amino terminus. Similarly, most of the trefoil knots were also on the boarder-line of being knotted with the exception of a very clear knot in an RNA methyltransferase. (Figure 5).

Clear knots in the protein chain are rather rare and it is always of interest to examine closely those that are identified. Sometimes these must be treated with caution as knots almost always involve surface loops, which with their greater exposure to solvent are more mobile than other parts of the protein chain and hence less well resolved. This can lead to errors in chain tracing and the erroneous creation of a knot. It is always better if there is more than one independent solution of the structure (preferably at as high a resolution as possible) and fortunately, this was the situation for both the clearly knotted structures described here.

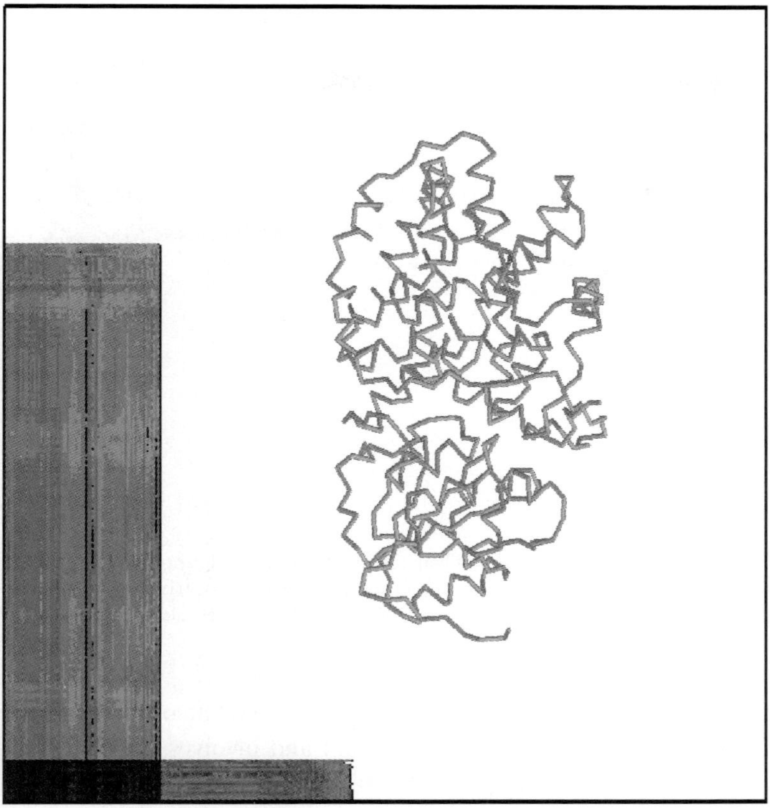

Fig. 4. To illustrate the effect of deleting the termini of proteins to find the minimal knot, an interesting test example was constructed by joining a protein with a trefoil knot and a protein with a figure-of-eight knot end-to-end (inset figure). The proteins (`1ipa` and `1yve`) are described later in the text. Each combination of amino and carboxy terminal deletion defines a cell in the plot with the bottom left corner being the full length chimeric protein. The X-axis indicates increasing deleting from one terminus while the Y-axis is deletions from the other terminus. For each deletion pair, the knotted state of the fragment was measured by the number of remaing points: white, no knot; black, two knots; grey, one knot. For the latter shade, it is difficult to distinguish the trefoil and the figure-of-eight because of noise, however, the latter is the smaller slightly darker rectangular area on the bottom.

2.2.1. *A protein trefoil knot*

A trefoil knot has been found in two homologous structures of an RNA methyltransferase [10,11] (PDB codes: `1IPA` and `1GZO`, respectively). These have 1.9Å root-mean-square (RMS) deviation over the 145 equivalent

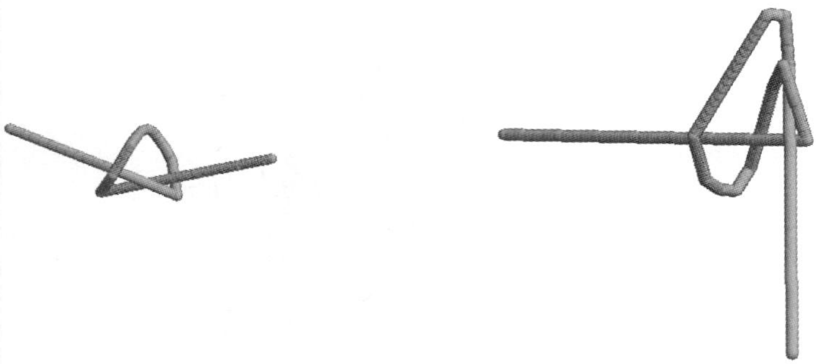

Fig. 5. Knots in proteins include a clear trefoil (left) and a figure-of-eight knot (right). The proteins in which these knots occur (`1ipa` and `1yve`, respectively) are discussed in more detail in the text. The figures show the smoothed trace close to the end of the iterations.

residues in the knotted domain. The knotted segment occurs towards the end of the chain in the C-terminal domain and involves the chain path at the edge of the β-sheet. (Figure 6).

Applying the current knot algorithm to a series of deletions from the termini, the minimal knot was found to lie between residues 187 and 238, well away from the amino (N) terminus and 25 residues away from the carboxy (C) terminus. It seems likely that this knot has arisen by the diffusion of the 25 residue C-terminal part of the chain through its loop.

2.2.2. *A protein figure-of-eight knot*

The figure-of-eight knot found in the two acetohydroxy acid isoisomerase structures, `1gz0`[12] and `1yve`[13], is much deeper than the trefoil in the methyltransferases and requires almost 80 residues to be "fed-through" the knotting loop. Although this loop would be large before it folds, it is difficult to imagine such a large directed motion occurring by chance in a reproducible way during the folding of the protein.

Clues to the folding of this protein can be found in a clear internal duplication within the domain (Figure 8). If it is assumed that the two

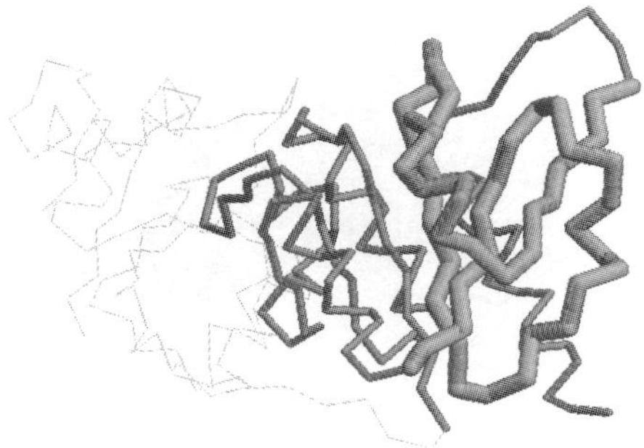

Fig. 6. The trefoil knot in an RNA methyltransferase (`1ipaA`). The backbone fold of the protein is shown as a trace linking α-carbon atoms. The amino terminal domain is drawn with a fine line while the carboxy terminal domain is drawn with medium thickness and within this the minimal knotted core is drawn thickest.

most deeply buried symmetrically equivalent helices initially pack together, then the remaining parts of each repeat can wrap around this core requiring only that the carboxy terminal segment must pass through the large loop between the repeats before this folds.

2.3. *Protein Pseudo-knots*

In the nucleic acid field, a class of knot called pseudo-knots is defined by the hydrogen-bonded links formed by base-pairing. If protein hydrogen bonds were to be given the same 'status' as base pairing in RNA or other covalent crosslinks (such as disulphide bonds) then proteins would provide a rich source of topological features. The analysis of the hydrogen-bond 'knot' can be treated more exactly using methods developed over many years to analyse knots formed by disulphide (or other) crosslinks [24]. Having reduced the protein chain (plus crosslinks) to a simple form, similar to the smoothed knots in Figure 5, then the two termini (which are assummed to be unambiguously clear of the knot), must be joined. In this configuration, if the resulting network can be placed flat on a plane without any lines crossing (including crosslinks) then there is no knot. Computationally, this can be

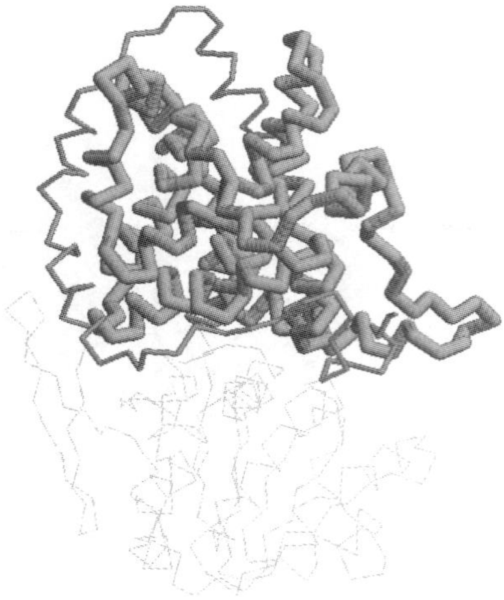

Fig. 7. The figure-of-eight knot in acetohydroxy acid isoisomerase (`1yveI`). The backbone representation of the complete native protein structure with the minimal knotted region drawn thickened. This region is preceded by a complete nucleotide binding domain (fine line) and followed by a long loop that wraps around the knotted domain (medium line).

solved by representing the protein (plus links) as a graph and using the algorithm of Kuratowski to check for planarity[15,16].

The smoothing algorithm described above can also be extended to deal with cross-links which can just be added as another set of lines that cannot pass through each other or through the backbone chain. These were implemented in a limited way, allowing a few links to be introduced. However, if too many are added, the increased number of bumps and tangles makes the algorithm much slower. The extended algorithm was tested on a small protein (called the "SET" domain) that has a potential pseudo-knot and has been variously described in the literature as a knot or not a knot[17].

2.3.1. *A pseudo-knot in a SET domain*

To examine the topology of the SET domain, the smoothing described above was applied to one of the recent crystal structures: the histone methyltransferase protein SET7/9 [18] (residues 135–343). Without cross-links, the domain was completely reduced to a straight line by the the application of the algorithm in 43 cycles of iteration, which is typical for a globular protein of

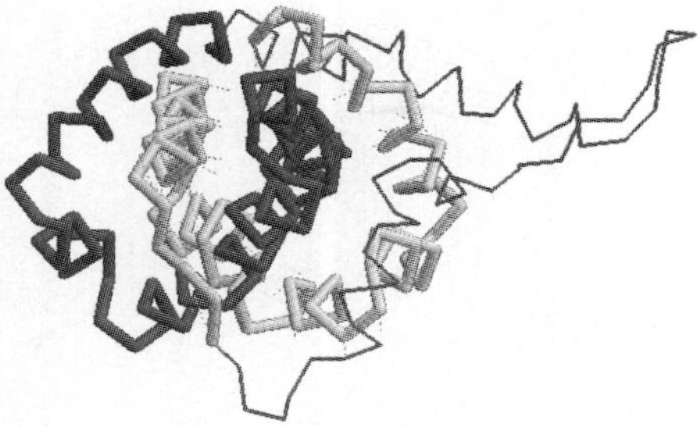

Fig. 8. Internal symmetry in acetohydroxy acid isoisomerase (`1yveI`). The backbone of the knotted C-terminal domain is drawn as a fine line with the two symmetrically equivalent portions drawn more thickly with one half grey and the other half white. The way these two halves wrap around each other may explain how the knot is formed. The RMS deviation is 3.2Å over the 80 superposed residues in the two halves.

this size. The speed of this reduction indicates that the fold does not even contain a slip-knot or tangle as these take longer to resolve than a simple fold. It was clear from this result that the SET domain does not contain a topological knot in the above sense of the term, where only the backbone topology is considered.

Further investigation was carried out using the limited cross-link algorithm by inserting hydrogen bond cross-links. Extensive trials inserting just one crosslink failed to create a knot and it was found that another pin was necessary to "lock" the structure. These trials have shown that the exact location of the two crosslinks are not important, provided that they both link the relatively parallel segments between residues 231–246 and 316–330 (which are hydrogen-bonded over most of their length). (Figure 9).

2.3.2. *Generalised protein knots*

The analysis of the SET domain has illustrated that, in general, proteins exhibit a progression of increasingly solid cross-links from close approach (van der Waals 'bonds') through hydrogen-bonds to disulphide (or other)

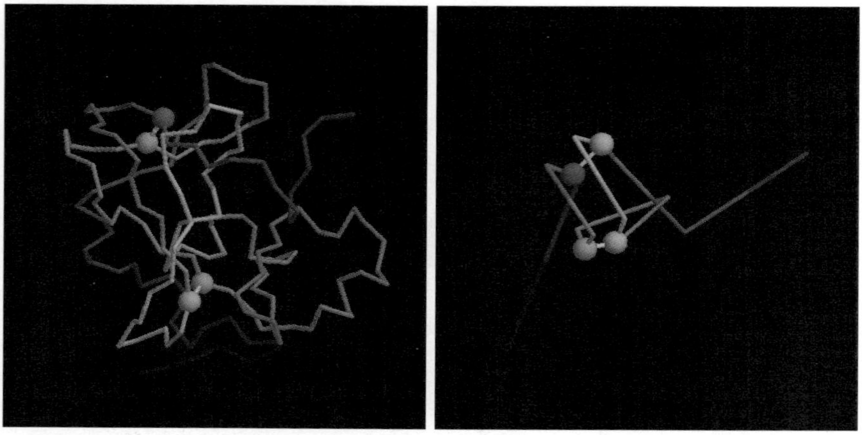

Fig. 9. The modified algorithm was applied with a pair of cross-linked hydrogen bonds as indicated by the linked spheres on the native structure (left). This reduced to a simplified form (right). If the two termini were fixed to a surface, the structure cannot be laid flat without either the chain or the links being crossed.

covalent-bonds. Unlike mathematical knots, the point at which a loop can be said to be a knot becomes a matter of energy, not simply logic. In principle, it would be possible to devise an algorithm to calculate the minimum energy required to undo a cross-linked knot in terms of the number of hydrogen bonds that must be broken, Van der Waals energy lost and even solvent effects.

To clarify the description of knot-like structures the term *threaded-loop* should be retained for un-bonded 'knots' while for the cross-linked knots, the term *covalent-knot* should be used to distinguish them from the true *topological-knot* in which the backbone fold alone is sufficient. This leaves the 'knot' found in the SET domain and, by analogy with the hydrogen bonded knots seen in RNA structures, the term *pseudo-knot* is used.

2.3.3. *Folding covalent and pseudo-knots*

Unlike the topological-knots, there is no mystery about how covalent-knots (or pseudo-knots) might form: they simply require cystine residues (or H-bonding groups) to lie close enough to become cross-linked either during or after folding. Like disulphide bonds in general, their function may be to give extra stability to the fold. This explanation raises the question of whether topological-knots might perform a similar function or are they just 'harmless' tangles that have arisen accidentally? Intriguingly, both deep topological knots considered above are in the catalytic domains of their proteins, with one even running through the active site [10,11]. However, there is little in the overall structure of a knot that can contribute any specific function that could not just as easily be constructed by an un-knotted piece of protein chain: so any advantage from their presence must derive from an indirect or entopic effect.

The distinction between topological knots and knots derived from cross-links, is not simply one of nomenclature but reflects a fundamental difference in the 'gymnastics' that the protein must perform to fold. With cross-links, it is only necessary for a loop to fold across another part of the chain in order to 'trap'. With this event, it is irrelevant how many residues lie before or after the knot site. To form a true topological knot, however, the protein chain must form a loop and then pass one of its termini sufficiently through the loop for it to be counted as a knot. This is clearly a more demanding exercise and would be expected to occur much less frequently.

3. Protein fold complexity

It is clear that there are different degrees of complexity associated with protein folds which can range from a simple fold, (equivalent to lowering a rope onto a flat surface), to a ball-of-string fold where the fold is built up by accretion around a core, to folds where the latter part of the fold must disrupt (or requires the participation of) the earlier part for the structure to be formed. This latter complexity is equivalent to making a "tuck" or for knots, threading the chain through a loop.

Over this hierarchy, can be added the degrees of knottedness discussed above, from the almost knotted (loop-penetration) through H-bonded cross-links (pseudo-knots) to the true topological knots. The latter again constitute a further hierarchy of the type of knot ranked by the number of crossings (trefoil, figure-of-eight, etc.).

3.1. *Topological indices*

Measures of complexity have been expressed in terms of the balance of local and non-local contacts and a measure that captures this is the topological index. Topological indices are abstract mappings of graphs to (real) numbers, while achieving information compression, and have been the subject of intensive study in theoretical chemistry[19]. One application is the management of molecular structure information. For this, unique topological indices are required which give different values even for structural isomers. This allows the shape of a molecule to be correlated with its physical, chemical or biological properties[20,21].

For protein chains, a topological index measures the 'connectedness' of the folded molecule which is expressed through the notion of 'effective chain length' [22]. This approach regards the folded chain as a set of loops (or cycles) held together by several non-covalent connections, thus focusing on the main topological difference between the folded structure and the unfolded chain which contains no cycles. In calculating topological indices, the minimal amino acid distance for a given pair of amino acids is defined as the shortest Euclidean distance $D(X, Y)$ between all possible pairs of non-hydrogen atoms (including the main chain atoms) across the two amino acids X and Y, *i.e.*

$$D(X, Y) = \min_{i,j}(d_{ij}) \tag{2}$$

where d_{ij} is the distance between the i-th atom belonging to amino acid X and the j-th atom belonging to amino acid Y. This distance measure assumes low values for amino acids which are in close contact in the structure and is independent of the side chain sizes. Other amino acid distances, such as the Euclidean distance between the centres of mass of two amino acids or the distances between α-carbon atoms could have been used as well without affecting the results. The minimal amino acid distances were calculated for all possible pairs of sequentially non-adjacent amino acids in a given structure, and those pairs for which this distance was smaller than a threshold were regarded as being connected. The number of connections per amino acid was not restricted.

In a 1D string of consecutively numbered nodes, the i-th node is separated from the j-th node by $j - i \quad (i \leq j)$ arcs. When additional arcs (connections) are added to this string, these will provide shortcuts so that the graph topological distance (the shortest route from node i to node j (measured by the number of arcs traversed) will be $d \leq j - i$. To measure connectedness in a graph, the minimal graph topological distances between all possible pairs of nodes were summed. For a 1D string with N nodes, this sum can be obtained analytically:

$$Conn_{1D} = \sum_{k=1}^{N-1} k(N-k) = \frac{N}{6}(N^2 - 1) \tag{3}$$

For general graphs, first a matrix of minimal topological distances between each pair of nodes was obtained [23], and then the elements of this matrix above the main diagonal were summed (*i.e.* counting each pair only once). Backbone and non-backbone connections were regarded as equivalent when calculating the connectedness index for 3D protein structures. The value of the connectedness index depends on the choice of the cutoff: if the cutoff is very large so that every amino acid is regarded as connected to all the others, then the value of the index for this maximally connected graph will obviously be $\frac{N}{2}(N-1)$, the lowest value possible. On the other hand, it reaches its maximal value when no non-backbone connections are present, as in the case of the corresponding 1D completely unfolded string (Eq. 3). The connectedness numbers for graphs of folded 3D structures should fall between these two limits and a value of 8Ångstroms was found to be best.

To obtain a measure for the degree of foldedness, the notion of the effective chain length was introduced. This is the (usually non-integral) number of nodes in a 1D string which would give the same connectedness index as the 3D graph in question. If the index is denoted as $Conn_{3D}$ and the effective length as N_{eff}, then from the definition and from Eq. 3 the following formula is obtained:

$$Conn_{3D} = \frac{N_{eff}}{6}(N_{eff}^2 - 1) \tag{4}$$

which can be solved for N_{eff}.

3.1.1. *"Tornado" plots*

A convenient way of visualising the non-backbone connections is to plot the connection spans versus the sequential positions (Figure 2). If a connection was made between the i-th and j-th amino acids in the sequence, then a horizontal straight line was drawn from the point (i, s) to the point (j, s) on the graph, s being the span of the connection ($i < j, s = j - i$). Local interactions show up on this plot as short lines at the bottom, whereas long-range interactions are indicated by long lines at the top. Connections with spans less then 5 were not plotted to avoid cluttering at the bottom of the graphs. Loops containing equivalent connections appear as vertical stacks of lines forming roughly equilateral triangles standing on their tips. As these structures are reminiscent of tornadoes, these plots have been called "tornado plots".

By their definition, non-backbone connections are indicative of closely packed regions in the protein structure, therefore tornadoes correspond to locally compact hairpin-like antiparallel substructures. Parallel strands are

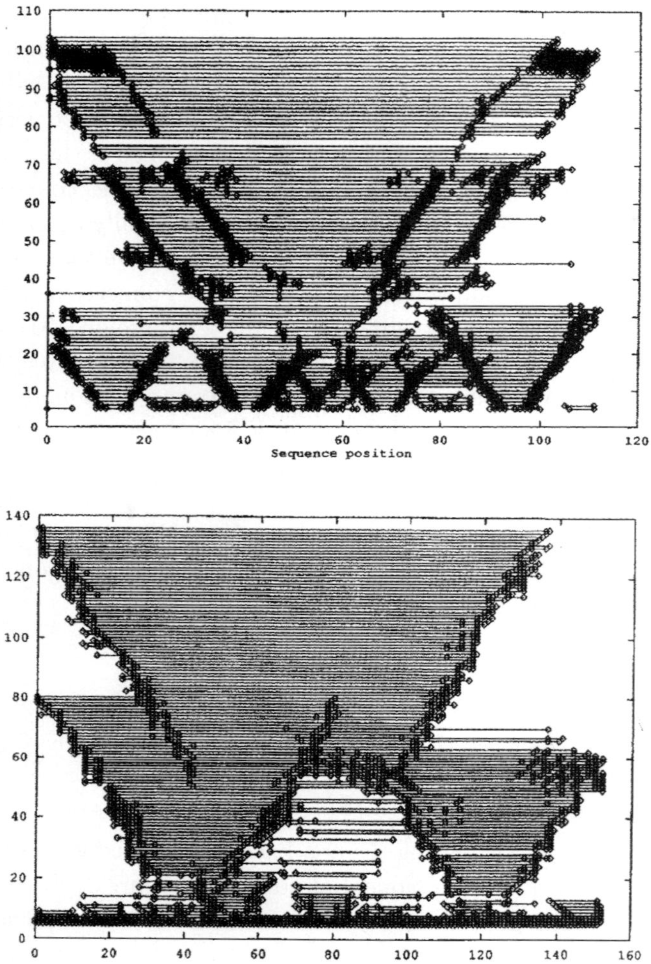

Fig. 10. **Tornado plots** for (Top) an all-*beta* type protein (the immunoglobulin domain 2rhe) and (Below) an all-α type protein (sperm whale myoglobin, 1mbo). Each span connecting two sequence positions (X-axis) is plotted on the Y-axis at a height equal to its length.

implicit in this representation, indicated by the presence of antiparallel motifs necessary for connecting parallel stretches of the chain. Although tornadoes often overlap, the inspection of tornado plots reveals the structural organisation of the molecule in terms of domains and sub-domains (Figure 10).

3.2. *Topological accessibility*

The 'fold' of a ball of string can be reproduced only by starting at the beginning and copying each turn of the string. It would be difficult to start at the outside and add layers of string towards the centre. If the chain must always remain locally packed, then a ball-of-string fold is accessible only from the amino terminus. A fold like the TIM-barrel is accessible from every point (adding in both directions) whereas Rossmann folds are accessible only up to a point before the chain reversal (Figure 11). If a measure of fold complexity is taken as the fraction of residues from which the fold is accessible then the TIM barrel is twice as accessible as the Rossmann fold. This measure seems to capture the intuitive sense of complexity for these folds which are poorly differentiated by other measures of local chain contact.

To apply this approach to real structures (as distinct from topology diagrams) requires some definitions of what constitutes a contact. In the above discussion, β-structure contacts were considered more important than α-structure contacts and in dealing with protein models a 'soft' cutoff is

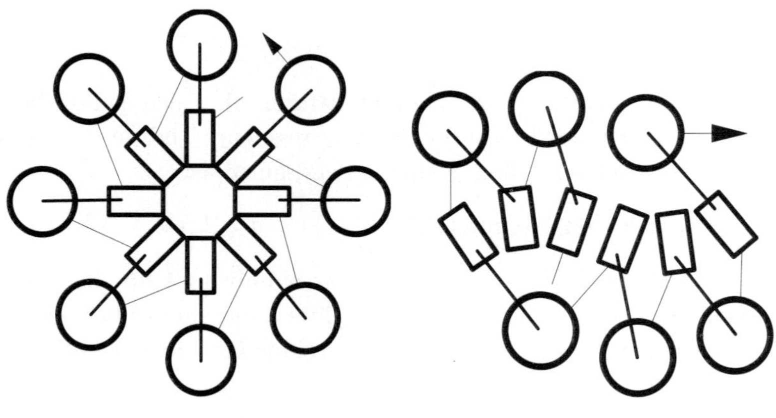

Fig. 11. Folding options for tandem $\beta - \alpha$ units. β-strands are represented by rectangles and α-helices are represented as bold circles. All strands run parallel and progress towards the viewer. (left) With eight units the sheet can form a barrel and the different radii of this circular form at the β and α level accommodate their different size. The barrel structure is found in many (unrelated) enzymes, typified by triosephosphate isomerase (TIM) and is referred to as a *TIM-barrel.* (right) Six units cannot form a barrel forcing an inversion in one half of the sheet to allow helices to be placed both above and below. The resulting arrangement has two-fold symmetry and occurs widely among di-nucleotide binding proteins. It is typified by the dehydrogenases where it is referred to as a *Rossmann fold.*

desirable. These requirements were captured in the following exponential switch function:

$$t_{ij} = 1 - 1/(1 + \exp(r + s_{ij} - d_{ij})), \tag{5}$$

giving a score t for residues i and j which are separated by a distance d_{ij}. The switch point occurs when $d_{ij} = r + s_{ij}$ which is determined by a base level r and a secondary structure component s. If either residues i and j are β then $s_{ij} = 5$; unless, if either i and j are α then $s_{ij} = 10$ and if neither condition applies, then $s_{ij} = 15$. By trial, a value of $r = 3$ was found to be reasonable which means that less than 12% is contributed by $\beta\beta$ interactions over 10Å, α interactions over 15Å and coil interactions over 20Å.

Given these measures, then starting at any residue, a segment can be "grown" by testing amino and carboxy terminal extensions on either side. If the current segment runs from residue n to c (in a chain of length N), then the best interactions are found for the extensions, as:

$$T_{n-1} = \max\{t_{n-1,j}\}, \forall\ n \leq j \leq c), \tag{6}$$

and similarly for the carboxy terminal extensions:

$$T_{c+1} = \max\{t_{c+1,j}\}, \forall\ (n \leq j \leq c), \tag{7}$$

The larger score ($\max\{T_{n-1}, T_{c+1}\}$) then determines which end of the current segment is extended to form the new segment. The process is then repeated. With each extension, a sum is maintained as:

$$S_i = \sum \log(1 + \max\{T_{n-1}, T_{c+1}\}), \forall\ n \geq 1, \forall\ c \leq N. \tag{8}$$

(A sum was taken of the log of the values simply to keep the numbers in a reasonable range). The sum S_i then gives a measure of how good the packing is when extensions are grown from residue i to encompass the whole chain. This value can be plotted for every residue, as shown for the small β/α protein Che-Y (Figure 12) and a larger TIM-barrel (Figure 13).

The S_i values can provide a measure of fold complexity but it is unclear what the best aspect might be. The obvious measure is the sum of S_i over all residues (S_{sum}). By this measure a TIM barrel would have a large sum while a Rossmann (or flavodoxin-like) fold would score less, being accessible only from the N-terminal portion. (Compare the areas below the curves in Figures 12 and 13.) Alternatively, the maximum value of S_i (S_{max}) might be of interest as this gives a measure of how easy it is for the structure to zip-up from an optimal starting point.

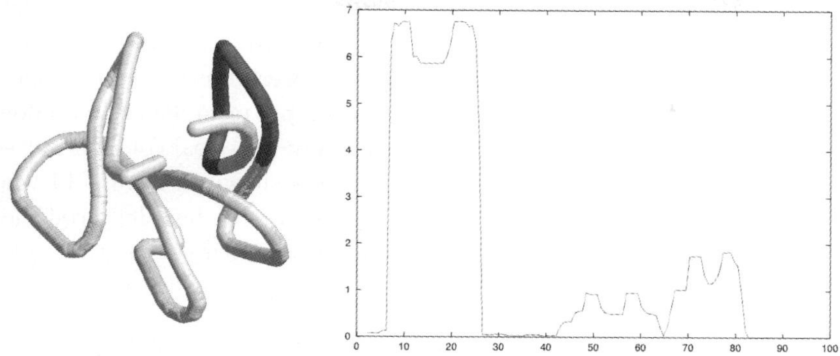

Fig. 12. Topological accessibility for the chemotaxis-Y protein (PDB: `3chy`). It has a Rossman-like fold which is accessible only from the amino terminal segment shaded grey (to the right of the smoothed fold). The values of S_i are plotted (right).

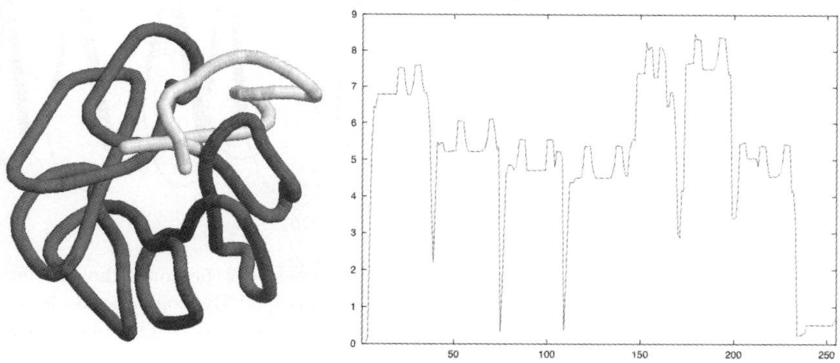

Fig. 13. Topological accessibility for the TIM-barrel protein (PDB: `1egz`) which is accessible from everywhere except a small amino terminal segment and the C-terminal helix. (Shaded grey on the smoothed fold). The values of S_i are plotted (right). The sharp dips in this plot occur where it is difficult to initiate packing in the middle of an extended β-strand.

The value S_{sum} was used as a measure of complexity and calculated for a non-redundant sample of single domain proteins. The simplest folds were highly repetitive structures and, overall, the all-α structures dominated. This may reflect a bias in the balance of the fairly *ad-hoc* values chosen for Equation 5 or may be because the chain is 'compacted' within an α-helix, giving less effective length to construct complex folds. The $\beta\alpha$ class, as might be expected, were completely dominated by TIM-barrels at the top of the list of simple folds. The all-β class, however, were much more complex than either the all-α or the $\beta\alpha$ classes, again possibly because the extended β-strand gives more chain length to 'play' with and create complex folds. Among the simpler all-β folds were the repeated propellor folds and simple barrels in which the chain simply meanders up and down. However complicated all-β folds can have many non-local contacts yet still have a simple underlying basis. An example is shown in Figure 14 where winding a double helix into a cylinder gives rise to the "Jelly-roll" fold, which in terms of non-local contacts, is quite complicated.

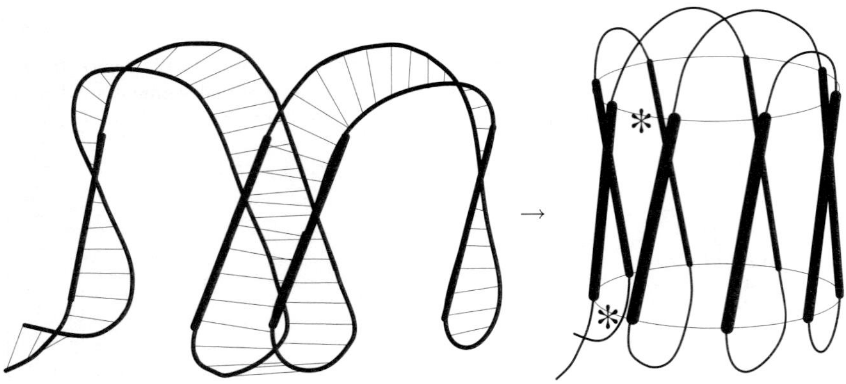

Fig. 14. Representations of an all-β protein (left) emphasising the double helical nature of the chain which may have played a rôle in folding. (right) Hydrogen-bonds between vertical strands creates a cylindrical β-sheet. (Joined at the asterisks). The fold is referred to as a *Jelly Roll.*

4. Random proteins

To this point the methods considered have been applied to the analysis of 'real' proteins. In this section, a different approach will be adopted and the methods discussed will be aimed at making model or 'fake' proteins. The philosophy pursued in this approach is that if a simple algorithm can create 'fake' proteins that reproduce the topological features seen in 'real' proteins, then we will have captured some of the underlying principles. Or in the words of Richard Feynman: "*What I cannot create for myself, I do not fully understand.*"

4.1. *Constructing Semi-random Folds*

4.1.1. *Self-avoiding Random Walks*

Random proteins have been generated many times in the past for testing purposes and a simple algorithm to do this is a self-avoiding random walk from one Cα to the next constrained to lie inside a sphere or ellipsoid [25,26]. The algorithm used here was implemented by selecting the next position in a growing chain to be preferentially in contact with its predecessors. Each residue was added to the chain in a random direction that did not seriously clash with any other. For each accepted position the number of packing contacts were counted and if they fell short of a target number the position was rejected. After a good number of attempts, the target height was successively reduced until an acceptable position was found.

The density of the 'fake' structures are dependent on the number of target neighbours: if this is set to find too few contacts then the resulting structures are not sufficiently compact while if set to find too many neighbours, then the time taken increases. Trials indicated that aiming for four or more neighbours produced compact structures while aiming for only two neighbours was not sufficient. As a compromise, the models described below aimed for three contacts (in addition to chain link to the previous monomer). The resulting structures are remarkably 'life-like', even incorporating elements reminiscent of secondary structure and, if allowed to grow large enough, 'breaking-up' into domain-like regions (Figure 15).

4.1.2. *Secondary structure lattice folds*

The hydrogen-bonded links across a β-sheet impose a layer structure onto the arrangement of secondary structures in a protein domain [27,28]. These layers can consist of either α-structure (packed α-helices) or β-structure (hydrogen-bonded β-sheets). There are seldom more than four layers in any one domain and each layer tends to be exclusively composed of one of these two types of secondary structure. The spacing between the axes of packed α-helices is typically 10Å, as is the spacing between β-sheets and between helices and sheets [29,30], while the spacing between the hydrogen-bonded strands in a sheet is close to 5Å. This makes 10 Ångstroms a convenient unit with which to 'digitise' protein structure onto a secondary structure lattice.

Folds for the β/α class can be generated by enumeration over the secondary structure lattices [30] (Figure 16). Before being realised as full 3D

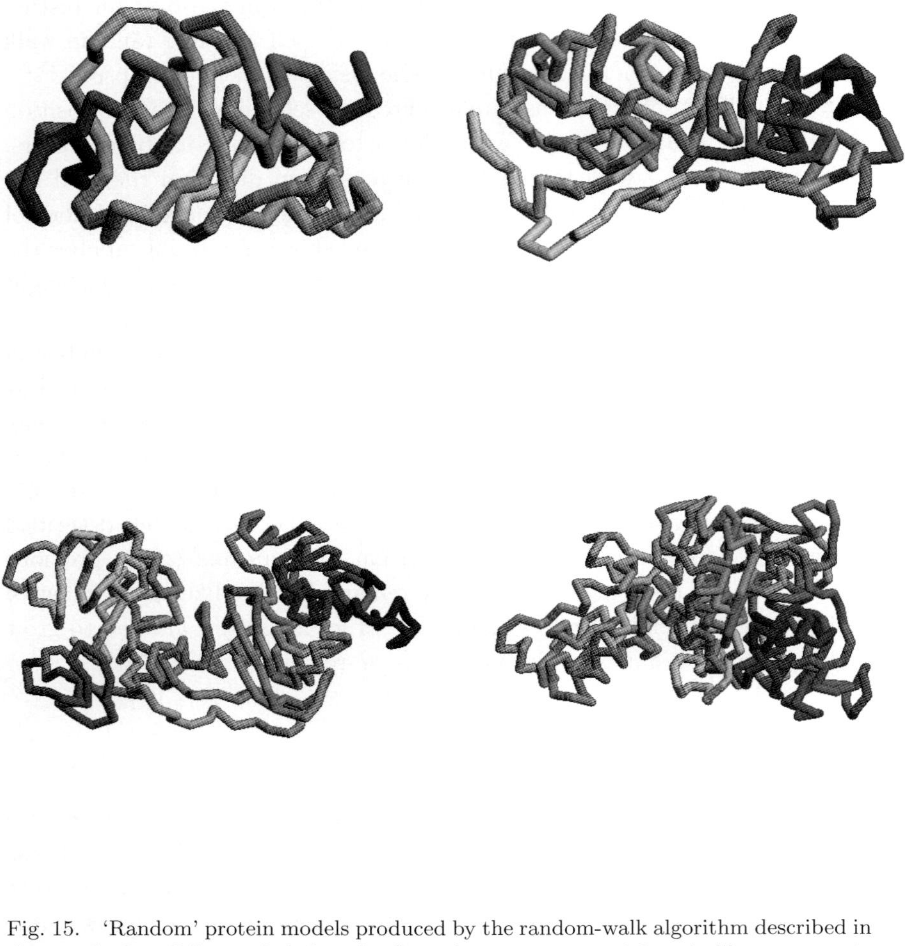

Fig. 15. 'Random' protein models produced by the random-walk algorithm described in the text for four different chain lengths. Secondary structure and domain-like components can be seen.

models, some constraints can be imposed on this class of fold including the handedness of connection of $\beta/\alpha/\beta$-units and the observation that loops rarely cross on the same face. Like the all-α models, these models were then also 'expanded' into Cα-models as described previously [31]. Typical examples are shown in Figure 17. In these models, all α-helices have the correct chirality and β-sheets have the correct twist.

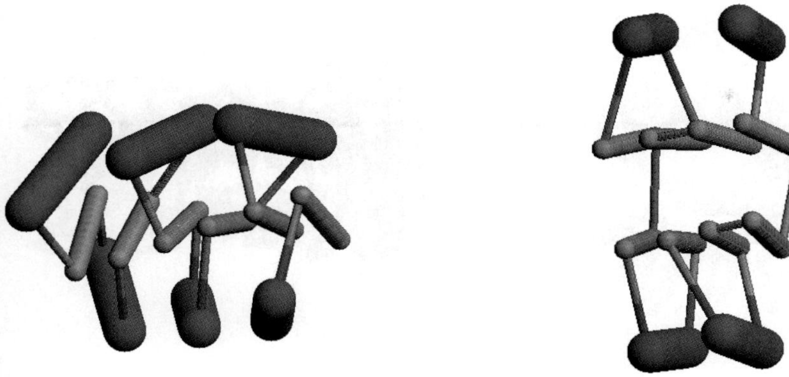

Fig. 16. Secondary structure lattice model. (left) A three-layer ($\alpha/\beta/\alpha$) lattice and (right) a four-layer ($\alpha/\beta/\beta\alpha$) lattice. α-helices are drawn thicker than β-strands. Both lattices can be extended indefinitely at either edge.

4.1.3. *Off-lattice folds*

For the analysis of the folds produced by these two methods, the random walk models are too unstructured and, without regular secondary structures, are difficult to classify. By contrast, the lattice-based folds come ready classified but may be too regular to be truly realistic. An intermediate situation would be more ideal and this can be realised by combining the two approaches as a random-walk constrained over a regular lattice.

The basic random walk algorithm described above can easily be modified to implement this by adding constraints to the conditions for accepting a new residue position. Besides the original packing target, the additional constraint for the new point to lie close to a point on a secondary structure line segment can be imposed. If the line represents an α-helix then the torsion angle with the preceding residues should also be in the range expected for a helix or the range for a strand if the edge is a β-strand.

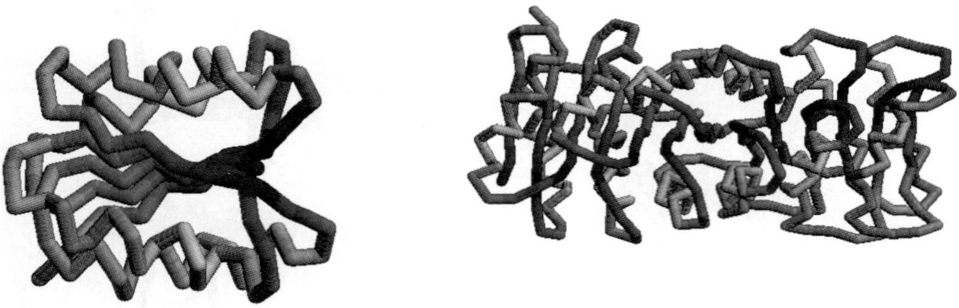

Fig. 17. Lattice-based protein models ranging from a small structure with a 5-stranded sheet (left) to large structures with 13 strands in the sheet (right). The former is viewed across the sheet while the latter is viewed end-on to the sheet.

4.2. *Analysis of semi-random folds*

4.2.1. *Semi-random Walks*

The chain smoothing algorithm of knot detection (2.1) was applied to a collection of structures generated by the semi-random walk algorithm outlined above. These structures, which ranged in length from 50–500 "residues" were constructed with a preference for maintaining local packing. They therefore should all be simple folds of the "dropping a rope on the floor" kind and would not be expected to contain knots. Of the 180 structures examined, none were knotted, with only 10 of the more complicated folds taking more than 100 cycles to reduce.

4.2.2. *Off-lattice folds*

The more protein-like folds generated with the lattice biased walk, while still maintaining a preference for local folding, have a greater propensity for knotting as the chain makes larger excursions through space to connect secondary structure positions. A collection of over 1000 structures based on a 125 residue β/α structure were analysed for knots. Three knotted conformations were found, including two trefoils and a figure-of-eight knot. Like

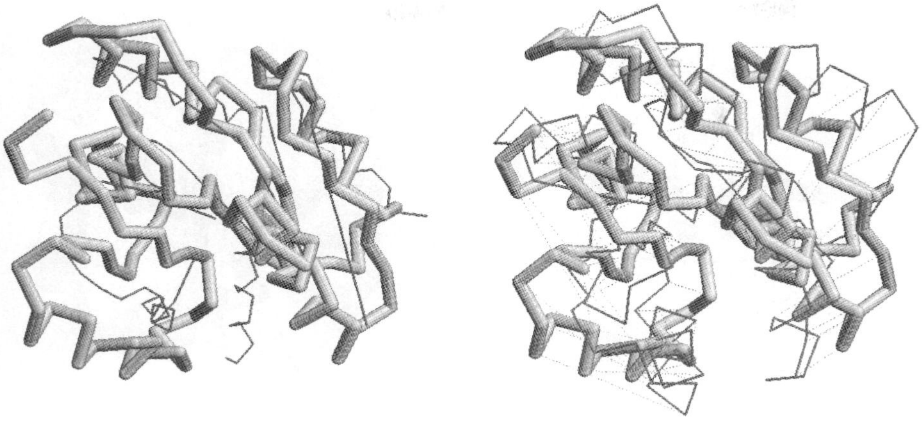

Fig. 18. program semi-random walks were generated for the small protein `3chy` over a secondary structure stick-figure with (left) helices expanded to half-size and (right) helices expanded to full-size. The guide path is shown as a fine line and the model backbone as a tube.

most protein "knots" two of these were threaded by only a few residues but one of the trefoil knots had 8 residues through the knotting loop (Figure 19).

Both in quality and quantity, these knots are similar to those found in real proteins, indicating that the off-lattice folds have a balance between local and non-local packing that is typical of native protein structure. However, when analysed in terms of the idea of topological accessibility (described above), these folds will all be accessible from some point in their structure and so all have a "ball of string" type of fold. They are nevertheless more complex than a typical ball-of-string since they have been wound from both ends. As a sizeable proportion of native proteins are not accessible in this way, these off-lattice models still constitute a restricted set of folds.

4.2.3. *Secondary structure lattice folds*

The folds generated by combinatorial enumeration over a secondary structure lattice are less of an unknown quantity as it is certain that some of the combinations will result in knotted topologies. Some example topologies are shown for a small β/α structure in Figure 20. If it is assumed that two

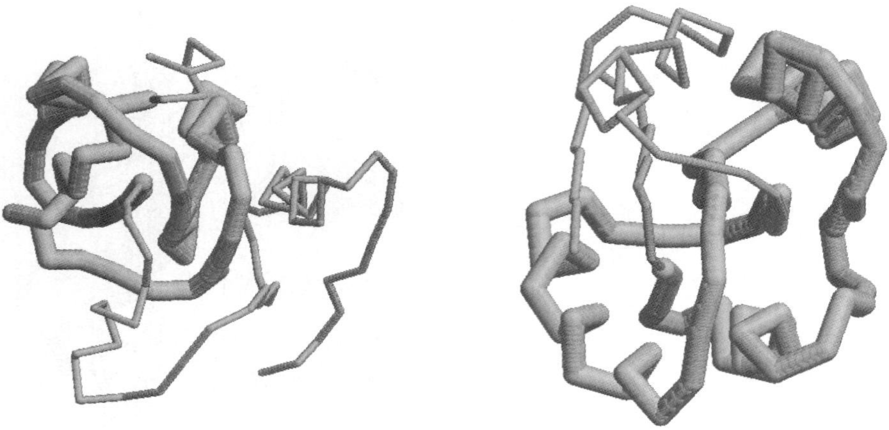

Fig. 19. Knots in "fake" proteins. (left) in a semi-random chain generated as lattice biased walk guided by secondary structure packing. (right) in a protein chain based closely on a secondary structure lattice. The backbone trace is shown with the minimal knotted region drawn thicker.

of the five β-strands (second and last) must lie on the edge of the sheet, then there are only six possible folds yet these exhibit many of the features discussed above, including a topologically inaccessible fold and a knot.

Some more realistic folds were generated for an alternating sequence of five β-strands and four α-helices. A random sample of 500 structures were drawn from the exhaustive enumeration of chain paths over the lattice and analysed for knots. Among these folds, 21 knotted topologies were found. Half of these knots had less than eight points[b] and were almost certainly trefoils. The remainder were then examined to check if they were trefoil knots that did not have sufficient iterations to be reduced to the minimum possible six points. All were found to be trefoils, which with hindsight, could have been anticipated. Although it is possible to construct a figure-of-eight knot on a grid of five β-strands with two α-helices above and below (visiting strand positions in the order 2,5,3,1,4), this topology includes a left-handed connection between positions 5 and 3 (strand 2 and 3 in the

[b]Although a figure-of-eight knot can be constructed with six sticks (seven points) this is not a likely result with the current algorithm. The minimum for a trefoil is six points (five sticks) [5].

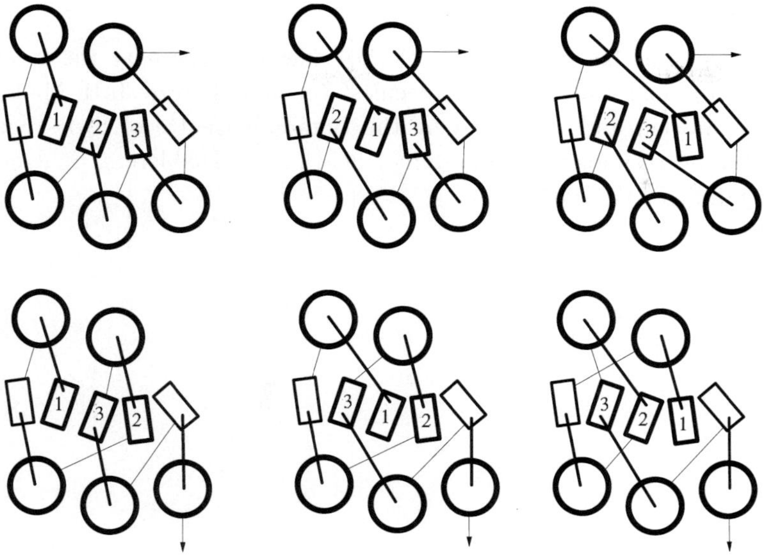

Fig. 20. Some possible folds for a small β/α protein. Labelling the three core strands as 1,2,3, the arrangements shown are: 123, 213, 231; 132, 312, 321. The folds are constructed so as to preserve a right-handed topology of connection between strands. Orders 123 and 213 are found in known proteins (flavodoxin and adenylate kinase), folds 231 and 132 infringe no known constraint (but fold 132 is topologically inaccessible), fold 321 has an unavoidable crossover of connections and fold 312 forms a knot.

sequence) which was forbidden by the rules of connection used to generate the folds.

5. Conclusions

In this chapter, I hope some indication has been given of the topological complexity of protein folds. Although the strict topological interpretation of protein structure provides little of any interest, by relaxing the constraints on the definition of cross-linking and admitting a definition of knots in an open string, some interesting results can be obtained. From a biological perspective, the occurrence of knots probably has little relevance. However, by considering their frequency of occurrence, some constraints may be placed on the ways in which proteins fold and, in particular, on their relative balance of local and non-local contacts along the sequence. Such constraints may prove to be of use in the construction of model proteins.

Acknowledgements

This chapter is based on a talk of the same title given at the meeting: "Knots, random walks and biomolecules" (co-organised by J. H. Maddocks and A. Stasiak and sponsored by the Bernoulli Centre of Swiss Federal Institute of Technology in Lausanne) held in Les Diablerets, Switzerland during July 14-17, 2003.

References

1. M. L. Connolly, I. D. Kuntz, and G. M. Crippen. Linked and threaded loops in proteins. *19*, Biopolymers:1167–1182, 1980.
2. M. L. Mansfield. Are there knots in proteins. *Nature Struc. Biol.*, 1:213–214, 1994.
3. W. R. Taylor. A deeply knotted protein and how it might fold. *Nature*, 406:916–919, 2000.
4. M. L. Mansfield. Fit to be tied. *Nature Struc. Biol.*, 4:116–117, 1997. News and Views.
5. J. A. Calvo, and K. C. Millett, Minimal edge picewise linear knots. In A. Stasiak, V. Katritch and L. H. Kauffman (ed.s), *Ideal Knots*, World Scientific, Singapore, 1998.
6. Pierański, P. In search of ideal knots. In A. Stasiak, V. Katritch and L. H. Kauffman (ed.s), *Ideal Knots*, World Scientific, Singapore, 1998.
7. Pierański, P., Przybyl, S. and Stasiak, A. Tight open knots. *Eur. Phys. J.*, E 6:123–128, 2001.
8. C. C. Adams. *The knot book: an elementary introduction to the mathematical theory of knots.* W. H. Freeman, New York, 1994.
9. W. R. Taylor. Searching for the ideal forms of proteins. *Biochem. Soc. Trans.*, 28:264–269, 2000.
10. O. Nureki, M. Shirouzu, K. Hashimoto, R. Ishitani, T. Terada, M. Tamakoshi, T. Oshima, M. Chijimatsu, K. Takio, D. G. Vassylyev, T. Shibata, Y. Inoue, S. Kuramitsu, and S. Yokoyama. An enzyme with a deep trefoil knot for the active-site architecture. *Acta Cryst., Sect.D*, 58:1129–1139, 2002.
11. G. Michel, V. Sauve, R. Larocque, Y. Li, A. Matte, and M. Cygler. The structure of the RlmB 23S rRNA methyltransferase reveals a new methyltransferase fold with a unique knot. *Structure*, 10:1303–1315, 2002.
12. K. Thomazeau, R. Dumas, F. Halgand, E. Forest, R. Douce, and V. Bi. Structure of spinach acetohydroxyacid isomeroreductase complexed with its product of reaction dihydroxy-methylvalerate, manganese and ADP-ribose. *Acta Cryst.Sect.D*, 56:389–399, 2000.
13. V. Biou, R. Dumas, C Cohen-Addad, R Douce, D. Job, and E. Pebay-Peyroula. The crystal structure of plant acetohydroxy acid isomeroreductase complexed with NADPH, two magnesium ions and a herbicidal transtition state analog at 1.65Å resolution. *EMBO J.*, 16:3405–3415, 1997.
14. C. Liang and K. Mislow. Topological chirality of proteins. *J. Am. Chem. Soc.*, 116:3588–3592, 1994.
15. G. M. Crippen. Topology of globular proteins. *J. Theor. Biol.*, 45:327–338, 1974.
16. M. H. Klapper and I. Z. Klapper. The 'knotting' problem in proteins: loop penetration. *Biochim. Biophys. Acta*, 626:97–105, 1980.
17. W. R. Taylor and K. Lin. A tangled problem. *Nature*, 421:25, 2003.
18. J. R. Wilson, C. Jing, P. A. Walker, S. R. Martin, S. A. Howell, G. M. Blackburn, S. J. Gamblin, and B. Xiao. Crystal structure and functional

analysis of the histone methyltransferase SET7/9. *Cell*, 111:105–115, 2002.
19. M. Randić. In search of structural invariants. *J. Math. Chem.*, 9:97–146, 1992.
20. G. Klopman and R. V. Henderson. A graph theory-based 'expert system' methodology for structure-activity studies. *J. Math. Chem.*, 7:187–216, 1991.
21. M. Mercier, O. Mekenyan, J. E. Dubois, and D. Bonchev. DARC/PELCO and OASIS methods: I. methodological comparison; modeling purine pk_a and antitumor activity. *Eur. J. Med. Chem.*, 26:575–592, 1991.
22. A. Aszódi and W. R. Taylor. Connection topology of proteins. *Comp. App. Bio. Sci.*, 9:523–529, 1993.
23. R. Sedgewick. *Algorithms in C.* Addison-Wesley, 1990.
24. C. Liang and K. Mislow. Topological features of protein structures: knots and links. *J. Am. Chem. Soc.*, 117:4201–4213, 1995.
25. F. E. Cohen and M. J. E. Sternberg. On the prediction of protein structure: the significance of the root-mean-square deviation. *J. Molec. Biol.*, 138:321–333, 1980.
26. J.M. Thornton and B.L. Sibanda. Amino and carboxy-terminal regions in globular proteins. *J. Molec. Biol.*, 167:443–460, 1983.
27. C. Chothia and A. V. Finkelstein. The classification and origins of protein folding patterns. *Ann. Rev. Biochem.*, 59:1007–1039, 1990.
28. A. V. Finkelstein and O. B. Ptitsyn. Why do globular proteins fit the limited set of folding patterns? *Prog. Biophys. Molec. Biol.*, 50:171–190, 1987.
29. F. E. Cohen, M. J. E. Sternberg, and W. R. Taylor. Analysis and prediction of protein β-sheet structures by a combinatorial approach. *Nature*, 285:378–382, 1980.
30. F. E. Cohen, M. J. E. Sternberg, and W. R. Taylor. Analysis and prediction of the packing of α-helices against a β-sheet in the tertiary structure of globular proteins. *J. Molec. Biol.*, 156:821–862, 1982.
31. W. R. Taylor. Towards protein tertiary fold prediction using distance and motif constraints. *Prot. Engng.*, 4:853–870, 1991.

CHAPTER 11

TYING DOWN OPEN KNOTS: A STATISTICAL METHOD FOR IDENTIFYING OPEN KNOTS WITH APPLICATIONS TO PROTEINS

Kenneth C. Millett
Department of Mathematics, University of California, Santa Barbara
Santa Barbara, CA 93016, USA
Email: millett@math.ucsb.edu
Homepage: http://www.math.ucsb.edu/~millett/KM.html

Benjamin M. Sheldon
Department of Mathematics, University of California, Santa Barbara
Santa Barbara, CA 93016, USA
Email: bensheldon@umail.ucsb.edu

A mathematical knot is simply a closed curve in three-space. Classifying open knots, or knots that have not been closed, is a relatively unexplored area of knot theory. In this note, we report on our study of open random walks of varying length, creating a collection of open knots. Following the strategy of Millett, Dobay and Stasiak, an open knot is closed by connecting its two open endpoints to a third point, lying on a large sphere that encloses the random walk deeply within its interior. The resulting polygonal knot can be analyzed and its knot type determined, up to the indeterminacy of standard knot invariants, using the HOMFLY polynomial. With many closure points uniformly distributed on the large sphere, a statistical distribution of knot types is created for each open knot. We use this method to continue the exploration of the knottedness of linear random walks and apply it also to the study of several protein chains. One new feature of this work is the use of an Eckert IV planar projection, preserving area, of the knotting distribution on the sphere to characterize the spatial properties of the distribution.

1. Introduction

Proteins are linear polymeric chains of amino acids connected in a specific sequence and folded into a specific spatial structure. While the linearity of proteins precludes them from forming real (closed) knots, their spatial structure has been shown to contain knotted conformations in a number of distinct methods developed for this purpose. Mansfield[6,7] used a method in which the protein structure is extended from the endpoints to two different points on a sphere that enclosed the protein at a relatively far distance. These two endpoints were then connected by a length of the generically unique great circle on the sphere and the knot type determined using knot invariants. Mansfield performed this operation 100 times for a protein in order to find a dominant knot in the resulting data. Taylor[9] described a continuous deformation of the protein structure in which the ends were held fixed and the rest of the protein is deformed around them until no further smoothing is possible. From the final configuration Taylor expects that the knot type can be identified. Applications of this method on a given protein can, however, yield different results depending upon order in which the deformations take place.[8] Both of these methods have something in common: in defining knots in proteins there is a degree of uncertainty.

The method we used, developed by Millett, Dobay and Stasiak,[8] is similar to Mansfield's as it attempts to find the dominant knot type from multiple closures of the protein structure. In their method a large sphere, relative to the size of the open knot, is created and centered at the center of the smallest sphere that encloses the protein (small-ball). A point is then picked at random with respect to the uniform distribution on this sphere and the two endpoints of the open knot are connected to this point, effectively closing the knot as shown in Figure 1.

The resulting knot, now closed, is analyzed by computing the HOMFLY polynomial.[4,3] When multiple random points are chosen on the sphere, a spherical distribution of knot types is thereby created that is associated to the single open knot. These distributions provide the information, i.e. the knotting spectrum of the configuration, that is used to identify the dominant knot type of the configuration. In this study, we have first tested the method on sets of 1000 random walks, each set consisting of walks of a specific length. This analysis provides a some insight into the structure of knotting. We wish, however, to have a deeper spatial sense of the distribution of knot types as they occur on the surface of the sphere containing the closure points. In order to capture the relative proportion and the spatial distribution, we selected the Eckert IV equal area projection of the sphere on which we have coded the knot types by color.

The analysis of the data has provided some insight into the spatial structure of the random walks and how this structure is reflected in the

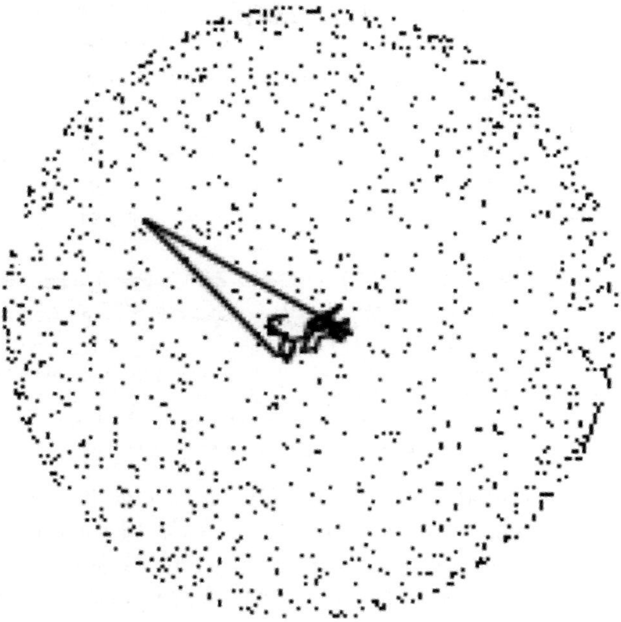

Fig. 1. A random walk of length 50 is closed to a random point on an enclosing sphere, generating a unique knot type.

knotting spectrum. In addition, we applied this method to eight proteins, those studied by Taylor, in order to test the extent of agreement between the two methods. In all cases, except the instance of 1kopA, the results are in agreement. In 1kopA,[1] Taylor[9] finds a figure-eight knot while we find no dominant type but, rather, a balance between the unknot and the trefoil knot.

2. Random Walks and Knotting

Before analyzing the knottedness of protein chains we extended that initial work of Millett, Dobay, and Stasiak[8] by applying the method to sets of 1000 classical random walks composed of uncorrelated equilateral segments with varying numbers of steps (50, 100, 150, 200, 250, and 300 steps). For each linear random walk in a set, the smallest enclosing sphere was determined and 10,000 random points on the large enclosing sphere were chosen

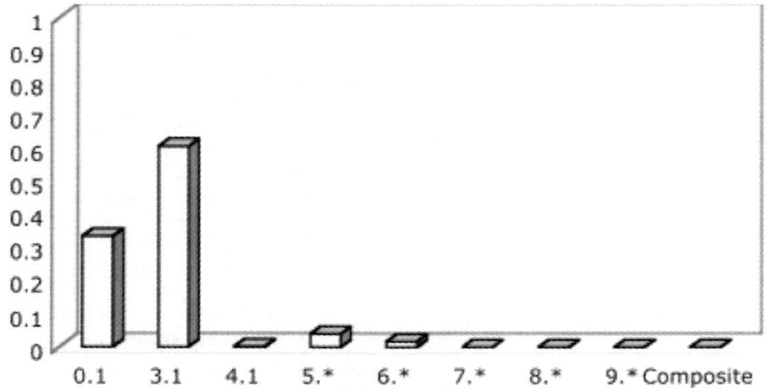

Fig. 2. The knot spectrum associated to a single linear random walk of 50 steps. 5* denotes the collection of knots with 5 crossings. Similarly for 6*, etc.

and analyzed via the HOMFLY knot polynomial invariant computation.[3] Typically an open knot will produce different closed knot types depending upon the choice of closure point on the sphere. In some regions, this knot type will be the same as the end-to-end closure of the configuration. As the choice of closure point moves about the sphere, the two closing segments will pass through segments of the linear random walk. Such passages may or may not have the effect of changing the closed topological knot type of a specified open knot.

The first task is translating the 10,000 point knotting distribution, an approximation of the continuous distribution, into a histogram representing the knot spectrum of the knot. Of great interest is not only determining the dominant knot type in the spectrum, but identifying a standard by which a certain knot type should be considered as dominant and, therefore, considered as the knot type associated to the open walk. One standard, "strong dominance", could be to require its occurrence in over 90% of the closures while, alternatively, one could require that its occurrence be twice as large as its nearest competitor, described as "dominance". Similarly, one could require that the knot type appear in 50% or more instances, giving "weak dominance". In Figure 2 we show the spectrum associated to a single linear random walk having 50 steps. In Figure 3 (Color Plate 1) we have collected 1,000 such spectra to give a sense of the presence of knots in such linear random walks. Figures 3 through 6 provide visual evidence of the increasing complexity of the spectra as the number steps in the random walk increases from 50 through 300 steps.

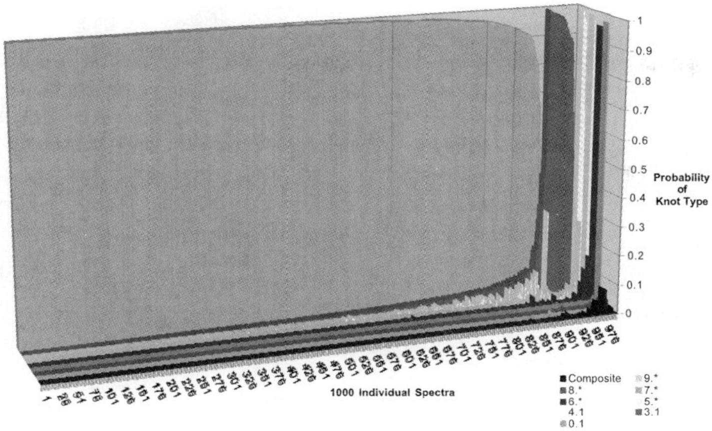

Fig. 3. The collected knot spectra for 1000 random walks of length 50.

For random walks of length 50, 95.9% (959 out of 1000) showed a certain knot type for over 90% of its closures, see Figure 3 (Color Plate 1). Of those, 32 had a single closure type with, therefore, a probability of one. All of these closures were the unknot. Applying the "weak dominance" criteria, one finds that in 99.1% of the analyzed random walks showed a dominant knot type that was formed in more than 50% of the random closures for each individual random walk. In a walk of this relatively short length, 50 steps, the dominant knot type appears to be easy to recognize. For longer walks one expects a higher degree of uncertainty as to the dominant type. For a random walk of length 300, see Figure 8 (Color Plate 6), only 86.1% (861 out of 1000) showed a certain knot type for over 90% of its closures. In this case, no knots had a probability of one but 18.6% had a knot type of probability 0.99 or more. Out of 1000 cases, only 4 random walks of length 300, 0.4%, had no knot type appearing in more than 50% of its closures. This provides concrete evidence that while longer walks produce greater complexity in determining a dominant knot type, this method has significant identification power at scales of interest in the natural sciences.

We have explored the relationship between the spatial character of the linear random walks and the observed knot types. For example, one of these spatial properties is the end-to-end length, i.e. the distance between the start of the walk and its end. One knows that the average end-to-end length is proportional to the square root of the length of the walk. There are

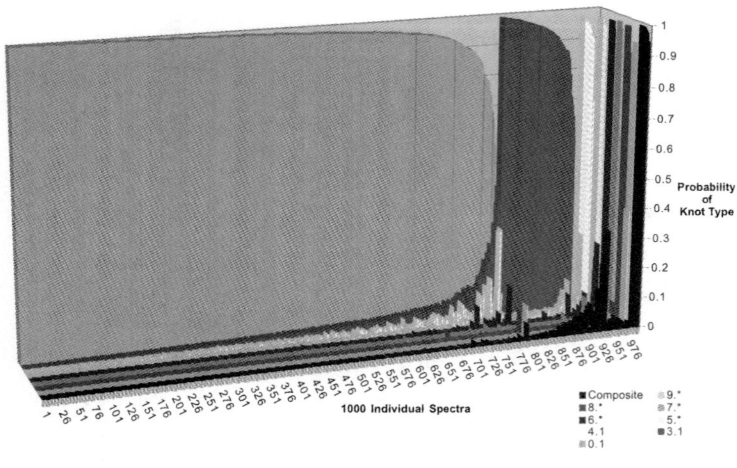

Fig. 4. Knot spectra for 1000 random walks of length 100.

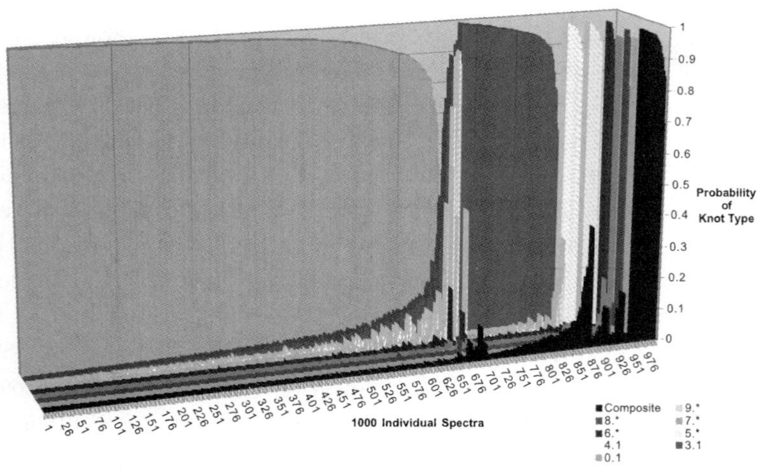

Fig. 5. Knot spectra for 1000 random walks of length 150.

statistically very few instances of random walks with extreme end-to-end lengths (either very short or very long) in relation to the number of steps in the random walk, see Figure 9. In either extreme, the distribution of knot types is more likely to have a strongly dominant knot. For small end-to-end lengths, this will be the knot type of the direct closure, and for large end-to-end lengths, one expects the trivial knot to dominate. Indeed, behavior of

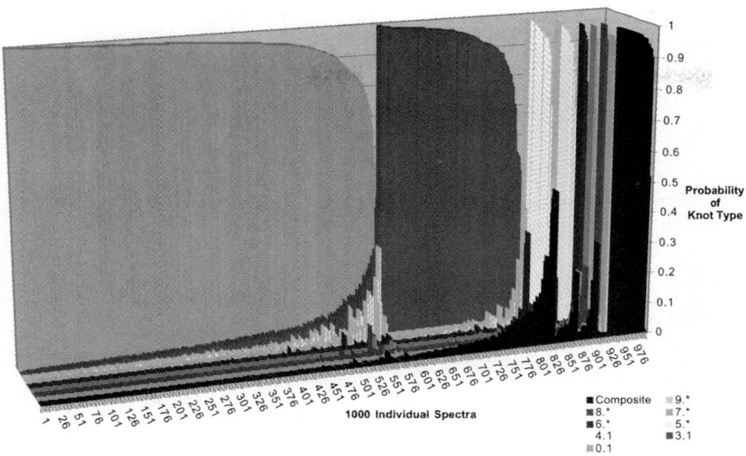

Fig. 6. Knot spectra for 1000 random walks of length 200.

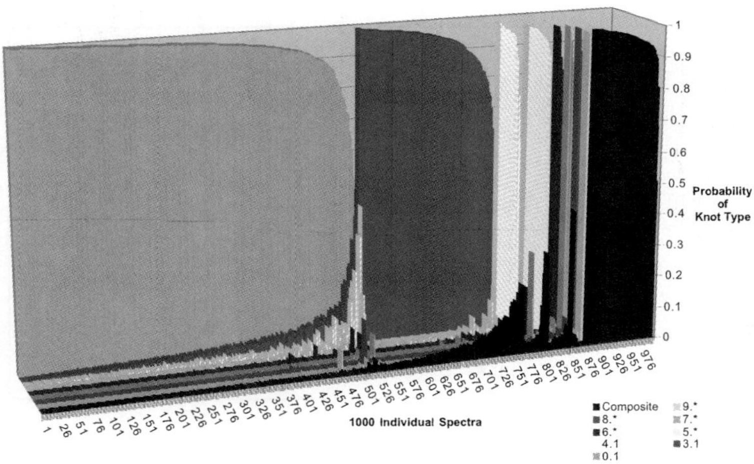

Fig. 7. Knot spectra for 1000 random walks of length 250.

this character is reflected in the data. For example, consider the case of 50 step linear random walks shown in Figure 10. The relationship between the end-to-end length, d, of a random walk of length L and the complexity of the knotting, as measured by the number of knot types observed, N(d), seems to take the form of a power function: $N(d) = Constant * L/\sqrt{d}$. While this functional relationship is surprisingly simple, we believe that it warrants

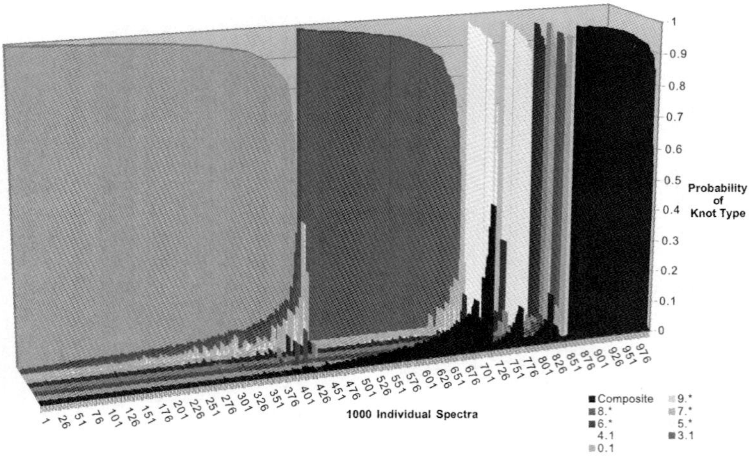

Fig. 8. Knot spectra for 1000 random walks of length 300.

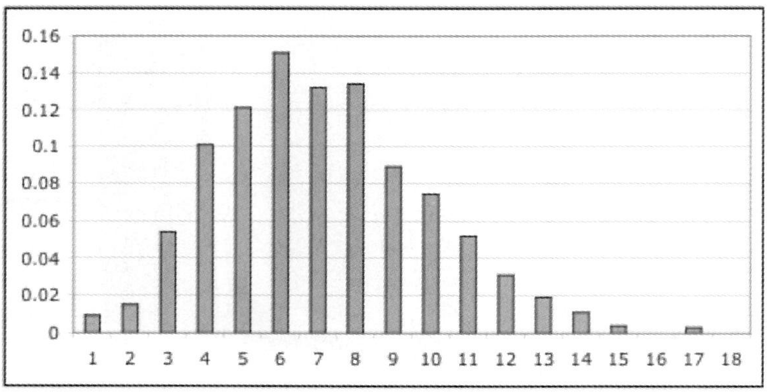

Fig. 9. The probability distribution of end-to-end lengths for 1000 random walks of length 50

further investigation in view of data developed in this project. What it tells one is certainly quite intuitive: the shorter the end-to-end length, the more knot types will appear up to a certain threshold point at which time most of the knots will be of the type determined by the end-to-end closure.

Another possibly interesting spatial characteristic is the compactness of the linear random walk. We calculated the ratio of the diameter of the smallest ball containing the walk and its end-to-end length distance. A

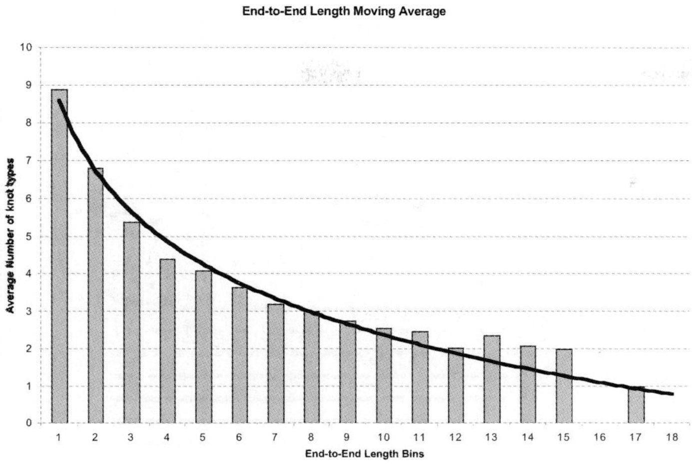

Fig. 10. The distribution of knot types as a function of the end-to-end distance for 1000 random walks of 150 steps. This function, the number of knot types, appears to be a simple function of the end-to-end distance, d, and the number of steps in the walk, L: $Constant * L/\sqrt{d}$.

smaller ratio signals a relatively compact walk. The general trend observed was that the more compact the random walk, the greater number of distinct knot types it would likely contain. See Figure 11.

We also looked at the distribution of the probabilities of the unknot knot. We observed that the distribution of the unknot in random walks was strongly bimodal. For a random walk, it is most likely that either almost all closures of a linear random walk are unknots, or almost none of the closures are unknots. For example, see Figure 12.

3. Visualization of the Knotting Spectrum

The principal objective of this project is to develop a method to visualize the distribution of knotting that occurs with the spherical closure of the linear knot under investigation. To do so, we graphically reproduce the sphere and indicate the type of the knot closure at each point by means of a color assigned to each point on the sphere depending upon the type of knot created by closure from that point The resulting image provides a richer picture of the knotting spectrum of the random walk. This procedure determines, on the closure sphere, a collection of regions areas defined by the equally colored points within them. The regions of a given color may be disconnected or not simply connected, as a given knot type may show

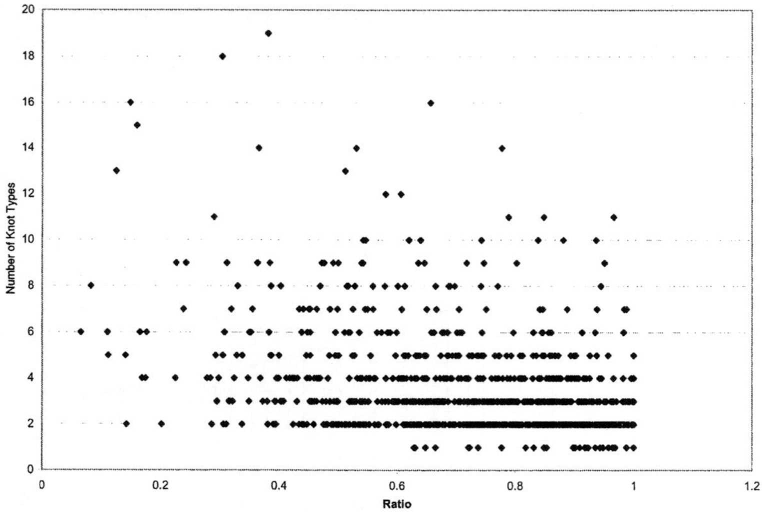

Fig. 11. A scatter plot of the number of knot types versus the ratio of the end-to-end length to the small ball radius for 1000 random walks of length 150.

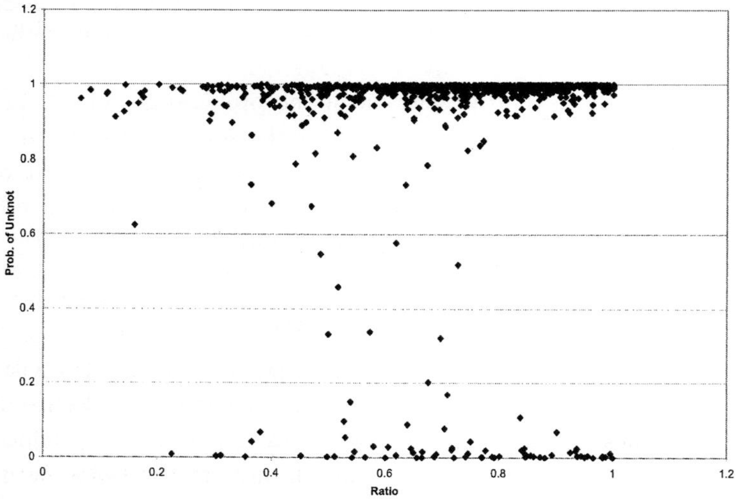

Fig. 12. The bimodal distribution of trivial knots versus the ratio of the end-to-end length to the small ball diameter for 1000 random walks of length 150.

up in distant regions of the sphere.

Key to attaining our principal objective is to employ an area preserv-

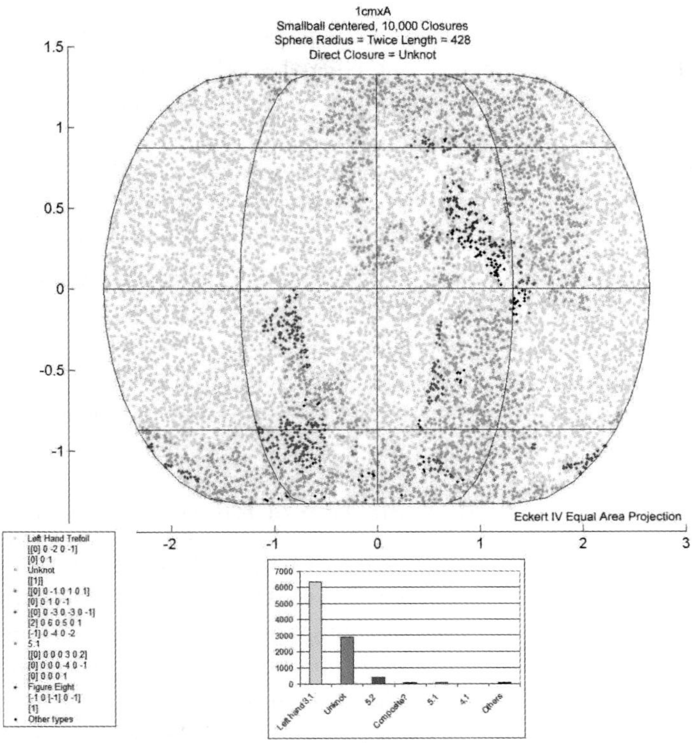

Fig. 13. An Eckert IV projection of the spectrum of the protein 1cmxA

ing planar representation of the sphere so as to link the statistics of the knot spectrum with the visual qualities of the image. The Eckert IV equal area projection[10] of the closure sphere appears to be an excellent vehicle to accomplish this purpose. This is a pseudocylindrical projection that is non-conformal, but presents the area with minimal distortion, one of its most important properties. See Figure 13 (Color Plate 7).[2] Using this projection greatly simplifies the analysis of a sphere distribution and an equal area projection was chosen in order to visually reproduce the statistical prevalence of each knot type: a knot type occurring in 30% of closures will have an area equal to 30% of the projection of the sphere. In view of its finer presentation of the knotting spectrum of the random walk, we call this the spectral sphere.

We believe that the local and global structure of the regions associated to the knot types presented in the spectral sphere are potential reflections of the spatial properties of the random walk. From a topological perspective, generically, the borders between two regions on the spectral sphere's knot types represent the passage of one edge through another. Such a passage can result in the change of the knot type but may knot always do so. In the spectral sphere we only record those instances in which there is a change in the knot type. Typically, such a change is reflected by a change of exactly one crossing in the minimal knot diagram of one of the associated knots. This may lead to substantial simplification of the diagram of the subsequent knot showing extensive global topological change between two closures. Cases in which two contiguous knot types in the spectral sphere can only be connected by a topological process involving more than one crossing change provide evidence of an insufficient resolution of closures necessary to accurately render the detail of the local properties.

4. Applications to the Identification of Knotting in Proteins

While we tested the spectral sphere method of analysis on linear random walk, the purpose for its development was its application to protein structure. The existence of classical knots in cyclic DNA, forming a closed path, is already well established. Knots in protein molecules have been studied but by taking into consideration the entire network of the protein,[5] by simplifying the the structure,[9] or by a closure method.[6,7] To test our spectral sphere method, we selected the 8 protein segments which had been studied by Taylor[9] from the Protein Data Bank.[1] This data consists of coordinates of the alpha-carbons in the backbone of the selected proteins. When the number of carbon atoms was too large for our HOMFLY programs, e.g. 1yve, 1znc, and 2btv, it was necessary to reduce the number of points by taking every other carbon atom in the backbone. In these cases, the data for both the full and reduced protein is shown when possible, Table 1. The case of $1yve_{256}$ is included as we were unable to get good data for $1yve_{513}$ (where the second number is the number of carbon atoms analyzed). We call attention to the fact that there are significant differences in the results between these two cases. For that reason, we are unable to draw any conclusions from the spectral calculations for those examples. Due to the computational constraints arising in our HOMFLY programs, we are limited to under 1,000 crossings is the knot projections. As a consequence, we were unable to apply our method to $2btvB_{885}$ because this data produced too many crossings. Fortunately, taking every other carbon atom did produce useable data.

Table 1. The analysis of seven protein segments, labeled with the number of carbon atoms employed. The dominant knot is **bold** and the direct closure is indicated by *italics. In the table, SMD is small ball diameter, EEL is end-to-end length, Ratio is EEL/SMD, EPsn is the termini distance from the small ball center divided by the small ball radius, Types is the number of distinct knot types observed, and the remaining data gives the proportion of knots of the indicated type.*

Protein	*SMD*	*EEL*	*Ratio*	*EPsn*	*Types*	0_1	3_1	4_1	5_1	5_2	6_+
$1cmxA_{214}$	74.6	33.4	0.45	0.48	13	*0.29*	**0.64**	0.00	0.01	0.04	0.02
$1dmxA_{237}$	74.9	30.1	0.40	0.68	11	*0.25*	**0.65**	0.02	0.05	0.01	0.02
$1fugA_{383}$	66.7	29.3	0.44	0.91	10	0.11	***0.85***	0.01	0.01	0.01	0.01
$1hcb_{248}$	63.6	41.4	0.65	0.70	11	0.28	**0.60**	*0.02*	0.07	0.02	0.02
$1kopA_{223}$	48.1	36.9	0.77	0.82	10	***0.45***	**0.45**	0.01	0.06	0.02	0.01
$1yve_{256}$	68.8	43.2	0.63	0.64	14	0.09	0.01	***0.66***	0.00	0.00	0.23
$1zncA_{262}$	55.1	35.9	0.65	0.65	14	*0.35*	**0.51**	0.02	0.06	0.02	0.04
$2btvB_{442}$	260.7	29.9	0.11	0.92	21	0.22	0.02	***0.64***	0.00	0.00	0.12

We note that, in the protein data, the expected termini behavior was not uniformly analogous to that observed in the linear random walks on which we tested our methods. To explore the consequences of this difference, we calculated the distance of the termini from the small-balls center and divided it by the small-balls radius. This provides one measure by which we can quantify the relative positon of the termini of the protein. A smaller ratio suggests that the termini maybe closer to the center of the protein mass. For the proteins investigated in this project (excluding 1yve) the average ratio was 0.751. Anticipating that the termini would lie close to the surface of the small ball enclosing the protein,[7] this number is smaller than expected with 1cmxA having the lowest ratio of 0.479. The consequences this might have upon the observation of knotting encountered in the protein may be worthy of further investigation.

The types of knots found through our analysis were not surprising in view of previous analysis of the presence of knotting in these proteins.[6,7,9] The dominant knots are relatively simple ones: unknot, trefoils and figure-eight. Only one protein (1fugA) had a dominant knot type above 80%.

1kopA is especially interesting because it has two essentially equally likely knot types, the unknot and the right trefoil, whereas Taylor[9] found the right trefoil. In all other cases we find agreement with Taylor's results. We note, however, that in only two of the eight cases did the direct end-to-end closure result in the same knot type.

5. Conclusions and Speculations

This project principally consisted of an effort in numerical analysis, the development of a spectral sphere data visualization method, applied to linear random walks and, ultimately, to eight proteins. Our findings strengthen the proposal that certain properties of classical closed knots can be identified and quantified in open knots. The relationship between end-to-end length and the number of knots in a random walk was experimentally discovered and, we believe, could be explained by theoretical mathematical analysis. An observation that is attractive for further investigation is the bimodal nature of distributions of the unknot; it suggests the exploration of this property for other knot types. An analytical proof of this would provide a strong foundation on which to build a theoretical understanding of the knotting spectrum of an open chain.

We suggest that the geometry of the spectral sphere is a rich arena for further research. The mechanics of changes in knot type and crossings of an open knot is closely related to the structure of the boundaries and connectivity of the knot regions on the spectral sphere. What is the fractal dimension of the boundary of the knotting regions on the sphere? How is the geometry of the knotting regions reflected in the spatial structure of the open knot or in the properties of the protein?

While it may appear that our explorations have raised more questions than it has answered, we believe that it has provided new evidence that this method of analysis of the presence of knotting in open chains is a powerful and potentially important new approach to a quantitative measure of knotting.

References

1. H. M. Berman, J. Westbrook, Z. Feng, G. Gilliland, T. N. Bhat, H. Weissig, I. N. Shindyalov, and P. E. Bourne. The protein data bank. *Nucleic Acids Rersearch*, 28:235–242, 2000.
2. G. I. Evendem. Cartographic release procedures. *Rel. 4, Second Int. Report, USGS*, 2003.

3. Bruce Ewing and Kenneth C. Millett. Computational algorithms and the complexity of link polynomials. In *Progress in knot theory and related topics*, pages 51–68. Hermann, Paris, 1997.
4. P. Freyd, D. Yetter, J. Hoste, W. B. R. Lickorish, K. Millett, and A. Ocneanu. A new polynomial invariant of knots and links. *Bull. Amer. Math. Soc. (N.S.)*, 12(2):239–246, 1985.
5. C. Liang and K. Mislow. Topological chirality of proteins. *J. Am. Chem. Soc*, 116(1):3588–3592, 1994.
6. Marc L. Mansfield. Are their knots in proteins? *Structural Biology*, 41:213, 1994.
7. Marc L. Mansfield. Fit to be tied. *Structural Biology*, 43:166, 1997.
8. Kenneth C. Millett, Akos Dobay, and Andrzej Stasiak. Linear random knots and their scaling behavior. *Macromolecules*, 38(2):601–606, 2005.
9. W. R. Taylor. A deeply knotted protein structure and how it might fold. *Nature*, 406:916–919, 2000.
10. Eric W. Weisstein. Eckert iv projection. *From MathWorld–A Wolfram Web Resource: http://mathworld.wolfram.com/EckertIVProjection.html*, 2004.

CHAPTER 12

SCALING OF THE AVERAGE CROSSING NUMBER IN EQUILATERAL RANDOM WALKS, KNOTS AND PROTEINS

Akos Dobay, Jacques Dubochet and Andrzej Stasiak
Laboratory of Ultrastructural Analysis, University of Lausanne
Lausanne, CH 1015, Switzerland
E-mail: Andrzej.Stasiak@unil.ch

Yuanan Diao
Department of Mathematics, University of North Carolina at Charlotte
Charlotte, NC 28223, USA
E-mail: ydiao@uncc.edu

We compare here the scaling behaviour of the mean average crossing number $\langle \text{ACN} \rangle$ of equilateral random walks in linear and closed form with the corresponding scaling of natural protein structures. We have shown recently that the scaling of $\langle \text{ACN} \rangle$ of equilateral random walks of length n follows the relation $\langle \text{ACN} \rangle = \frac{3}{16} n \ln n + bn$ and that a similar result holds for equilateral random polygons [14]. Furthermore, our earlier numerical studies indicated that when random polygons of length n are divided into individual knot types, the $\langle \text{ACN}(\mathcal{K}) \rangle$ for each knot type $\mathcal{K}$ can be described by a function of the form $\langle \text{ACN}(\mathcal{K}) \rangle = a(n - n_0) \ln(n - n_0) + b(n - n_0) + c$ where a, b and c are constants depending on $\mathcal{K}$ and n_0 is the minimal number of segments required to form $\mathcal{K}$ [14]. Here we analyze in addition natural protein structures and observe that the relation $\langle \text{ACN} \rangle = \frac{3}{16} n \ln n + bn$ also describes accurately the scaling of $\langle \text{ACN} \rangle$ of protein backbones.

1. Introduction

Random walks are frequently used to model the behaviour of polymers at thermodynamic equilibrium [16,17,19]. The simplest but also the most fundamental type of random walks is represented by chains composed of freely jointed segments of equal length (equilateral) where the individual segments have no thickness. Such random walks are known as ideal random walks and are used to model the behaviour of polymers under so-called theta conditions where polymer segments neither attract nor repel each other [27]. The behaviour of ideal random walks is thoroughly researched and it is well established, for example, that such measures of overall dimensions of ideal random walks like the average end to end distance or the average radius of gyration scale with the number of segments n as n^ν where $\nu = 0.5$ [16,17,19]. Although the overall dimensions provide important information about the modelled polymers, frequently additional characteristics of polymers are of interest. Among these characteristics are such that can tell whether a polymer is knotted or such that can measure the extent of polymer entanglement. In the case of circular polymers their knot type can be rigorously determined [1,24]. However, the information about the knot type tells us only the minimal crossing number that one could see upon elimination of all nugatory crossings. The determination of the knot type does not tell us, for example, how many nugatory crossings there are in the analyzed polymer trajectories. In addition, in the case of linear polymers it is difficult to define their knotedness [25,32]. There are no difficulties, however, with defining a directional crossing number that corresponds to the perceived number of crossings that can be observed on an orthogonal projection of a given non-perturbed random walk in open or cyclised form. To be independent of the choice of a particular projection, the average crossing number ACN is defined as the average of directional crossing numbers over all possible orthogonal projections of a given rigid or momentary configuration. Here we will consider $\langle \mathrm{ACN} \rangle$, i.e. the average of ACN over the whole statistical ensemble of ideal random walks (or polygons) with a given number of segments. Of course $\langle \mathrm{ACN} \rangle$ corresponds also to the time averaged ACN of randomly fluctuating polymer in solution. $\langle \mathrm{ACN} \rangle$ provides an interesting measure of physical behaviour of knotted polymers since it shows a strong correlation with the experimentally observed speed of electrophoretic migration of knotted DNA molecules of the same size but of various knot types [30]. Furthermore, $\langle \mathrm{ACN} \rangle$ correlates well with the expected sedimentation coefficient of different types of DNA knots formed on the same size DNA molecules [33]. In the case of protein chains, $\langle \mathrm{ACN} \rangle$ provides an interesting measure of their compactness [4] and several studies were devoted to investigate how $\langle \mathrm{ACN} \rangle$ in proteins scales with the length of polypeptide chain [2,3,5,6,7,20]. Another scaling aspect of ACN was discussed in the case of ideal geometric representations of knots [8,9,10,11,22]. In this paper we compare the scaling of $\langle \mathrm{ACN} \rangle$ in random walks in open and closed forms with $\langle \mathrm{ACN} \rangle$ scaling determined for protein backbones.

2. Simulation methods

The average crossing number of a random walk or polygon W can be calculated by the modified Gauss formula

$$\frac{1}{4\pi}\int_W\int_W \frac{|(\dot{\gamma}(t),\dot{\gamma}(s),\gamma(t)-\gamma(s))|}{|\gamma_(t)-\gamma(s)|^3}\,dt\,ds \qquad (1)$$

where γ is an arclength parameterization of W. However, the application of this formula leads to problems when the analyzed configurations have some non-consecutive segments very close to each other. For this reason our numerical determination for the ACN of simulated random chains or of protein backbones is based on detecting the number of crossing points in numerous projections of the analyzed configurations. We calculated the number of crossings in individual projections and then averaged over 50 randomly chosen directions of projections in case of random walks for which we had large statistical samples (10^5 configurations per analysed chain size) and over 1000 randomly chosen directions in case of analysed protein stuctures. To generate random equilateral walks (open walks) we first created a set of random unit vectors with the same origin that uniformly sampled the surface of the unit sphere. The vectors were then joined sequentially while maintaining their original directions. To generate random equilateral polygons (closed walks) we followed the approach of Klenin *et al.* [23]. To construct a 100 segment long random polygon, for example, we first create a set of 50 random unit vectors uniformly sampling the surface of a unit sphere. Subsequently, we add to this set another 50 unit vectors that are opposite to the original set. This procedure assures that the sum of the 100 vectors is zero and that the trajectory obtained by any random permutation of sequential joining of all 100 vectors will be always closed. To eliminate the correlation between anti-parallel vectors the entire set of vectors was de-correlated by multiple rotations of randomly chosen pairs of vectors around their respective sum vectors. The number of independent rotations was 10-times bigger than the number of segments in a given random equilateral walk. Finally, all 100 randomized vectors were sequentially joined to create an equilateral random polygon. Knot types of the resulting random polygons were recognized by calculation of their HOMFLY polynomials [18]. The trajectories traced by backbones of proteins were entered as segmented chains where each segment corresponded to the distance between sequential α-Carbons in a given protein structure taken from protein structure databank (PDB). For good statistics, it is important to analyze a maximal number of "independent" proteins and eliminate from the data set all "duplicates", i.e. more than one entries defining the same protein structure under somewhat different crystallization conditions or a very similar protein structure from another organism [31]. The set of 2230 filtered and thus "independent" protein trajectories was kindly provided by William Taylor.

3. Results

3.1. ⟨ACN⟩ *scaling in linear and closed random walks*

Figure 1(A) shows the ⟨ACN⟩ values obtained in numerical simulations of ideal random walks in a linear (open) form and in a closed form. We have analyzed walks with up to 1000 segments and each of the ⟨ACN⟩ data points was obtained by averaging the ACN values from 10^5 independent random configurations of open or closed random walks of the corresponding size. To fit the data points we have used the analytically predicted scaling function $an \ln n + bn$ [14], leaving the two parameters a and b free. In both cases (closed and open walks) we have obtained an excellent fit (with the correlation coefficient $R = 1$) where the prefactor a was practically equal to the analytically predicted 3/16 ($3/16 \pm 0.0007$ and $3/16 \pm 0.001$, respectively). Therefore, we have proceeded with another fit where the prefactor a was set to 3/16, leaving just one free parameter b. These fits are presented in Figure 1 and it is visible that the fitted functions pass almost perfectly through all the data points. These results therefore confirm earlier theoretical prediction that the ⟨ACN⟩ of open and closed random walks scales with the number of segments n as $\frac{3}{16} n \ln n + bn$. Comparing the ⟨ACN⟩ of open and closed random walks of the same chain length one notices that closed walks have higher ⟨ACN⟩ than open walks. This result is a consequence of the fact that closed walks have smaller overall dimensions than open walks of the same chain length and this increases the number of perceived crossings in a random projection. As can be expected from the scaling function, the difference of the ⟨ACN⟩ between closed and open random walks of the same chain length n can be described by the linear relation. In fact the difference of ⟨ACN⟩ between closed and open walks with n segments can be accurately described by the formula $\langle \text{ACN} \rangle_{\text{closed}} - \langle \text{ACN} \rangle_{\text{open}} = \frac{3}{16}(n - c)$ where c is the ⟨ACN⟩ of three segment-long open random walks. Figure 1(B) shows that in closed chains each segment contribution to ⟨ACN⟩ is bigger by 3/16 than the corresponding contribution of each segment in open chains.

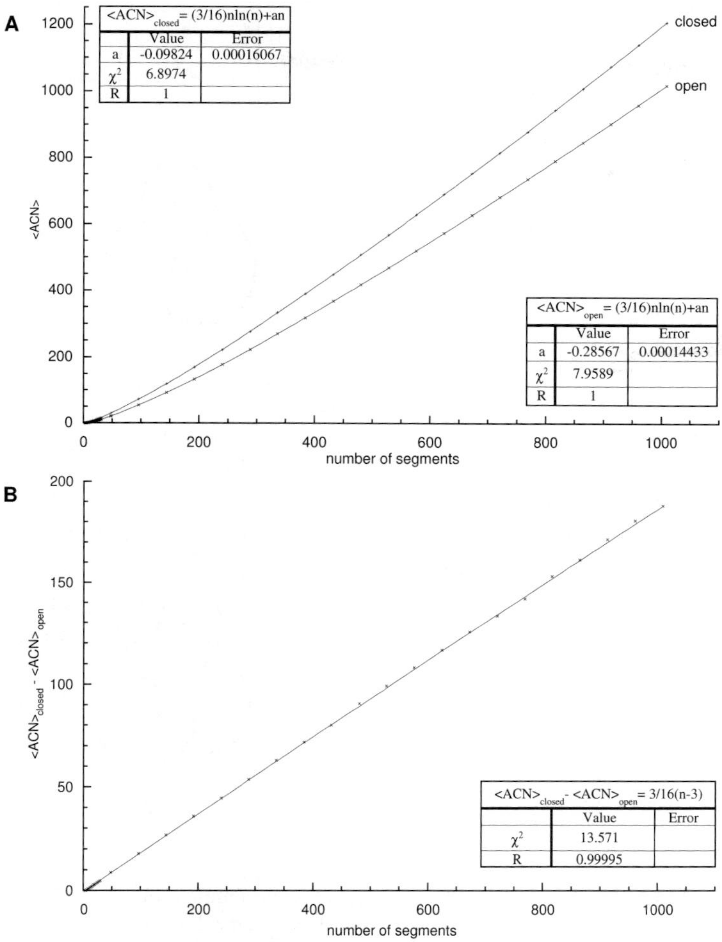

Fig. 1. Comparison of the mean average crossing number values ⟨ACN⟩ for corresponding chain lengths of closed and open ideal random walks (equilateral random polygons and equilateral random walks). The standard deviation is about the size of the data points. The values of the correlation coefficient (R) and chi-squared test (χ^2) are given in the inset. A. The ⟨ACN⟩ values obtained in numerical simulations of closed and open random walks are marked as data points and the fitting functions are listed. The analyzed sample sizes of simulated configurations were bigger than 10^5 for each chain length. B. The difference of ⟨ACN⟩ between closed and open ideal random walks of the same length.

3.2. ⟨ACN⟩ *scaling in the individual knot types*

The ⟨ACN⟩ data points for closed walks presented in Figure 1 were obtained for all closed random walks of a given chain, i.e. the averaging was

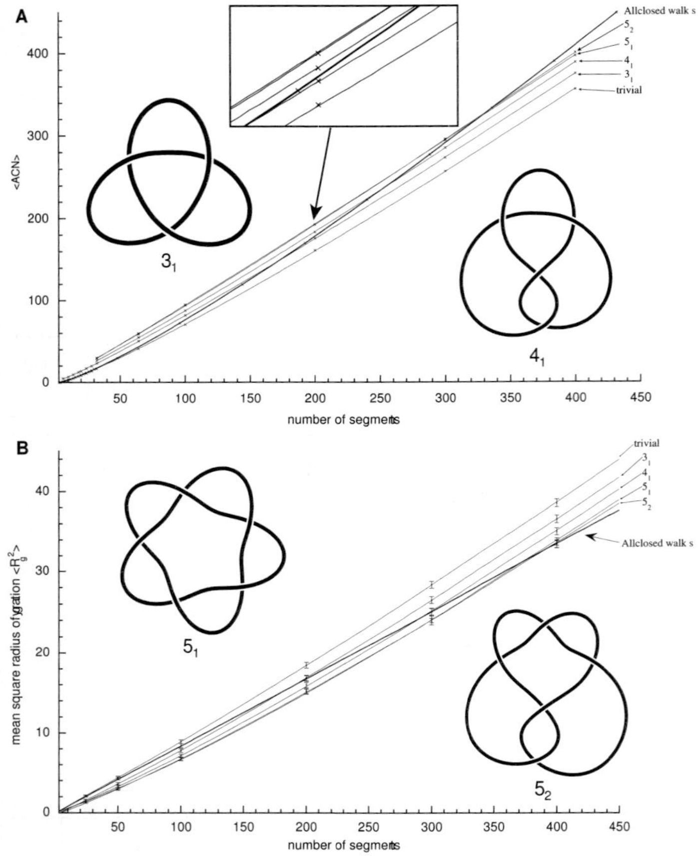

Fig. 2. Effect of the topology on the average crossing number values $\langle \mathrm{ACN} \rangle$ and $\langle \mathrm{R}_\mathrm{g}^2 \rangle$ for different knot types of closed random walk. The standard deviation is about the size of the data points. A. The $\langle \mathrm{ACN} \rangle$ values obtained in numerical simulations of closed random walks representing different knot types (trivial, 3_1, 4_1, 5_1 and 5_2) and all closed walks grouped together. The data points are marked and the fitting functions are listed in Table 1. The statistical sets for different knots and different chain size were not the same. Highest quality data are for unknots with 2,319,455 configurations analyzed in total and the smallest data set was that of 5_1 knots with 11,406 configurations analyzed in total. Note that $\langle \mathrm{ACN}(\mathcal{K}) \rangle$ profiles for individual knot types intersect with the $\langle \mathrm{ACN} \rangle$ profile for all closed random chains of the same length. B. The profiles for radius of gyration $\langle \mathrm{R}_\mathrm{g}^2 \rangle$ of individual random knots intersect with the $\langle \mathrm{R}_\mathrm{g}^2 \rangle$ profile of all closed random walks. The points of intersections define an R_g-based equilibrium length (the data here are taken from [15]).

performed over configurations forming unknots and different types of knots. It is possible, however, to divide closed walks into individual knot types

Table 1. Fitting values of the formula (2) used in Fig. 2

Fitting parameters	0_1	3_1	4_1	5_1	5_2
n_0	3	6	7	8	8
a	0.116±0.002	0.083±0.001	0.077±0.003	0.04±0.01	0.063±0.008
b	0.21±0.01	0.442±0.008	0.5±0.2	0.78±0.06	0.62±0.05
c	-0.17±0.13	4.9±0.1	7.9±0.4	8.2±1.4	10.5±1.1
χ^2	1.2685	0.41756	0.25687	2.5491	1.6548
R	1	1	1	0.99999	0.99999

and to analyze how the $\langle \text{ACN} \rangle$ scales when a given type of knot increases its chain length. Figure 2 illustrates how the $\langle \text{ACN} \rangle$ values scale with the chain length of random polygons representing various types of knots. It is clearly visible that random configurations of more complex knots have higher $\langle \text{ACN} \rangle$ values than the random configurations of simpler knots. This is a consequence of the fact that more complex knots have smaller overall dimensions than less complex knots with the same chain length [15,26,28] and therefore in an average projection the chain has to cross with itself more frequently in more complex knots. Interestingly, the $\langle \text{ACN} \rangle$ scaling of individual knot types can be accurately fitted with the formula:

$$\langle \text{ACN}(\mathcal{K}) \rangle = a(n - n_0)\ln(n - n_0) + b(n - n_0) + c \qquad (2)$$

where n_0 is the minimal number of segments needed to form a given knot [14] and a, b and c are free parameters (the function is defined for $n > n_0$). As can be seen in Table 1 the prefactor a decreases with the complexity of knots indicating that the contribution of $n \ln n$ part of the formula decreases with the complexity of the knot. However, the contribution of the linear component increases with the complexity of knots. The constant c also increases with the complexity of the knot and its value is close to the ACN of ideal geometric representation of a given knot [22,29]. For short chain lengths, the difference between the $\langle \text{ACN} \rangle$ of random configurations of a given knot type and the $\langle \text{ACN} \rangle$ of random configurations of unknots with the same chain length is well approximated by the actual ACN value of ideal (ropelength minimizing) geometric representation of a given knot type [22]. For example, for 6 and 14 segment-long random knots the difference between $\langle \text{ACN} \rangle$ of trefoil knots and unknots amounts to 4.14 and 5.21, while the ACN of an ideal trefoil is close to 4.26. This relation between the ACN of ideal knots and the $\langle \text{ACN} \rangle$ of random walks with relatively small chain length was noted earlier [22]. However, as the length of the analyzed random walks increases the $\langle \text{ACN} \rangle$ values of random knots of different types diverge from each other as observed in [21].

Comparing the fitted parameters of $\langle \text{ACN} \rangle$ scaling profiles of individual knot types (see Table 1) with these obtained for all closed walks grouped together irrespectively of their knot type (see Figure 1a) it is striking that the $n \ln n$ part has a much weaker contribution in the case of individual knot types while the linear contribution is much stronger. Consequently, the $\langle \text{ACN}(\mathcal{K}) \rangle$ profiles of random polygons of each knot type $\mathcal{K}$ show a lower growth rate than the $\langle \text{ACN} \rangle$ profile of all closed walks grouped together independently of their knot type (see Figure 2).

3.3. *The equilibrium length of a knot*

Comparing the $\langle \text{ACN}(\mathcal{K}) \rangle$ profiles for random polygons of knot type $\mathcal{K}$ with that of all closed walks grouped together, it is visible that (with the exception of the unknots) the $\langle \text{ACN}(\mathcal{K}) \rangle$ profiles intersect with the profile for all closed walks (see Figure 2). This is due to the fact that individual knot types show a smaller growth rate than all closed walks grouped together (see the discussion above), while each individual knot type (with an exception of the unknots) initially has higher $\langle \text{ACN}(\mathcal{K}) \rangle$ values than the $\langle \text{ACN} \rangle$ values for the ensemble of all closed walks. The more complex the knot, the later its $\langle \text{ACN}(\mathcal{K}) \rangle$ profile intersects with the $\langle \text{ACN} \rangle$ profile of all closed walks. The points of intersection between profiles $\langle \text{ACN}(\mathcal{K}) \rangle$ and $\langle \text{ACN} \rangle$ determine $\langle \text{ACN} \rangle$-based equilibrium length of the corresponding knot type $\mathcal{K}$. Below the equilibrium length of a given knot shows an excess of $\langle \text{ACN} \rangle$ as compared with the $\langle \text{ACN} \rangle$ of all possible walks realized with the same chain length. Therefore if one would cut a knot realized with a chain shorter than its equilibrium length, let it equilibrate and then promote reclosure of the ends, one would observe a tendency to form simpler knots than the starting knot type. Above the equilibrium length, the situation reverses. If one would cut a knot realized with the chain length longer than its equilibrium length, one would observe after a reclosure of the chain its tendency to form more complex knots. At the equilibrium length, however, a knot would show no tendency to decrease or increase its $\langle \text{ACN} \rangle$ after cutting, equilibration and reclosure. Interestingly, the equilibrium lengths determined by the intersections of $\langle \text{ACN} \rangle$ profiles of individual knot types with $\langle \text{ACN} \rangle$ profile of all closed walks grouped together practically coincide with the equilibrium lengths determined by intersections of the corresponding profiles of the mean radius of gyration $\langle R_g \rangle$ (see Figure 2). So for example, the equilibrium length for trefoils and figure of eight knots based on measurements of $\langle ACN \rangle$ amounted to 176 ± 10 and 258 ± 10 segments, respectively, while the equilibrium length of these knots based on measurements of $\langle R_g \rangle$ amounted to 174 ± 14 and 270 ± 17 segments, respectively [15] (the bigger error range for R_g-based equilibrium length is caused by a smaller statistical set of data used to determine $\langle R_g \rangle$).

3.4. *Scaling of* $\langle$ACN$\rangle$ *in natural protein structures*

To describe the structure of proteins, one can follow their backbones and analyze conformations determined by the positions of sequential C_α atoms (α-Carbons). These representations of proteins structure are formally equivalent to segmented linear chains and can be studied as such. There has been a substantial interest in scaling of overall dimensions of protein structures as well as in the entanglement of protein backbones including studies of $\langle ACN \rangle$ scaling [2,3,5,6,20]. Interestingly, native protein structures were shown to have a complex scaling behaviour where small single domain proteins were behaving like collapsed walks while longer proteins resembled more ideal random walks such as those analysed in Fig. 2 [5,7]. There is a fundamental difference between collapsed and ideal walks. The former increase their overall dimensions (like the radius of gyration $\langle R_g \rangle$) proportionally to the cubic root of the number of residues in the polypeptide chain while the later increase their overall dimensions proportionally to the square root of their chain length [19]. It was postulated earlier that in the maximally tightly packed, i.e. collapsed, walks $\langle ACN \rangle$ should scale according to a simple power law where the entanglement exponent β should be equal to 4/3 [6,8,9,10,11,13]. As we have discussed above, the $\frac{3}{16} n \ln n + bn$ scaling of $\langle ACN \rangle$ was proven to apply to ideal random walks that behave like chains in which segments neither attract nor repulse each other (see Figure 1). We decided therefore to check what is the scaling of $\langle ACN \rangle$ in proteins when one analyses the whole range of determined protein structures. Figure 3 presents a dot plot where ACN values, measured for backbones of 2230 independent proteins, are plotted against the number of $C_\alpha - C_\alpha$ segments in the respective protein structures. All points are fitted using either the formula $\langle ACN \rangle = \frac{3}{16} n \ln n + bn$ or the formula $\langle ACN \rangle = an^{4/3}$. Notice that in both cases we just have one free parameter and that in both cases the fit is excellent. Although the formula $\langle ACN \rangle = an^{4/3}$ gives a slightly better fit, one can certainly say that both formulas describe very accurately the scaling behaviour of ⟨ACN⟩ in natural proteins.

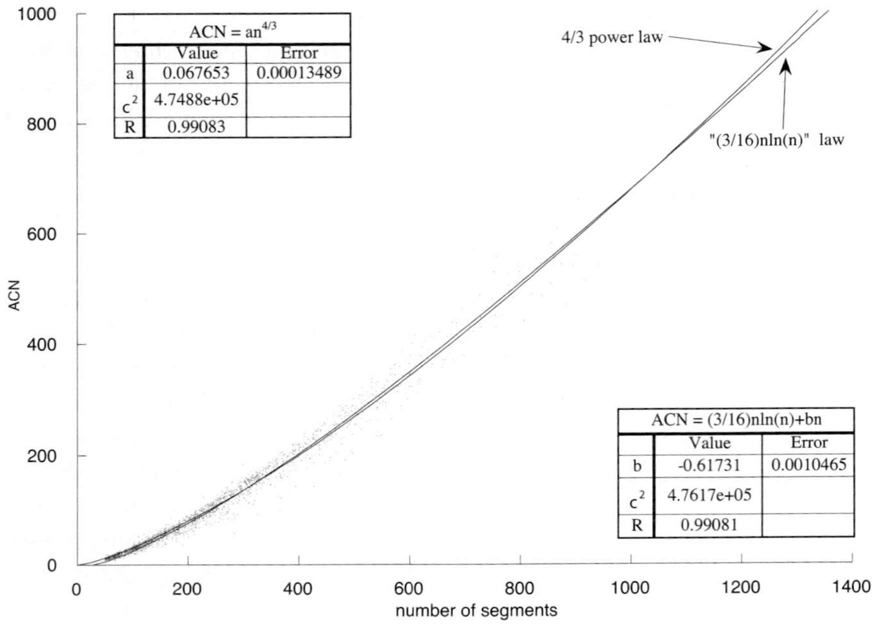

Fig. 3. Scaling of ACN in natural protein structures. ACN values calculated for the backbones of 2230 independent proteins were fitted using two formulas: $\langle \mathrm{ACN} \rangle = \frac{3}{16} n \ln n + bn$ and $\langle \mathrm{ACN} \rangle = an^{4/3}$. The quality of the fit and the fitted value of respective free parameter are indicated in the insets.

4. Conclusions and Outlook

We have compared here the scaling of ACN in ideal random walks in which segments neither repulse nor attract each other with the ACN scaling in backbones of proteins and observed that in both cases the scaling of ⟨ACN⟩ can be accurately described by the formula $\langle \mathrm{ACN} \rangle = \frac{3}{16} n \ln n + bn$ where the free parameter b amounts to -0.29 ± 0.0001 in the case of ideal random walks and -0.62 ± 0.001 in the case of protein backbones. While the relation $\langle \mathrm{ACN} \rangle = \frac{3}{16} n \ln n + bn$ was analytically proven to apply to ideal random walks [14] there is no formal proof that the same relation should also hold for protein structures. Therefore, one should take into account that ⟨ACN⟩ scaling in protein backbones can be also accurately described by a power law relation $\langle \mathrm{ACN} \rangle = an^{4/3}$ where the 4/3 exponent was predicted theoretically for tightly packed walks [6,8,9,10,11,13]. In fact the power law relation fits the data slightly better (see Fig. 3). Why would the two different fitting formulas give both a very good fit? The answer lies probably in the fact that protein structures show a complex scaling behaviour where the relatively small single domain proteins behave more like compact walks and where larger proteins behave more like ideal random walks [5,7]. To verify this possible explanation we have reanalysed our data by dividing all proteins into two groups: smaller and bigger than 300 residues. We noticed that for smaller proteins the 4/3-power law scaling gave a better fit than $n \ln n$ scaling of $\langle \mathrm{ACN} \rangle$ while the opposite was the case for bigger proteins (data not shown). We confirmed therefore, using a much larger statistic sample than before, that proteins show a complex scaling behaviour of $\langle \mathrm{ACN} \rangle$ [5,7]. However, for most practical purposes both scaling relations: $n \ln n$ and 4/3-power law provide a very accurate description of $\langle \mathrm{ACN} \rangle$ scaling in proteins over the whole size range of known protein structures.

The relation $\langle \mathrm{ACN} \rangle = \frac{3}{16} n \ln n + bn$ applies also to ideal random polygons but the linear term is less important than in the case of ideal random walks and the parameter b amounts to circa -0.1. Interestingly, the difference between $\langle \mathrm{ACN} \rangle$ for closed and open ideal random walks with the same number of segments n follows a linear relation $\langle \mathrm{ACN} \rangle_{\mathrm{closed}} - \langle \mathrm{ACN} \rangle_{\mathrm{open}} = \frac{3}{16} n$. We have also analyzed the scaling of $\langle \mathrm{ACN} \rangle$ for individual knot types K and observed that in each case the observed relation can be described by a formula $\langle \mathrm{ACN}(\mathcal{K}) \rangle = a(n - n_0) \ln(n - n_0) + b(n - n_0) + c$, where n_0 is the minimal number of equilateral segments needed to form the given knot type $\mathcal{K}$ and where the constant c approximates the ACN value of the ideal knot of type $\mathcal{K}$. The $\langle \mathrm{ACN}(\mathcal{K}) \rangle$ profiles show slower growth rates than the corresponding $\langle \mathrm{ACN} \rangle$ profiles of all closed walks grouped together. As the complexity of the knot type $\mathcal{K}$ increases, the coefficient a in our fitting formula decreases while the coefficient b increases. This scaling behaviour of ACN in individual knot types causes the $\langle \mathrm{ACN} \rangle$ scaling profiles of each knot (with exception of unknot) to intersect the scaling profile of $\langle \mathrm{ACN} \rangle$ of all closed walks grouped together. The points of intersection define the equilibrium length of a given knot type, i.e. the chain length at which a given knot type is not under- or over-knotted and would therefore show no tendency to increase or decrease its $\langle \mathrm{ACN} \rangle$ upon closure, equilibration and re-closure.

Acknowledgments

We thank G. Arteca, R. Kusner, K. Millett and W. Taylor for discussions. We also thank W. Taylor for providing us with the filtered set of coordinates of independent protein structures. This work was supported by Swiss National Science Foundation grants 31-68151.02, PA00A-105000 and 3100A0-103962 and Y. Diao is partially supported by NSF Grant #DMS-0310562. A preliminary version of this paper was presented at the conference 'Knots, random walks and biomolecules' (co-organized by J. H. Maddocks & A. Stasiak and sponsored by the Bernoulli Centre of the Swiss Federal Institute of Technology in Lausanne) held in Les Diablerets, Switzerland during 14-17 July 2003.

References

1. C. C. Adams, *The Knot Book*, W.H. Freeman and Company, New York, 1994.
2. G. A. Arteca, *Overcrossing spectra of protein backbones: characterization of three-dimensional molecular shape and global structural homologies, Biopolymers*, **33**, 1829-1841, 1993.
3. G. A. Arteca, *Scaling behaviour of some molecular shape descriptors of polymer chains and protein backbones*, Phys. Rev. E, **49** 2417-2428, 1994.
4. G. A. Arteca, *Scaling regimes of molecular size and self-entanglements in very compact proteins*, Phys. Rev. E, **51**, 2600-2610, 1995.
5. G. A. Arteca, *Self-similarity in entanglement complexity along the backbones of compact proteins*, Phys. Rev. E, **56**, 4516-4520, 1997.
6. G. A. Arteca, *Analytical estimation for the entanglement complexity of a bond network*, J. Chem. Inf. Comp. Sci., **42**, 326-330, 2002.
7. G. A. Arteca, *Different molecular size scaling regimes for inner and outer regions of proteins*, Phys. Rev. E, **54**, 3044-3047, 1996.
8. G. Buck, *Four-thirds power law for knots and links*, Nature, **392**, 238-239, 1998.
9. G. Buck and J. Simon, *Thickness and crossing number of knots*, Topology and its Applications, **69**, 1-12, 1999.
10. J. Cantarella, R. B. Kusner and J. M. Sullivan, *Tight knot values deviate from linear relations*, Nature, **392**, 237-238, 1998.
11. J. Cantarella, R. B. Kusner and J. M. Sullivan, *On the minimum ropelength of knots and links*, Invent. math., **150**, 257-286, 2002.
12. J. A. Calvo and K. C. Millett, in *Ideal Knots*, Eds. A. Stasiak, V. Katritch and L.H. Kauffman, World Scientific, Singapore, 107-128, 1998.
13. Y. Diao and C. Ernst, *The complexity of lattice knots*, Topology and its Applications, **90**, 1-9, 1998.
14. Y. Diao, A. Dobay, R. B. Kusner, K. Millett and A. Stasiak, *The average crossing number of equilateral random polygons*, J. Phys. A: Mathematical and General, **36**, 11561-11574, 2003.

15. A. Dobay, J. Dubochet, K. Millett, P.-E. Sottas and A. Stasiak, *Scaling behaviour of random knots*, Proc. Natl. Acad. Sci. USA, **100**, 5611-5615, 2003.
16. M. Doi and S. F. Edwards, *The Theory of Polymer Dynamics*, Oxford University Press, Oxford, 1986.
17. P. J. Flory, *Principles of Polymer Chemistry*, Cornell University Press, Ithaca, NY, 1953.
18. P. Freyd, D. Yetter, J. Hoste, W. Lickorish, K. Millett and A. Ocneau, *A new polynomial invariant of knots and links*, Bull. Amer. Math. Soc., **12**, 239-246, 1985.
19. P. G. de Gennes, *Scaling Concepts in Polymer Physics*, Cornell University Press, Ithaca, N.Y., 1979.
20. P. Grassberger, *Opacity and entanglement of polymer chains*, J. Phys. A: Math. Gen., **34**, 9959-9963, 2001.
21. J.-Y. Huang and P.-Y. Lai, *Crossings and writhe of flexible and ideal knots*, Phys. Rev. E, **63**, 21506, 2001.
22. V. Katritch, J. Bednar, D. Michoud, R. G. Scharein, J. Dubochet and A. Stasiak, *Geometry and physics of knots*, Nature, **384**, 142-145, 1996.
23. K. V. Klenin, A. V. Vologodskii, V. V. Anshelevich, D. A.M. and M. D. Frank-Kamenetskii, *Effect of excluded volume on topological properties of circular DNA*, J. Biomol. Struct. Dyn., **5**, 1173-1185, 1988.
24. M. A. Krasnow, A. Stasiak, S. J. Spengler, F. Dean, T. Koller and N. R. Cozzarelli, *Determination of the absolute handedness of knots and catenanes of DNA*, Nature, **304**, 559-560, 1983.
25. M. L. Mansfield, *Are there knots in proteins?*, Nat. Struct. Biol., **1**, 213-214, 1994.
26. E. Orlandini, M. C. Tesi, E. J. Janse Van Rensburg and S. G. Whittington, *Asymptotics of knotted lattice polygons*, J. Phys. A: Math. Gen., **31**, 5935-5967, 1998.
27. D. Poland and H. A. Scheraga, *Theory of Helix-Coil Transitions in Biopolymers*, Academic Press, New York, 1970.
28. M. K. Shimamura and T. Deguchi, *Anomalous finite-size effect for the mean-squared gyration radius of Gaussian random knots*, J. Phys. A: Math. Gen., **35**, 241-246, 2002.
29. A. Stasiak, J. Dubochet, V. Katritch and P. Pieranski, in *Ideal Knots*, Eds. A. Stasiak, S. Katritch and L.H. Kauffman, World Scientific, Singapore, 1-19, 1998.
30. A. Stasiak, V. Katritch, J. Bednar, D. Michoud and J. Dubochet, *Electrophoretic mobility of DNA knots*, Nature, **384**, 122, 1996.
31. W. R. Taylor, *Multiple sequence threading: an analysis of alignment quality and stability*, J. Mol. Biol., **269**, 902-943, 1997.
32. W. R. Taylor, *A deeply knotted protein structure and how it might fold*, Nature, **406**, 916-919, 2000.
33. A. Vologodskii, N. Crisona, B. Laurie, P. Pieranski, V. Katritch, J. Dubochet and A. Stasiak, *Sedimentation and electrophoretic migration of DNA knots and catenanes*, J. Mol. Biol., **278**, 1-3, 1998.

CHAPTER 13

FOLDING COMPLEXITY IN A RANDOM-WALK COPOLYMER MODEL

Gustavo A. Arteca

Département de Chimie et Biochimie, Laurentian University
Ramsey Lake Road, Sudbury, Ontario, Canada P3E 2C6
E-mail: gustavo@laurentian.ca

Basic folding features of a macromolecular chain can be conveyed by using a geometrical measure of chain entanglement, e.g., the mean overcrossing number (or "average crossing number") $\overline{N}$. In this work, we study the dependence of the configurationally-averaged $\langle\overline{N}\rangle$ value with the monomer number n, for a simple random-walk copolymer model involving two constant bond lengths b_1 and b_2. The results support the validity of the scaling $\langle\overline{N}\rangle \approx a(z)\, n\, \ln(n)$, over the entire range of bond-length ratios $z = b_1/(b_1 + b_2)$. The results serve as a benchmark for recognizing specific effects on polymer shape caused by chemical composition, or charging scheme, in realistic heteropolymers.

1. Introduction

Important aspects of polymer shape cannot be captured by using descriptors that depend exclusively on monomer geometry, e.g., the mean chain size and anisometry.[26,22,3] For instance, protein conformers both in the unfolded and in the collapsed state can have similar mean sizes yet distinct secondary and tertiary structural content.[20,24,30,32,12,11,16,17] Some of these differences are accounted by bringing the chain connectivity (or "topology"), in addition to the chain geometry, into the analysis of polymer shape. This expanded geometrical description serves to quantify how complex a chain's *self-entanglements* are; entanglement complexity can be used as a measure of a protein's folding features.[15,29,25] In this note, we use the mean overcrossing number (or "average crossing number") $\overline{N}$ as a descriptor of chain entanglement. For a chain consisting of a sequence of connected linear bonds, $\overline{N}$ is the number of projected bond-bond crossings (or "overcrossings") averaged over all possible spatial directions.[9,23,1] It can be computed as a sum over pair contributions:[5]

$$\overline{N} = 2\sum_{i=1}^{n-2}\sum_{j=i+2}^{n} \overline{N}_{i,j}\,, \tag{1}$$

where $\overline{N}_{i,j}$ is the average crossing number for two nonadjacent bonds i and j. This latter contribution gives the probability that the bonds i and j would produce *one* crossing when projected along an arbitrary direction in space. Such probability is also equivalent to the area of the unit sphere spanned by the image of the Gauss mapping defined by the particular pair of bonds. In turn, this area is computed as a double line integral:[5]

$$\overline{N}_{i,j} = \frac{1}{4\pi}\int_0^1\int_0^1 \frac{|(\dot{\gamma}_i(s)\times\dot{\gamma}_j(t))\cdot(\gamma_i(s)-\gamma_j(t))|}{\|\gamma_i(s)-\gamma_j(t)\|^3}\,ds\,dt \tag{2}$$

with $\gamma_i(s) = \mathbf{R}_i + s(\mathbf{R}_{i+1} - \mathbf{R}_i)$ being the bond between the i- and $(i+1)$-monomers, and $\dot{\gamma}_i(s) = \mathbf{R}_{i+1} - \mathbf{R}_i$ the corresponding parametric derivative.

Numerical simulations and analytical estimations have been used to study the behaviour of the mean overcrossing number averaged over accessible (random) configurations, denoted by $\langle\overline{N}\rangle$, in the case of *homopolymers*, i.e., with a constant bond length $\|\mathbf{R}_{i+1} - \mathbf{R}_i\| = b$. (For a review, see Refs. [6,7] and others quoted therein.) From recent work, the emerging consensus is that $\langle\overline{N}\rangle$ satisfies a scaling law with the number of monomers n which changes with the universality classes associated with the dominant monomer-monomer interactions: *i*) Collapsed polymers (e.g., compact or self-attracting chains, ideal knots) follow $\langle\overline{N}\rangle \approx an^{4/3}$.[14,13,2,8] *ii*) Chains in an ideal-solvent condition (e.g., random walks) appear to satisfy $\langle\overline{N}\rangle \approx a\,n\ln(n)$.[6,21,18] *iii*) Swollen polymers (e.g., self-avoiding walks or chains with excluded-volume interaction) are conjectured to follow a different power law: $\langle\overline{N}\rangle \approx a\,n$, for sufficiently large n values.[21] Differentiating these regimes requires long chains because the behaviours for medium-size chains are well represented by an *apparent* power law $\langle\overline{N}\rangle \approx a\,n^{\beta_{\mathit{eff}}}$, with an effective exponent β_{eff} between 1.4 and 1.1 for compact to swollen chains, respectively.[6,7] In this work, we explore the validity of the scaling regime for random walks from a different viewpoint.

Whereas the entanglement complexity of open and closed homopolymer chains with excluded volume interaction are discussed in the literature, the effects on polymer shape caused by variable chain composition are less well characterized. Here, we are interested in understanding the simplest of shape effects on a *regular copolymer*. For the purposes of this work, a "regular copolymer" is defined as a chain *effectively* equivalent to an AB diblock polymer, i.e., one whose structure is determined by the presence of two different monomers in regular sequence ABABAB.... In our context, each basic unit A and B will be represented by a *single* bond length. Each of these effective bond lengths may arise from a number of distinct situations, for instance: (a) A and B may simply be the labels for two distinct types of bonds that alternate along the backbone of a homo- or heteropolymer, (b) A and B can be two monomers with different size that are freely articulated, yet internally rigid, or (c) one of the units can correspond to a short linker between two larger sections of the polymer characterized by a much larger persistence length. An example of the first type can be found in protein backbones, where one can simplify the sequence of distinct bonds N–C_α, C_α–C, and C–N, into two effective ones, e.g., the peptide bond C–N and another "effective" bond centred about the α-carbon (N——C). Examples of the second kind can be found in homopolymers such as polysaccharides or perylene-containing polyimides.[28] In both cases, the polymer may exhibit a sequence of large rigid units (e.g., the sugar ring or the fused aromatic rings in perylene) connected by a short linker. Finally, examples of the third kind could be found in the simplified models of α-helical proteins that consider the helices as rigid rods connected by short equivalent links (e.g., the β-turns). In this case, a complex heteropolymer is described as a sequence of two distinct entities, clearly differentiated in terms of effective length. Other examples could be generated by making use of various strategies for backbone rigidization,[27] by considering a coarse-grained model of diblock polymers with alternating persistence lengths,[19] or regions with distinct charge distributions in polyampholytes.[31]

To our knowledge, the only available entanglement characterization for a *nonequilateral* polymer backbone corresponds to the experimental native states of proteins. Here, the entanglement measure associated with the α-carbon chains of short globular proteins supports the law $\overline{N} \approx a\, n^{4/3}$ expected for compact polymers.[2] However, although the C_α–C_α distance between consecutive monomers varies with composition, this effect can be considered simply a small random noise about $b = 3.8$Å. As mentioned before, it is possible that the mean shape description be different if one takes into account all the main chain atoms (N, C_α and C) along the protein backbone, and then models the chain as a *regular* copolymer following the sequence N——C–N'——C'–N''—— etc. In our context, this latter sequence is characterized by the alternance of two distinct bond lengths b_1 and b_2.

In the next sections, we study the configurationally-averaged $\langle\overline{N}\rangle$ value as a function of the number of monomers in a simple regular copolymer model with a random distribution of configurations. The model consists of a random walk (freely-jointed chain) with fixed bonds which alternate in length (e.g., $b_1 - b_2 - b_1 - b_2 -$ etc). Below, we focus on the scaling behaviour of $\langle \overline{N} \rangle$ and its dependence the parameter z, representing the ratio between the two bond lengths, $z = b_1/(b_1 + b_2)$.

2. Simulation method and shape characterization

The copolymer model consists of a chain of $n - 1$ ($n \gg 1$) bonds which alternate between two constant values of length, b_1 and b_2; the first bond corresponds to the b_1 value. Each model will be characterized by the parameter $z = (1 + b_2/b_1)^{-1}$. Except for the variable bond lengths, the chain conformation is chosen to be an ideal random walk (RW), i.e., a freely-jointed chain without excluded-volume interaction. In the limit $z = 1/2$, we find a homopolymer chain of n monomers; similarly, $z = 0$ and $z = 1$ correspond also to a homopolymer chain, but one with only $(n/2) + 1$ monomers (for even values of n).

Configurations for the above chain model are generated by using a "naïve" Monte Carlo approach [2] with $n \leq 500$. The procedure generates strictly independent conformers, all with the same statistical weight; conformational attrition is not a major limiting factor for the present range of n values. Statistical averages are computed using 5000 and 1000 conformers for $n < 200$ and $n \geq 200$, respectively. The quality of the conformational search is assessed by using the known scaling for the configurationally-averaged mean radius of gyration $\langle R_g^2 \rangle^{1/2}$:

$$\langle R_g^2 \rangle^{1/2} \approx n^{\nu}, \tag{3}$$

and estimating the scaling exponent ν by a linear regression $\ln\langle R_g^2 \rangle^{1/2}$ vs $\ln n$. The results obtained over a range of z values appear in Table 1, and are consistent with the well-known value $\nu = 1/2$ for RWs. Other average shape properties will thus be found in the ideal-solvent regime.

The mean overcrossing number $\langle \overline{N} \rangle$ for a given conformer is computed using 10^4 random projections;[1] this ensures a statistical error in $\overline{N}$ below the error in the configurational average $\langle \overline{N}(z) \rangle$. We have tested two possible scaling laws for $\langle \overline{N} \rangle$ as a function of the z ratio:

$$\langle \overline{N}(z) \rangle \approx a\, n\, \ln(n)\,, \tag{4a}$$

$$\langle \overline{N}(z) \rangle \approx a\, n^{\beta_{\text{eff}}}\,, \tag{4b}$$

Table 1. Scaling exponent ν for the configurationally-averaged mean radius of gyration, $\langle R_g^2 \rangle^{1/2}$, Eq. (3) computed as a function of the bond length ration z for $n \in [75, 500]$. Error bars are 95% confidence intervals; the third column is the correlation coefficient. The value $\nu \approx 1/2$ is consistent with having reached an exhaustive sampling for random-walk configurations.

z	ν	C
1/20	0.49 ± 0.01	0.99992
3/40	0.48 ± 0.04	0.99921
1/10	0.50 ± 0.01	0.99986
3/20	0.48 ± 0.03	0.99880
1/6	0.49 ± 0.06	0.99840
1/4	0.51 ± 0.01	0.99984
1/3	0.51 ± 0.01	0.99979
3/8	0.49 ± 0.04	0.99912
10/23	0.50 ± 0.03	0.99879
1/2	0.50 ± 0.01	0.99986

corresponding to the proposed exact behaviour (Eq. (4a))[21,18] and the apparent effective behaviour (Eq. (4b)) for intermediate chain lengths.[6] Below, we compare Eqs. (4) and as a representation of the entanglement complexity for copolymers models having different bond-length ratios, $z \in [0, 1]$.

3. Scaling behaviour as a function of the bond-length ratio

For each copolymer model, we have fitted Eqs. (4) using linear regressions with $n = 75$, 100, 150, 200, 300, 400, and 500. The homopolymer cases were derived from the results already available in the literature for $n = 70$, 94, 150, 198, 296, 390, and 500 using in a random-walk hydrocarbon model.[4] (These values were also used for the ν-estimates in Table 1.) A similar range of n was used in Ref. [18] to verify the exact scaling behaviour for *equilateral random homopolymers*, which is expected to be $\langle \overline{N} \rangle \approx (3/16)n\ \ln(n) + c\,n$ for both open and closed walks. Numerical estimates indicate a negative correction to scaling, given by $c \approx -0.3$ and $c \approx -0.1$ for open and closed chains, respectively.[18]

In the case of our copolymer model, the results emerging from a fitting according to the $(n \ln n)$-law are shown in Figure 1; the corresponding correlation coefficients are given in Table 2. (The case $z = 0$ is handled differently by re-fitting the data for $z = 1/2$ as $\langle \overline{N}(z) \rangle$ vs. "$2n \ln(2n)$.") Clearly, the results appear to follow the $(n \ln n)$-scaling law equally well for all z values, without any systematic deviation.

The remaining columns of Table 2 give the results from the power-law

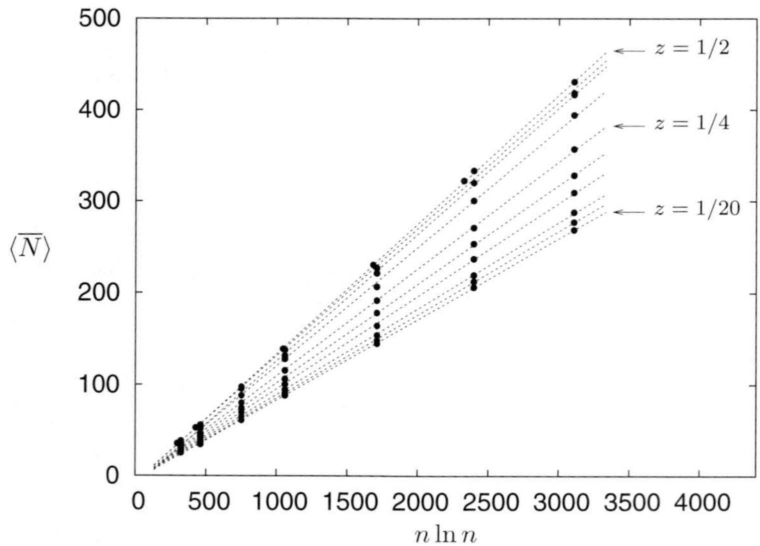

Fig. 1. Variation of the configurationally-averaged mean overcrossing number $\langle \overline{N} \rangle$ with the monomer number n. For copolymer random walks, a "$n \ln n$" scaling law describes well the behaviour for all values of the bond-length ratio $z = b_1/(b_1 + b_2)$.

fitting (4b). The data produce an effective scaling exponent $\beta_{eff} \approx 1.26$ that is valid for all z; consistently with the lack of monomer attraction in ideal RWs, this value is below the one corresponding to collapsed chains ($\beta = 4/3$). The fact that there are no systematic deviations of β_{eff} as a function of z reinforces the notion that all copolymer chains are found in the same scaling regime. However, a comparison of the correlation coefficients in Table 2 indicates that Eq. (4a) provides a more accurate representation of the data. This result is consistent with the scaling behaviour recently uncovered for homopolymers.[6,7,21,18]

According to these observations, all dependence with the bond-length ratio will be confined to the scaling amplitude $a(z)$. Figures 2 and 3 show the results for $a(z)$ according to the scaling laws (4a) and (4b), respectively. Both laws produce a qualitatively similar $a(z)$ functions, with a maximum range $a(1/2)/a(0)$ that is slightly larger than 2; the resulting scaling amplitudes $a(z)$ resemble bell-shaped curves with a rather flat maximum; the function obtained with Eq. (4a) is better defined than the one in Fig. 3. From this analysis, we arrive to the conclusion that a simple polynomial fitting of the data in Fig. 2, together with the scaling law (4a), should provide

Table 2. Estimation of scaling behaviour for the configurational average of the mean overcrossing number, $\langle \overline{N}(z) \rangle$, over a range of bond-length ratios z. The second and third columns give the correlation coefficients for the least-square fittings to Eqs. (4a) and (4b), respectively. The last column give the estimates for an effective scaling exponent according to (4b) with 95% confidence intervals. The data are better represented by Eq. (4a) for most z values.

z	C, Eq. (4a) $n \ln n$-law	C, Eq. (4b) n^{β}-law	β_{eff}, Eq. (4b)
0	0.99993	0.99983	1.27 ± 0.03
1/20	0.99982	0.99965	1.26 ± 0.04
3/40	0.99996	0.99992	1.25 ± 0.03
1/10	0.99994	0.99992	1.25 ± 0.04
3/20	0.99990	0.99993	1.26 ± 0.02
1/5	0.99994	0.99988	1.26 ± 0.02
1/4	0.99995	0.99985	1.27 ± 0.03
1/3	0.99984	0.99979	1.27 ± 0.03
3/8	0.99993	0.99988	1.28 ± 0.03
10/23	0.99957	0.99962	1.27 ± 0.04
1/2	0.99995	0.99983	1.27 ± 0.03

a complete and accurate description of the mean entanglement complexity of a regular copolymer model with no excluded-volume interaction.

It should also be noted that the numerical results for $a(z)$ about $z = 1/2$ agree well with the estimates deduced from Ref. [18] for equilateral random walks. Upon considering the negative correction to scaling in open walks, a scaling law with a single dominant term should produce an *underestimate*: $a(1/2) \approx (3/16) - (0.3/\ln(n))$. Given the range of chain lengths used here (with a mean-ln$(n) \approx 5.3$), the expected maximum is $a \approx 0.131$, in good agreement with the results in Fig. 2. In contrast, note that the results in Fig. 3 for the maximum in a agree better with the dominant term expected from the asymptotics of equilateral walks, i.e., $\max a = 3/16 = 0.1875$. The better agreement is consistent with the fact that Fig. 3 corresponds to an *effective* scaling law, and therefore should include finite-size and correction-to-scaling effects associated with intermediate n values.

Although we cannot account at present for the detailed form of the amplitude $a(z)$ as a function of z, a rough estimate can be obtained using mean-field arguments. Here, we follow the approach in Ref. [6], which predicts a "$n \ln n$" scaling law for $\overline{N}$ in a homopolymer chain. To this end, let us consider a random chain with n-bonds of equal length b, embedded in a cubic lattice and having a span $R \approx n^{\nu} b$. With Eq. (6) in Ref. [6], the

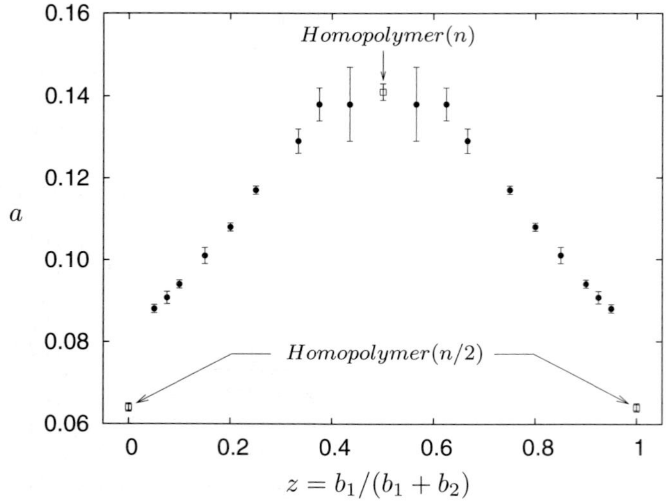

Fig. 2. Scaling amplitude $a = \lim_{n\to\infty} \langle \overline{N}\rangle/(n \ln n)$ for copolymer random walks with variable bond-length ratio $z = b_1/(b_1+b_2)$. Error bars represent 95% confidence intervals. The white squares correspond to the results obtained with freely-jointed homopolymer random walks (i.e., those with constant bond length); the values for $z = 0$ and $z = 1$ correspond to chains that are half-as-long as the ones for $z = 1/2$.

configurationally-averaged $\overline{N}$ value is represented as:

$$\langle \overline{N}\rangle = 2\sum_{i=1}^{n-2}\sum_{j=i+2}^{n}\langle \overline{N}_{i,j}\rangle$$

$$\approx \frac{n^2}{2}\int g^{(3D)}(r)r^2\langle \overline{N}_{i,j}(r,\phi)\rangle_\phi \, dr, \; n \gg 1, \tag{5}$$

where $g^{(3D)}(r) = const \cdot r^{-3+1/\nu}$ is the distribution function for the $n^2/4$ pairs of perpendicular bonds that overcross; with the normalization condition $\int_b^R r^2 g^{(3D)}(r)dr = 1$, the distribution function becomes $g^{(3D)}(r) \approx 2rR^2$ for an RW chain. The mean value $\langle \overline{N}_{i,j}(r,\phi)\rangle_\phi$ gives the contribution of a single pair of bonds separated by a mean distance r to the total overcrossing number, averaged over all possible orientations represented by the angle $\phi \in [0, 2\pi]$.

Using Eq. (9) in Ref. [6] for $\langle \overline{N}_{i,j}(r,\phi)\rangle_\phi$, we find an expression for $\langle \overline{N}\rangle$ in a homopolymer RW chain:

$$\langle \overline{N}\rangle \approx const \cdot \frac{n^2}{R^2} b^2 O[\ln(R/b)]\,, \tag{6}$$

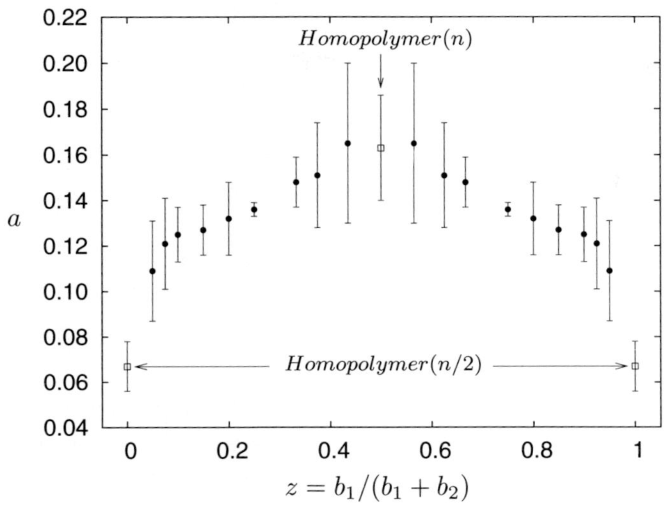

Fig. 3. Re-estimation of the scaling amplitude using a power-law $\langle\overline{N}\rangle = an^{\beta_{eff}}$. The effective exponent β_{eff} is given in the last column of Table 2. Symbols are the same as those in Fig. 2. Qualitatively, the amplitude a varies with the bond-length ratio parameter $z = b_1/(b_1 + b_2)$ in a similar fashion as the scaling amplitude in Fig. 2, although with a larger statistical error.

where $O[\ln(R/b)] = O(\ln n)$. Since $R^2 \approx nb^2$, Eq.(6) produces a "$n \ln n$" scaling. If we now consider our regular copolymer model with two bonds $b_1 < b_2$, we can modify Eq. (6) by noting that there are three pair contributions to $\overline{N}$, associated with: *i*) overcrossings between "short" bonds, $\langle\overline{N}_{b_1,b_1}\rangle$, *ii*) between "long" bonds, $\langle\overline{N}_{b_2,b_2}\rangle$, and *iii*) between bonds of different lengths $\langle\overline{N}_{b_1,b_2}\rangle$. The total $\langle\overline{N}\rangle$ value for the copolymer can now be computed by adding all pairs:

$$\langle\overline{N}_{TOT}\rangle = 2\sum_i\sum_{j>i}\langle\overline{N}_{b_1,b_1}\rangle + 2\sum_i\sum_{j>i}\langle\overline{N}_{b_2,b_2}\rangle + 2\sum_i\sum_j\langle\overline{N}_{b_1,b_2}\rangle \quad (7)$$

The first two terms in (7) can be estimated using Eq. (6) for two *half-chains* with n_1 and n_2 bonds (or monomers), occupying the spatial span R for the complete chain with $n = n_1 + n_2$ bonds. The last term in Eq. (7) is more difficult because the bonds have different length, but it must scale similarly with b and n, yet vanish for $b_1 = 0$. Moreover, only perpendicular bonds overcross in a cubic lattice, and there are twice as many contributions to $\langle\overline{N}_{b_1,b_2}\rangle$ than to the other two cases (if $n_1 = n_2 = n/2$). Putting all these considerations together, we arrive at the following expression:

$$\langle \overline{N}_{TOT} \rangle \approx const \cdot \left\{ \frac{n_1^2}{R^2} b_1^2 + \frac{n_2^2}{R^2} b_2^2 + 2x \frac{n_1 n_2}{R^2} b_1 b_2 \right\} O[\ln(R/b)] \quad (8)$$

with an unknown constant x. Since the chain contour length is $L = n_1 b_1 + n_2 b_2 = n_2 b_2/(1-z)$, the span of a copolymer RW can be represented as $R^2 \approx n_2 b_2^2/(1-z)^2$. Using $n_1 = n_2 = n/2$, we get:

$$\begin{aligned} \langle \overline{N}_{TOT} \rangle &\approx const \cdot \frac{n}{2}(1-z)^2 \left\{ \left(\frac{b_1}{b_2}\right)^2 + 1 + 2x\frac{b_1}{b_2} \right\} O[\ln(n)] \\ &\approx const \cdot \left\{ (1-z)^2 + z^2 + 2xz(1-z) \right\} nO[\ln(n)], z \in [0, 1/2] \,. \end{aligned} \quad (9)$$

The z-function in Eq.(9) represents well the $a(z)$ in Fig. 2 up to $z = 1/2$, if we take x as a fitting parameter with values $x \in [3, 4]$; the particular case $x = 3$ ensures a *concave* $a(z)$ function, with a maximum at $z = 1/2$ and a ratio $a(1/2)/a(0) = 2$, similar to that found in the numerical simulations.

4. Conclusions

The numerical results in this work support the notion that the averaged mean number of overcrossings $\langle \overline{N} \rangle$ scales as "$n \ln n$" for the entire range of regular copolymers (i.e., nonequilateral walks with two distinct, alternating monomers), as expressed by the bond-length ratio z. It would also appear that the main qualitative features observed for the $\langle \overline{N}(z) \rangle$ function can be accounted by the rough mean-field argument leading to Eq.(9).[6,7]

It should be noted that mean-field approach incorporates *no* information on the *ordering* of the two different types of bonds b_1 and b_2 along the chain (that is, their "regularity" along the chain). Therefore, we conjecture that the observed scaling should also apply to copolymers with *nonregular, random sequences.* We believe that this law would remain valid for random-walk chains constructed as a sequence of i-bonds whose respective lengths b_i are chosen from a random distribution $b_i \in [0, b]$. In the latter case, we find a *general heteropolymer in an ideal solvent*; our results thus provide an estimation of how random variations of bond lengths or monomer types (e.g., variations in sequence composition) may affect the polymer's average folding features. The notion of the general validity of scaling laws for irregular chains is also supported by the fact that the backbones of experimental native states of maximally compact globular proteins resemble random nonequilateral walks, yet they are found in the same scaling regime as collapsed equilateral walks.[2]

As mentioned earlier, the observed scaling behaviour should be not only typical of simple copolymers, but also of other systems that are equivalent to polymers with two effective "bond lengths scales" or persistence lengths. An interesting case is found in randomly-charged *quenched* polyampholytes.[31] The ground state of these polymers is spatially organized as a series of locally-neutral compact globules connected by long strings. (Similar structures are found in unfolded protein ions.[10]) Interestingly, the mean size of the quenched polyelectrolytes [31] scales as $n^{1/2}$, thus providing an excellent example of a system with two distinct local size scales, yet the same global size scaling of a random walk.

Acknowledgments

This work is related to a presentation at the conference "Knots, random walks and biomolecules" (co-organized by J.H. Maddocks and A. Stasiak and sponsored by the Bernoulli Centre of the Swiss Federal Institute of Technology in Lausanne) held in Les Diablerets, Switzerland, during July 14-17, 2003. I thank the organizers for their wonderful hospitality and for providing a fantastic atmosphere for discussions. I would also like to thank A. Stasiak and Y. Diao for supplying the preprint [18], and for several useful comments that improved the present manuscript. This research was supported by NSERC (Canada) and the Canada Research Chairs' Program. I am indebted to the Department of Physical Chemistry and Prof. Tapia (Uppsala, Sweden), and Gamla Teatern (Östersund, Sweden) for the generous use of their facilities while completing part of this work.

References

1. G.A. Arteca. Overcrossing spectra of protein backbones: characterization of 3d molecular shape and global structural homologies. *Biopolymers*, 33:1829–1841, 1993.
2. G.A. Arteca. Scaling behavior of some molecular shape descriptors of polymer chains and protein backbones. *Phys. Rev. E*, 49:2417–2428, 1994.
3. G.A. Arteca. Molecular shape descriptors. In K. Lipkowitz and D.B. Boyd, editors, *Reviews in Computational Chemistry*, volume 9, chapter 5, pages 191–253. VCH Publishers, 1996.
4. G.A. Arteca. Dynamic molecular shape analysis of configurational transitions associated with melting and premelting of n-alkane chains. *J. Phys. Chem. B*, 101:4097–4104, 1997.
5. G.A. Arteca. Path integral calculation of the mean number of overcrossings in an entangled polymer network. *J. Chem. Inf. Comput. Sci.*, 39:550–557, 1999.

6. G.A. Arteca. Analytical estimation of scaling behavior for the entanglement complexity of a bond network. *J. Chem. Inf. Comput. Sci.*, 42:326–330, 2002.
7. G.A. Arteca. A measure of folding complexity for d-dimensional polymers. *J. Chem. Inf. Comput. Sci.*, 43:63–67, 2003.
8. G.A. Arteca, R.E. Cachau, and K. Veluri. Structural complexity of hydrogen-bonded networks. *Chem. Phys. Lett.*, 319:719–724, 2000.
9. G.A. Arteca and P.G. Mezey. The shapes of macromolecular backbones: three-dimensional characterization by spherical shape maps. *Biopolymers*, 32:1609–1621, 1992.
10. G.A. Arteca, C.T. Reimann, and O. Tapia. Transitions in chain entanglement and compactness associated with in vacuo unfolding of lysozyme ions. *J. Phys. Chem. B*, 105:4992–4998, 2001.
11. G.A. Arteca, K. Veluri, and O. Tapia. Pathways for folding and re-unfolding transitions in denatured conformations of anhydrous proteins. *Chem. Phys. Lett.*, 370:555–562, 2003.
12. Gustavo A. Arteca and O. Tapia. A simulation protocol to study proteins in vacuo. controlled re-folding and re-unfolding transitions. *Chem. Phys. Lett.*, 365:148–156, 2002.
13. G. Buck. Four-thirds power law for knots and links. *Nature*, 392:238–239, 1998.
14. J. Cantarella, R.B. Kusner, and J.M. Sullivan. Tight knot values deviate from linear relations. *Nature*, 392:237–238, 1998.
15. C. Chothia. One thousand families for the molecular biologist. *Nature*, 357:543–544, 1992.
16. V. Daggett. Molecular dynamics simulations of the protein unfolding/folding reaction. *Acc. Chem. Res.*, 35:422–429, 2002.
17. V. Daggett and A.R. Fersht. Is there a unifying mechanism for protein folding? *Trends Biochem. Sci.*, 28:18–25, 2003.
18. Y. Diao, A. Dobay, R.B. Kusner, K. Millett, and A. Stasiak. The average crossing number of equilateral random polygons. *J. Phys. A*, 36:11561–11574, 2003.
19. T. Edvinsson, C. Elvingson, and G.A. Arteca. Effect of compression on the molecular shape of polymer mushrooms with variable stiffness. *J. Chem. Phys.*, 116:9510–9517, 2002.
20. A.V. Finkelstein. Can protein unfolding simulate protein folding? *Protein Eng.*, 10:843–845, 1997.
21. P. Grassberger. Opacity and entanglement of polymer chains. *J. Phys. A: Math. Gen.*, 34:9959–9963, 2001.
22. O. Jagodzinski, E. Eisenriegler, and K. Kremer. Universal shape properties of open and closed polymer chains: renormalization group analysis and Monte Carlo experiments. *J. Phys. I France*, 2:2243–2279, 1992.
23. E.J. Janse van Rensburg, D.A.W. Sumners, E. Wasserman, and S.G. Whittington. Entanglement complexity of self-avoiding walks. *J. Phys. A*, 25:6557–6566, 1992.
24. L. Mirny and E. Shakhnovich. Protein folding theory: From lattice to all-atom models. *Annu. Rev. Biophys. Biomol. Struct.*, 30:361–396, 2001.

25. A.G. Murzin. How far divergent evolution goes in proteins. *Curr. Opin. Struct. Biol.*, 8:380–387, 1998.
26. J. Rudnick and G. Gaspari. The shapes of random walks. *Science*, 237:384–389, 1987.
27. A.R. Sanford, K. Yamato, X. Yang, L. Yuan, Y. Han, and B. Gong. Well-defined secondary structures. *Eur. J. Biochem.*, 271:1416–1425, 2004.
28. P.R. Sundararajan, G. Sacripante, and Z.Y. Wang. The possibility of intra-molecular nematic order via chain folding in perylene-containing polyimides: a molecular dynamics study. *Comp. Theoret. Polymer Sci.*, 10:219–220, 2000.
29. J.M. Thornton, C.A. Orengo, A.E. Todd, and F.M.G. Pearl. Proteins folds, functions and evolution. *J. Mol. Biol.*, 293:333–342, 1999.
30. V.N. Uversky and A.L. Fink. The chicken-egg scenario of protein folding revisited. *FEBS Lett.*, 515:79–83, 2002.
31. S. Wolfling and Y. Kantor. Model ground state of polyampholytes. *Phys. Rev. E*, 57:5719–5725, 1998.
32. B. Zagrovic, C.D. Snow, S. Khaliq, M.R. Shirts, and V.S. Pande. Native-like structure in the unfolded state of proteins. *J. Mol. Biol.*, 323:153–164, 2002.

CHAPTER 14

UNIVERSAL CHARACTERISTICS OF POLYGONAL KNOT PROBABILITIES

Kenneth C. Millett

Department of Mathematics,
University of California, Santa Barbara
Santa Barbara, CA 93016, USA
Email: millett@math.ucsb.edu
Homepage: http://www.math.ucsb.edu/~millett/KM.html

Eric J. Rawdon

Department of Mathematics and Computer Science,
Duquesne University
Pittsburgh, PA 15282, USA
Email: rawdon@mathcs.duq.edu
Homepage: http://www.mathcs.duq.edu/~rawdon

There is a striking qualitative similarity among the graphs of the relative probabilities of corresponding knot types across a wide range of random polygon models. In many cases one has theoretical results describing the asymptotic decay of these knot probabilities but, in the finite range, there is little theoretical knowledge and a variety of functional models have been used to fit the observed structures. In this paper we compare a selection of these models and study the extent to which each provides a successful fit for five distinct random knot models. One consequence of this study is that while such models are quite successful in this finite range, they do not provide the theoretically predicted asymptotic structure. A second result is the observed similarity between the global knot probabilities and those arising from small perturbations of three ideal knots.

1. Introduction

The probability that a random curve in 3-space is unknotted as a function of its length has long been the subject of experimentation, conjecture, and mathematical analysis in the context of lattice polygons, random 3-space polygons, and smooth curves. The graphs of these probability functions, independent of the context, have many fundamental similarities, only some of which are often captured in proposed equations to calibrate the dependence of the knot probability on the length of the curve. For example, Figs. 1-5 show the graphs of length to probability for five different knot models. In view of the fundamental similarity between the overall shape of these and other graphs, we wish to capture the key features of the functional dependence whose universal nature is illustrated in these graphs and which is the subject of this paper. One purpose is to explore a variety of proposed functional models seeking a single one that has the property of fitting the data at "finite scales" as well as in the "asymptotic scales". In addition, we seek a single functional model that works for unknots as well as any non-trivial knot type.

For example, we propose that in the trivial knot probability data, the largest number of edges for which only trivial knots occur, within a given physical model, plays a role analogous to the number of edges for which a non-trivial knot attains its maximal probability. The objective of determining the features of the model from finite scale data has potential experimental importance due to the computational complexity associated with the generation and precise analysis of the knot data in the asymptotic regime.

We will first describe the various models for knots we consider in this paper, geometric, equilateral, and perturbations of ideal trivial, trefoil and figure-eight knots. Next we describe the Monte Carlo exploration of these knot ensembles or knot spaces and review the methods of analysis that are employed. We will then give a qualitative description of the knot probability data with attention to the critical features of the data. Many functional models have been proposed for such data, some which are intended to apply only in certain regimes. The data considered here principally concerns the "finite range", that is for numbers of edges in the polygonal models which are less than those necessary for the known asymptotic exponential decay to account for the data values. We will describe some of these functional models and compare the application of these models to our data sets in the finite scales and their relationship to expected asymptotic behavior. We propose two simple models that appear to be sufficiently robust, capturing the key features of the data for both trivial and non-trivial knots provided by our models in this finite range. These finite scales are the regime of many important scientific applications and, as a consequence, these model functions may capture important aspects of their structure. A clear distinction between the finite and asymptotic ranges is visible in the structure of the best fit model functions.

2. Models of Knotting

A geometric polygonal knot K is defined by an ordered set of *vertices* $\{v_1, ..., v_n\}$ in $\mathbb{R}^3$ which are cyclically connected by straight line segments, called the *edges* of K. The set of all such knots is denoted $Geo(n)$. The subset of $Geo(n)$, where one requires that the length of each edge is equal to one, is called the space of equilateral knots, denoted $Equ(n)$. These spaces are considered as subsets of $3n$-dimensional Euclidean space and, as a consequence, inherit a natural distance function between two knots given by the maximum distance between the corresponding vertex coordinates. A third set of knots that we will consider in this paper consists of those knots in $Geo(n)$ each of whose vertices are, respectively, within a given distance r of the corresponding vertex of a knot K in $Equ(n)$. These sets are denoted $N(K, r)$. In particular, we will analyze the sets $N(K, r)$ where K is ropelength-minimized and r is the value of the thickness radius for K.

The thickness radius $r(K)$ of a polygonal knot employed in this paper is that defined by Rawdon in Refs. [22,23,24] which corresponds to the radius determined by polygonal analogs of the radius of curvature and the doubly-critical self-distance for smooth knots [15]. It provides an embedded solid torus neighborhood, whose radius we call the *thickness of* K, having K as its central curve. The *ropelength* is the ratio of the length of the polygon to the thickness radius. Given a fixed number of vertices and topological knot type that can be realized with the given number of edges, one can seek the minimum ropelength configuration of this type. These thickest knots have also been called *ideal knots* [13,18,19,25]. They are useful representatives for models as they provide spatial information that model physical phenomena. The family of knots created by a perturbation of the vertices of such an ideal knot K bounded by the radius of the thickest tube $r(K)$ determines the subset of $Geo(n)$ which we denote by $N(K, r(K))$. In particular, we will be analyzing $N(K, r(K))$ where K is a ropelength-minimized unknot, trefoil, or figure-eight knot. We call these spaces $Tube(0_1)$, $Tube(3_1)$, and $Tube(4_1)$, respectively, and suppress the number of edges n in the notation. Note that every polygon in $N(K, r(K))$ lies within the thick solid torus of radius $r(K)$ surrounding K and therefore are satellites of K.

3. Generation and Analysis of Knot Probability Data

The estimation of the relative proportion of a subset of a knot space, for example $Geo(n)$, consisting of a given knot type requires the uniform generation or sampling of the members of the entire space. Then the occurance of the knot type can be identified and the number of these in the sample gives an estimation of their frequency in $Geo(n)$. In this section we will describe how this sampling was accomplished for each of the knot spaces in which we are interested and, in addition, describe how the analysis of the knot type was achieved.

We define $Geo(n)$ to be the space of knots determined by the ordered set of n vectors uniformly distributed within the unit ball. These vectors are the vertices of the random knot. We propose this definition because the true set of geometric polygonal knots in 3-space, even those rooted at the origin, is a non-compact space. As a consequence, it is not possible to determine a proportion of the entire knot space, per se, determined by a given knot type. One must employ a secondary method. For the purposes of the present project, we have employed a spatial contraction that requires that the most distant vertex from the origin lie within the unit ball. The result is equivalent to a compactification of the knot space and allows one to estimate the knot probabilities in this knot space. We note, however, that this strategy to estimate geometric knot probabilities disrupts the spatial and physical structure of the knots and the average physical characteristics of the population. As a consequence, these knots may not provide a model for any biological or physical situation such as DNA knots or ring polymers in solution or melt. Rather, they are proposed as a vehicle to respond to the mathematical requirements of estimating the relative proportion of knots of various topological types.

The generation of equilateral knots requires a different strategy. Beginning with a regular polygon of n edges, we perform a sequence of random crankshaft rotations, which do not preserve the knot type, given by selecting two of the n vertices randomly, selecting one of the two arcs of segments of the polygon spanned by these vertices, and performing a random rotation of the segment about the axis determined by the two selected vertices. The number of pivot transformations, corresponding to steps in a random walk in $Equ(n)$, vary from 31,207,000, for the octagonal knots, to approximately 1,000,000 for some of the 150 estimates of knotting probabilities for n between 6 and 500.

While the regular equilateral polygon of n edges is the thickest unknot of n edges, the corresponding question as to what is the thickest configuration for a given number of edges and knot type is still an open question. There are, however, excellent candidates for these thickest knots. Our "thick" knots were computed by TOROS [21], which uses simulated annealing to minimize the ropelength (i.e. the ratio of length to thickness radius) of polygonal knots. We cannot assume that the resulting knots are truly ideal due to the finite capabilities of computers and the nature of simulated annealing. Therefore, we will refer to these knots as *tight* or *thick*, although we do believe that they are excellent approximations of the ideal conformations. The perturbations of one of the resulting tight knots K is determined by adding to each vertex a perturbation vector that is uniformly distributed in a ball of radius $r(K)$.

Once generated, the data is encoded so as to facilitate the calculation of the HOMFLY [8] polynomial using the Ewing-Millett program [7]. This polynomial is one of the very useful, but not entirely faithful, means to characterize the knot populations [10,11]. We employ them as a surrogates for the knot type and, thereby, create knot population distributions which are, in fact, distributions of the associated HOMFLY polynomials. For the range of edge numbers and knots we analyze here, the polynomial is an effective tool for measuring the probabilities of the trivial, trefoil, and figure-eight knots. One feature, however, that affects the quality of the resulting data for large numbers of edges derives from the complexity of the resulting knot presentations. We were unable to analyze presentations with more than 999 crossings in our preprocessing knot simplification program. Also, for those with less than 1,000 crossings, we were unable to analyze knots whose presentations could not be simplified to fewer than 250 crossings. These are the current limitations of the HOMFLY program. We believe that these two factors could lead to a reduced accuracy in the probability calculations for equilateral knots with larger numbers of edges.

4. Knot Probabilities and Associated Functional Models

In our study of knot probabilities, we begin with geometric polygonal knots, whose probability distribution as a function of the number of edges is shown in Fig. 1. The data starts at five edges since six edges are required for non-trivial knotting to occur. The proportion of unknots monotonically decreases from one to, asymptotically, zero with a characteristically exponential decay observed for large numbers of edges, i.e. in the aysmptotic regime. We note that, for the unknot data, the graph is concave down immediately following the number of edges required to create a non-trivial knot and that a change of concavity occurs later in the graph. The initial concavity resembles the behavior that is observed at a local maximum. This suggests the potential of a structure analogous to that observed at the point of maximal probability in non-trivial knot probability. Only for "large n" does one observe the characteristic exponential decay.

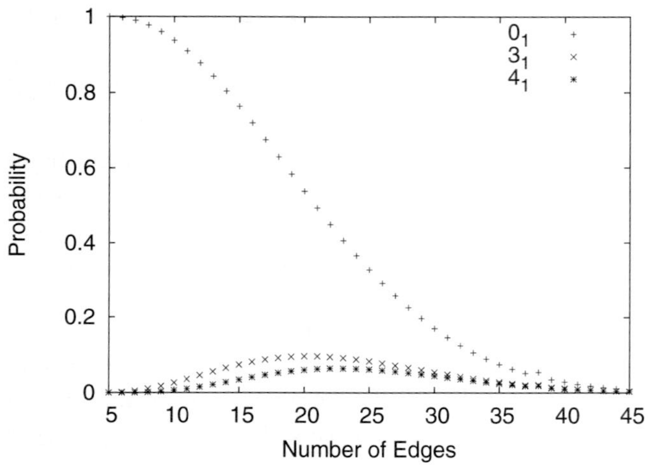

Fig. 1. Probability data for geometric polygonal knots.

For the trefoil and figure-eight knots, one observes a maximum and two changes in concavity: one between the initial data points and the point of maximum proportion and one following the maximum beyond which point the asymptotic range begins. The "asymptotic scale" dependence appears to be one of an exponential decay.

In Fig. 2, one can observe similarities and differences between data from equilateral polygonal knots and that of geometric polygonal knots. Figs. 3, 4, and 5 show the corresponding data for perturbations of the thick trivial, trefoil, and figure-eight knots respectively.

For $Equ(n)$ at larger numbers of edges (greater than 100), the data sets used to compute the proportions are sometimes relatively small, on the order of 1 million, compared to those employed for the smaller numbers of edges, on the order of 30 million for numbers of edges less than 20. At large numbers of edges, a generic projection has a large number of

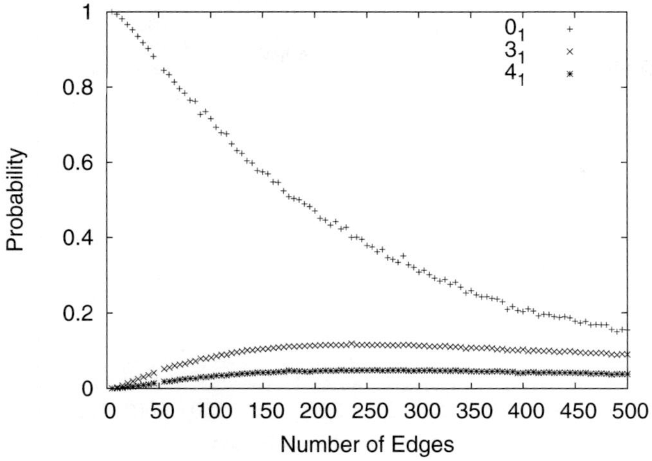

Fig. 2. Probability data for equilateral polygonal knots.

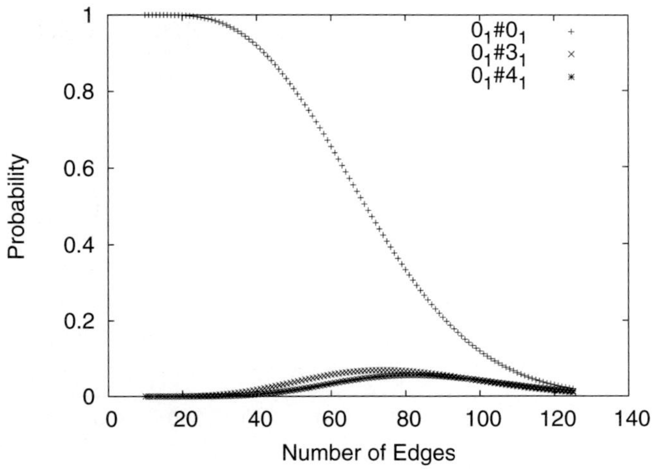

Fig. 3. Probability data for perturbations of the thick trivial knot.

crossings, often exceeding the current limitations of our program. While in theory the complexity of computing the HOMFLY polynomial increases exponentially on the number of crossings, for random knots there are many nugatory crossings which can be eliminated through careful preprocessing.

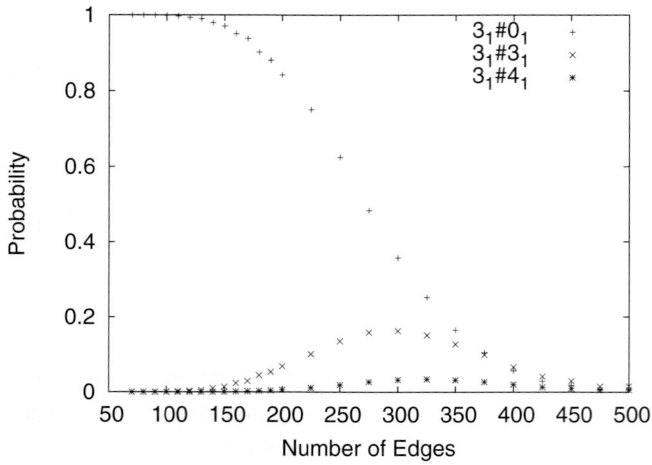

Fig. 4. Probability data for perturbations of the thick trefoil knot.

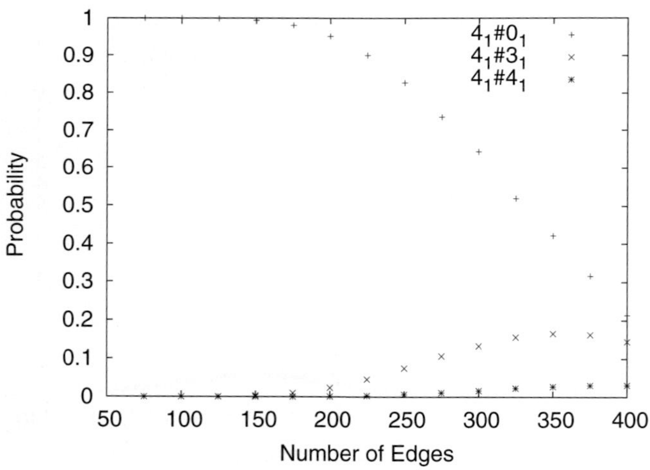

Fig. 5. Probability data for perturbations of the thick figure-eight knot.

Still, computing probabilities from a larger sample set for the larger numbers of edges in $Equ(n)$ was beyond our computational capacity [7].

For the probabilities within the thick trefoil tube, note that for $n < 200$, we have data points for every 10 edges. After $n = 200$, we have data points separated by 25 edges. This may slightly bias the fit towards the "small n" regime, although it does not appear have created any anomalous behavior.

We fit these different data sets with several functional models which will be defined in the following sections. The fitting was done using a combination of two programs: `gnuplot` and random perturbation programs written in Maple and Mathematica. `Gnuplot` uses the Marquardt-Levenberg method to flow the fitting parameters to "better values". This is done to minimize the sum of the squared residuals (which we will refer to as the SSR), i.e. the sum of the squared distances between the data points and the function values at these points

$$SSR = \sum_{i=1}^{n} (f(x_i) - y_i)^2$$

where f is the fitting function and (x_i, y_i) are the data points. Like any sort of gradient-type descent, this algorithm is vulnerable to becoming stuck in local minima within the parameter space. Furthermore, the algorithms for nonlinear graph fitting can be extremely sensitive to the starting values for the parameters. We tried several different starting parameter values to minimize the impact of these effects. Also, we employed large-scale perturbations to confirm the Marquardt-Levenberg parameter values and to reduce the risk of these local minima. Furthermore, noting that the fitting could be improved by using the Maple or Mathematica perturbation programs, we used decreasing scales of perturbations to improve the final SSR values. The extent of improvement we saw varied with each case.

4.1. *The Exponential Decay Model (ED)*

Although somewhat similar in nature, apparently distinct functions have been employed as models for the kinds of data shown in these figures or, at least, certain aspects of such data. First, Hammersley [9] shows that the number of self-avoiding polygons grows as $e^{\kappa n+o(n)}$, where n is the number of edges. Sumners and Whittington [26] and Pippenger [20] show that the number of those that are unknotted grows as Eqn. (1) so that the probability of unknotted polygons decreases as $e^{-\alpha n+o(n)}$. Diao et al. extended this result to Gaussian random polygons [5]. This has been refined, numerically in Refs. [4,14,16], to propose that the probability of the unknot is given by Eqn. (1) with n_c being a model dependent characteristic length of random knotting and n_0 being the minimal number of edges required to construct a non-trivial knot in this model. The exponential decay model is given by

$$P_0(n) = e^{-(n-n_0+1)/n_c} . \tag{1}$$

Because functions of this form are always decreasing and never equal to zero, this model will not achieve the goal of providing a single functional model that is valid for non-trivial as well as trivial knots. Furthermore, this functional model does not apply simultaneously to both the finite range and the asymptotic range as no change in concavity can occur for such functions. Parenthetically, we note that the location and size of the transition region between "finite range" and "asymptotic range" data appears uncertain. The data that serves as a basis for our study is consists principally in values within the "finite range". This plays a critical role in analysis of the fitting functions. As a consequence, we seek a functional model that addresses the change of concavity objective, an objective that appears to concern finite range behavior and not that of the asymptotic range. For an example of a study that might extend into the asymptotic range, we call attention to Moore, Lua, and Grosberg [17], who have reported an extensive analysis of data for knots from 15 to 3000 edges. Modulo large scale data collection and analysis concerns, their data fits this model very well at the larger scales. While not significant for their purposes, there is a less visible, but present, systematic departure from the model for small numbers of edges, i.e. in the finite range.

Fig. 6 shows a best fit of the probability that a knot in $Geo(n)$ is an unknot as a function of the number of edges for the exponential decay function model: $n_0 = 15.2901$, $n_c = 9.0695$, and the sum of the square residuals SSR is 0.3448.

4.2. *The Deguchi-Tsurusaki Model (DT)*

For non-trivial knots, Deguchi and Tsurusaki [1,2,3] have proposed Eqn. (2) where p_K is called the topological exponent of the knot, n_c is a variable to be fit, and n_K is the minimal number of edges required to create the knot. So long as p_K is greater than one, such functions satisfy the initial concavity requirement discussed earlier for non-trivial knots. They are, therefore, attractive functional model candidates with which to model this data. They also provide the desired asymptotic exponential decay and exhibit the changes in concavity observed in the data for both trivial and non-trivial knots. This model has been successfully used by Katritch et al. [12] (using a slightly different, yet equivalent, form) to model knotting

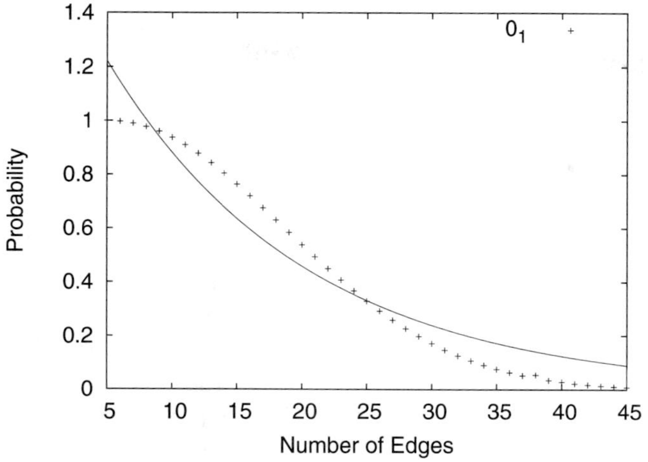

Fig. 6. Unknots in $Geo(n)$: fitting with the exponential decay model.

data beyond the finite range. The Deguchi-Tsurusaki model is given by

$$P_K(n) = C_K((n - n_K + 1)/n_c)^{p_K} e^{(-(n-n_K+1)/n_c)} . \tag{2}$$

While in fitting the data we did not require that p_K be greater than one, the fitting process often provides this property in the Deguchi-Tsurusaki model.

First, we ask if it is possible to model the trivial knot using the same functional model. Since the unknot has a different character than the non-trivial knots, we allow the n_K parameter to be a free parameter when fitting to the unknot data. Fig. 7 shows a best fit of the probability that a random knot in $Geo(n)$ is a trivial knot as a function of the number of edges for the Deguchi-Tsuruski model: $C_0 = 5.7752e - 31$, $n_0 = -53.4664$, $n_c = 2.0820$, $p_0 = 29.2866$, and $SSR = 0.002249$. While the Deguchi-Tsurusaki function fits this data quite well, there is a systematic deviation apparent in the graph. One might suspect that this is a consequence of the finite range constraint on the data set. An objective of the next models will be to provide a better fit, both numerically and visually, with the data.

4.3. *The Dobay et al. Model (DSDS)*

For non-trivial knots, Dobay, Sottas, Dubochet, and Stasiak [6] have proposed Eqn. (3) where p_K, called the topological exponent of the knot, n_c,

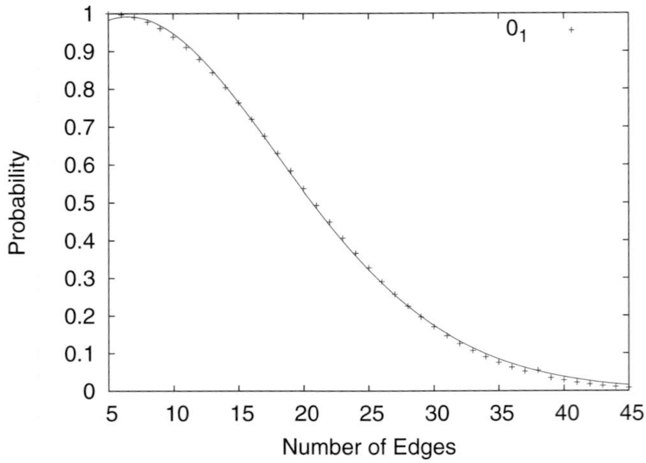

Fig. 7. Unknots in $Geo(n)$: fitting with the Deguchi-Tsurusaki model.

and q_K are variables to be fit, and n_K is the minimal number of edges required to create the knot. So long as p_K is greater than one and n_c and q_K are positive, such functions satisfy the initial concavity requirement discussed earlier for non-trivial knots and, therefore, are attractive functional model candidates with which to model this data. Dobay et al. [6] chose q_K to be an empirical constant equal to 0.18. The values of q_K arising from fitting our data do not suggest the existence of a single value that will work for all data. Functions of this nature do not provide the desired asymptotic exponential decay but do exhibit the changes in concavity observed in the data for both trivial and non-trivial knots. This model was proposed in the study of random walks in which the ends of the walk come in close vicinity (with distance less than or equal to two steps) to each other and thereby provides an opportunity for closure. The model is proposed to take into account the probability of closure in the course of a random walk. The Dobay et al. model is given by

$$P_K(n) = C_K(n - n_K + 1)^{p_K} e^{(-(n^{q_K})/n_c)} . \tag{3}$$

In fitting the data we did not require that p_K be greater than one although the fitting process often provides this property.

As earlier, we ask if it is possible to model the trivial knot using the same functional model. Since the unknot has a different character than the non-trivial knots, we allow the n_K parameter to be a free parameter when fitting to the unknot data. This is also true in the models proposed in the next two sections when applied to the trivial knot. Fig. 8 shows a best fit of the probability that a random knot in $Geo(n)$ is a trivial knot as a function of the number of edges for the Dobay et al. model: $C_0 = 0.3780$, $n_0 = -11.8976$, $n_c = 774.6570$, $p_0 = 0.3519$, $q_0 = 2.1763$, and $SSR = 0.0001348$.

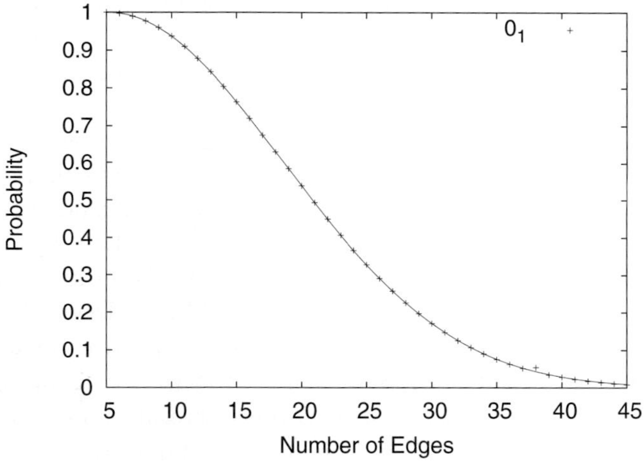

Fig. 8. Unknots in $Geo(n)$: fitting with the Dobay et al. model.

4.4. *The Quadratic Variation (QV)*

In the search for models that could better fit the observed finite range data, we use the Deguchi-Tsurusaki model as our point of departure by giving the function an equivalent but slightly different format and including quadratic factors to test the potential importance of second order effects in the exponential term in this range. The fact that theoretical results show that quadratic decay does not occur asymptotically does not preclude the presence of such effects in the finite range. The quadratic variation model is given by

$$P_K(n) = C_K((n - n_K + 1)^{p_K})e^{(-k_K n - l_K n^2)} . \tag{4}$$

As in the previous models n_K represents the number of edges required to create the knot type in the given model, except in the case of trivial knots when it is allowed to be a free parameter.

The remaining values are fitted to the data in an attempt to capture critical features of the data such as inflection points, the maximum, etc. For non-trivial knots, at $n = n_K - 1$, $P_K(n) = 0$, implying that p_K is greater than zero. In view of the initial concavity of the probability function data for non-trivial knots, one might also require that p_K be greater than one for non-trivial knots, although we have not done so here. For the unknot, we have allowed the values for n_0 and p_0 to be free parameters determined by the data. $P_0(n)$ is a monotonically decreasing function for $n \geq n_0$ with a single inflection point. Theoretical results, Sumners and Whittington [26] and Pippenger [20], show that the decrease is a linear exponential decay. To insure this, we require the parameters k_0 and l_0 to be positive. For non trivial knots, $P_K(n)$ is a monotonically decreasing function for n greater than the unique value for which $P_K(n)$ attains its maximum, and has a single inflection point. For the trivial geometric random knots, Fig. 9 shows the best fit for the quadratic variation model: $C_0 = 1.1614e - 51$, $n_0 = -53.4692$, $p_0 = 29.2908$, $k_0 = 0.4803$, $l_0 = 2.0825e - 40$, and $SSR = 0.002249$, roughly the same quality of fit achieved by the Deguchi-Tsurusaki function. The size of "l" may suggest an artifact of the fitting process rather than the presence of a significant second order term. A significant second order term would be inconsistent with theoretical results on the exponential decay and provide evidence of the domination of finite range data.

Further evidence that the data's dominant features reside in the finite range is clearly provided by allowing the fitting function to have fuller freedom in the exponent, for example by allowing k and/or l to have negative values. Let QV_- denote this relaxed model. Allowing k and l to have negative values gives an order of magnitude improvement in the fit. This results in a rather different function having the parameter values: $C_0 = 0.9145614$, $n_0 = 4.0371$, $p_0 = 0.0001229$, $k_0 = -0.03179$, $l_0 = 0.002920$, and $SSR = 0.0001853$. This fit is of the same order of magnitude as that given by the Dobay et al. model where $SSR = 0.0001348$.

4.5. *The Full Variation Model (FV)*

Finally, we test for a second order variation in the power law factor by introducing an independent power term. The full variation model is given by

$$P_K(n) = C_K(n - n_K + 1)^{p_K}(1 + b_K(n - n_K + 1)^{q_K})e^{(-k_K n - l_K n^2)} \,. \quad (5)$$

For non-trivial knots, in order that P_K be defined at $n = n_K - 1$, we require that both p_K and q_K be greater than 0 and we invoke the same

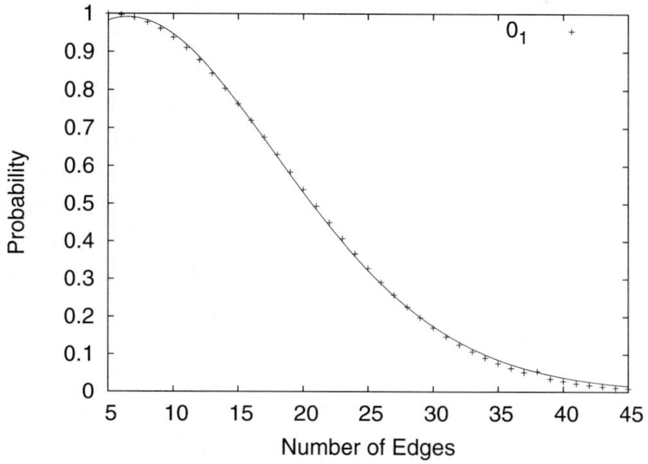

Fig. 9. Unknots in $Geo(n)$: fitting with the quadratic variation model.

constraint on the functional models of trivial knots.

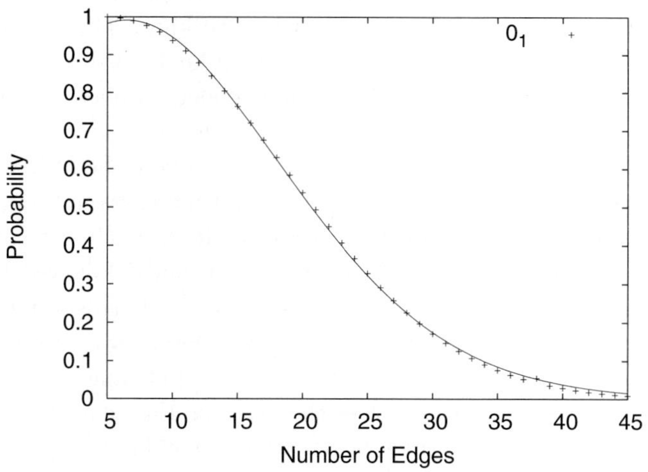

Fig. 10. Unknots in $Geo(n)$: fitting with the full variation model.

For the case of geometric random unknots, Fig. 10 shows the fit using the full variation model: $C_0 = 1.1600e - 51$, $n_0 = -53.4702$, $p_0 = 29.2910$,

$q_0 = 0.2263$, $b_0 = 0.000002700$, $k_0 = 0.4803$, $l_0 = 2.0823e - 40$, and $SSR = 0.002249$.

The evidence of the finite range character of the data appearing in the quadratic variation model is also present in the full variation. By allowing the fitting function to have negative k and l, we define the FV_- relaxed model. Fitting this model results in a rather different function having the parameter values: $C_0 = 0.9123$, $n_0 = 3.3827$, $p_0 = 0.00009720$, $q_0 = 2.2828$, $b_0 = 0.0002283$, $k_0 = -0.03465$, $l_0 = 0.003423$, and $SSR = 0.0001399$. This fit is essentially the same as that given by the Dobay et al. model where $SSR = 0.0001348$.

4.6. *Applications to Unknot Data*

In addition to geometric polygonal unknot data, we have applied these same functional models to the case of equilateral polygonal trivial knots and the vertex perturbations of the tight unknot, trefoil knot, and the figure-eight knots which are less than the radius of the thick tube surrounding the core knot. In the case of the perturbations within the tight knots, one always gets a satellite knot, usually in the form of a connected sum with the core knot. Thus, trivial knots within these emsembles correspond to perturbations which result in the same knot type as the core knot. The results of our analysis are similar to that for geometric trivial knots. Table 1 reports the parameters for each of the four functional models applied to the trivial knots in the five data sets.

For the geometric unknots, shown in Fig. 11, where the SSR values are plotted on a logarithmic scale, we see significant increases in the quality of fit as one goes from the exponential decay to the Deguchi-Tsurusaki and Dobay et al. models to the quadratic variation model followed by a less significant improvement in the full variation model. This is not uniformly the case with all the unknot data sets as one observes in Table 2. Here we see that all models give equivalent fits for the full range of data of the equilateral unknots while this is less the case for the geometric unknots, and the perturbations of the thick trivial knot, trefoil knot, and figure-eight knot.

Since the Dobay et al. and the QV_- and FV_- (in which the exponential term is not linear or the linear part has a positive coefficient) most often give the best fit to the data, one is lead to conclude that the nature of the finite range is quite different from that of the asymptotic range where theoretical results prove a linear exponential decay. This conclusion is supported by

Table 1. The fitting parameters for the unknot probability.

				Probability of 0_1				
Model	C_K	n_K	n_c	p_K	k_K	l_K	b_K	q_K
				$Equ(n)$				
ED		1.31e+01	2.50e+02					
DT	1.07e+00	4.44e-10	2.44e+02	9.24e-03				
DSDS	9.51e-01	5.95e-01	1.17e+02	4.75e-02				8.92e-01
QV	1.06e+00	6.00e+00		4.30e-03	4.10e-03	4.31e-29		
QV_-	1.04e+00	5.98e+00		4.46e-03	4.06e-03	4.47e-29		
FV	1.04e+00	5.98e+00		4.27e-03	4.06e-03	4.45e-29	1.00e-08	8.65e-09
FV_-	1.04e+00	5.98e+00		4.27e-03	4.06e-03	4.45e-29	1.00e-08	8.65e-09
				$Geo(n)$				
ED		9.07e+00	1.53e+01					
DT	5.78e-31	-5.35e+01	2.08e+00	2.93e+01				
DSDS	3.78e-01	-1.19e+01	7.75e+02	3.52e-02				2.17e+00
QV	1.16e-51	-5.35e+00		2.93e+01	4.80e-01	2.08e-40		
QV_-	9.15e-01	4.04e+00		1.23e-04	-3.18e-02	2.92e-03		
FV	1.16e-51	-5.35e+01		2.93e+01	4.80e-01	2.08e-40	2.65e-06	2.26e-01
FV_-	9.12e-01	3.38e+00		9.72e-05	-3.47e-02	3.42e-03	2.28e-04	2.28e+00
				$Tube(0_1)$				
ED		2.51e+01	5.61e+01					
DT	2.97e-13	-1.05e+02	8.09e+00	1.62e+01				
DSDS	1.07e-33	-4.25e+01	8.10e-01	2.06e+01				6.77e-01
QV	1.16e-33	-1.06e+02		1.62e+01	1.23e-01	2.60e-06		
QV_-	8.32e-01	9.61e-01		3.87e-04	-1,67e-02	3.51e-04		
FV	1.15e-33	-1.06e+02		1.62e+01	1.23e-01	2.64e-06	7.14e-05	2.32e-02
FV_-	1.17e-01	8.49e-01		3.36e-04	-1,67e-02	3.51e-04	6.07e+00	3.93e-07
				$Tube(3_1)$				
ED		1.15e+02	2.01e+02					
DT	2.01e+00	-8.46e+00	6.40e+01	1.97e+00				
DSDS	6.89e-21	-2.08e+02	4.65e+01	8.54e+00				1.03e+00
QV	1.69e-03	3.07e-02		1.71e+00	1.47e-02	6.54e-09		
QV_-	6.36-01	7.88e-01		3.33e-07	-7978e-03	3.34e-05		
FV	9.53e-02	4.25e-03		5.59e-01	7.69e-06	2.06e-05	5.69e-05	4.32e-02
FV_-	4.16e-01	8.75e-07		2.39e-10	-7.28e-03	3.11e-05	6.022-01	1.51e-14
				$Tube(4_1)$				
ED		1.12e+02	3.79e+02					
DT	2.85e-07	-3.77e+02	4.59e+01	1.09e+01				
DSDS	1.28e-29	-8.96e+01	5.59e-01	1.59e+01				4.87e-01
QV	4.04e-29	-3.78e+02		1.10e+01	2.19e-02	3.54e-09		
QV_-	7.89e-01	-6.84e+00		2.23e-06	-4,42e-03	1.76e-05		
FV	3.01e-29	-3.79e+02		1.10e+01	2.19e-02	3.59e-09	6.43e-09	2.60e-02
FV_-	1.27e-01	9.94e+00		1.93e-04	-4.41e-03	1.76e-05	5.21e+00	3.87-04

the corresponding SSR values for the trefoil and figure-eight knots in Table 2.

Deguchi and Tsurusaki [4] propose that the parameter, p_K "should be universal: it is independent upon the models of random polygon(s) and is

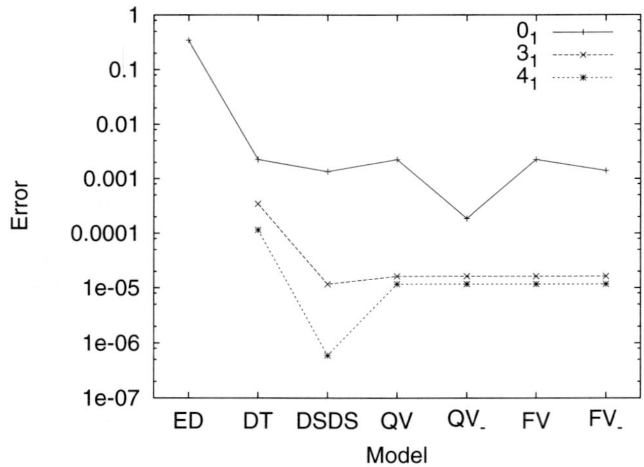

Fig. 11. Comparison of SSR values of the exponential decay, Deguchi-Tsurusaki, Dobay et al., quadratic variation, and full variation models for trivial, trefoil, and figure-eight knots in $Geo(n)$.

determined only by the knot K". While our data may concern models for random polygons outside the category they had in mind, we have explored the degree to which such universality holds. It appears that the different data sets require rather different values for p_K as shown in Fig. 12.

Comparison of p_K for the full variation model is shown in Fig. 13. Here also, one observes significant variation in the values of p_K for the four polygonal trivial knot models.

Tables 1, 3, and 4 report the parameter values for the different models and knot space data sets for the equilateral, geometric, and perturbations of the tight trivial knot, trefoil knot, and figure-eight knots. Table 2 gives the associated SSR values for each set of parameters.

5. Analysis of Functional Models of Non-trivial Knot Probability

With the exception of the exponential decay model, the functional models have been developed to study knotting probability of non-trivial knots, and therefore, one expects a better fit with the data will be possible. We will not include the exponential decay model in our analysis as it is of interest only in the asymptotic range. Furthermore, in this note, we explore the data for two nontrivial knot types, the trefoil knot and the figure-eight knot,

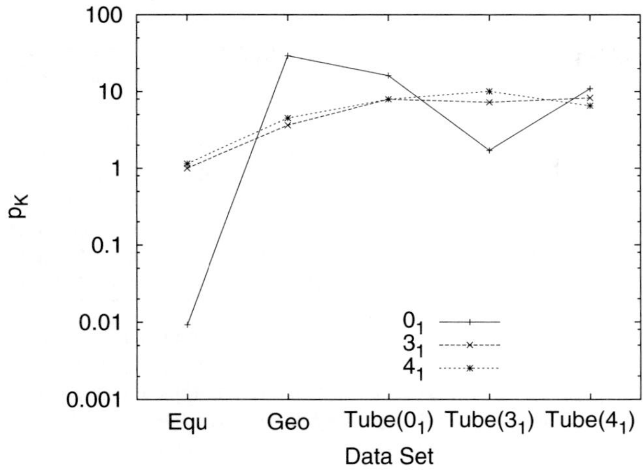

Fig. 12. Comparison of p_K of the Deguchi-Tsurusaki model for all knots and data sets.

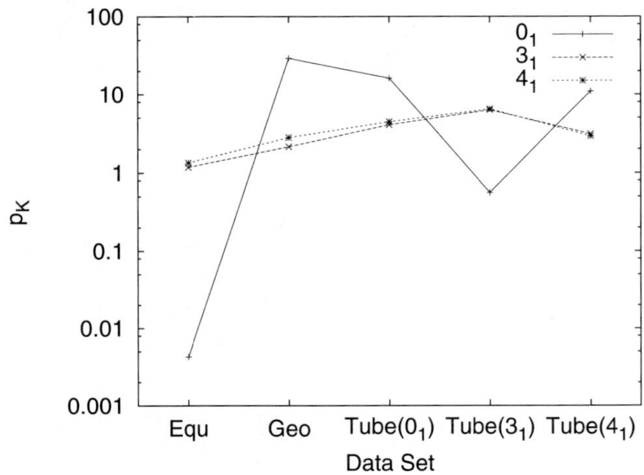

Fig. 13. Comparison of p_K of the full variation model for all knots and data sets.

as they occur in five polygonal knot contexts: equilateral, geometric and perturbations of the tight unknot, trefoil knot, and figure-eight knot.

Table 2. SSR values for all knots, data sets, and models.

	SSR values				
Model	$Equ(n)$	$Geo(n)$	$Tube(0_1)$	$Tube(3_1)$	$Tube(4_1)$
			0_1		
ED	1.2117e-02	3.4482e-01	2.0215e+00	6.1453e-01	2.5860e-01
DT	9.0422e-03	2.2490e-03	7.5266e-02	7.8516e-02	3.1228e-02
DSDS	5.1867e-03	1.3483e-03	2.4599e-02	3.6708e-02	1.8822e-02
QV	9.1813e-03	2.2488e-03	7.4494e-02	8.3402e-02	3.1181e-02
QV_-	4.9453e-03	1.8529e-04	2.1224e-02	2.5167e-02	1.6621e-02
FV	4.9450e-03	2.2488e-03	7.4469e-02	2.7480e-02	3.1149e-02
FV_-	4.9450e-03	1.3987e-03	2.1217e-02	1.2541e-02	1.6647e-02
			3_1		
DT	4.0408e-04	3.4432e-04	2.5416e-04	7.2692e-04	5.7876e-05
DSDS	1.4319e-04	1.1645e-05	1.1827e-06	1.2210e-04	3.0108e-05
QV	4.0422e-04	1.6298e-05	9.8789e-07	7.2692e-04	5.4966e-05
QV_-	4.0422e-04	1.6298e-05	9.8789e-07	1.1109e-04	1.1787e-05
FV	3.4764e-04	1.6298e-05	9.8788e-07	8.9632e-05	5.4966e-05
FV_-	3.4764e-04	1.6298e-05	9.8788e-07	8.9632e-05	1.1767e-05
			4_1		
DT	1.1416e-04	1.1504e-04	1.2672e-04	1.8650e-05	1.4368e-06
DSDS	1.8545e-05	5.9223e-07	2.9904e-06	9.7717e-06	5.3741e-07
QV	1.1750e-04	1.1693e-05	7.8559e-07	9.7804e-06	7.9669e-07
QV_-	1.1750e-04	1.1693e-05	4.3770e-07	9.7440e-06	5.1171e-07
FV	1.0769e-04	1.1692e-05	5.9955e-07	7.1188e-06	7.9141e-07
FV_-	1.0769e-04	1.1692e-05	4.3343e-07	7.1188e-06	5.1171e-07

5.1. *The Trefoil Knot*

As in the case of the equilateral unknots, the Deguchi-Tsurusaki, quadratic variation, and full variation functional models are essentially equally successful for trefoil knots. This is also the case for the perturbations of the tight figure-eight knot but not for the geometric trefoil knots nor for the perturbations of either the tight unknot or trefoil knot. The quadratic variation is essentially as good a model as the full variation and both are substantially better than the Deguchi-Tsurusaki model as shown in Fig. 14. Note that because the different data sets have different numbers of data points, one should compare the quality of the fitting models within a particular data set, not across the data sets. We include the connecting lines only to simplify the reading of the graph.

Comparison of p_{3_1} for the Deguchi-Tsurusaki model with that of the full variation model is shown in Figs. 12 and 13. Here also, one observes significant variation in the values of p_{3_1} for the five polygonal models.

Table 3. The fitting parameters for the trefoil probability.

Probability of 3_1								
Model	C_K	n_K	n_c	p_K	k_K	l_K	b_K	q_K
$Equ(n)$								
DT	3.16e-01	6.00e+00	2.47e+02	9.99e-01				
DSDS	4.25e-04	6.00e+00	7.09e+00	1.55e+00				5.49e-01
QV	1.31e-03	6.00e+00		9.99e-01	4.04e-03	6.96e-10		
QV_-	1.31e-03	6.00e+00		9.99e-01	4.04e-03	6.96e-10		
FV	2.55e-03	6.00e+00		1.18e+00	3.43e-03	3.62e-12	-5.90e-01	6.43e-02
FV_-	2.55e-03	6.00e+00		1.18e+00	3.43e-03	3.62e-12	-5.90e-01	6.43e-02
$Geo(n)$								
DT	3.44e-02	6.00e+00	4.05e+00	3.63e+00				
DSDS	1.30e-03	6.00e+00	2.25e+02	2.11e+00				1.92e+00
QV	1.36e-03	6.00e+00		2.15e+00	1.80e-02	2.97e-03		
QV_-	1.36e-03	6.00e+00		2.15e+00	1.80e-02	2.97e-03		
FV	1.36e-03	6.00e+00		2.15e+00	1.80e-02	2.97e-03	9.34e-16	1.11e-08
FV_-	1.36e-03	6.00e+00		2.15e+00	1.80e-02	2.97e-03	9.34e-16	1.11e-08
$Tube(0_1)$								
DT	7.77e-05	1.10e+01	8.79e+00	7.09e+00				
DSDS	4.45e-08	1.10e+01	1.76e+03	4.00e+00				1.94e+00
QV	4.30e-08	1.10e+01		4.05e+00	5.53e-03	3.90e-04		
QV_-	4.30e-08	1.10e+01		4.05e+00	5.53e-03	3.90e-04		
FV	4.30e-08	1.10e+01		4.05e+00	5.53e-03	3.90e-04	9.34e-06	1.11e-02
FV_-	4.30e-08	1.10e+01		4.05e+00	5.53e-03	3.90e-04	9.34e-06	1.11e-02
$Tube(3_1)$								
DT	1.28e-04	8.00e+01	2.84e+01	7.27e+00				
DSDS	3.55e-09	8.00e+01	7.97e+05	3.65e+00				2.51e+00
QV	5.73e-14	8.00e+01		7.27e+00	3.53e-02	1.22e-26		
QV_-	1.25e-08	8.00e+01		2.95e+00	-1.73e-02	5.27e-05		
FV	2.06e-11	8.00e+01		6.31e+00	4.24e-02	7.20e-06	5.28e-10	4.33e+00
FV_-	2.06e-11	8.00e+01		6.31e+00	4.24e-02	7.20e-06	5.28e-10	4.33e+00
$Tube(4_1)$								
DT	1.54e-05	7.50e+01	3.38e+01	8.30e+00				
DSDS	6.53e-10	7.50e+01	5.96e+09	3.65e+00				3.86e+00
QV	7.01e-17	7.50e+01		8.09e+00	2.88e-02	1.16e-26		
QV_-	5.66e-10	7.50e+01		3.09e+00	-2.27e-02	4.75e-05		
FV	7.02e-17	7.50e+01		8.09e+00	2.88e-02	1.16e-26	9.37e-06	1.11e-02
FV_-	5.60-10	7.50e+01		3.10e+00	-2.26e-02	4.75e-05	9.99e-10 0	1.00e+00

5.2. *The Figure-Eight Knot*

As in the case of the equilateral unknots and the trefoil knot, the Deguchi-Tsurusaki, Dobay et al., quadratic variation, and full variation functional models are essentially equally successful for equilateral figure-eight knots. This is also the case for the perturbations of the tight figure-eight knot but not for the geometric trefoil knots nor for the perturbations of either the tight unknot or trefoil knot. For the geometric figure-eight knots and for

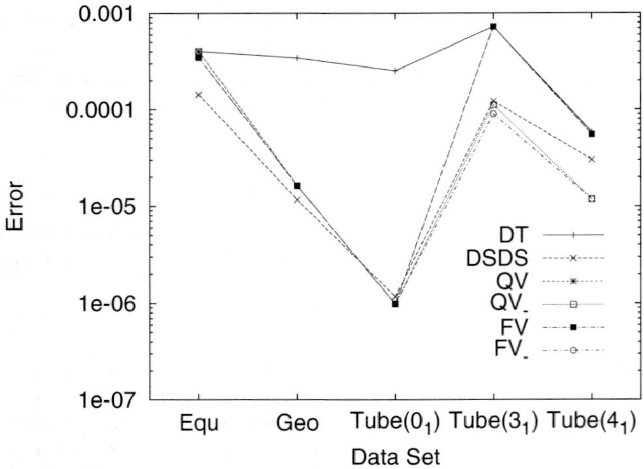

Fig. 14. Comparison of SSR values of the Deguchi-Tsurusaki, Dobay et al., quadratic variation, and full variation models for all trefoil knots.

the perturbations of the tight unknot and the trefoil knot, the quadratic variation is essentially as good a model as the full variation and both are substantially better than the Deguchi-Tsurusaki model as shown in Fig. 15.

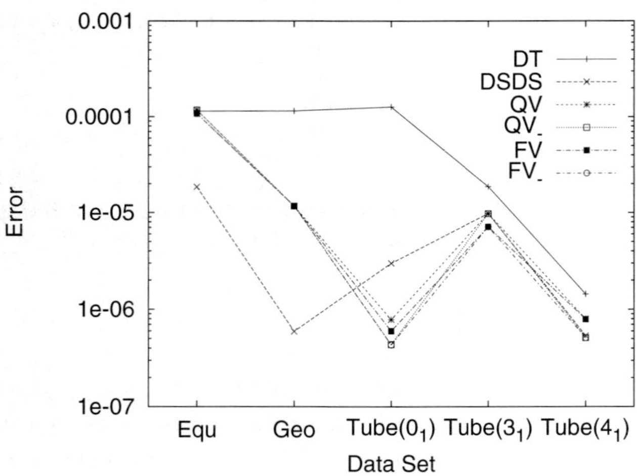

Fig. 15. Comparison of SSR values of the Deguchi-Tsurusaki, Dobay et al., quadratic variation, and full variation models for all figure-eight knots.

Table 4. The fitting parameters for the figure-eight probability.

				Probability of 4_1				
Model	C_K	n_K	n_c	p_K	k_K	l_K	b_K	q_K
				$Equ(n)$				
DT	1.32e-01	7.00e+00	2.24e+02	1.15e+00				
DSDS	6.89e-05	7.00e+01	3.69e+00	1.92e+00				4.88e-01
QV	2.80e-04	7.00e+00		1.14e+00	4.43e-03	1.26e-10		
QV_-	2.80e-04	7.00e+00		1.14e+00	4.43e-03	1.26e-10		
FV	1.28e-03	7.00e+00		1.35e+00	3.82e-03	2.04e-12	-8.27e-01	2.31e-02
FV_-	1.28e-03	7.00e+00		1.35e+00	3.82e-03	2.04e-12	-8.27e-01	2.31e-02
				$Geo(n)$				
DT	6.37e-03	7.00e+00	3.54e+00	4.54e+00				
DSDS	2.08e-04	7.00e+01	2.14e+02	2.72e+00				1.93e+00
QV	2.34e-04	7.00e+00		2.80e+00	3.13e-02	3.02e-03		
QV_-	2.34e-04	7.00e+00		2.80e+00	3.13e-02	3.02e-03		
FV	2.34e-04	7.00e+00		2.80e+00	3.13e-02	4.13e-04	4.79e-08	1.27e+00
FV_-	2.34e-04	7.00e+00		2.80e+00	3.13e-02	4.13e-04	4.79e-08	1.27e+00
				$Tube(0_1)$				
DT	1.10e-05	1.60e+01	8.21e+00	7.95e+00				
DSDS	2.99e-06	1.60e+01	1.26e+03	4.78e+00				1.88e+00
QV	3.83e-09	1.60e+01		4.63e+00	2.93-03	4.07e-04		
QV_-	5.82e-09	1.60e+01		4.43e+00	-4.26-03	4.31e-04		
FV	4.28e-09	1.60e+01		4.58e+00	1.01e-03	4.13e-04	8.31e-08	1.27e-03
FV_-	5.73e-09	1.60e+01		4.44e+00	-3.97e-03	4.30e-04	9.97e-08	1.11e-03
				$Tube(3_1)$				
DT	6.06e-08	8.00e+01	2.36e+01	1.01e+01				
DSDS	9.77e-06	8.00e+01	1.95e+04	6.21e+00				1.96e+00
QV	4.32e-15	8.00e+01		6.17e+00	6.37e-04	3.85e-05		
QV_-	1.45e-14	8.00e+01		5.79e+00	-3.25e-03	4.19e-05		
FV	2.43e-13	8.00e+01		6.49e+00	3.78e-02	1.06e-05	5.04e-10	4.49e+00
FV_-	2.43e-13	8.00e+01		6.49e+00	3.78e-02	1.06e-05	5.04e-10	4.49e+00
				$Tube(4_1)$				
DT	1.01e-04	1.25e+02	4.27e+01	6.53e+00				
DSDS	5.37e-07	1.25e+02	2.41e+07	4.06e+00				2.96e+00
QV	1.84e-12	1.25e+02		4.84e+00	3.72e-04	2.20e-05		
QV_-	1.47e-10	1.25e+02		2.90e+00	-2.60e-02	4.71e-05		
FV	1.93e-10	1.25e+02		4.82e+00	1.14e-04	2.22e-05	-7.55e-11	1.64e-07
FV_-	1.47e-10	1.25e+02		290e+00	-2.60e-03	4.71e-05	-1.25e-11	1.06e-02

Comparison of p_{4_1} for the Deguchi-Tsurusaki model with that of the full variation model is shown in Figs. 12 and 13. Here also, one observes significant variation in the values of p_{4_1} for the five polygonal models.

6. Conclusions and Speculations

This search for a universal functional model for the probability of specific knot types in a range of models for knots has shown that there may exist such functions, but only with some constraint on the range of application and the expectation that the model be effective in both the finite and asymptotic ranges. For example, the Dobay et al. model is usually the most successful in fitting the data in the range of the data used in this article. Similarly, the QV_- and FV_- models are more successful than the remaining models, but these three models suffer from the defect that are not consistent with the theoretical results giving the specific nature of the asymptotic decay. Among those models that more closely match the asymptotic requirements, the results of this project suggest that the quadratic variation of the exponential portion of the Deguchi-Tsurusaki function is sufficient to capture much of the structure of the data. One is still motivated to relax the expectation that the fitting function be consistent with the theoretically proven linear exponential decay in the asymptotic range. Even with some additional flexibility, however, the model is not entirely successful in capturing all the desired finite range structure of critical behaviour as is shown in Fig. 16. Here the model fails to closely follow the slow decay near the initial data points. This same problem is encountered with efforts to fit the trivial knot distributon in other data sets. Other challenging features, such as the concavity of the function for non-trivial knots, provide additional problems as illustrated in Fig. 17. Notice how the data values wander back and forth across the graph of the model function. This suggests a "higher order" behavior that is not captured in the model function assuming, of course, that the data are sufficiently accurate to demonstrate this structure.

Overall, however, one is struck by the qualitative fit achieved by the quadratic variation in almost all cases when the requirement of proper asymptotic behavior is relaxed. This provides evidence of the existence of a phase transition separating the "finite scale" from the "asymptotic scale" in which the linear exponential decay is sufficient to give a good fit of the data, e.g. as given by the Deguchi-Tsurusaki function. Similarly, one observes a close fit of the data using the Dobay et al. model when we relax the condition of linear exponential decay. The various functions do not uniformly provide the same quality of fit in the finite range, even when one allows full freedom in the exponential coefficients. This can be observed through a comparison of the SSR values achieved for various models, even those which are not consistent with the asymptotic decay to those achieved which are consistent. Note that, in the case of the Dobay et al. model, there is an exceptional improvement over the fit achieved by the other models when considering fitting figure-eight knot data.

Thus, we are led to two conclusions. First, there is no known single function that adequately models the data in both the finite and asymptotic scales. Second, while the Deguchi-Tsurusaki function may be the optimal fitting function in the asymptotic regime, our quadratic variation is a better option in the finite scale regime.

Consideration of the potential universality of p_K across the five models shows that there is good agreement between the equilateral and geometric trefoil and figure-eight knot values, but that this is no longer the case for the perturbations as shown in Figs. 12 and 13.

The data analysis undertaken here provides information to test the extent of universality of these significant functional parameters for each of the random knot models: the equilateral knots, the geometric knots, and the perturbations of the thick trivial, trefoil, and figure-eight knots. We propose that the similarity one observes in the nature of the knot probability distribution functions is a reflection of much larger similarity in the structure of the various knot spaces, perhaps at differing scales, that are reflected in the changing parameters. One word of caution is necessary, however, in that there are subtle artifacts of the numerical limitations of the accuracy of the data generation and the software employed to analyze the data to extract the parameters. The extraction of the key information is a very delicate process indeed as the number of edges increases.

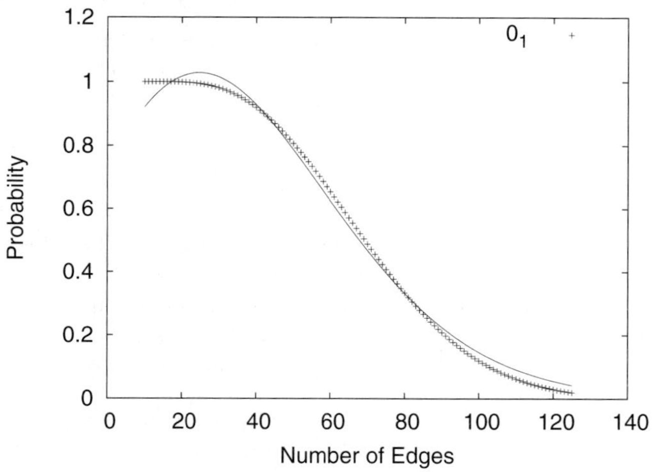

Fig. 16. Perturbation of the thick trivial knot: application of the full variation model to trivial knots.

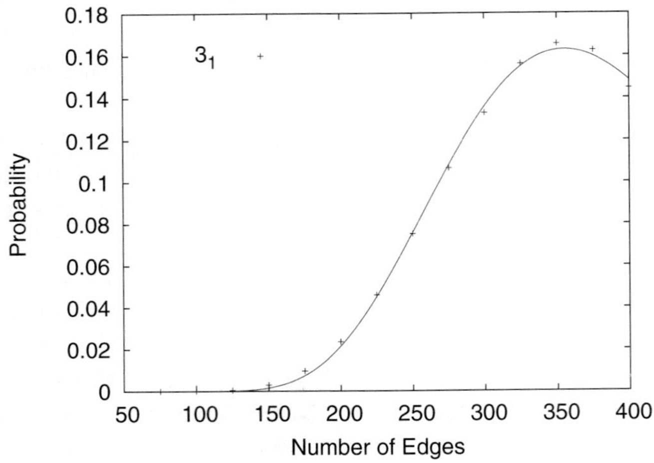

Fig. 17. Perturbation of the thick figure-eight knot: application of the full variation model to trefoil knots.

Acknowledgements

We wish to thank Akos Dobay, Tetsuo Deguchi, Yuanan Diao, Andrzej Stasiak, and Stuart Whittington for their review of an early draft and recommendations that have contributed to the work reported here. We also thank Michael Piatek for generating the tube perturbation data and Johnny Zarate for writing the program used to analyze this data. Rawdon was supported by NSF Grant No. 0311010.

References

1. Tetsuo Deguchi and Kyoichi Tsurusaki. Topology of closed random polygons. *J. Phys. Soc. Japan*, 62(5):1411–1414, 1993.
2. Tetsuo Deguchi and Kyoichi Tsurusaki. A statistical study of random knotting using the Vassiliev invariants. In *Random knotting and linking (Vancouver, BC, 1993)*, volume 7 of *Ser. Knots Everything*, pages 89–121. World Sci. Publishing, River Edge, NJ, 1994.
3. Tetsuo Deguchi and Kyoichi Tsurusaki. Universality of random knotting. *Phys. Rev. E*, 55(5):6245–6248, August 1997.
4. Tetsuo Deguchi and Kyoichi Tsurusaki. Numerical application of knot invariants and universality of random knotting. In *Knot theory (Warsaw, 1995)*, volume 42 of *Banach Center Publ.*, pages 77–85. Polish Acad. Sci., Warsaw, 1998.
5. Yuanan Diao, Nicholas Pippenger, and De Witt Sumners. On random knots.

J. Knot Theory Ramifications, 3(3):419–429, 1994. Random knotting and linking (Vancouver, BC, 1993).
6. Akos Dobay, Pierre-Edouard Sottas, Jacques Dubochet, and Andrzej Stasiak. Predicting optimal lengths of random knots. *Lett. Math. Phys.*, 55(3):239–247, 2001. Topological and geometrical methods (Dijon, 2000).
7. Bruce Ewing and Kenneth C. Millett. Computational algorithms and the complexity of link polynomials. In *Progress in knot theory and related topics*, pages 51–68. Hermann, Paris, 1997.
8. P. Freyd, D. Yetter, J. Hoste, W. B. R. Lickorish, K. Millett, and A. Ocneanu. A new polynomial invariant of knots and links. *Bull. Amer. Math. Soc. (N.S.)*, 12(2):239–246, 1985.
9. J. M. Hammersley. The number of polygons on a lattice. *Proc. Cambridge Philos. Soc.*, 57:516–523, 1961.
10. J. Hoste, 2000. personal communication.
11. Jim Hoste, Morwen Thistlethwaite, and Jeff Weeks. The first 1,701,936 knots. *Math. Intelligencer*, 20(4):33–48, 1998.
12. Vsevold Katrich, Wilma K. Olsen, Jacques Vologodskii, Alexander Dubochet, and Andrzej Stasiak. Tightness of random knotting. *Phys. Rev. E*, 61(5):5545–5549, 2000.
13. V. Katritch, W. K. Olson, P. Pieranski, J. Dubochet, and A. Stasiak. Properties of ideal composite knots. *Nature*, 388:148–151, July 1997.
14. Kleanthes Koniaris and M. Muthukumar. Self-entanglement in ring polymers. *J. Chem. Phys.*, 95(4):2873–2881, May 1991.
15. R. A. Litherland, J. Simon, O. Durumeric, and E. Rawdon. Thickness of knots. *Topology Appl.*, 91(3):233–244, 1999.
16. J. P. J. Michels and F. W. Wiegel. On the topology of a polymer ring. *Proc. Roy. Soc. London Ser. A*, 403(1825):269–284, 1986.
17. Nathan T. Moore, Rhonald C. Lua, and Alexander Y. Grosberg. Topologically driven swelling of a polymer loop. *PNAS*, 101(37):13431–13435, 2004.
18. J. O'Hara. Energy of knots. In *Ideal knots*, pages 288–314. World Sci. Publishing, River Edge, NJ, 1998.
19. Piotr Pierański. In search of ideal knots. In *Ideal knots*, pages 20–41. World Sci. Publishing, River Edge, NJ, 1998.
20. Nicholas Pippenger. Knots in random walks. *Discrete Appl. Math.*, 25(3):273–278, 1989.
21. E. Rawdon. TOROS. http://www.mathcs.duq.edu/∼rawdon. Program for visualizing, manipulating, and thickness maximizing knots.
22. E. Rawdon. *The Thickness of Polygonal Knots*. PhD thesis, University of Iowa, 1997.
23. Eric J. Rawdon. Approximating the thickness of a knot. In *Ideal knots*, pages 143–150. World Sci. Publishing, River Edge, NJ, 1998.
24. Eric J. Rawdon. Approximating smooth thickness. *J. Knot Theory Ramifications*, 9(1):113–145, 2000.
25. Andrzej Stasiak, Jacques Dubochet, Vsevolod Katritch, and Piotr Pieranski. Ideal knots and their relation to the physics of real knots. In *Ideal knots*, pages 1–19. World Sci. Publishing, River Edge, NJ, 1998.

26. D. W. Sumners and S. G. Whittington. Knots in self-avoiding walks. *J. Phys. A*, 21(7):1689–1694, 1988.

CHAPTER 15

THE AVERAGE CROSSING NUMBER OF GAUSSIAN RANDOM WALKS AND POLYGONS

Yuanan Diao

Department of Mathematics
University of North Carolina at Charlotte
Charlotte, NC 28223

Claus Ernst

Department of Mathematics
Western Kentucky University
Bowling Green, KY 42101

In this paper, we extend results about the average crossing number of equilateral random walks and polygons to the average crossing number of the Gaussian random walks and polygons. We show that the asymptotical behavior of the ACN for the two models are very similar. More precisely, we show that the mean average crossing number $\langle \text{ACN} \rangle$ of Gaussian random walks and polygons of length n is of the form $\frac{1}{2\pi} n \ln n + O(n)$.

1. Introduction

Random walks and random polygons are frequently used to model linear long-chain molecules or ring polymers. They are useful in the study of the behavior of polymers at thermodynamic equilibrium. The two most frequently used continuum models are the equilateral random walks and the Gaussian random walks. An equilateral random walk is composed of freely jointed line segments of equal length, whereas a Gaussian random walk is composed of freely jointed line segments whose lengths follow the Gaussian distribution. Numerous studies are devoted to the topological properties of random polygons. For example, it is investigated what types of knots are formed on polymer chains [1,11,14], and it is theoretically proven that knotting becomes inevitable when the length of a random polygon approaches infinity

[2,5,12,15]. Furthermore, for the equilateral and Gaussian random polygons, it has been shown that knots in the global sense do exist with a high probability if the length of the polygon is large [4,10]. However, while the overall dimensions of equilateral and Gaussian random walks (polygons) are well understood (they scale with the number of segments n as $\sqrt{n}$ [6,7,9], the problem is much harder and less is known for random polygons with certain topological constraints (such as a fixed knot type).

A Gaussian random vector $X = (x, y, z)$ is a random point whose coordinates x, y and z are independent standard normal random variables (with mean $= 0$ and variance $= 1$). The pdf (probability density function) of X is the joint pdf of x, y and z, which is

$$\begin{aligned} f(X) &= \left(\frac{1}{\sqrt{2\pi}}e^{-\frac{x^2}{2}}\right)\left(\frac{1}{\sqrt{2\pi}}e^{-\frac{y^2}{2}}\right)\left(\frac{1}{\sqrt{2\pi}}e^{-\frac{z^2}{2}}\right) \\ &= \left(\frac{1}{\sqrt{2\pi}}\right)^3 e^{-\frac{x^2+y^2+z^2}{2}} \\ &= \left(\frac{1}{\sqrt{2\pi}}\right)^3 e^{-\frac{|X|^2}{2}}. \end{aligned}$$

A Gaussian random walk of n steps consists of $n+1$ consecutive points $X_0 = (0,0,0) = O$, X_1, X_2, ... , X_n such that $Y_{k+1} = X_{k+1} - X_k$ ($k = 0, 1, ..., n-1$) are independent Gaussian random vectors. It follows that the joint pdf for all the vertices is

$$\begin{aligned} f(X_1, X_2, ..., X_n) &= \left(\frac{1}{\sqrt{2\pi}}\right)^{3n} e^{-\frac{1}{2}(|Y_1|^2+|Y_2|^2+\cdots+|Y_n|^2)} \\ &= \left(\frac{1}{\sqrt{2\pi}}\right)^{3n} e^{-\frac{1}{2}(|X_1|^2+|X_2-X_1|^2+\cdots+|X_n-X_{n-1}|^2)}. \end{aligned}$$

In this paper, we are interested in a particular measure of a polymer entanglement, i.e., the average crossing number of a random walk or a random polygon. A Gaussian random polygon is a conditioned random walk such that the last vertex coincides with the starting point, which can be always assumed to be the origin. For the readers who are not familiar with probability theory, we need to point out that the probability that a Gaussian random walk or polygon has self-intersections is zero. Furthermore, almost all projections of any random walk or polygon are regular projections. Thus, for a given random walk (or random polygon), with probability one, the crossing number associated with a randomly chosen projection of the random walk (the number of crossings one observes when the walk or polygon is projected to a plane under the given projection direction) is well defined. Therefore, we may define the average crossing number of the walk (or polygon) as the average of this crossing number over all possible projection directions (which is equivalent to the unit sphere). It is known [3] that the mean average crossing number ($\langle$ACN$\rangle$) of all equilateral random walks and polygons of length n is of the form $\frac{3}{16} \cdot n \cdot \ln n + O(n)$. We will study the same problem for the Gaussian random walks and polygons in this paper. Since the end to end distance behavior of two vertices along an equilateral random walk is approximately Gaussian [13], it is probably not surprising that the mean ACN for the Gaussian random walks and polygons would also follow the $n \ln n$ growth rule (although the difficulties one encounter in the two cases are different technically). As in the case of the equilateral random walks and polygons, we are also able to determine the coefficient of the $n \ln n$ term for the Gaussian random walks and polygons. More precisely, we will show that the mean ACN's for all the Gaussian random walks and polygons of length n are both of the form $\frac{1}{2\pi} n \ln n + O(n)$.

2. The Gaussian Random Walks and Polygons

In this section, we will discuss some basic probability properties of Gaussian random walks and polygons. For readers not familiar with probability theory, chapters 2, 4 and 5 in [16] will be of help. As defined in the last section, a Gaussian random vector $X = (x, y, z)$ is a random point whose coordinates x, y and z are independent standard normal random variables (with mean $= 0$ and variance $= 1$) and a Gaussian random walk of n steps consists of $n + 1$ consecutive points $X_0 = (0, 0, 0) = O$, X_1, X_2, ... , X_n such that $Y_{k+1} = X_{k+1} - X_k$ ($k = 0, 1, ..., n-1$) are n independent Gaussian random vectors. A Gaussian random walk of n steps is denoted by GW_n. The line segment that joins X_{k-1} and X_k (which has length $|Y_k| = |X_k - X_{k-1}|$) is called the k-th step of GW_n. A Gaussian random polygon GP_n is a conditioned GW_n of n steps such that the last vertex X_n coincides with the starting point $X_0 = O$. Thus, if we let $g_n(X_n)$ be the pdf of X_n for a GW_n, then the joint pdf of X_1, X_2, ..., X_{n-1} of a GP_n is

$$h(X_1, X_2, ..., X_n) = f(X_1, X_2, ..., X_n)/g_n(O).$$

Let A be a three by three real orthonormal matrix. Under the transformation $(x, y, z) \longrightarrow (x, y, z)A^T = (x', y', z')$ (where A^T is the transpose of

A), the new variables x', y' and z' are apparently also independent standard normal random variables. So the vector $X' = (x', y', z')$ is also a Gaussian random vector. If we let $r = |X| = |X'|$, θ be the angle between X' and the z'-axis ($0 \leq \theta \leq \pi$) and ϕ be the angle between the projection of X' to the $x'y'$-plane and the positive x'-axis ($0 \leq \phi < 2\pi$), then one can show that the random variables r, θ and ϕ are also independent. Furthermore, the probability density functions of r, θ and ϕ are given by

$$f_1(r) = \sqrt{\frac{2}{\pi}} r^2 e^{-\frac{r^2}{2}}, \tag{1}$$

$$f_2(\theta) = \frac{1}{2} \sin\theta, \text{ and} \tag{2}$$

$$f_3(\phi) = \frac{1}{2\pi} \tag{3}$$

respectively.

In the following, we will list several probability density functions that we will be using in the next section. These results are all well known and can be calculated directly. Please refer to [16].

The pdf of the k-th vertex X_k (not to be confused with $Y_k = X_k - X_{k-1}$) in a GW_n is given by

$$g_k(X_k) = \left(\frac{1}{\sqrt{2\pi k}}\right)^3 e^{-\frac{|X_k|^2}{2k}}. \tag{4}$$

Notice that if we let x_j be the x-coordinate of X_j ($1 \leq j \leq k$), then $x_k = x_1 + (x_2 - x_1) + \cdots + (x_k - x_{k-1}$ is the sum of k identical independent Gaussian random variables. It follows that x_k is also a Gaussian random variable. For a Gaussian random variable, one simply needs to know its mean and variance in order to write down its pdf. In this case, it is easy to see that the mean of x_k is 0 and the variance of x_k is n. Repeating this for the y and z coordinates of X_k then leads to the above equation.

Similarly, the pdf of the random vector $X_{j+k} - X_j$ (where X_j and X_{j+k} are vertices from the same GW_n) is given by

$$g_k(X_{j+k} - X_j) = \left(\frac{1}{\sqrt{2\pi k}}\right)^3 e^{-\frac{|X_{j+k} - X_j|^2}{2k}}. \tag{5}$$

On the other hand, if the above vertices are from a GP_n, then the pdf of X_k may be obtained from its integral form

$h_k(X_k) = \int\int\dots\int \frac{1}{g_n(O)} f(X_1)f(X_2 - X_1)f(X_3 - X_2)\dots f(O - X_{n-1})dX_1\dots dX_{k-1}dX_{k+1}\dots dX_{n-1}$, which is,

$$h_k(X_k) = \left(\frac{1}{\sqrt{2\pi}\sigma_{n,k}}\right)^3 e^{-\frac{|X_k|^2}{2\sigma_{n,k}^2}}, \tag{6}$$

where $\sigma_{n,k} = \sqrt{\frac{k(n-k)}{n}}$. Notice that in the integral above (and the one below), the integration is to be taken over bfR^3 for each vector involved. Similarly, the pdf of $X_{j+k} - X_j$ in a GP_n is:

$$h_k(X_{j+k} - X_j) = \left(\frac{1}{\sqrt{2\pi}\sigma_{n,k}}\right)^3 e^{-\frac{|X_{j+k}-X_j|^2}{2\sigma_{n,k}^2}}. \tag{7}$$

Finally, the joint pdf of X_1, X_{k+1} and X_{k+2} in a GP_n is

$$\begin{aligned} &h(X_1, X_{k+1}, X_{k+2}) \qquad (8)\\ &= \int\int\dots\int \frac{f(X_1)f(X_2 - X_1)\dots f(O - X_{n-1})}{g_n(O)} dX_2\dots dX_k dX_{k+3}\dots dX_{n-1}\\ &= \left(\sqrt{2\pi n}\right)^3 f(X_1)f(X_{k+2} - X_{k+1})g_k(X_{k+1} - X_1)g_{n-k-2}(X_{k+2})\\ &= \left(\sqrt{2\pi n}\right)^3 \left(\frac{1}{\sqrt{2\pi}}\right)^3 e^{-\frac{|X_1|^2}{2}} \cdot \left(\frac{1}{\sqrt{2\pi}}\right)^3 e^{-\frac{|X_{k+2}-X_{k+1}|^2}{2}} \cdot\\ &\quad \left(\frac{1}{\sqrt{2\pi k}}\right)^3 e^{-\frac{|X_{k+1}-X_1|^2}{2k}} \cdot \left(\frac{1}{\sqrt{2\pi(n-k-2)}}\right)^3 e^{-\frac{|X_{k+2}|^2}{2(n-k-2)}}. \end{aligned}$$

3. The Main Results and their Proofs

The following two theorems are the main results of this paper.

Theorem 1: Let χ_n be the ACN of a Gaussian random walk of n steps and $E(\chi_n)$ be its expected value; then

$$E(\chi_n) = \frac{1}{2\pi} n \ln n + O(n).$$

Theorem 2: Let χ'_n be the ACN of a Gaussian random polygon of n steps and $E(\chi'_n)$ be its expected value; then

$$E(\chi'_n) = \frac{1}{2\pi} n \ln n + O(n).$$

The following lemma is key to the proofs of the theorems.

Lemma 3: *Let P, Q, P_1 and Q_1 be four points in $\mathbf{R}^3$ such that P, Q are fixed and $P_1 - P$, $Q_1 - Q$ are two independent Gaussian random vectors. Let $a(\ell_1, \ell_2)$ be the average crossing number between the two line segments $\ell_1 = PP_1$ and $\ell_2 = QQ_1$, then we have*

$$E(a(\ell_1, \ell_2)) = \frac{1}{2\pi r^2} + O(\frac{1}{r^{2.5}}), \tag{9}$$

where $r = |P - Q|$.

Proof: Without loss of generality, let us assume that $P = O$ and Q is on the positive z-axis (recall from last section that a rotation of 3-space transforms independent Gaussian random vectors to independent Gaussian random vectors). Let θ_1 be the angle between $V_1 = \overrightarrow{PP_1}$ and the z-axis and θ_2 be the angle between $V_2 = \overrightarrow{QQ_1}$ and the z-axis. Furthermore, let φ be the angle between the projections of V_1 and V_2 on the xy-plane. See Figure 1 below.

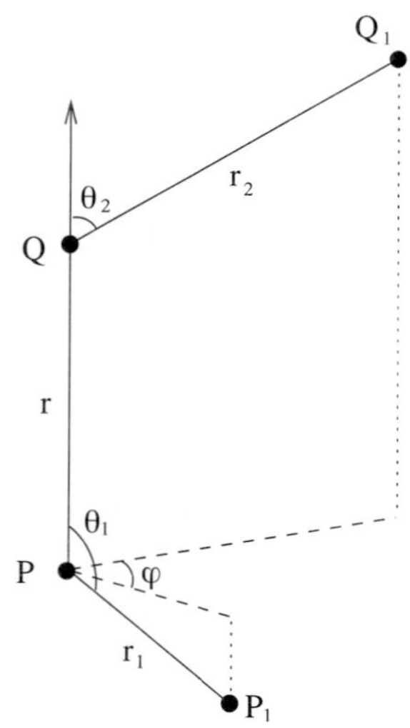

Fig. 1. The ACN of two random edges

Notice that φ is the angle between two plane vectors and its measure is between 0 and π. We leave it to the reader to show that φ is a random variable uniformly distributed on the interval $[0, \pi]$. In [8], it is shown that for fixed P_1 and Q_1, the average crossing number $a(\ell_1, \ell_2)$ between the edges ℓ_1 and ℓ_2 is given by

$$a(\ell_1, \ell_2) = \frac{1}{2\pi} \int_{\ell_1} \int_{\ell_2} \frac{|(\dot{\gamma}_1(t), \dot{\gamma}_2(s), \gamma_1(t) - \gamma_2(s))|}{|\gamma_1(t) - \gamma_2(s)|^3} dtds, \tag{10}$$

where γ_1 and γ_2 are the arclength parameterizations of ℓ_1 and ℓ_2 respectively, and $(\dot{\gamma}_1(t), \dot{\gamma}_2(s), \gamma_1(t) - \gamma_2(s))$ is the triple scalar product of $\dot{\gamma}_1(t)$, $\dot{\gamma}_2(s)$, and $\gamma_1(t) - \gamma_2(s)$. We can write

$$\gamma_1(t) = t \cdot \frac{V_1}{|V_1|}, \quad 0 \le t \le |V_1|,$$
$$\gamma_2(s) = \overrightarrow{OQ} + s \cdot \frac{V_2}{|V_2|}, \quad 0 \le s \le |V_2|.$$

By an elementary calculation, we have

$$\begin{aligned} &\int_{\ell_1} \int_{\ell_2} |(\dot{\gamma}_1(t), \dot{\gamma}_2(s), \gamma_1(t) - \gamma_2(s))| dtds \\ &= \int_0^{|V_1|} \int_0^{|V_2|} |(\frac{V_1}{|V_1|}, \frac{V_2}{|V_2|}, \overrightarrow{OQ})| dtds \\ &= |(V_1, V_2, \overrightarrow{OQ})| = r r_1 r_2 \sin\varphi \sin\theta_1 \sin\theta_2, \end{aligned}$$

where $r = |\overrightarrow{OQ}|$, $r_1 = |V_1|$ and $r_2 = |V_2|$. Recall from the last section that θ_1 and θ_2 are independent and their pdf's are $\frac{1}{2}\sin\theta_1$ and $\frac{1}{2}\sin\theta_2$ respectively. For r large enough, if r_1, $r_2 \le \sqrt{r}$, then we have $r - 2\sqrt{r} \le |\gamma_1(t) - \gamma_2(s)| \le r + 2\sqrt{r}$. From this we obtain the following:

$$\frac{1}{|\gamma_1(t) - \gamma_2(s)|^3} = \frac{1}{r^3} + O(\frac{1}{r^{3.5}}).$$

It follows that

$$a(\ell_1, \ell_2) = \frac{1}{2\pi r^2} r_1 r_2 \sin\varphi \sin\theta_1 \sin\theta_2 + r_1 r_2 O(\frac{1}{r^{2.5}}) \tag{11}$$

and

$$\begin{aligned} &E(a(\ell_1, \ell_2)) \\ &= \int\int a(\ell_1, \ell_2) f(V_1) f(V_2) dV_1 dV_2 \\ &= \int_0^{\pi} \int_0^{\pi} \int_0^{\pi} \int_0^{\infty} \int_0^{\infty} a(\ell_1, \ell_2) p(\theta_1, \theta_2, \varphi, r_1, r_2) dr_1 dr_2 d\theta_1 d\theta_2 d\varphi, \end{aligned} \tag{12}$$

where $p(\theta_1, \theta_2, \varphi, r_1, r_2)$ is the joint pdf of θ_1, θ_2, φ, r_1 and r_2. Since the random variables involved here are independent of each other, $p(\theta_1, \theta_2, \varphi, r_1, r_2)$

is simply the product of the individual pdf's of these random variables. Recall from (2.2) that the pdf's of θ_1 and θ_2 are $\sin\theta_1/2$ and $\sin\theta_2/2$ respectively, and that the pdf of φ (this is not to be confused with the φ defined in the last section) is $1/\pi$. The pdf of r_1 is simply the integration of the pdf of V_1 over the sphere $|V_1| = r_1$. This turns out to be

$$4\pi r_1^2 \left(\frac{1}{2\pi}\right)^{\frac{3}{2}} e^{-\frac{r_1^2}{2}} = \sqrt{\frac{2}{\pi}} r_1^2 e^{-\frac{r_1^2}{2}}.$$

Similarly, the pdf of r_2 is

$$\sqrt{\frac{2}{\pi}} r_2^2 e^{-\frac{r_2^2}{2}}.$$

It follows that

$$p(\theta_1, \theta_2, \varphi, r_1, r_2) = \frac{1}{2\pi^2} \sin\theta_1 \sin\theta_2 r_1^2 r_2^2 e^{-\frac{r_1^2+r_2^2}{2}} \tag{13}$$

Thus, $E(a(\ell_1, \ell_2))$ can be estimated by splitting the integral in (12) in two parts: One with both r_1, $r_2 \leq \sqrt{r}$ (which we will call I_1) and the other with either $r_1 > \sqrt{r}$ or $r_2 > \sqrt{r}$, which we will call I_2. Since $a(\ell_1, \ell_2) \leq 1$, and if one assumes that $r_1 > \sqrt{r}$ one can bound I_2 from above as follows:

$$\begin{aligned}
I_2 &\leq \int_0^\pi \int_0^\pi \int_0^\pi \int\int_{r_1 \text{ or } r_2 > \sqrt{r}} p(\theta_1, \theta_2, \varphi, r_1, r_2) dr_1 dr_2 d\theta_1 d\theta_2 d\varphi \\
&= \frac{2}{\pi} \int\int_{r_1 \text{ or } r_2 > \sqrt{r}} r_1^2 r_2^2 e^{-\frac{r_1^2+r_2^2}{2}} dr_1 dr_2 \\
&< \frac{4}{\pi} \int_{\sqrt{r}}^\infty r_1^2 e^{-\frac{r_1^2}{2}} dr_1 \int_0^\infty r_2^2 e^{-\frac{r_2^2}{2}} dr_2 \\
&= 2\sqrt{\frac{2}{\pi}} \int_{\sqrt{r}}^\infty r_1^2 e^{-\frac{r_1^2}{2}} dr_1 \\
&= 2\sqrt{\frac{2}{\pi}} \int_{\sqrt{r}}^\infty e^{-\frac{r_1^2}{4}} (r_1^2 e^{-\frac{r_1^2}{4}}) dr_1 \\
&\leq 2\sqrt{\frac{2}{\pi}} e^{-\frac{r}{4}} \int_{\sqrt{r}}^\infty r_1^2 e^{-\frac{r_1^2}{4}} dr_1 \\
&= O(e^{-\frac{r}{4}})
\end{aligned}$$

Hence I_2 is bounded above by $O(\frac{1}{r^{2.5}})$. On the other hand, using (11) for $a(\ell_1, \ell_2)$, we split I_1 into two integrals. The error term $O(\frac{1}{r^{2.5}})$ gives the following:

$$O(\frac{1}{r^{2.5}})\int_0^{\pi}\int_0^{\pi}\int_0^{\pi}\int_0^{\sqrt{r}}\int_0^{\sqrt{r}} r_1 r_2 p(\theta_1,\theta_2,\varphi,r_1,r_2)dr_1dr_2d\theta_1d\theta_2d\varphi = O(\frac{1}{r^{2.5}}).$$

The remaining integral

$$\int_0^{\pi}\int_0^{\pi}\int_0^{\pi}\int_0^{\sqrt{r}}\int_0^{\sqrt{r}} \frac{1}{4\pi^3 r^2}\sin\varphi\sin^2\theta_1\sin^2\theta_2 r_1^3 r_2^3 e^{-\frac{r_1^2+r_2^2}{2}}dr_1dr_2d\theta_1d\theta_2d\varphi$$

can be written as the product of the following five integrals:

$$\frac{1}{4\pi^3 r^2}\int_0^{\pi}\sin\varphi d\varphi = \frac{1}{2\pi^3 r^2},$$

$$\int_0^{\pi}\sin^2\theta_1 d\theta_1 = \int_0^{\pi}\sin^2\theta_2 d\theta_2 = \frac{\pi}{2},$$

and

$$\int_0^{\sqrt{r}} r_1^3 e^{-\frac{r_1^2}{2}}dr_1 = \int_0^{\sqrt{r}} r_2^3 e^{-\frac{r_2^2}{2}}dr_2 = 2-(2+r)e^{-r/2} = 2+O(\frac{1}{r^{2.5}}).$$

The result now follows.

We will now give a proof for Theorem 1.

Proof: Let ℓ_k be the k-th segment of a GW_n, that is, $\ell_k = \overline{X_{k-1}X_k}$ ($1 \le k \le n$). Let $a(\ell_i,\ell_j)$ be the average crossing number between ℓ_i and ℓ_j; then we have for the average crossing number χ_n of GW_n

$$\chi_n = \frac{1}{2}\sum_{1\le i,j\le n} a(\ell_i,\ell_j),$$

and

$$E(\chi_n) = \frac{1}{2}\sum_{1\le i,j\le n} E(a(\ell_i,\ell_j)) = \sum_{1\le i<j\le n} E(a(\ell_i,\ell_j)).$$

By symmetry, $E(a(\ell_{i_1},\ell_{j_1})) = E(a(\ell_{i_2},\ell_{j_2}))$ whenever $|j_1-i_1| = |j_2-i_2|$. It follows that

$$E(\chi_n) = \sum_{3\le j\le n}(n-j+1)E(a(\ell_1,\ell_j)), \tag{14}$$

where j starts at 3 since $a(\ell_1,\ell_2)=0$. Letting $r_j = |X_{j-1}-X_1|$, $P = X_1$, $P_1 = O$, $Q = X_{j-1}$ and $Q_1 = X_j$, we obtain

$$E(a(\ell_1,\ell_j)|X_{j-1}-X_1) = \frac{1}{2\pi r_j^2} + O(\frac{1}{r_j^{2.5}})$$

for any fixed X_j by Lemma 3. Since r_j is a random variable depending only on $X_{j-1}-X_1$, and since $X_{j-1}-X_1$ has the same pdf as X_{j-2}, it follows that

$$\begin{aligned} E(a(\ell_1,\ell_j)) &= \int E(a(\ell_1,\ell_j)|X_{j-1}-X_1)g_{j-2}(X_{j-1}-X_1)d(X_{j-1}-X_1) \\ &= \int_0^\infty E(a(\ell_1,\ell_j)|X_{j-1}-X_1)4\pi r_j^2(\frac{1}{\sqrt{2\pi(j-2)}})^3 e^{-\frac{r_j^2}{2(j-2)}}dr_j \\ &= I_1+I_2, \end{aligned}$$

where

$$I_1=\int_{10}^\infty E(a(\ell_1,\ell_j)|X_{j-1}-X_1)4\pi r_j^2(\frac{1}{\sqrt{2\pi(j-2)}})^3 e^{-\frac{r_j^2}{2(j-2)}}dr_j$$

and

$$I_2=\int_0^{10} E(a(\ell_1,\ell_j)|X_{j-1}-X_1)4\pi r_j^2(\frac{1}{\sqrt{2\pi(j-2)}})^3 e^{-\frac{r_j^2}{2(j-2)}}dr_j.$$

The choice of the number 10 here is quite arbitrary, we simply want to avoid the singularity in the integration caused by the error term $O(\frac{1}{r_j^{2.5}})$. Since $E(a(\ell_1,\ell_j)|X_{j-1}-X_1)$ is bounded above by one, it is easy to see that I_2 is bounded above as follows:

$$I_2 \le \int_0^{10} 4\pi r_j^2(\frac{1}{\sqrt{2\pi(j-2)}})^3 e^{-\frac{r_j^2}{2(j-2)}}dr_j = O(\frac{1}{j^{1.5}})\int_0^{10} r_j^2 e^{-\frac{r_j^2}{2(j-2)}}dr_j = O(\frac{1}{j^{1.5}}).$$

On the other hand, using

$$E(a(\ell_1,\ell_j)|X_{j-1}-X_1)=\frac{1}{2\pi r_j^2}+O(\frac{1}{r_j^{2.5}})$$

from Lemma 3, we can split I_1 into two integrals. The one containing the error term $O(\frac{1}{r_j^{2.5}})$ is bounded above as follows:

$$\begin{aligned} & 4\pi(\frac{1}{\sqrt{2\pi(j-2)}})^3\int_{10}^\infty O(\frac{1}{\sqrt{r_j}})e^{-\frac{r_j^2}{2(j-2)}}dr_j \\ &\le O(\frac{1}{(j-2)^{1.5}})\int_0^\infty O(\frac{1}{\sqrt{r_j}})e^{-\frac{r_j^2}{2(j-2)}}dr_j \\ &= O(\frac{1}{(j-2)^{1.5}})O((j-2)^{.25}) \\ &= O(\frac{1}{j^{1.25}}). \end{aligned}$$

Finally, the remaining integral is estimated as:

$$
\begin{aligned}
&2\left(\frac{1}{\sqrt{2\pi(j-2)}}\right)^3 \int_{10}^{\infty} e^{-\frac{r_j^2}{2(j-2)}}\, dr_j \\
&= \left(\frac{1}{\sqrt{2\pi(j-2)}}\right)^3 \int_{-\infty}^{\infty} e^{-\frac{r_j^2}{2(j-2)}}\, dr_j - \left(\frac{1}{\sqrt{2\pi(j-2)}}\right)^3 \int_{-10}^{10} e^{-\frac{r_j^2}{2(j-2)}}\, dr_j \\
&= \left(\frac{1}{\sqrt{2\pi(j-2)}}\right)^3 \sqrt{2\pi(j-2)} + O\left(\frac{1}{j^{1.5}}\right) \\
&= \frac{1}{2\pi(j-2)} + O\left(\frac{1}{j^{1.5}}\right) = \frac{1}{2\pi j} + O\left(\frac{1}{j^{1.5}}\right).
\end{aligned}
$$

Combining the above bounds, we get

$$E(a(\ell_1, \ell_j)) = \frac{1}{2\pi j} + O\left(\frac{1}{j^{1.25}}\right).$$

So

$$E(\chi_n) = \frac{1}{2\pi}(n+1) \sum_{3 \le j \le n} \frac{1}{j} - \frac{1}{2\pi}(n-3) + \sum_{3 \le j \le n} O\left(\frac{n-j+1}{j^{1.25}}\right)$$

by (14). Since

$$
\begin{aligned}
&\sum_{3 \le j \le n} O\left(\frac{n-j+1}{j^{1.25}}\right) \\
&= (n+1) O\left(\sum_{3 \le j \le n} \frac{1}{j^{1.25}}\right) - O\left(\sum_{3 \le j \le n} \frac{1}{j^{0.25}}\right),
\end{aligned}
$$

$\sum_{3 \le j \le n} \frac{1}{j^{1.25}}$ is convergent and $\sum_{3 \le j \le n} \frac{1}{j^{0.25}}$ is of the order of $n^{0.75}$, it follows that $\sum_{3 \le j \le n} O(\frac{n-j+1}{j^{1.25}}) = O(n)$. Theorem 1 now follows from the fact that $\sum_{3 \le j \le n} \frac{1}{j} - \ln n$ converges to a constant independent of n. This finishes the proof of Theorem 1.

Finally, we will prove Theorem 2.

Proof: Let χ'_n be the ACN of a Gaussian random polygon of $n \ge 4$ steps. As we did in the proof of Theorem 1, let ℓ_k be the k-th segment of a GP_n, that is, $\ell_k = \overline{X_{k-1} X_k}$ $(1 \le k \le n)$. Let $a(\ell_i, \ell_j)$ be the average crossing number between ℓ_i and ℓ_j; then we have

$$\chi'_n = \frac{1}{2} \sum_{1 \le i,j \le n} a(\ell_i, \ell_j),$$

and

$$E(\chi_n') = \frac{1}{2} \sum_{1 \le i,j \le n} E(a(\ell_i, \ell_j)).$$

We have $E(a(\ell_{i_1}, \ell_{j_1})) = E(a(\ell_{i_2}, \ell_{j_2}))$ whenever $|j_1 - i_1| = |j_2 - i_2|$ or $|j_1 - i_1| = n - |j_2 - i_2|$ by symmetry. It follows that

$$E(\chi_n') = n \sum_{3 \le j \le (n+1)/2} E(a(\ell_1, \ell_j)).$$

j starts at 3 in the above formula since $a(\ell_1, \ell_2)$ is always 0. Let $r_j = |X_{j-1} - X_1|$. We have

$$E(a(\ell_1, \ell_j)) = \int_{\mathbf{R}^3} \int_{\mathbf{R}^3} \int_{\mathbf{R}^3} a(\ell_1, \ell_j) h(X_1, X_{j-1}, X_j) dX_1 dX_{j-1} dX_j, \quad (15)$$

where

$$h(X_1, X_{j-1}, X_j) = \left(\sqrt{2\pi n}\right)^3 f(X_1) f(X_j - X_{j-1}) g_{j-2}(X_{j-1} - X_1) g_{n-j}(X_j) \quad (16)$$

is the joint pdf of X_1, X_{j-1} and X_j (see (8)). In the proof of Theorem 1, Lemma 3 was used to estimate $E(a(\ell_1, \ell_j))$. However we cannot apply Lemma 3 directly since the random vectors X_1, $X_j - X_{j-1}$ and X_j are not independent Gaussian random vectors in the case of a polygon GP_n. In the following, we will show that the pdf $h(X_1, X_{j-1}, X_j)$ in (15) can be replaced by $f(X_1) f(X_j - X_{j-1}) h_j(X_j)$ (which is the pdf of the three corresponding independent random vectors X_1, $X_j - X_{j-1}$ and X_j), while keeping the error term small enough so the result of the theorem will still hold.

By (6), the pdf of X_j is

$$h_j(X_j) = \left(\frac{1}{\sqrt{2\pi}\sigma_{n,j}}\right)^3 e^{-\frac{|X_j|^2}{2\sigma_{n,j}^2}}, \quad (17)$$

where $\sigma_{n,j} = \sqrt{j}\sqrt{\frac{n-j}{n}} = O(j^{0.5})$ for $j \le (n+1)/2$. It follows that $(\frac{1}{\sqrt{2\pi}\sigma_{n,j}})^3 = O(j^{-1.5})$ and therefore $P(|X_j| \le 10) = O(j^{-1.5})$. By the fact that $a(\ell_1, \ell_j) \le 1$, we get the following inequality:

$$\begin{aligned}
&\int_{|X_j| \le 10} \int_{\mathbf{R}^3} \int_{\mathbf{R}^3} a(\ell_1, \ell_j) h(X_1, X_{j-1}, X_j) dX_1 dX_{j-1} dX_j \\
&\le \int_{|X_j| \le 10} \int_{\mathbf{R}^3} \int_{\mathbf{R}^3} h(X_1, X_{j-1}, X_j) dX_1 dX_{j-1} dX_j \\
&= P(|X_j| \le 10) = O(j^{-1.5}).
\end{aligned}$$

Notice again that the use of the number 10 above is quite arbitrary and it plays no significant role in the proof. On the other hand, we have

$$\begin{aligned}
P(|X_j| \geq j^{5/8}) &= \int_{|X_j|\geq j^{5/8}} \left(\frac{1}{\sqrt{2\pi}\sigma_{n,j}}\right)^3 e^{-\frac{|X_j|^2}{2\sigma_{n,j}^2}} dX_j \\
&= O(j^{-1.5}) \int_{j^{5/8}}^{\infty} r^2 e^{-\frac{r^2}{2\sigma_{n,j}^2}} dr \\
&\leq O(j^{-1.5}) \int_{j^{5/8}}^{\infty} r^2 e^{-\frac{r^2}{2j}} dr \\
(\text{let } r = \sqrt{j}u) \quad &= O(j^{-1.5}) \int_{j^{1/8}}^{\infty} j^{1.5} u^2 e^{-\frac{u^2}{2}} du \\
&= O(1) \int_{j^{1/8}}^{\infty} u^2 e^{-\frac{u^2}{4}} e^{-\frac{u^2}{4}} du \\
&\leq O(e^{-\frac{j^{0.25}}{4}}) \int_0^{\infty} u^2 e^{-\frac{u^2}{4}} du \\
&\leq O(j^{-1.5}).
\end{aligned}$$

Notice that the choice of the power 5/8 in the above integral is also somewhat arbitrary. It can actually be chosen from a small range of numbers without affecting the result of the theorem. Its use will become clear as the end of the proof. Thus

$$\begin{aligned}
&\int_{|X_j|\geq j^{5/8}} \int_{\mathbf{R}^3} \int_{\mathbf{R}^3} a(\ell_1, \ell_j) h(X_1, X_{j-1}, X_j) dX_1 dX_{j-1} dX_j \\
&\leq P(|X_j| \geq j^{5/8}) \leq O(j^{-1.5})
\end{aligned}$$

as well. Now consider $P(|X_1| \geq r^{1/4})$ where $r = |X_j|$. The joint pdf of X_1 and X_j is

$$\begin{aligned}
&h(X_1, X_j) \\
&= \int\int \cdots \int \frac{f(X_1)f(X_2 - X_1)\ldots f(O - X_{n-1})}{g_n(O)} dX_2 \ldots dX_{j-1} dX_{j+1} \ldots dX_{n-1} \\
&= \left(\sqrt{2\pi n}\right)^3 f(X_1) g_{j-1}(X_j - X_1) g_{n-j}(X_j) \\
&= \left(\sqrt{2\pi n}\right)^3 \left(\frac{1}{\sqrt{2\pi}}\right)^3 e^{-\frac{|X_1|^2}{2}} \left(\frac{1}{\sqrt{2\pi(j-1)}}\right)^3 e^{-\frac{|X_j - X_1|^2}{2(j-1)}} \left(\frac{1}{\sqrt{2\pi(n-j)}}\right)^3 e^{-\frac{|X_j|^2}{2(n-j)}} \\
&\leq \left(\frac{1}{\sqrt{2\pi(j-1)}}\right)^3 \left(\sqrt{\frac{n}{2\pi(n-j)}}\right)^3 e^{-\frac{|X_1|^2}{2}}.
\end{aligned}$$

Using the facts that $j \le (n+1)/2$, $\int_{\mathbf{R}^3} e^{-\frac{|X_1|^2}{4}} dX_j = 8\pi^{3/2}$ and that $\int_{\mathbf{R}^3} e^{-\frac{|X_j|^{0.5}}{4}} dX_j$ converges to a constant independent of j, we get

$$\begin{aligned}
&P(|X_1| \ge r^{1/4}) \\
&\le \int_{\mathbf{R}^3}\int_{\mathbf{R}^3} \left(\frac{1}{\sqrt{2\pi(j-1)}}\right)^3 \left(\sqrt{\frac{n}{2\pi(n-j)}}\right)^3 e^{-\frac{|X_1|^2}{4}} e^{-\frac{|X_j|^{0.5}}{4}} dX_1 dX_j \\
&= \int_{\mathbf{R}^3} \left(\frac{1}{\sqrt{j-1}}\right)^3 \left(\sqrt{\frac{n}{\pi(n-j)}}\right)^3 e^{-\frac{|X_j|^{0.5}}{4}} dX_j \\
&= O(j^{-1.5})
\end{aligned}$$

Similarly,

$$P(|X_j - X_{j-1}| \ge r^{1/4}) \le O(j^{-1.5}).$$

It now follows that

$$E(a(\ell_1, \ell_j)) = I + O(j^{-1.5}), \tag{18}$$

where

$$I = \int_\Omega a(\ell_1, \ell_j) h(X_1, X_{j-1}, X_j) dX_1 dX_{j-1} dX_j \tag{19}$$

and Ω is the region defined by $10 < r < j^{5/8}$, $|X_1| < r^{1/4}$ and $|X_j - X_{j-1}| < r^{1/4}$. In Ω, we have

$$r - 2r^{1/4} < |X_{j-1} - X_1| < r + 2r^{1/4}.$$

This leads to

$$\frac{|X_{j-1} - X_1|^2}{2(j-2)} = \frac{r^2}{2j} + O(j^{-7/32})$$

and

$$e^{-\frac{|X_{j-1}-X_1|^2}{2(j-2)}} = (1 + O(j^{-7/32})) e^{-\frac{r^2}{2j}}$$

since $e^{O(j^{-7/32})} \le 1 + O(j^{-7/32})$. Thus the function $g_{j-2}(X_{j-1} - X_1)$ (which is part of $h(X_1, X_{j-1}, X_j)$) in (19) can be written as

$$\begin{aligned}
g_{j-2}(X_{j-1} - X_1) &= \left(\frac{1}{\sqrt{2\pi(j-2)}}\right)^3 e^{-\frac{|X_{j-1}-X_1|^2}{2(j-2)}} \\
&= (1 + O(j^{-7/32})) \left(\frac{1}{\sqrt{2\pi(j-2)}}\right)^3 e^{-\frac{r^2}{2j}} \\
&= (1 + O(j^{-7/32})) \left(\frac{1}{\sqrt{2\pi j}}\right)^3 e^{-\frac{|X_j|^2}{2j}}.
\end{aligned}$$

Hence in (19) the function $h(X_1, X_{j-1}, X_j)$ can be written as

$$\begin{aligned}
&\left(\sqrt{2\pi n}\right)^3 f(X_1)f(X_j - X_{j-1})g_{j-2}(X_{j-1} - X_1)g_{n-j}(X_j) \\
&= (1 + O(j^{-7/32})) \cdot f(X_1)f(X_j - X_{j-1}) \cdot \\
&\left(\frac{1}{\sqrt{2\pi j}}\right)^3 e^{-\frac{|X_1|^2}{2j}} \cdot \left(\frac{1}{\sqrt{2\pi(n-j)}}\right)^3 e^{-\frac{|X_j|^2}{2(n-j)}} \\
&= (1 + O(j^{-7/32})) \cdot f(X_1)f(X_j - X_{j-1}) \cdot \left(\frac{1}{\sqrt{2\pi}\sigma_{n,j}}\right)^3 e^{-\frac{|X_j|^2}{2\sigma_{n,j}^2}} \\
&= (1 + O(j^{-7/32}))f(X_1)f(X_j - X_{j-1})h_j(X_j). \qquad (20)
\end{aligned}$$

Using (18), (19), and (20) we now have

$$E(a(\ell_1, \ell_j)) = O(j^{-1.5}) + (1 + O(j^{-7/32})) \cdot J_\Omega \qquad (21)$$

where

$$J_\Omega = \int_\Omega a(\ell_1, \ell_j)f(X_1)f(X_j - X_{j-1})h_j(X_j)dX_1dX_{j-1}dX_j.$$

Now observe that the integral

$$J = \int_{\mathbf{R}^3}\int_{\mathbf{R}^3}\int_{\mathbf{R}^3} a(\ell_1, \ell_j)f(X_1)f(X_j - X_{j-1})h_j(X_j)dX_1dX_{j-1}dX_j \quad (22)$$

is the mean average crossing number of the two edges ℓ_1, ℓ_j under the conditions that X_1, X_j, $X_{j-1} - X_j$ are independent Gaussian random vectors such that the variance of X_1 and $X_{j-1} - X_j$ is one and the variance of X_j is $\sigma_{j,n}^2$. Therefore, we can apply Lemma 3 and the argument used in the proof of Theorem 1 to estimate J. More precisely, we have

$$\begin{aligned}
J &= \int_{\mathbf{R}^3} E(a(\ell_1, \ell_j)|X_j)\left(\frac{1}{\sqrt{2\pi}\sigma_{n,j}}\right)^3 e^{-\frac{|X_j|^2}{2\sigma_{n,j}^2}} dX_j \\
&= \int_0^\infty \left(\frac{1}{2\pi r^2} + O(\frac{1}{r^{2.5}})\right) 4\pi r^2 \left(\frac{1}{\sqrt{2\pi}\sigma_{n,j}}\right)^3 e^{-\frac{r^2}{2\sigma_{n,j}^2}} dr
\end{aligned}$$

As in the proof of Theorem 1 this integral can be split into two:

$$\int_0^{10} \left(\frac{1}{2\pi r^2} + O(\frac{1}{r^{2.5}})\right) 4\pi r^2 \left(\frac{1}{\sqrt{2\pi}\sigma_{n,j}}\right)^3 e^{-\frac{r^2}{2\sigma_{n,j}^2}} dr = O(j^{-1.5})$$

and

$$\int_{10}^\infty \left(\frac{1}{2\pi r^2} + O(\frac{1}{r^{2.5}})\right) 4\pi r^2 \left(\frac{1}{\sqrt{2\pi}\sigma_{n,j}}\right)^3 e^{-\frac{r^2}{2\sigma_{n,j}^2}} dr$$

The second integral has two parts. The one containing the error term $O(\frac{1}{r_j^{2.5}})$ can be bounded above as

$$\begin{aligned}
&4\pi\left(\sqrt{\frac{n}{2\pi j(n-j)}}\right)^3\int_{10}^{\infty}O(\frac{1}{\sqrt{r}})e^{-\frac{nr^2}{2j(n-j)}}dr\\
&=O(\frac{1}{j^{1.5}})\int_{10}^{\infty}\frac{1}{\sqrt{r}}e^{-\frac{nr^2}{2j(n-j)}}dr\\
&=O(\frac{1}{j^{1.5}})O(j^{.25})=O(j^{-1.25})
\end{aligned}$$

and the other part is estimated as follows:

$$\begin{aligned}
&2\left(\sqrt{\frac{n}{2\pi j(n-j)}}\right)^3\int_{10}^{\infty}e^{-\frac{nr^2}{2j(n-j)}}dr\\
&=\left(\sqrt{\frac{n}{2\pi j(n-j)}}\right)^3\int_{-\infty}^{\infty}e^{-\frac{nr^2}{2j(n-j)}}dr-\left(\sqrt{\frac{n}{2\pi j(n-j)}}\right)^3\int_{-10}^{10}e^{-\frac{nr^2}{2j(n-j)}}dr\\
&=\left(\sqrt{\frac{n}{2\pi j(n-j)}}\right)^3\sqrt{\frac{2\pi j(n-j)}{n}}+O(j^{-1.5})\\
&=\frac{1}{2\pi}\frac{n}{j(n-j)}+O(j^{-1.5})=\frac{1}{2\pi}(\frac{1}{j}+\frac{1}{n-j})+O(j^{-1.5}).
\end{aligned}$$

Combining the above bounds, we get

$$J=\frac{1}{2\pi}\left(\frac{1}{j}+\frac{1}{n-j}\right)+O(j^{-1.25}). \tag{23}$$

Following similar arguments used in the beginning of this proof, we also have

$$J=O(j^{-1.25})+J_{\Omega}. \tag{24}$$

Combining (21), (23) and (24), we obtain the following inequality:

$$\begin{aligned}
E(a(\ell_1,\ell_j))&=O(j^{-1.5})+\left(1+O(j^{-7/32})\right)\cdot(\frac{1}{2\pi}\left(\frac{1}{j}+\frac{1}{n-j}\right)+O(j^{-1.25}))\\
&=\frac{1}{2\pi}\left(\frac{1}{j}+\frac{1}{n-j}\right)+O(j^{-39/32}).
\end{aligned}$$

Similar to the proof of Theorem 1, we now estimate

$$\begin{aligned} E(\chi'_n) &= n \sum_{3\le j\le (n+1)/2} E(a(\ell_1,\ell_j)) \\ &= n \sum_{3\le j\le (n+1)/2} \frac{1}{2\pi}\left(\frac{1}{j}+\frac{1}{n-j}\right) + nO(\sum_{3\le j\le (n+1)/2} j^{-39/32}) \\ &= \frac{1}{2\pi} n\ln n + O(n), \end{aligned}$$

since $\sum_{3\le j\le (n+1)/2}\frac{1}{n-j}$ and $\sum_{3\le j\le (n+1)/2} j^{-39/32}$ both converge.

Acknowledgments

The authors are partially supported by NSF Grant #DMS-0310562.

References

1. T. Deguchi and K. Tsurusaki, *A statistical study of random knotting using the Vassiliev invariants*, Journal of Knot Theory and Its Ramifications, Vol.3, No.3 (1994), 321–353.
2. Y. Diao, *The Knotting of Equilateral Polygons in $\boldsymbol{R}^3$*, Journal of Knot Theory and its Ramifications, Vol.4, No.2 (1995), 189–196.
3. Y. Diao, A. Dobay, R. B. Kusner, K. Millett and A. Stasiak, *The Average Crossing Number of Equilateral Random Polygons*, Journal of Physics A: Mathematical and General, Vol. 36, No. 46 (2003), 11561 - 11574.
4. Y. Diao, J. Nardo and Y. Sun, *Global Knotting in Equilateral Random Polygons*; Journal of Knot Theory and its Ramifications, Vol.10, No.4 (2001), 597–607.
5. Y. Diao, N. Pippenger and D. W. Sumners, *On random knots*, Journal of Knot Theory and its Ramifications, Vol.3, No.3 (1994), 419–429.
6. M. Doi and S.F. Edwards, *The theory of polymer dynamics*, Oxford University Press, 1986.
7. J. P. Flory, *Principles of Polymer Chemistry*, Cornell University Press, 1953.
8. M.H. Freedman and Z. He, *Divergence-free Fields: Energy and Asymptotic Crossing Number,* Annals of Math., 134 (1991), 189–229.
9. P. G. de Gennes, *Scaling Concepts in Polymer Physics*, Cornell University Press, 1979.
10. D. Jungreis, *Gaussian Random Polygons Are Globally Knotted,* Journal of Knot Theory and its Ramifications, Vol.3, No.4 (1994), 455–464.
11. V. Katritch, W. K. Olson, A. V. Vologodskii, J. Dubochet and A. Stasiak, *Tightness of random knotting*, Phys. Rev. E. 61 (2000), 5545–5549.
12. N. Pippenger, *Knots in Random Walks,* Discrete Appl. Math. 25 (1989), 273-278.
13. L. Rayleigh, *On the problems of random vibrations, and of random flights in one, two, or three dimensions*, Phil. Mag. S. 6. Vol.37, No.220 (1919), 321–347.

14. V. V. Rybenkov, N. R. Cozzarelli and A. V. Vologodskii, *The probability of DNA knotting and the effective diameter of the DNA double helix*, Proc. Nat. Acad. Sci. USA, Vol.90 (1993), 5307–5311.
15. D.W. Sumners and S.G. Whittington, *Knots in Self-Avoiding Walks,* J.Phys.A: Math.Gen. 21(1988), 1689-1694.
16. V. K. Rohatgi and A. K. MD. Ehsanes Saleh, *An Introduction to Probability and Statistics*, John Wiley & Sons, 2001.

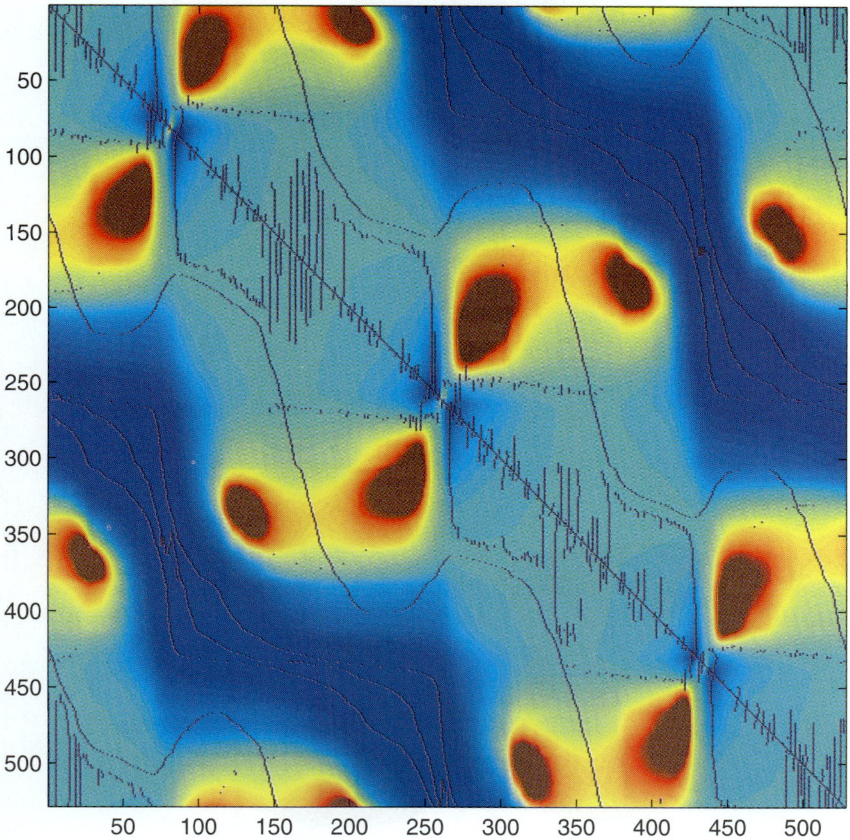

Plate 5.1. Trefoil knot: A colour plot of the function $\mathsf{pt}(i, j) := \min_{\sigma \in \mathbf{a}_j} \mathsf{pt}(\mathbf{M}_i, \sigma)$ evaluated on each arc of the approximately ideal 264-biarc trefoil knot, with blue for low and red for high values. In addition there is a dark blue dot superposed at indices j with local minima of $\mathsf{pt}(s, \cdot)$ realised in the arc $\mathbf{a}_j$. Note that the plot is a uniformly spaced function of indices corresponding to a non-uniform mesh in arc length.

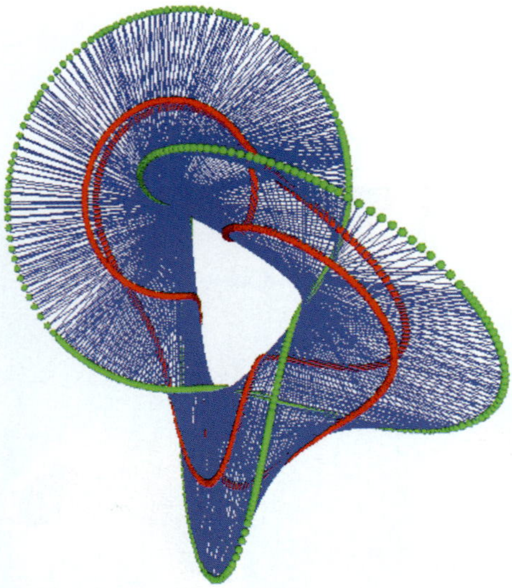

(a)

(b)

Plate 5.2. Trefoil knot: Two visualisations of the approximately ideal, biarc trefoil knot in three dimensional space. In part (a) the green spheres are centred at the mid-points $\mathbf{M}_i$ of the arcs $\mathbf{a}_i$, the red balls are the set of contact points $\mathbb{C}_{\mu^*}$ with $\mu^* = 8.1861 \cdot 10^{-6}$, and the blue line segments are the associated contact chords. At this resolution the red spheres overlap to form a tube, whose centreline is another trefoil. The contact chords appear to form a surface, which is confirmed in the visualisation of part (b), which is a shaded triangulation of the blue contact chords.

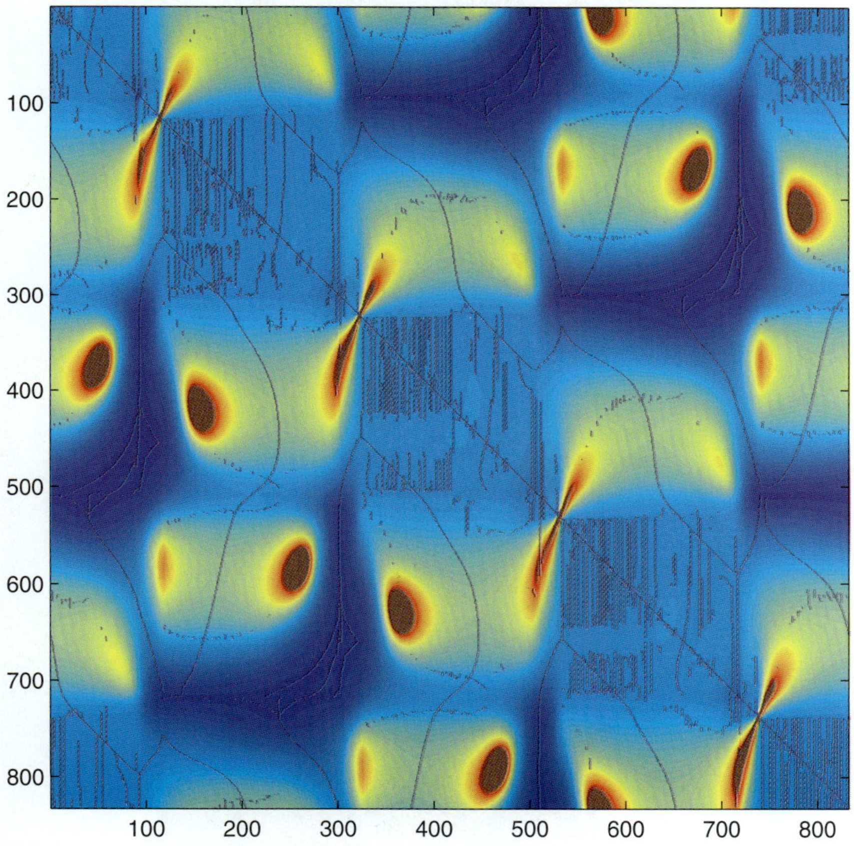

Plate 5.3. Figure-eight knot: A colour plot of the function $\mathsf{pt}(i,j) := \min_{\sigma\in\mathbf{a}_j} \mathsf{pt}(\mathbf{M}_i,\sigma)$ evaluated on each arc of the approximately ideal biarc figure-eight knot, with blue for low and red for high values. In addition there is a dark blue dot superposed at each index j associated with a local minimum.

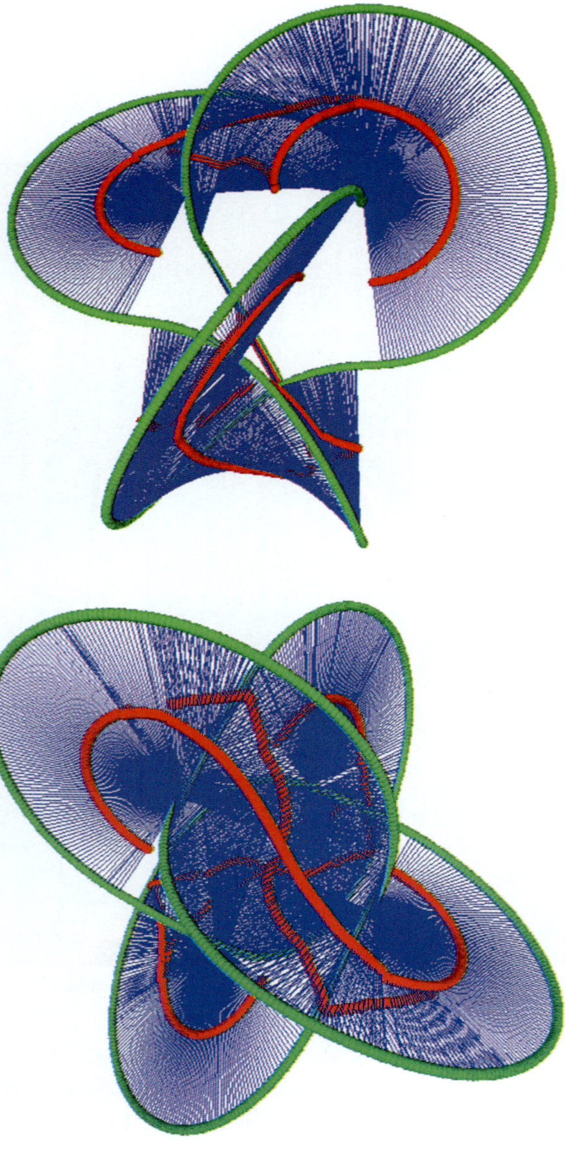

Plate 5.4. Figure-eight knot: Two images of the approximately ideal, figure-eight knot (the green balls are centred at the mid-points $\mathbf{M}_i$ of the arcs $\mathbf{a}_i$), the set of contact points $\mathbb{C}_{\mu^*}$ in three dimensional space (red balls), and associated contact chords (blue line segments), all for $\mu^* = 4.2115 \cdot 10^{-6}$.

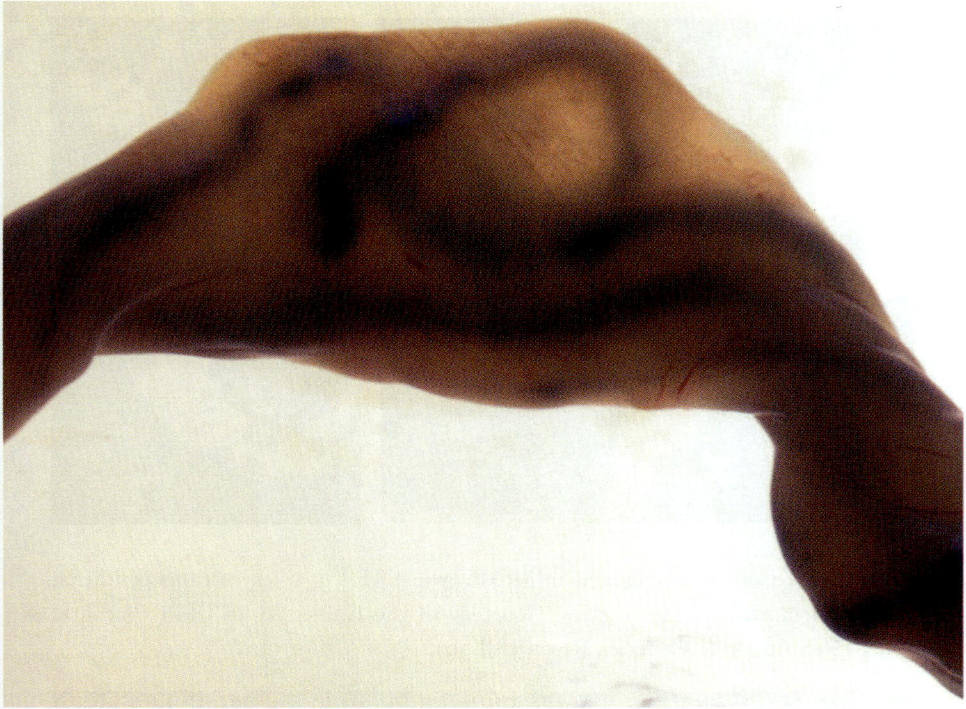

Plate 6.1a. A false knot is an enlargement of the umbilical cord due to excessive vascular growth. Picture Courtesy of the University of Utah Placental Bank (Reproductive Genetics Research Lab).

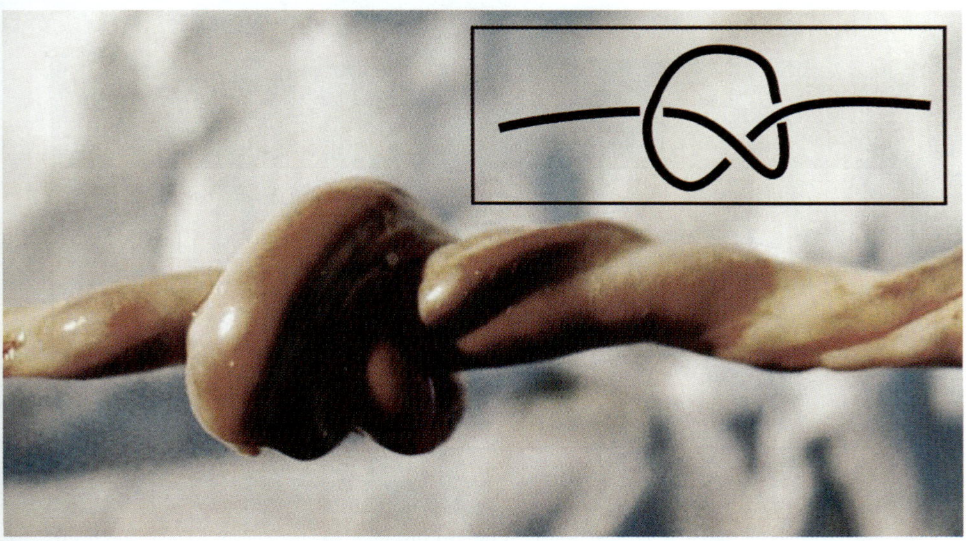

Plate 6.1b. A true knot occurs when the fetus passes through a loop and forms a knot. Copyright ©1996 Massachusetts Medical Society. All rights reserved.

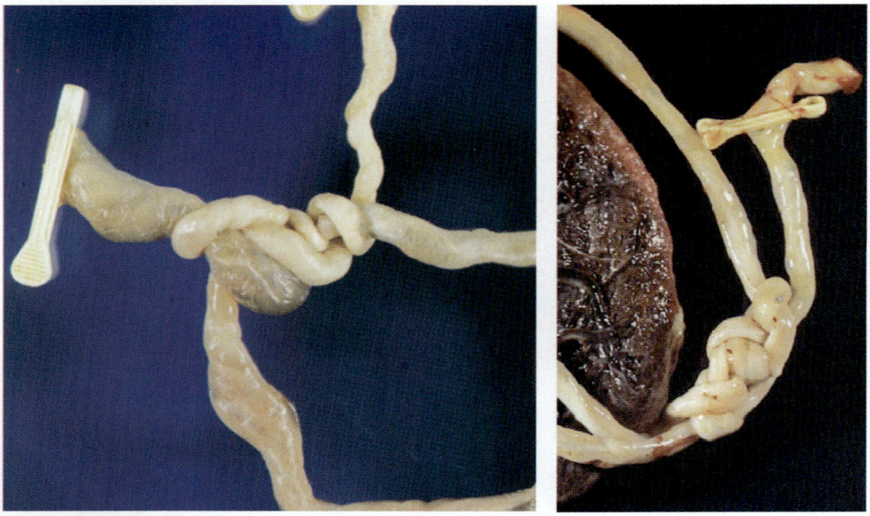

Plate 6.2a. Cord entanglement in MoMo twins. This is a very serious condition with high mortality rate. Picture Courtesy of the University of Utah Placental Bank (Reproductive Genetics Research Lab).

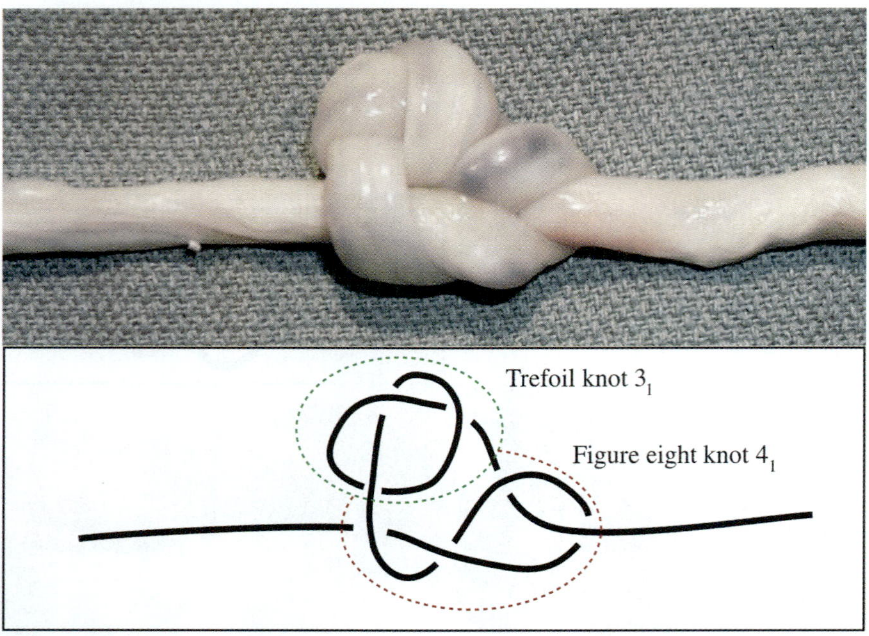

Plate 6.2b. An example of complex knot in an umbilical cord, a composition of an eight-knot with a trefoil knot. The infant was in good condition, without any clinical evidence of intrapartum asphyxia. Copyright ©1996 Massachusetts Medical Society. All rights reserved.

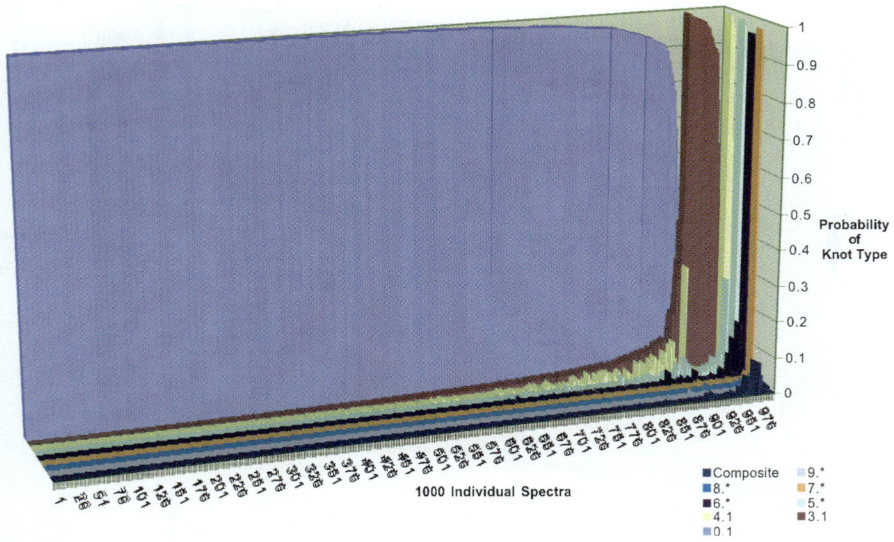

Plate 11.1. Knot spectra for 1000 random walks of length 50.

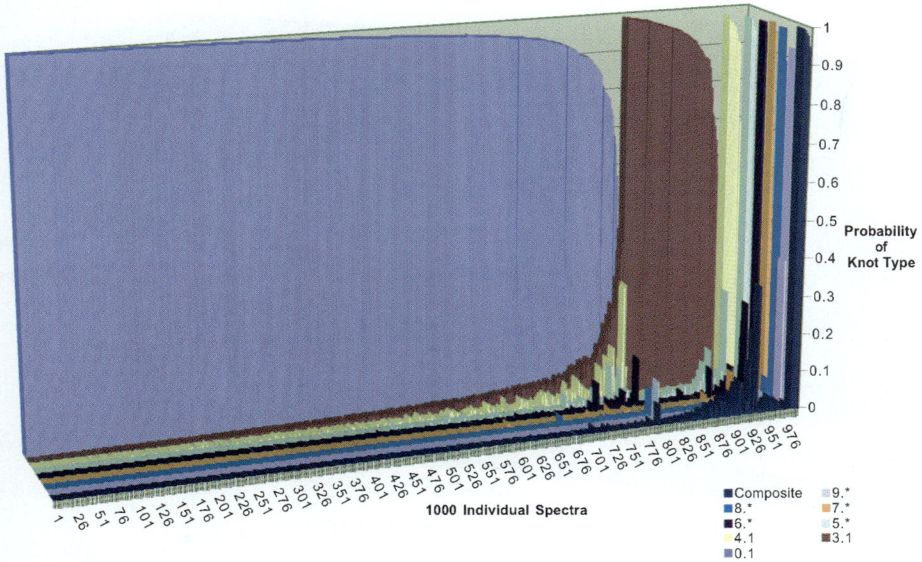

Plate 11.2. Knot spectra for 1000 random walks of length 100.

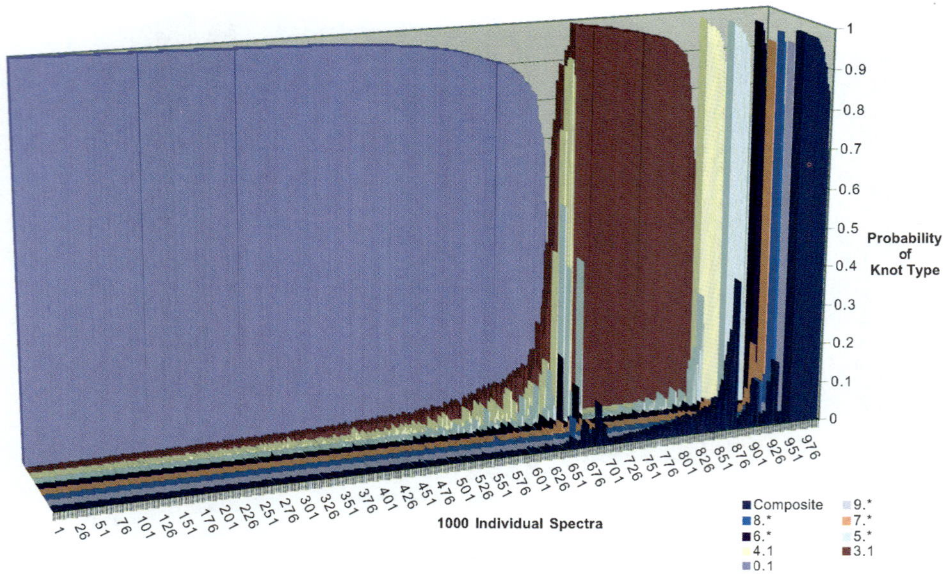

Plate 11.3. Knot spectra for 1000 random walks of length 150.

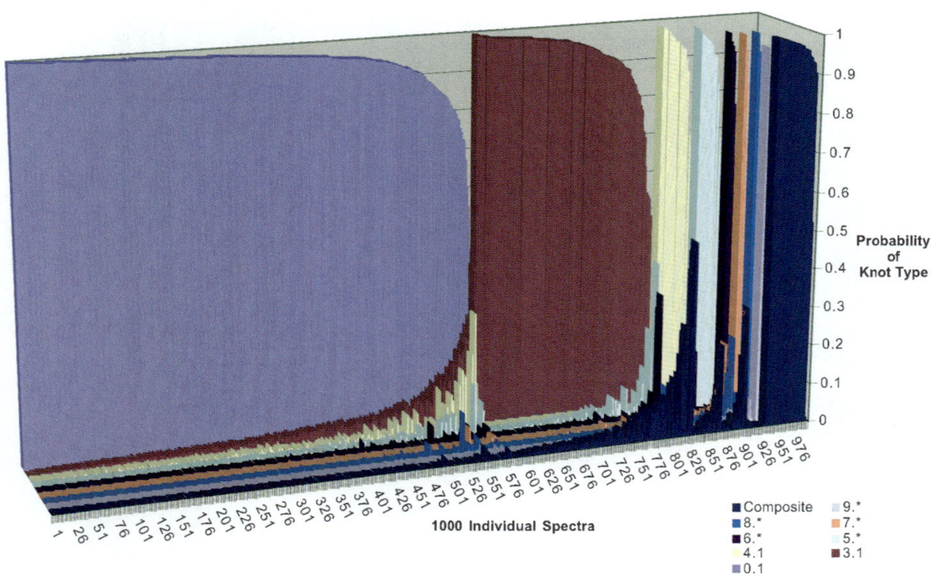

Plate 11.4. Knot spectra for 1000 random walks of length 200.

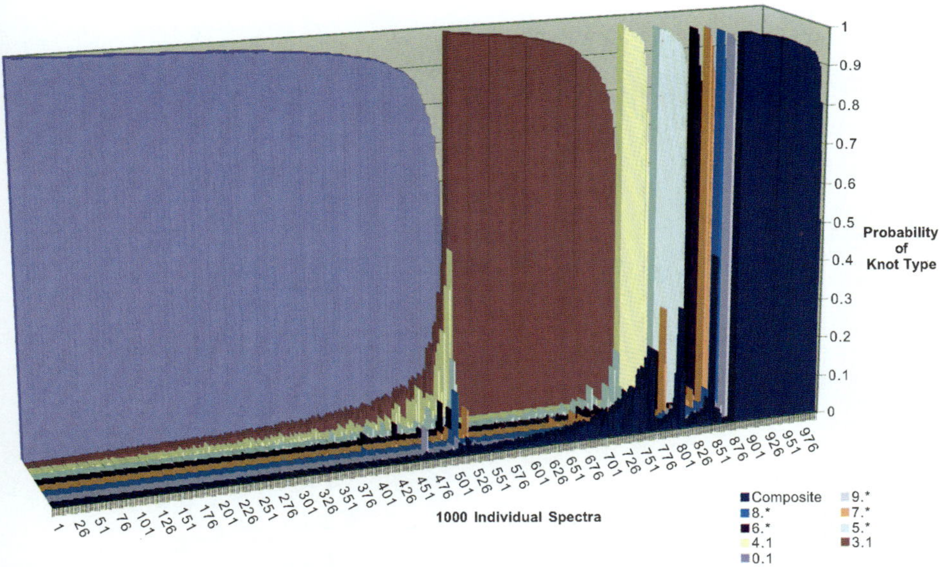

Plate 11.5. Knot spectra for 1000 random walks of length 250.

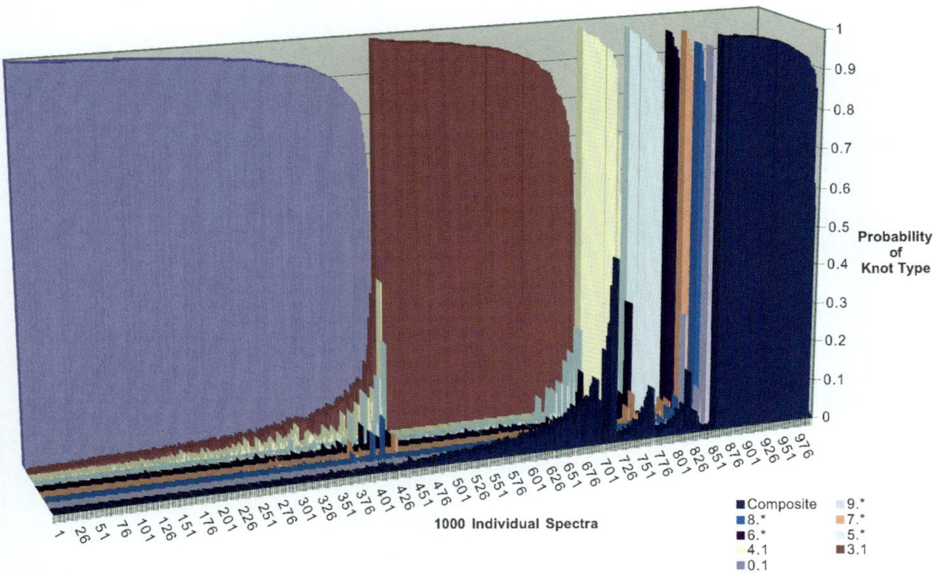

Plate 11.6. Knot spectra for 1000 random walks of length 300.

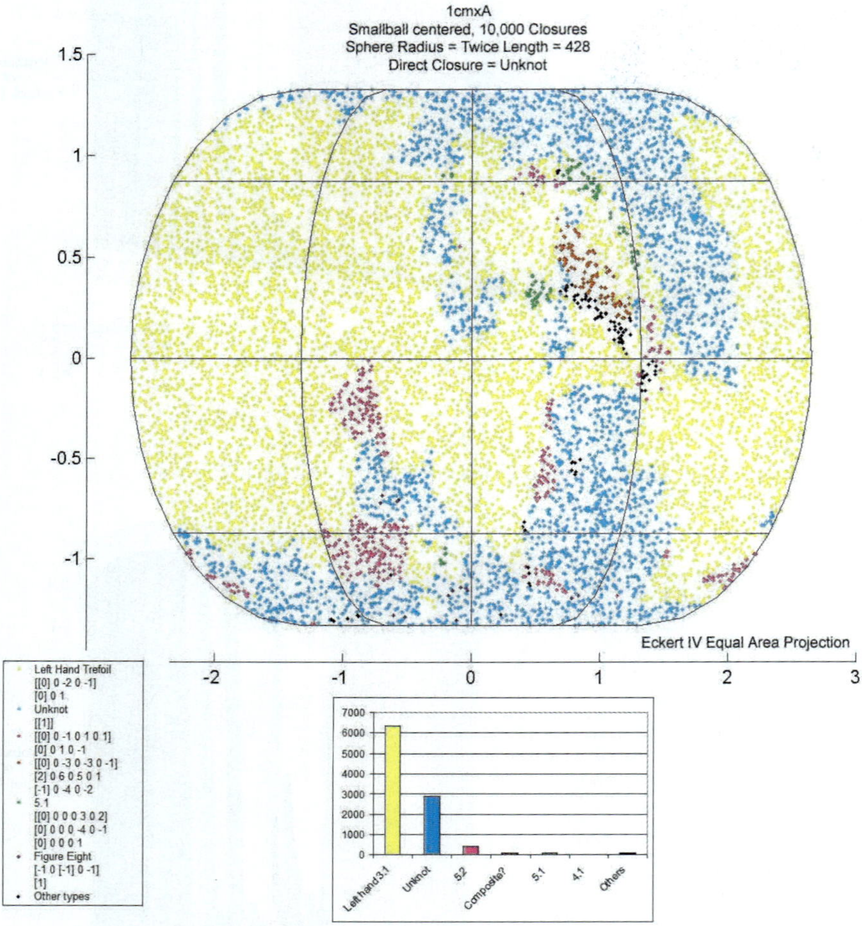

Plate 11.7. An Eckert IV projection of the spectrum of the protein 1cmxA.

CHAPTER 16

ROPELENGTH OF TIGHT POLYGONAL KNOTS

Justyna Baranska, Piotr Pieranski
Poznan University of Technology
Laboratory of Computational Physics and Semiconductors
Nieszawska 13A, 60 965 Poznan, Poland
E-mail: pieransk@man.poznan.pl, jmarch@phys.put.poznan.pl

Eric J. Rawdon
Department of Mathematics and Computer Science
Duquesne University
Pittsburgh, PA 15282, USA
E-mail: rawdon@mathcs.duq.edu

A physical interpretation of the rope simulated by the SONO algorithm is presented. Properties of the tight polygonal knots delivered by the algorithm are analyzed. An algorithm for bounding the ropelength of a smooth inscribed knot is shown. Two ways of calculating the ropelength of tight polygonal knots are compared. An analytical calculation performed for a model knot shows that an appropriately weighted average should provide a good estimation of the minimum ropelength for relatively small numbers of edges.

1. Introduction

Knots tied on material objects, e.g. a rope of a finite thickness, are called physical knots [20]. It is well known from everyday life that open knots can be tightened by pulling apart the loose ends of the rope. Tightening a closed knot, i.e. a knot tied on a rope whose ends are spliced, is more difficult since one has to engage processes in which the rope will reduce its length while keeping its diameter intact. Such a process, difficult to create in the laboratory, can be simulated numerically and we find in the literature various examples of such simulations [3,5,14,16,19,21,22]. From the intuitive point of view, it is rather obvious that in tightening a knot tied on a rope of a finite diameter, one must arrive at a conformation at which the tightening process stops. This conformation is called *tight.* Any small change in the shape of a tight conformation needs an elongation of the rope. Obviously, tight conformations of knots tied on different types of rope, with different physical properties, will be different. Thus, to make the notion of the tight knot unambiguous, one must specify the physical model of the rope on which the knot is tied. The simplest of such models is the *perfect rope.* Perfect rope is *perfectly floppy* – needs no force to be bent, but, at the same time, it is *perfectly hard* – needs an infinite force to be squeezed, its diameter always remains intact. It is also *perfectly slippery.* Its formal definition will be given in the next section.

Tight conformations can be seen as minimizers of the ropelength function [2,13]. The minima can be only local. While it has been shown that global minima exist [4,8,9], with the exception of the trivial knot and some simple links, the minimizing conformations are not known. Knots in the conformations at which the ropelength reaches its global minimum are called *ideal.* As long as the ideal conformations are not known analytically, we cannot be certain that the conformation at which one arrives is ideal.

In simulations of the processes in which knots are tightened, the knots are represented by a finite sequence of points in $\mathbb{R}^3$. When connected, the points can be seen as vertices of a self-avoiding polygon. Most often the simulation programs either start from a non-equilateral conformation and aim at making it equilateral (e.g. SONO) or start from an equilateral conformation and try to keep it such (crankshaft rotations). Let P be such an equilateral polygon. The essential problem we face is to construct, using the coordinates of the vertices of P, a smooth knot K_P which can be seen as the axis of the knot tied on the perfect rope. In constructing K_P, we make some assumptions concerning the shape of the pieces of which it is built. In what follows we assume that the knot K_P is built from smoothly connected arcs inscribed into P, which makes the knot piecewise C^2 smooth. As a result of this assumption and provided the radius of the rope is properly chosen, we show that the knot K_P can indeed be tied with perfect rope leaving it self-avoiding. On the other hand, we also know that the inscribed knot is not ideal, thus, its ropelength can serve only as an upper bound for the minimum ropelength of the unknown ideal conformation. In finding tight polygonal knots P with an increasing number of vertices, we see that the ropelengths of the inscribed smooth knots K_P converge to a value, which we believe is the ropelength of the ideal conformation, although we will not prove that here. This convergence is slow however, and one might want an approximation using relatively few edges. This is especially important for complicated knots where the number of edges needed to adequately approximate the smooth ropelength would not be practical. With this aim, a second, weighted-average, approach is explored. This technique provides surprisingly accurate approximations with relatively few edges. Together, we provide both a theoretically-sound approach with slow convergence and a practical approach with fast convergence. Depending on the application, we expect that both techniques will be useful.

2. Computing upper bounds for smooth knots inscribed in tight polygons

2.1. *Ropelength of knots tied on the perfect rope*

Smooth ropelength models perfect rope as a non self-intersecting tube with a $C^{1,1}$ knot as its core. For completeness, we include these definitions and some properties of this knot energy.

Definition 1: For a C^1 knot K and $x \in K$, let $D_r(x)$ be the disk of radius r centered at x lying in the plane normal to the tangent vector at x. Let

$$R(K) = \sup\{r > 0 \,:\, D_r(x) \cap D_r(y) = \emptyset \text{ for all } x \neq y \in K\}\,.$$

The quantity $R(K)$ is called the *thickness radius* of K. Define the *ropelength* of K to be

$$Rope(K) = Length(K)/R(K)\,,$$

where $Length(K)$ is the arclength of K.

The thickness radius $R(K)$ is the radius of a thickest tube that can be placed about the knotted core K without self-intersection. Note that in some of the literature, the thickness refers to the *diameter* of this thickest tube, in which case the ropelength is half of what we use here. The problem we face is how, knowing the shape of K, to find $R(K)$.

Suppose we are given a concrete knot K and we are trying to reproduce it with an impenetrable tube. Thus, K serves here as the core of the knotted tube. We have to pay attention to two types of interactions between the tube's normal disks that restrict the radius R of the tube.

(A) The interactions between the disks centered on pairs of points located infinitesimally close to the diagonal of $K \times K$ – these normal disks will start overlapping when the radius of curvature is smaller than R.

(B) The interactions between disks centered on points located away from the diagonal of $K \times K$ – the disks will overlap when the Euclidean distance between the points is smaller than $2R$.

Fig. 1 illustrates the problem. These intuitive observations are captured by the quantities below and the subsequent lemma.

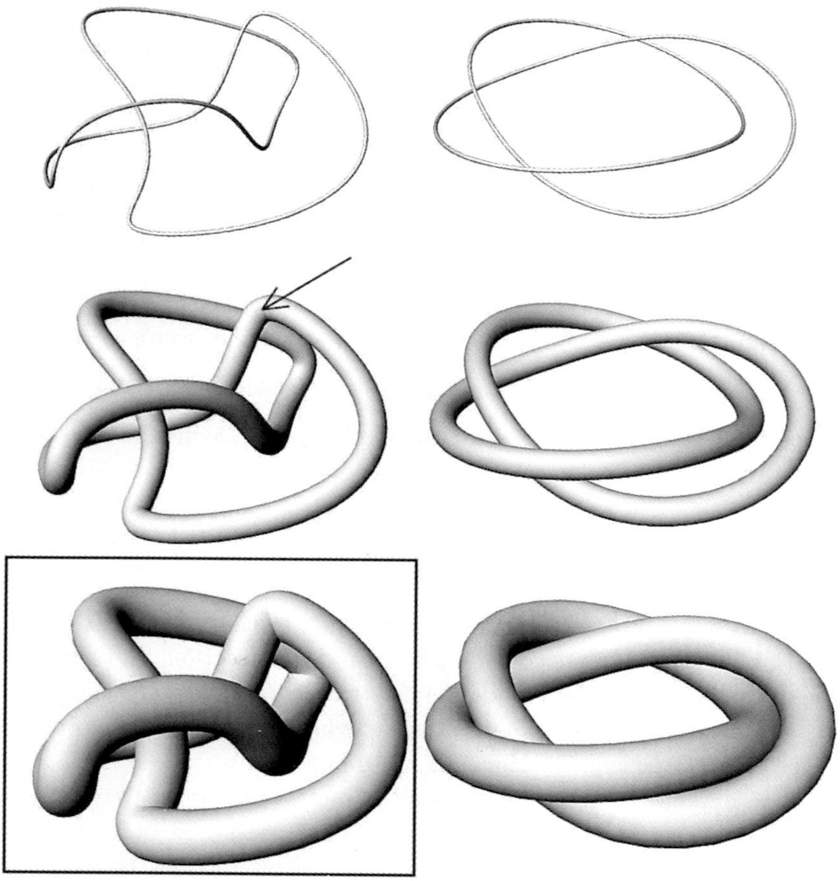

Fig. 1. Two different conformations of the trefoil knot are inflated. The inflation of the conformation shown on the left stops first because of its sharp bends at which interactions of type A intervene. Further inflation would lead to the development of singularities in the tube surface (see the framed picture). Inflation of the conformation shown on the right stops at the point above which interactions of type B intervene.

Definition 2: For a C^2 knot K with unit tangent map T, let $MinRad(K)$ be the minimum of the radius of curvature measured at each of the points on the knot. The *doubly-critical self-distance* is the minimum distance between pairs of points on the knot whose connecting chord is perpendicular to the tangent vectors at both of the points. In other words, let

$$DC(K) = \{(x, y) \in K \times K \, : \, T(x) \perp \overline{xy} \perp T(y), x \neq y\}$$

where $\overline{xy}$ is the chord connecting x and y. Define the *doubly-critical self-distance* by

$$DCSD(K) = \min\{|x - y| \, : \, (x, y) \in DC(K)\},$$

where $| \cdot |$ is the standard $\mathbb{R}^3$ norm.

There is a fundamental relationship between $R(K)$, $MinRad(K)$, and $DCSD(K)$.

Lemma 3: *Suppose K is a C^2 knot. Then*

$$R(K) = \min\{MinRad(K), DCSD(K)/2\}\ .$$

Proof: See [13]. □

In Refs. [4,8,9], it is shown that ropelength minima exist as (at worst) $C^{1,1}$ curves. After reworking the definition of $MinRad$, one can extend Lemma 3 to include C^1 (and thus $C^{1,1}$ curves). The following is taken from Ref. [7]. For a C^0 function $f : \mathbb{R} \to \mathbb{R}^3$, define the *dilation* of f by

$$dil(f) = \sup\left\{\frac{|f(s) - f(t)|}{|s - t|} \, : \, s, t \in \mathbb{R}, s \neq t\right\}.$$

Note that if K is a C^2 knot parameterized by arclength with unit tangent map T, then $MinRad(K) = 1/dil(T)$. Thus, the dilation gives a generalization for $MinRad$ to knots that are C^1. Since $MinRad$ and $1/dil(T)$ are equal for C^2 knots, we use $MinRad$ to denote $1/dil(T)$ for all C^1 knots. If a knot is C^1 but not $C^{1,1}$, then $dil(T) = \infty$, in which case $MinRad$ is assumed to be 0. In this paper, we are mainly interested in $C^{1,1}$ knots, in which case $dil(T)$ is finite and $MinRad$ is positive. Alternate, but equivalent, approaches for defining the ropelength of $C^{1,1}$ knots are explored in Refs. [4,6,10].

Lemma 4: *Suppose K is a C^1 knot. Then*

$$R(K) = \min\left\{MinRad(K), \frac{DCSD(K)}{2}\right\}.$$

Furthermore, if K is $C^{1,1}$, then $R(K) > 0$.

Proof: See Refs. [4,7]. □

2.2. *Ropelength of polygonal knots*

As mentioned in the introduction, the problem we face in analyzing knots provided by numerical simulations is that they are not given in a smooth continuous form. For instance, programs based on the SONO algorithm deliver nearly equilateral polygonal knots. Let P be such a knot. It is represented by a sequence of n points whose positions are indicated by vectors $\vec{v}_1, \vec{v}_2, \ldots, \vec{v}_n$. We shall refer to them as the *vertices* or *beads* of the polygonal knot P that one obtains by joining consecutive vertices. Because the knots that we are considering are closed, we implicitly take all subscripts modulo n. We define

$$\begin{aligned}\vec{e}_i &= \vec{v}_{i+1} - \vec{v}_i \\ e_i &= |\vec{e}_i| \\ \vec{r}_{i,j} &= \vec{v}_j - \vec{v}_i \\ r_{i,j} &= |\vec{r}_{i,j}| .\end{aligned}$$

The vectors $\vec{e}_i$ will be called *edges*. When referring to the vertices as points on a polygonal knot, we shall omit the vector notation.

In what follows, we will use the notion of the index distance ID between the vertices:

$$ID(v_i, v_j) = \begin{cases} |i-j|, & \text{when } |i-j| \le n/2 \\ n - |i-j|, & \text{otherwise} \end{cases}.$$

In the SONO simulations, the vertices of a knot interact with each other as if they were spheres, the properties of which will be presented later. Taking into account the representation, we shall extend the notion of the thickness radius in terms of virtual spheres. We do this by defining notions of $MinRad$ and $DCSD$ for polygons similar to the characterization theorem (Lemma 4) for smooth thickness radius. Alternate approaches have appeared in Refs. [5,6,10,11,12,17,18,19,21,23].

Definition 5: Suppose P is a polygonal knot. Let

$$Rad(v_i) = \frac{\min\{e_{i-1}, e_i\}}{2\tan(\theta_i/2)}$$

where θ_i is the turning angle of the tangent vectors at v_i and e_i is the length of the edge $\vec{e}_i$. Note that $Rad(v_i)$ is the radius of an arc α of a circle that can be inscribed in the corner at v_i so that α intersects the shorter of the adjacent edges at the midpoint and the tangents at the intersection points on P coincide with the tangents on α. Let

$$MinRad(P) = \min_i\{Rad(v_i)\}\,.$$

If P happens to be equilateral with edge length dl, this reduces to

$$MinRad(P) = \frac{dl}{2\tan(\theta_{max}/2)}\,,$$

where θ_{max} is the maximum turning angle. See Figure 2.

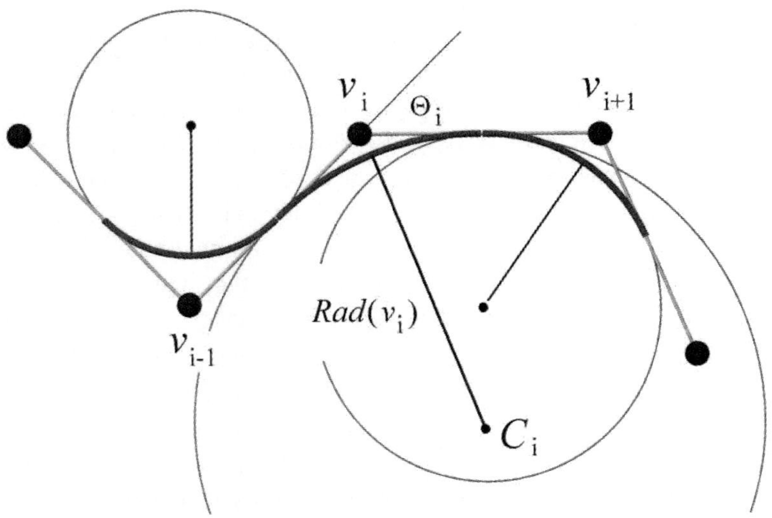

Fig. 2. Vertices and edges of a piece of an equilateral polygonal knot P. Inscribed arcs of radii $Rad(v_i)$ are also shown. To simplify the drawing we assumed that all of the vertices are located in the same plane.

Definition 6: Let P be a polygonal knot and let $MinEdge(P)$ and $MaxEdge(P)$ be the minimum and maximum of the edge lengths of P.

We define the *sphere thickness radius* to be

$$R_s(P) = \min\{\sqrt{MinRad(P)^2 + (MinEdge(P)/2)^2}, SR(P)\}$$

where $SR(P)$, the *sphere radius*, is half the minimum distance between vertices whose index distance is at least $\lceil \pi MinRad(P)/MaxEdge(P) - 1 \rceil$.

The construction of the sphere thickness radius needs a few words of explanation. The knots we will be considering are very close to being equilateral, so for now assume that P is an equilateral knot with common edge length dl. Imagine that about each of the vertices, there is a sphere of radius R. When R is larger than $dl/2$, the spheres about nearby vertices intersect. The case in which $R < dl/2$ need special attention, but since they never occur in the simulations we eventually analyze, we exclude them assuming implicitly in what follows that $R > dl/2$. The surface of the union of all spheres is bumpy, so we refer to this as the *corrugated tube* or *bead rope* about the knot. The problem we face is to inflate the spheres as much as possible without violating the condition of the self-avoidedness of the corrugated tube formed by their union.

First, let us explain what we mean by self-avoiding in the case of the corrugated tube. Consider three consecutive vertices v_{i-1}, v_i, and v_{i+1} and the spheres S_{i-1}, S_i, and S_{i+1}, around them all of radius R. The spheres S_{i-1} and S_i intersect at a circle C_{i-1} whose radius equals $R_D = \sqrt{R^2 - (dl/2)^2}$. The spheres S_i and S_{i+1} intersect at a circle C_i of the same radius. Let D_{i-1} and D_i be the disks filling the circles. The disks are inclined at an angle equal to the turning angle θ_i. We assume that the disks are hard and hence cannot overlap; thus, at most they may become tangent at a single point. The latter happens when the sphere radius is larger than $\frac{dl}{2\sin(\theta_i/2)}$. For a given equilateral polygonal knot P, the smallest of such radii is $\sqrt{MinRad(P)^2 + (dl/2)^2}$. When R is chosen larger than $\sqrt{MinRad(P)^2 + (dl/2)^2}$, we see the behavior shown in Figure 3a.

When $R \leq \sqrt{MinRad(P)^2 + (dl/2)^2}$, all portions of the corrugated tube are well-defined locally and one must check if index-distant vertices are too Euclidean-close. Let us explain what we mean by stating that the portions of the corrugated tube are well-defined. Each portion of the corrugated tube is a sphere with two pieces cut off. The cutting planes are perpendicular to the polygon edges and are located at the midpoints. Consecutive portions are thus connected with disks of radius R_D. As stated above, the disks are assumed to be hard and they are not allowed to overlap. A portion of the corrugated tube which fulfills this condition is said to be well-defined. If the Euclidean distance between two vertices is smaller than $2R$, the portions of the corrugated tube overlap. Thus, we check all the Euclidean distances between sufficiently index-distant vertices and find the smallest one. This will be the value of $2SR(P)$. This is a geometrical interpretation of the self-avoidedness conditions found in the definition of $R_s(P)$. When R is chosen larger then $SR(P)$, we see the behavior shown in Figure 3b.

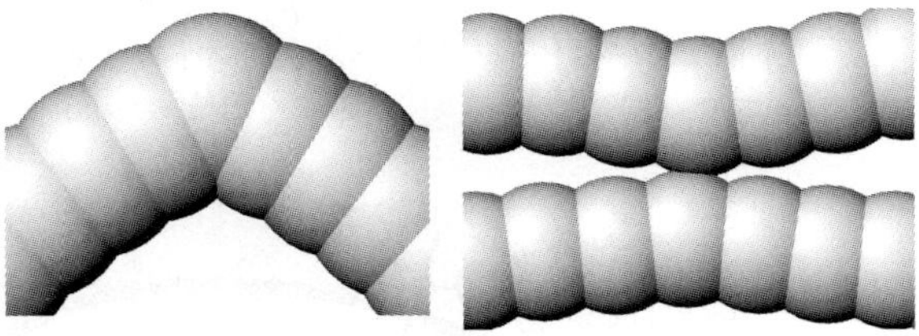

Fig. 3. Violation of the self-avoidedness of the bead tube. (a) The tube bends too quickly. (b) Two index-distant portions of the tube are too close to each other.

We will be mainly interested in optimizing R_s, but to do so we need a scale-invariant version of R_s. There are various ways to do this and we explore two normalizations in the next section. The corrugation will play a part in how the knot packs efficiently in these optimizing conformations. In the short term, our goal is to show that for a sufficiently "thick" polygon P, there exists a smooth knot inscribed in P whose thickness radius is close to $R_s(P)$. We first present the inscribing algorithm. A similar formulation for inscribing polygons appears in Ref. [19].

Proposition 7: *For a given n-edge polygonal knot P, a $C^{1,1}$ curve K_P can be inscribed in P in such a way that $MinRad(K_P) = MinRad(P)$. Furthermore, there exists a bijection between K_P and P so that for each pair $x' \in P$ and $x \in K_P$, we have $|x - x'| \leq R_s(P)\left(\sec\left(\frac{\theta_{max}}{2}\right) - 1\right)$.*

Proof: An arc α_i of a circle of radius $Rad(v_i)$ can be inscribed at v_i such that α_i is tangent to $\vec{e}_{i-1}$ and $\vec{e}_i$ and intersects the shorter of the adjacent edges at the midpoint. Let K_P be the result of inscribing arcs of radius $Rad(v_i)$ at each vertex v_i and removing the bypassed corners. The resulting knot K_P consists of arcs of circles and straight line segments (where the lengths of adjacent edges are not equal). Since there is no overlapping of adjacent inscribed arcs, K_P is well-defined as a (possibly self-intersecting) closed curve. The curve K_P has a piecewise constant radius of curvature, and $MinRad(K_P) = MinRad(P)$. The knot K_P is C^1 and piecewise C^2, and thus, lies in the category of $C^{1,1}$ curves.

For each x on the inscribed curve K_P, we define a unique point x' on P. If x is a point also on P, let $x' = x$. Otherwise, x lies on an arc, say α_i, whose center is C_i. Let x' be the intersection of the ray $\overrightarrow{C_i x}$ with $e_{i-1} \cup e_i$ (see Figure 4). Simple trigonometric calculations show that

$$\|x - x'\| \leq MinRad(P) \left(\sec \left(\frac{\theta_{max}}{2} \right) - 1 \right). \qquad \square$$

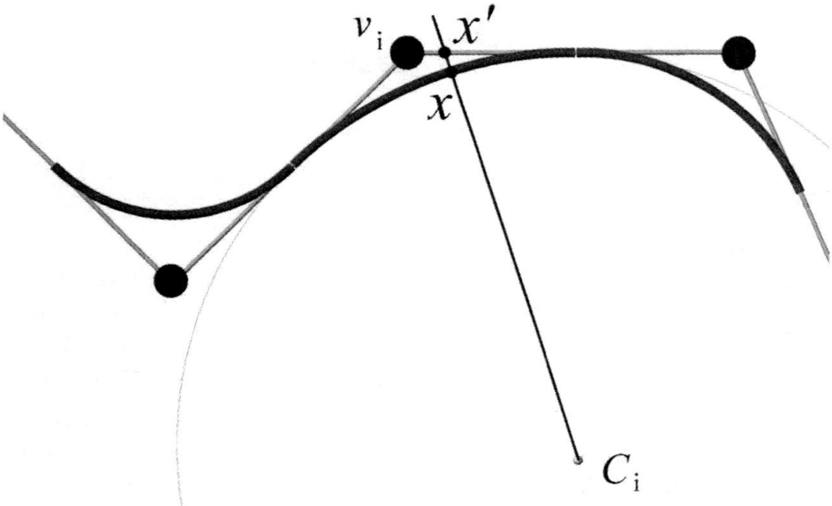

Fig. 4. The definition of the points x and x' used in Proposition 7.

For each $x' \in P$, the point x' has a closest vertex with respect to arc distance. In particular, for the points on edge e_i, half of the points are closer to v_{i-1} and half are closer to v_i. We call the closest vertex to x' its *corresponding vertex*. The midpoint of each edge has two corresponding vertices. It will be convenient to think of the midpoint as corresponding to each vertex in different situations and this will not pose any problems in this work. For each $x \in K_P$, we define its *corresponding vertex* to be the corresponding vertex of the point $x' \in P$. Note that the smaller θ_{max} is, the closer the inscribed curve K_P will be to the polygon P.

While the knots we eventually analyze are nearly equilateral, any knot realized on a computer is inherently not exactly equilateral. Because we wish to prove mathematically rigorous upper bounds, we will now assume that the knot need not be exactly equilateral. However, one will notice that this construction will yield the best upper bounds when the knots are close to equilateral.

The corrugated tube about the polygon is a union of many spheres, some of which are allowed to intersect. The intersection between the surface of two adjacent spheres S_i, S_{i+1} consists of a circle whose radius is $\sqrt{R_s(P)^2 - e_i^2/4}$ and whose center is the midpoint of the edge $\vec{e}_i$. This value will be important, so we define the smallest of these as $R_c(P) = \sqrt{R_s(P)^2 - MaxEdge(P)^2/4}$. We will show now that the smooth tube of radius $R_c(P)$ about K_P is non self-intersecting, i.e. that the thickness radius $R(K_P)$ of K_P is at least $R_c(P)$. See Figure 5. This is a bit delicate and will take some effort to prove. We will need the following technical lemma.

Lemma 8: *Let P be an polygonal knot, $R_s(P) > MaxEdge(P)/2$, and K_P be the smooth knot inscribed in P via the algorithm of Proposition 7. If $x, y \in K_P$ and $arc(x, y) \geq \pi MinRad(K_P) = \pi MinRad(P)$, then $ID(v_x, v_y) \geq \lceil \pi MinRad(P)/MaxEdge(P) - 1 \rceil$, where v_x and v_y are the vertices corresponding to x and y.*

Proof: Suppose $arc(x, y) \geq \pi MinRad(P)$. Now $arc(x, y)$ is the minimum arclength on K_P measured over the two paths connecting x and y. By the construction of the inscribed smooth knot, we know that $arc(x, y)$ is no larger than the arclength along P between the corresponding points x' and y'. In the worst case, both x and y are midpoints and their corresponding vertices v_x and v_y lie on the shorter arc between x and y. In such a case, we have

$$arc(x, y) \leq arc_P(v_x, v_y) + 2(MaxEdge(P)/2),$$

where arc_P denotes the minimum arclength when walking along the polygon P. Now $arc_P(v_x, v_y) \leq n\, MaxEdge(P)$ where n is the number of edges between v_x and v_y. Thus, $n\, MaxEdge(P) \geq \pi MinRad(P) - MaxEdge(P)$ or $n \geq \pi MinRad(P)/MaxEdge(P) - 1$. □

In other words, if $arc(x, y) \geq \pi MinRad(P)$, then we know that the corresponding vertices v_x and v_y will be a part of the set over which

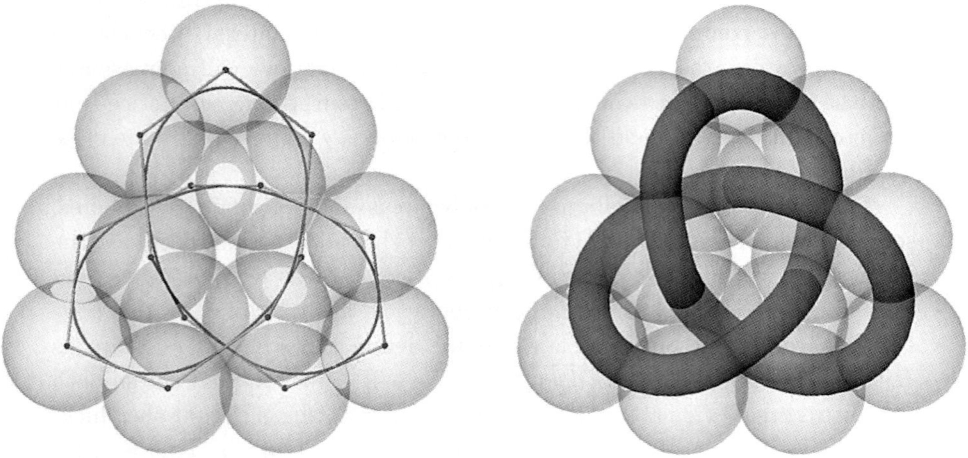

Fig. 5. The picture on the left presents the polygonal trefoil knot P tightened by SONO. The distance between its vertices is guarded by the virtual spheres of radius $SR(P)$. Since in the case shown in the figure $SR(P) < \sqrt{MinRad(P)^2 + (MinEdge(P)/2)^2}$, the value of $SR(P)$ serves as the sphere thickness radius $R_s(P)$. Consecutive spheres S_i, S_{i+1} meet at circles of radius $R_c(P) = \sqrt{R_s(P)^2 - MaxEdge(P)^2/4} \leq \sqrt{R_s(P)^2 - e_i^2/4}$. The smooth knot K_P shown also in the figure consists of arcs inscribed into K. The picture on the right shows the inscribed knot K_P inflated to radius $R_c(P)$. It is well visible, and we prove it in the text, that the smooth tube knot, hidden as a whole inside the corrugated tube made of the virtual spheres, remains self-avoiding.

$SR(P)$ is computed. This will be important in ensuring that $SR(P)$ bounds $DCSD(K_P)$.

We now have the tools to bound the thickness radius of the inscribed knot K_P.

Theorem 9: Let P be a polygonal knot, $R_s(P) > MaxEdge(P)/2$, and K_P be the smooth knot inscribed in P via the algorithm of Proposition 7. Then $R(K_P) \geq R_c(P) = \sqrt{R_s(P)^2 - \frac{MaxEdge(P)^2}{4}}$.

Proof: By the construction of K_P, we know that $MinRad(K_P) = MinRad(P)$. We split the proof into two cases: when $DCSD(K_P)/2 \geq MinRad(K_P)$ and when $DCSD(K_P)/2 < MinRad(K_P)$.

In the former case, we have

$$\begin{aligned} R(K_P) &= MinRad(K_P) \\ &= MinRad(P) \\ &\geq \sqrt{(MinRad(P)^2 + MinEdge(P)^2/4) - MaxEdge(P)^2/4} \\ &\geq \sqrt{R_s(P)^2 - MaxEdge(P)^2/4} \\ &= R_c(P). \end{aligned}$$

In the latter case, we assume $R(K_P) = DCSD(K_P)/2$. By Ref. [19], we know that when $DCSD(K_P)$ is realized at a pair of points, say a and b, then $arc(a, b) \geq \pi MinRad(K_P) = \pi MinRad(P)$. Therefore, the corresponding vertices v_a and v_b have $ID(v_a, v_b) \geq \lceil \pi MinRad(P)/MaxEdge(P) - 1 \rceil$ by Lemma 8. Let D_a and D_b be the disks of radius $R_c(P)$ normal to K_P at a and b respectively and let B_a and B_b be the spheres of radius $R_s(P)$ centered at v_a and v_b respectively. Now a and b lie inside B_a and B_b respectively, and by the definition of $R_s(P)$, we know that the interiors of B_a and B_b do not intersect. Furthermore, the normal disks D_a and D_b are contained within B_a and B_b respectively. Therefore, D_a and D_b can only intersect, in the worst case, on the boundary. This implies that $|a - b| \geq 2R_c(P)$ or $DCSD(K_P)/2 \geq R_c(P)$. □

The previous result is not sharp in general. However, for the sake of this paper, it does provide a computable lower bound for the thickness radius of K_P. Furthermore, the length of K_P, which is smaller than the length of P, can be computed explicitly. We call $Length(K_P)/R_c(P)$ the *inscribed ropelength* and denote it L_c. The value of L_c is an upper bound for the ropelength of the inscribed K_P, and as $MaxEdge(P) \to 0$ (note that changing $MaxEdge(P)$ affects n, P, and K_P), our bound on $R(K_P)$ approaches $R_s(P)$. Since $R_s(P) > dl/2$, the inscribed arcs lie within the corrugated tube about P, which guarantees that K_P has the same knot type as P. Since K_P is a $C^{1,1}$ smooth knot with the same knot type as P, we have a bound on the ropelength of one smooth knot within the knot type. Thus, we have an upper bound for the minimum smooth ropelength for the given knot type.

3. Interpretation of simulations performed with the SONO algorithm

Let us now describe the SONO algorithm using the notions defined above. First notice that in discussing the ropelength of smooth knots, we are considering knots K of a given length $Length(K)$ looking for their thickness radius $R(K)$. The value $R(K)$ is the maximum radius for which K could be inflated without violating the conditions that the surface of the tube must remain self-avoiding. The value $Rope(K) = Length(K)/R(K)$ is the ropelength of K. The application of the virtual inflation suggests numerical calculations in which the inflation process forces the knot to change its conformation. Such changes may maximize $R(K)$ and thus minimize $Rope(K)$. Working with polygonal knots, it proves more convenient to consider a different scheme in the simulations: keep the sphere radius SR fixed and shorten the edge lengths so that the knot eventually arrives at a conformation for which $R_s(P) = SR$. This is the approach we discuss below.

3.1. *Basic procedures of SONO*

There are two essential procedures on which SONO is based [15]. The basic goal of SONO is to reduce $Length(P)$ subject to the constraint that $R_s(P) \geq SR$, for some fixed rope radius SR, while keeping the polygon very close to being equilateral. In practice, we choose $SR = 1$. The reduction of the polygon length is achieved by reducing its goal edge length dl. We aim at simulating equilateral knots; thus one of these procedures called EqualizeEdges (EE), checks the length of the edges, and, if they differ from the desired value dl, introduces necessary corrections. Suppose the distance e_i between vertex i and $i+1$ is different from the desired dl value. Then the vertices are shifted toward new positions so that their distance is closer to dl:

$$\vec{v'}_i = \vec{v}_i - c_1(dl - e_i)\,\vec{e}_i/e_i$$
$$\vec{v'}_{i+1} = \vec{v}_{i+1} + c_1(dl - e_i)\,\vec{e}_i/e_i$$

$c_1 \in (0, 1/2]$. Usually we work with $c_1 = 1/2$.

As a result of multiple applications of the procedure, the dispersion of the edge lengths in the final conformations becomes very small, and thus, the knots we get from our simulations are very close to equilateral.

The second procedure called RemoveOverlaps (RO), checks the distances between the vertices. When the RO procedure finds that the minimum distance $2SR$ is violated, the vertices v_i and v_j are shifted away from each other to a distance equal to $2SR$ or, what proves to speed up the initial stage of the tightening process, exceeding $2SR$ by ϵ:

$$\vec{v'}_i = \vec{v}_i - c_2(2SR - r_{i,j} + \epsilon)\,\vec{r}_{i,j}/r_{i,j}$$
$$\vec{v'}_j = \vec{v}_j + c_2(2SR - r_{i,j} + \epsilon)\,\vec{r}_{i,j}/r_{i,j}$$

where $c_1 \in (0, 1/2]$ and is usually set equal to 1/2. The value of ϵ changes during the tightening process. Initially it is of the order 10^{-2} and reaches the level of 10^{-7} at its end.

The procedures RO and EE are in some circumstances contradictory, but their multiple application leads to simultaneous reduction of both the overlaps and the dispersion of the edge lengths. Both parameters are constantly monitored.

When the goal edge length *dl* is small, the tightened polygonal knots develop short regions where $Rad(v_i)$ tends to become small. An additional procedure ControlCurvature (CC) monitors this, never allowing θ_i to be larger than $2\arcsin\left(\frac{e_i}{2SR}\right)$. This, in particular in the final stages of the tightening, makes sure that the tight conformation of P will have $R_s(P) = SR$.

3.2. *Physical sense of the SONO algorithm and practical details of simulations*

Let us now discuss the physical sense of the simulation based on the SONO algorithm. Since our goal is to gain intuition, it is covenient to assume that the knot is equilateral at this point. The simulated knot P can be seen as tied on a closed necklace of beads. The necklace is unusual, because its beads are unusual: if their index distance $ID(v_i, v_j) \leq \lceil \frac{\pi MinRad(P)}{dl} \rceil$, they are allowed to overlap, otherwise they behave as hard spheres of radius SR and repel infinitely hard. The RO procedure simulates this interaction. Because of the action of the EE procedure, the centers of the consecutive beads can be seen as connected with nonextensible rods of controlled length *dl*. The rods are connected to each other by elasticity free but angle-limiting joints. Each of the beads has the shape of a sphere with two parts of it cut off. The cutting planes are located in the middle of the edges which lead to it. Consecutive beads are thus connected via disks of radius $R_c = \sqrt{SR^2 - dl^2/4}$. The necklace can be bent in any direction but only to some extent: the CC procedure limits the angle between the consecutive rods to $2\arcsin\left(\frac{dl}{2SR}\right)$. As a result, $MinRad(P) \geq R_c$, which prevents the disks separating consecutive beads to overlap. At most they are allowed to become tangent. Because of its construction, the surface of the bead rope is not smooth. It is corrugated and the corrugation is more pronounced, the larger the distance between consecutive beads. This distance is, of course, the edge length dl.

The tightening process runs as follows. Suppose we start with a loose conformation. The EE and RO procedures make the lengths of all edges nearly equal and remove overlaps between the beads. After multiple applications of the procedures, the dispersion of the edge lengths and the sum of all overlaps fall below a fixed level. Then SONO reduces the required edge length dl. Now, as the EE procedure tries to adjust the length of the edges to the new value by shortening the distance between consecutive beads, new overlaps may appear. Subsequently they are removed by the RO procedure. And so on.

In the final tight conformation, the dispersion of edge lengths and the sum of all overlaps are smaller than 10^{-6} which allows us to consider the closest beads as just touching. Keeping in mind that the CC procedure does not allow $\sqrt{MinRad(P) + dl^2/4}$ to be smaller than SR, the final tight knot is a nearly equilateral knot P whose $R_s(P) = SR$.

The shrinking rope forces the knot to change its conformation. At the end, we arrive at a conformation for which further shortening is no longer possible because it creates non-removable overlaps. Obviously, if the number of beads is small, the final value of dl is large and the bead rope is strongly corrugated. This may create problems, since the knot is more likely to become stuck in a local minimum. The reason for this is as follows: one part of the rope winding around another part may get into a groove between consecutive beads even if shifting it to another groove would allow further shortening of the rope. The shifting will not be achieved by SONO, because it would need a (rather large) temporary increase of the ropelength. Working with a small number of beads, we set ϵ to a higher value to minimize these effects. In general, the tightening process should be seen as a physical experiment in which the experimenter watches and adjusts the parameters.

3.3. *The problem of finding the right ropelength, an experimental approach*

In Figure 6, we have the tightest conformations of the trefoil knot found by SONO working with rope consisting of n =15, 30, 60, 120, 240, and 480 vertices.

It seems plausible that the best estimation of the minimal length of the rope needed to tie a particular knot will come from analyzing the polygonal conformation obtained for highest n. As seen in the figure, at $n = 480$

Fig. 6. Tight conformations of the trefoil knot tied on a rope consisting of 15, 30, 60, 120, 240 and 480 beads.

the surface of the bead rope is visually smooth – the corrugation of its surface, so visible at $n = 15, 30, 60$, becomes undetectable by the naked eye. The problem we face is that in tightening much larger knots, for instance the $(2, 99)$ torus knot, reaching this level of the surface smoothness would require a very large number of beads, which is both awkward and time-consuming. For practical reasons, it is reasonable to limit the number of beads as much as possible. We present a picture showing the details of the SONO tightened trefoil for such a small n in Figure 7.

Calculations run faster when n is small, but what about the accuracy of the ropelength we can extract from the analysis of the final, tight conformation?

4. Ropelength of SONO knots

We are mainly interested in understanding the optimal conformations of R_s. However, R_s varies with scale, so one could always increase R_s simply by scaling the polygon. In the second section, we explored one way to normalize R_s, namely by analyzing the ropelength L_c of the inscribed knot K_P. In doing so, we overestimate the minimum ropelength. In this section, we present a different normalization which appears to underestimate the ropelength. In the next section, we combine these two notions to obtain a reasonable approximation of the minimum ropelength with relatively few edges.

The simplest way to determine the ropelength is to sum the length of all edges of the knot and divide by $R_s(P)$:

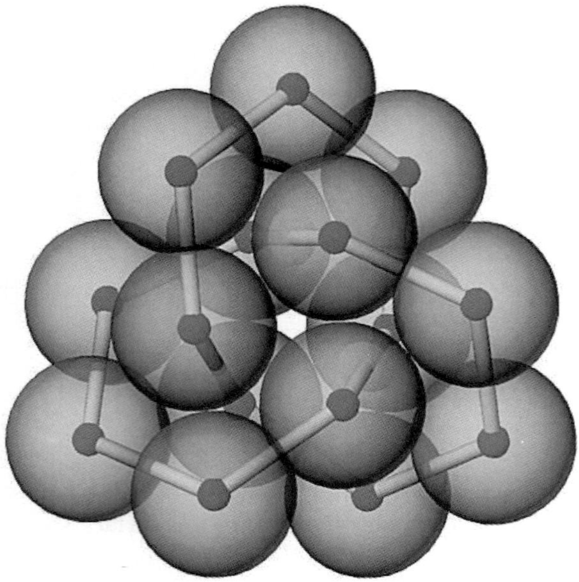

Fig. 7. Details of a trefoil knot tightened by SONO. The knot is tied on a piece of corrugated rope consisting of $n = 15$ beads. The centers of the beads make the vertices of the polygonal knot. The vertices are connected with straight cylindrical edges.

$$L_p = \left(\sum_{i=1}^{n} e_i \right) / R_s(P) \approx n\, dl / R_s(P) \, .$$

In what follows we shall refer to L_p as the *raw polygonal ropelength*. In calculating the ropelength in such a simplistic manner, we implicitly assume that the polygonal knot makes the axis of a rope of radius 1, which is obviously wrong. Certainly, if the cylindrical segments of the rope are not to overlap, the radius should be smaller. This becomes clear for large dl. However erratic the raw polygonal length may seem, it will prove to be very useful.

It is interesting to see how the values of the raw polygonal and inscribed ropelength behave in practice as n increases. For the sake of convenience, Figure 8 shows both values plotted versus the goal edge length dl.

First notice that as the edge length diminishes, the raw polygonal and the inscribed ropelengths apparently converge to a common value. We call

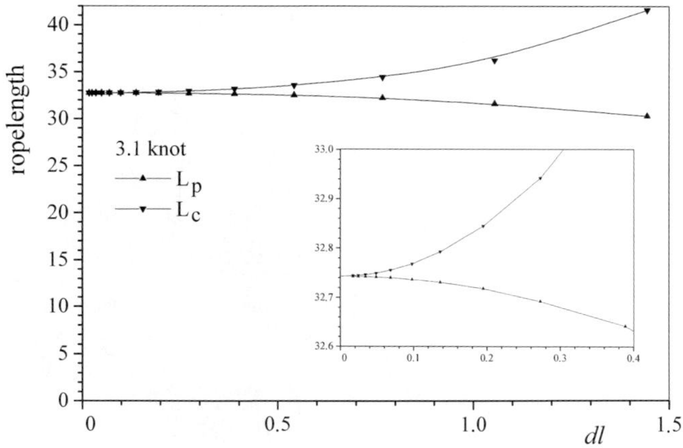

Fig. 8. The raw polygonal L_p and inscribed L_c ropelength for SONO tightened trefoil knot versus the edge length with n=21, 30, 42, 60, 84, 120, 168, 240, 336, 480, 672, 960, 1342, and 1920.

this value the *limiting ropelength*, denoted L_∞. In spite of its erratic nature, the raw polygonal ropelength converges faster than the inscribed ropelength. The raw polygonal ropelength L_p appears to underestimate while the inscribed ropelength L_c overestimates the ropelength. Can one find a weighted average that will accurately estimate the ropelength with relatively few edges? Below, we present a heuristic reasoning which provides the answer.

4.1. *The problem of finding the right ropelength, an analytic approach*

By looking at SONO tightened knots, see for instance Figure 9, one may notice that we often deal with pieces of a rope winding around other pieces of the rope. At smaller n, the rope is strongly corrugated. Thus, to be precise, we deal with pieces of a corrugated rope winding around other pieces of the corrugated rope. In a model situation shown also in the picture, the central piece is straight and the other piece winds tightly around it. By analyzing the raw polygonal and inscribed ropelength functions in this simplified model, we will shed some light on their functioning in the description of tight knots.

To analyze the problem, we consider a still simpler case of a corrugated rope wound toroidally around a corrugated cylinder. See Figure 10.

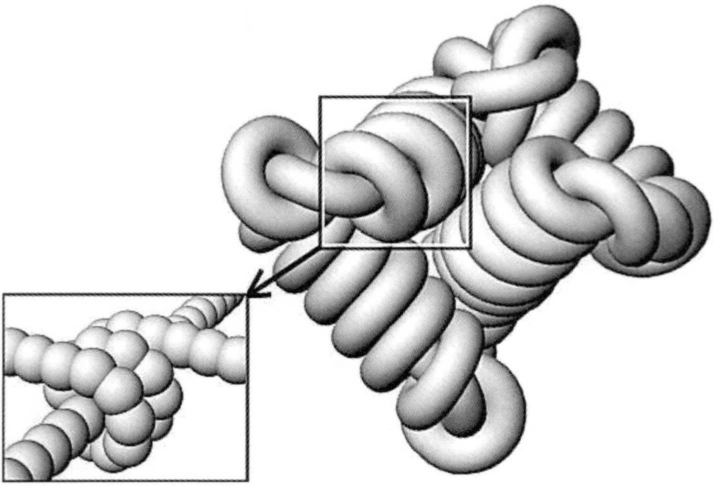

Fig. 9. A tight conformation of the (2,59) torus knot found by SONO. A typical situation where one piece of the rope is winding around another piece is shown in the frame. In simulations, the rope is corrugated as shown in the inset.

In the latter case, we may perform a rigorous analysis of the ways in which the raw polygonal and the inscribed ropelength estimate the limiting value of the ropelength. The latter is known since it is simply the length of a smooth torus winding tightly around a smooth cylinder. Assuming that both the cylinder and the torus are made of a tube whose radius $R = 1$, the value is 4π.

Now, let us consider the corrugated torus on the corrugated cylinder. Both of them can be seen as unions of spheres. We assume that all the spheres have radius $R = 1$. The distance between consecutive spheres is dl. Obviously, for an arbitrary dl there are problems with a clean closing of the torus. In what follows we shall concentrate on a short piece of the torus, so this is not problematic.

The best way to wind a piece of a corrugated rope around a straight piece of the same rope is to wind it inside the groove between two consecutive spheres. Figure 10 illustrates the situation. Our aim is to find the raw

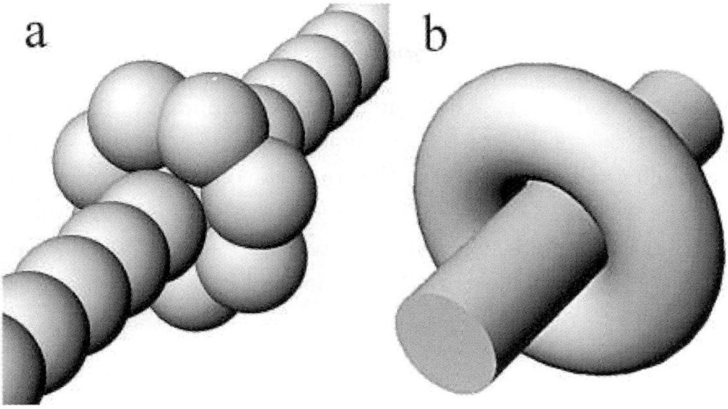

Fig. 10. (a) A piece of a corrugated rope winds toroidally around the corrugated cylinder (a straight piece of corrugated rope). (b) In the $n \to \infty$ limit, we deal with a torus winding around a cylinder.

polygonal length and the inscribed ropelength of a single piece of the rope and compare them with the limiting length of the analogous piece of the torus wound tightly around the cylinder. The relations we find here should apply to the situation we face in calculating the ropelength of the SONO tightened knots.

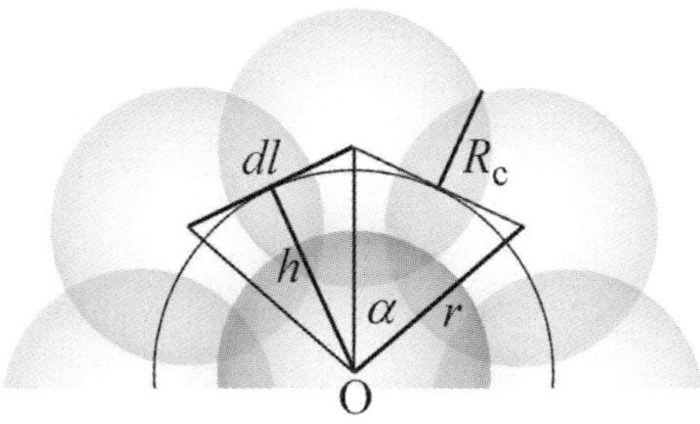

Fig. 11. The geometry of the corrugated rope wound within a groove of a straight piece of the same rope. The value of R_c is the radius of the circles at which consecutive spheres intersect. Thus, it is also the radius of the smooth rope which can be safely placed inside the corrugated rope. Its axis runs along the inscribed arcs.

Let us look at Figure 11. Assume the distance between the spheres, i.e. the polygon edge length, is dl. Since the corrugated rope is wound within a groove, its spheres are at a distance $r < 2$ from the center O of the axis of the corrugated cylinder. We get

$$r = \sqrt{4 - dl^2/4}\,.$$

The distance is different when the corrugated rope is wound not within the groove, but on the hill of the corrugated cylinder. Here $r = 2$. In real situations, such as this presented in Figure 9, the corrugated rope runs both in the grooves and on the hills. It is thus reasonable to assume an average r. It would be an error to calculate the average as an arithmetic average, since the shape of the hills is not saw-tooth. The hills have the form of circular arcs.

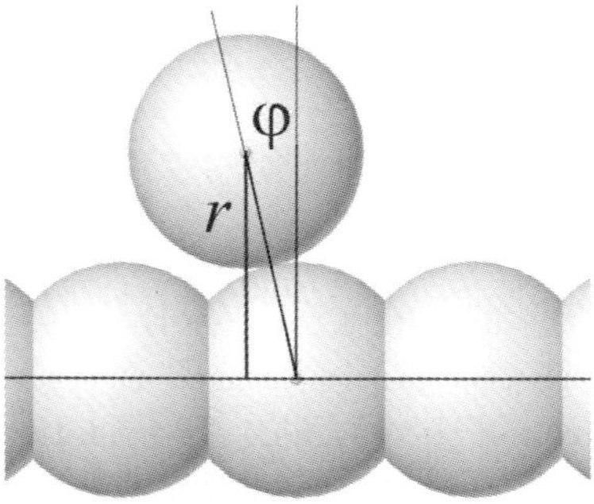

Fig. 12. The bead of the winding rope can be located anywhere on the circular hill between two consecutive grooves.

The position of the bead can be parametrized by the angle φ, whose maximum φ_{max} value is given by

$$\varphi_{max} = \arcsin\left(\frac{dl}{4}\right).$$

See Figure 12. The average value of r is found by integration:

$$r_{av} = \frac{1}{\varphi_{max}} \int_0^{\varphi_{max}} 2cos(\varphi)d\varphi = \frac{dl}{2\arcsin(\frac{dl}{4})}.$$

Let us return now to Figure 11. The angle α at which the edge is seen from the center O is given by

$$\alpha = 2\arcsin\left(\arcsin\left(\frac{dl}{4}\right)\right).$$

The middle point of the edge is found at the distance h from O, where

$$h = \frac{dl\sqrt{\arcsin(\frac{dl}{4})^{-2} - 1}}{2}.$$

The length of the inscribed arc joining the middle points of two consecutive edges is

$$d\lambda = dl \arcsin\left(\arcsin\left(\frac{dl}{4}\right)\right)\sqrt{\arcsin\left(\frac{dl}{4}\right)^{-2} - 1}\,.$$

Since the arc makes the axis of the smooth rope of radius

$$R_c = \sqrt{1 - dl^2/4}\,,$$

the normalized length $dL_c = d\Lambda/R_c$ of the rope segment is

$$dL_c = \frac{2\,dl \arcsin\left(\arcsin\left(\frac{dl}{4}\right)\right)\sqrt{\arcsin(\frac{dl}{4})^{-2} - 1}}{\sqrt{4 - dl^2}}\,.$$

The raw polygonal length of the segment is simply

$$dL_p = dl\,.$$

Now, let us ask the basic question: what is the limiting ropelength dL of the segment? It is the length of a piece of a smooth torus (wound tightly on the cylinder of unit radius) seen at angle α from the center O:

$$dL = 4 \arcsin\left(\arcsin\left(\frac{dl}{4}\right)\right)\,.$$

The relationship between the limiting length dL, its raw polygonal length dL_p, and its inscribed length dL_c estimations should be similar to those observed for the limiting L_∞, raw polygonal L_p, and inscribed L_c ropelengths found for knots tightened by SONO. Thus, let us see how the raw polygonal dL_p and inscribed rope dL_c approximations differ from the limiting value dL. To get a clear quantitative estimation of errors that we make using dL_p and dL_c, we plot the relative deviations $(dL_p - dL)/dL$ and $(dL_c - dL)/dL$ of the raw polygonal and inscribed segment length from its limiting dL value. See Figure 13.

By looking at the picture, one can clearly see that dL_p underestimates while dL_c overestimates the length of the segment. Perhaps an appropriately weighted average of the values could provide a better estimation of the limiting length. To check this, we solve the equation

$$(a dL_p + dL_c)/(a + 1) = dL\,.$$

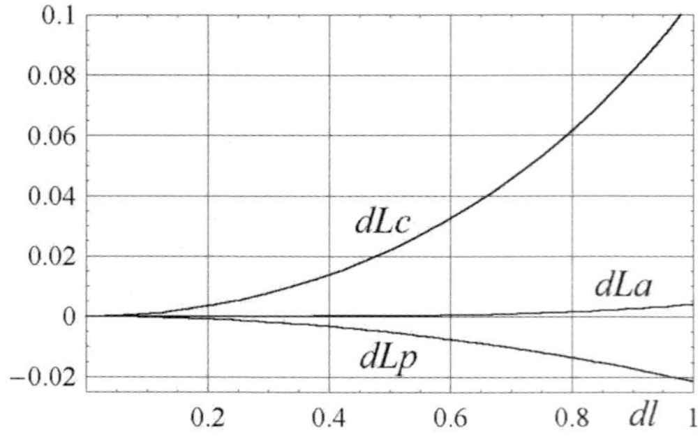

Fig. 13. The relative deviations of the raw polygonal dL_p and inscribed dL_c segment lengths from the limiting value dL. The value dL_a is the relative deviation of the weighted average $dL_a = (4dL_p + dL_c)/5$. Compare the plots with the plots shown in Figure 15.

Its solution is

$$a = \frac{2\left(2 - \frac{dl\sqrt{\arcsin(\frac{dl}{4})^{-2} - 1}}{\sqrt{4 - dl^2}}\right)\arcsin(\arcsin(\frac{dl}{4}))}{dl - 4\arcsin(\arcsin(\frac{dl}{4}))}.$$

The functional dependence of the weight a on the edge length dl looks rather complex, but its plot versus dl reveals that the dependence is very weak – at $dl = 1$, its value is close to 5, but as dl diminishes, it converges quickly to 4. See Figure 14.

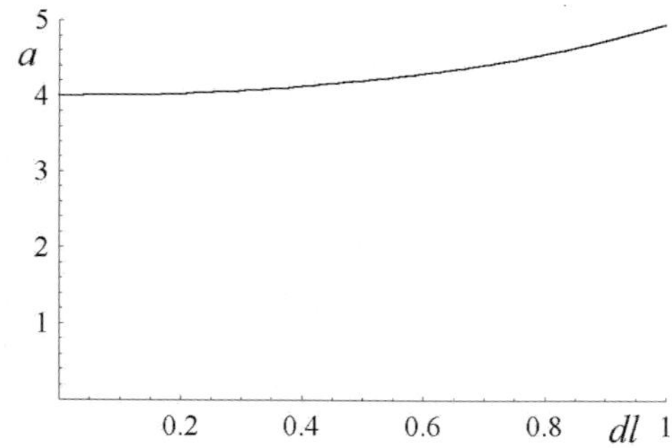

Fig. 14. Weight a versus dl.

By looking at the plot, one may suspect that the expansion of $a(dl)$ should have zero order term equal to 4 and much smaller higher order terms. We carried out the expansion to get:

$$a = 4 + \frac{353\, dl^2}{480} + O(dl^4)\,.$$

Since the relations between L_p, L_c, and L_∞ in tight knots should be similar to the relations between dL_p, dL_c, and dL, we arrive to the conclusion that the weighted average

$$L_a = (4L_p + L_c)/5$$

should be almost independent of the edge length dl and, thus, it should provide a good estimate of the L_∞ value. To check the hypothesis, we perform a series of tests. First of all, we calculate the weighted average for the trefoil knot, whose L_p and L_c are plotted in Figure 15. Looking at the plot of L_a, one clearly sees that, as expected, it is almost independent of dl.

Table 1. The raw polygonal, inscribed and weighted average ropelength of the trefoil knot tightened by SONO.

n	dl	L_p	L_c	$L_a = (4L_p + L_c)/5$
120	0.27243068	32.69168	32.94095	32.7415
240	0.13637831	32.73079	32.79243	32.7431
480	0.06820844	32.74005	32.75542	32.7431
960	0.03410665	32.74238	32.74622	32.7431
1920	0.01705362	32.74295	32.74391	32.7431

Similar tests have been performed for the next torus knots: 5_1, 7_1, and 9_1. See Tables 1–4.

Table 2. The raw polygonal, inscribed and weighted average ropelength of the 5_1 knot tightened by SONO.

n	dl	L_p	L_c	$L_a = (4L_p + L_c)/5$
180	0.26182883	47.12919	47.45869	47.20
360	0.13105733	47.18064	47.26225	47.20
720	0.06554618	47.19325	47.21360	47.20
1440	0.03277605	47.19751	47.20261	47.20

The applicability of the weighted average ropelength to other knots requires further analysis.

5. Discussion

Numerical experiments carried out by the SONO algorithm provide us with polygonal knots which can be seen as skeletons of tight knots tied on a

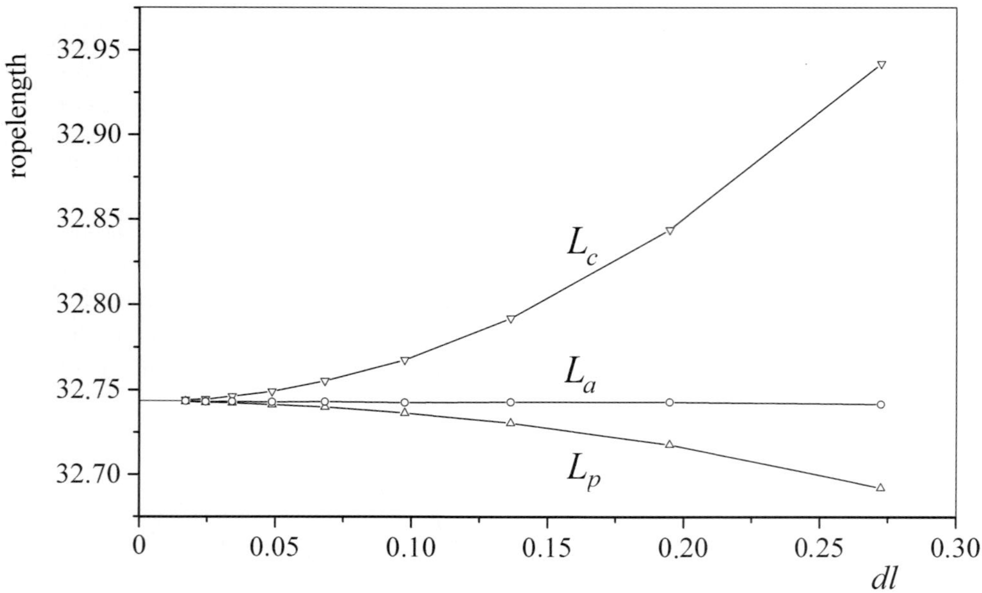

Fig. 15. Raw polygonal L_p, inscribed L_c, and weighted average ropelength L_a for SONO tightened trefoil knot versus the edge length for n= 240, 480, 960, and 1920. The data were fitted with with second order polynomial curves.

Table 3. The raw polygonal, inscribed and weighted average ropelength of the 7_1 knot tightened by SONO.

n	dl	L_p	L_c	$L_a = (4L_p + L_c)/5$
154	0.39748	61.21150	62.21861	61.41
308	0.19921	61.35573	61.60250	61.40
616	0.09966	61.39211	61.45352	61.40
1232	0.04984	61.40131	61.41663	61.40

Table 4. The raw polygonal, inscribed and weighted average ropelength of the 9_1 knot tightened by SONO.

n	dl	L_p	L_c	$L_a = (4L_p + L_c)/5$
188	0.40140	75.46436	76.72711	75.72
376	0.20118	75.64443	75.95430	75.71
752	0.10066	75.69522	75.77243	75.71
1504	0.05034	75.70902	75.72832	75.71

corrugated rope. The vertices of the polygonal knots are the centers of the spherical segments of the corrugated tube. As we demonstrated, one can place a smooth tube of a smaller radius inside the corrugated tube to obtain a smooth knot tied within a smooth tube. We know that L_c, the bound on its ropelength, is larger than the ropelength of the ideal knot. In observing the ropelength of the inscribed knots, we see that it converges with the increasing number of vertices to a value L_∞, which can be seen as an estimation for the ropelength of an ideal knot. Determining the exact value of L_∞ is a subtle problem and shall not be discussed here; however, by combining the results we obtained for finite n with the inscribing algorithm, we are able to find an upper bound for the ropelength of a few ideal knots. For the trefoil knot, the bound equals 32.744, for the 5_1 knot it equals 47.203, for the 7_1 knot it equals 61.417, and for the 9_1 knot it is 75.728. These are the smallest upper bounds obtained so far [a].

As said above, the L_c values we find in analyzing the polygonal knots delivered by SONO can be seen as upper bounds for the ropelength of their ideal smooth conformations. But what about the weighted average values L_a? To see their utility we must adopt the less rigorous point of view of an experimental physicist. The coordinates of the polygonal knots delivered by SONO can be seen as results of measurements carried out on knots tied on the corrugated rope. Analyzing the experimental data, one finds that the knot length calculated according to the raw polygonal length formula systematically increases with n while the length calculated according to the inscribed arcs formula systematically decreases with n. Plotting the values together, one notices that they converge to a common value: L_c from below, L_p from above. One also finds that, located between the two plots, the plot of their weighted average is almost flat and horizontal. By looking at the L_a values presented in the tables, one can clearly see that L_a provides a good estimate of the ropelength, even at a small number of vertices.

Acknowledgments

We thank Maciej Oszwaldowski for indicating the possibility of an analytical analysis of the weighted average algorithm. Pieranski and Baranska were supported under project TB 62-176/04-DS. Rawdon was supported by NSF Grant No. 0311010.

[a]Since the time this chapter was accepted for publication, we performed a new series of more extensive simulations ending at $n = 2544$. Analysis of their results provided a more precise (and slightly smaller) value of 32.7434. See Ref. [1].

References

1. J. Baranska, P. Pieranski, S. Przybyl, and E.J. Rawdon. Length of the tightest trefoil knot. *Phys. Rev. E*, 70:051810, 2004.
2. Gregory Buck and Jeremey Orloff. A simple energy function for knots. *Topology Appl.*, 61(3):205–214, 1995.
3. J. A. Calvo, K. C. Millett, and E. J. Rawdon, editors. *Physical Knots: Knotting, Linking, and Folding Geometric Objects in* $\mathbb{R}^3$, volume 304 of *Contemporary Mathematics*, Providence, RI, 2002. Amer. Math. Soc., Amer. Math. Soc. Physical Knotting and Linking (Las Vegas, NV, 2001).
4. Jason Cantarella, Robert B. Kusner, and John M. Sullivan. On the minimum ropelength of knots and links. *Invent. Math.*, 150(2):257–286, 2002.
5. M. Carlen, B. Laurie, J. H. Maddocks, and J. Smutny. Biarcs, global radius of curvature, and the computation of ideal knot shapes. In this volume, 2004.
6. Y. Diao, C. Ernst, and E. J. Janse van Rensburg. Thicknesses of knots. *Math. Proc. Cambridge Philos. Soc.*, 126(2):293–310, 1999.
7. O. Durumeric, R. A. Litherland, E. Rawdon, and J. Simon. Thickness of knots 2. preprint.
8. O. Gonzalez and R. de la Llave. Existence of ideal knots. *J. Knot Theory Ramifications*, 12(1):123–133, 2003.
9. O. Gonzalez, J. H. Maddocks, F. Schuricht, and H. von der Mosel. Global curvature and self-contact of nonlinearly elastic curves and rods. *Calc. Var. Partial Differential Equations*, 14(1):29–68, 2002.
10. Oscar Gonzalez and John H. Maddocks. Global curvature, thickness, and the ideal shapes of knots. *Proc. Natl. Acad. Sci. USA*, 96(9):4769–4773 (electronic), 1999.
11. Vsevolod Katritch, Jan Bednar, Didier Michoud, Robert G. Scharein, Jacques Dubochet, and Andrzej Stasiak. Geometry and physics of knots. *Nature*, 384(6605):142–145, 1996.
12. Robert B. Kusner and John M. Sullivan. On distortion and thickness of knots. In *Topology and geometry in polymer science (Minneapolis, MN, 1996)*, pages 67–78. Springer, New York, 1998.
13. R. A. Litherland, J. Simon, O. Durumeric, and E. Rawdon. Thickness of knots. *Topology Appl.*, 91(3):233–244, 1999.
14. K. C. Millett and Eric J. Rawdon. Energy, ropelength, and other physical aspects of equilateral knots. *J. Comput. Phys.*, 186(2):426–456, 2003.
15. Piotr Pierański. In search of ideal knots. In *Ideal knots*, pages 20–41. World Sci. Publishing, River Edge, NJ, 1998.
16. E. Rawdon. *The Thickness of Polygonal Knots*. PhD thesis, University of Iowa, 1997.
17. Eric J. Rawdon. Approximating the thickness of a knot. In *Ideal knots*, pages 143–150. World Sci. Publishing, River Edge, NJ, 1998.
18. Eric J. Rawdon. Approximating smooth thickness. *J. Knot Theory Ramifications*, 9(1):113–145, 2000.
19. Eric J. Rawdon. Can computers discover ideal knots? *Experiment. Math.*, 12(3):287–302, 2003.

20. Jonathan Simon. Physical knots. In *Physical knots: knotting, linking, and folding geometric objects in* $\mathbb{R}^3$ *(Las Vegas, NV, 2001)*, volume 304 of *Contemp. Math.*, pages 1–30. Amer. Math. Soc., Providence, RI, 2002.
21. Jana Smutny. *Global Radii of Curvature, and the Biarc Approximation of Space Curves: In Pursuit of Ideal Knot Shapes.* PhD thesis, EPFL, 2004.
22. A. Stasiak, V. Katritch, and L. H. Kauffman, editors. *Ideal knots.* World Scientific Publishing Co. Inc., River Edge, NJ, 1998.
23. Andrzej Stasiak, Jacques Dubochet, Vsevolod Katritch, and Piotr Pieranski. Ideal knots and their relation to the physics of real knots. In *Ideal knots*, pages 1–19. World Sci. Publishing, River Edge, NJ, 1998.

CHAPTER 17

A FAST OCTREE-BASED ALGORITHM FOR COMPUTING ROPELENGTH

Ted Ashton* and Jason Cantarella†

Department of Mathematics, University of Georgia, Athens, GA 30602

The *ropelength* of a space curve is usually defined as the quotient of its length by its *thickness*: the diameter of the largest embedded tube around the knot. This idea was extended to space polygons by Eric Rawdon, who gave a definition of ropelength in terms of doubly-critical self-distances (local minima or maxima of the distance function on pairs of points on the polygon) and a function of the turning angles of the polygon. A naive algorithm for finding the doubly-critical self-distances of an n-edge polygon involves comparing each pair of edges, and so takes $O(n^2)$ time. In this paper, we describe an improved algorithm, based on the notion of *octrees*, which runs in $O(n \log n)$ time. The speed of the ropelength computation controls the performance of ropelength-minimizing programs such as Rawdon and Piatek's TOROS. An implementation of our algorithm is freely available under the GNU Public License.

1. Introduction

For a C^2 curve in 3-space, *ropelength* is the quotient of the length of the curve by its *thickness*: the diameter of the largest embedded tube around the curve. Minimizing ropelength is the same as fixing the diameter of the tube and minimizing its length— if the tube is knotted, we are pulling the knot tight, and so the minimum ropelength curves in any knot type are often called *tight* knots. Since the problem is such a natural one, the definition of thickness has been discovered and rediscovered by several authors[1,3,14], with the earliest results known (to these authors) on the problem credited to Krötenheerdt and Veit in 1976[12].

In the past decade, there has been a great deal of interest in exploring the geometry of tight knots; the definition of thickness has been refined and fully

*email:ashted@uga.edu

†email:cantarel@math.uga.edu

understood[10], it has been shown that $C^{1,1}$ minimizers exist in each knot type[5,8,9], some minimizing links have been found[5], and a theory of ropelength criticality has started to emerge[4,21]. The development of this theory has been fueled by a steady stream of numerical data on ropelength minimizers, from Pieranski's original SONO algorithm[15] and Rawdon's TOROS[16], to second-generation efforts such as Smutny and Maddocks' biarc computations[6,19] and the RIDGERUNNER project of Cantarella, Piatek, and Rawdon. All of these algorithms have in their innermost loops a computation of the ropelength of a curve in 3-space.

Intuitively, the thickness of a tube is controlled locally by the curvature of the core curve, and globally by the approach of "distant" sections of the tube. Rawdon, in his thesis[17], defined a radius of curvature for a corner of a polygon. A given corner has two circles which are tangent to both incident edges and tangent to one of the edges at its center. He proved that we can define a sensible polygonal radius of curvature as the radius of the smaller of those two circles.

More precisely:

Definition 1: If P_n is a polygonal curve in $\mathbb{R}^3$ with edges $e_1, \ldots, e_n$, and α_i is the turning angle of the polygon made by edges e_i and e_{i+1}, then let

$$\operatorname{minRad}(P_n) = \min_{i \in 1,\ldots,n} \left\{ \frac{|e_i|}{2\tan\left(\frac{\alpha_i}{2}\right)}, \frac{|e_{i+1}|}{2\tan\left(\frac{\alpha_i}{2}\right)} \right\} \tag{1}$$

where we take $e_{n+1} = e_1$ if the polygon is a closed curve, and take $i \in 1, \ldots, n-1$ otherwise.

Definition 2: Using the distance function on $P_n \times P_n$ given by $D(x,y) = |x-y|$, we say that a pair xy of P_n (bounding the chord $\overline{xy}$) is a *pair of closest approach* of P_n if it is a non-trivial local minimum of the distance function. The length of the shortest such chord is denoted $\operatorname{POCA}(P_n)$ (and we take $\operatorname{POCA}(P_n) = \infty$ if no such chord exists).

Definition 3: We define the *thickness* of P_n by

$$\operatorname{Thi}(P_n) = \min\left\{2\operatorname{minRad}(P_n), \operatorname{POCA}(P_n)\right\}. \tag{2}$$

We note that the value which Rawdon uses in place of $\operatorname{POCA}(P_n)$ in his original definition of polygonal thickness[17] is different. In particular, it is always finite. But Rawdon reports that the equivalence of the two definitions follows from results in an upcoming paper[13].

As computing the radius of curvature at a given corner only involves the edges incident to that corner, computing $\operatorname{minRad}(P_n)$ requires only $O(n)$ time. On the other hand, all previous efforts to compute thickness have used some variant of Algorithm 1 for computing $\mathrm{POCA}(P_n)$. This algorithm is clearly $O(n^2)$. So we have focused our attention on improving the $\mathrm{POCA}(P_n)$ calculation.

```
for i = 1 to n do
    for j = i + 1 to n do
        check e_i and e_j for local min chords;
        compare to previous shortest local min chord;
    end
end
```

Algorithm 1: Standard Algorithm for $\mathrm{POCA}(P_n)$.

Our algorithm concentrates on reducing the total number of edge-edge checks performed by grouping the edges according to their positions in space into a data structure known in computer graphics as an *octree*. We will use the octree to optimize the inner loop of Algorithm 1, and show that we can isolate a constant-size set of candidate e_j's for any given e_i in time $O(\log n)$. The new algorithm will then perform $O(n \log n)$ edge-edge checks, and one octree construction (which will also require time $O(n \log n)$).

Before continuing, it is reasonable to ask whether such a complicated algorithm can be implemented in a way that provides a practical advantage over Algorithm 1. We believe that our implementation, `liboctrope`, does. We give performance data for some test problems in Section 6. And more importantly, we invite interested readers to download `liboctrope` and test the code themselves (`http://ada.math.uga.edu/research/software/octrope`).

2. Edge-Edge Checks

The quantity $\mathrm{POCA}(P_n)$ is defined to be the smallest nontrivial local minimum of the distance function $D(x, y)$ on pairs of points on the polygon P_n. To understand it, we first make an observation about the nature of these local minima.

Lemma 4: *If we orient the curve P_n and let $T^-(x)$, $T^+(x)$ denote the inward and outward tangent vectors of P_n at x (they are different if and only if x is a vertex with nonzero turning angle). Every pair xy which locally minimizes $D : P_n \times P_n \to \mathbb{R}$ has*

$$T^-(x) \cdot (y - x) \geq 0 \qquad T^+(x) \cdot (y - x) \leq 0 \tag{3}$$

$$T^-(y) \cdot (x - y) \geq 0 \qquad T^+(y) \cdot (x - y) \leq 0. \tag{4}$$

We note that if x is in the interior of an edge, then the above relations force $T^{\pm}(x) \cdot (x - y) = 0$.

Proof: There are three cases: either both x and y are on the interior of an edge, one is an edge point and one a vertex, or both are vertices, as shown in Figure 1. At x, the distance from y must not decrease to first order as one moves away from x in either direction along the curve: a computation verifies that this is equivalent to the first line of the statement of the Lemma. A similar argument at y completes the proof. □

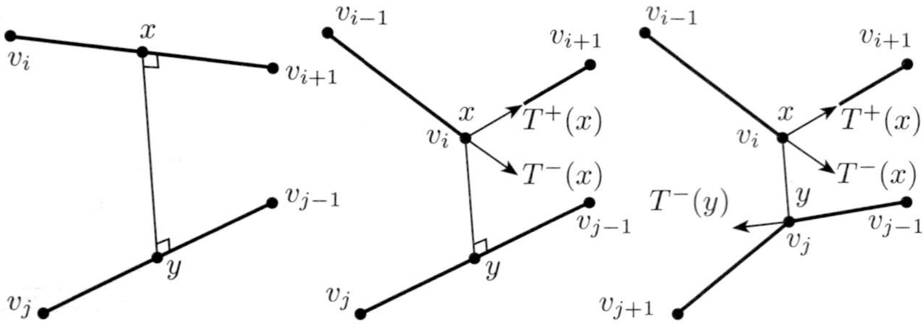

Fig. 1. We see the three cases in the proof of Lemma 4, from left to right an *edge-edge* pair, a *vertex-edge* pair, and a *vertex-vertex* pair. In the center and right figures we see $T^-(x)$ and $T^+(x)$, and in the righthand figure we also see $T^-(y)$. We have not drawn $T^+(y)$, but it would be colinear with $v_j v_{j+1}$ as with $T^+(x)$.

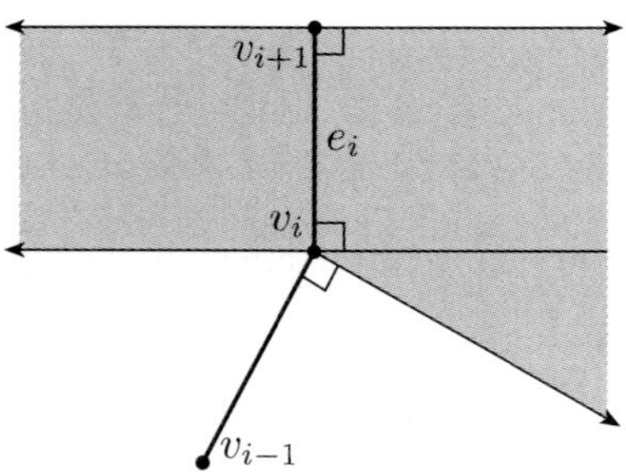

Fig. 2. The shaded area represents the region of space in which the second point y of a locally minimal pair xy can lie when x is on the edge e_i or is the vertex v_i. This region consists of the infinite slab of parallel planes normal to e_i which pass through e_i, together with the wedge extending from vertex v_i in the outward direction from the vertex.

We now make a definition:

Definition 5: The ith *ramp*, R_i of a polygonal curve P_n is the union of the planes through edge $e_i = v_i v_{i+1}$ with normal vector $v_{i+1} - v_i$, together with the wedge of vectors w defined by the inequalities

$$(w - v_i) \cdot T^-(v_i) \geq 0 \quad (w - v_i) \cdot T^+(v_i) \leq 0. \tag{5}$$

See Figure 2.

This leads naturally to the Lemma:

Lemma 6: *If xy is a pair of points on P_n which locally minimizes D, and x is on the half-open edge $e_i - \{v_{i+1}\}$, then y is in the ramp R_i.*

Proof: If x is in the interior of the edge, Lemma 4 implies that $\overline{xy}$ must be perpendicular to e_i, and hence that y is in the union of normal planes through e_i. If x is at v_i, the inequalities above are those of the statement of Lemma 4. □

It may seem like we have only rephrased Lemma 4. In fact, we have gained an important geometric insight about the problem– for any edge e_i, the ramp R_i will probably be very close to a thin slab which only intersects the remainder of P_n in a few places (see Figure 3). If we can isolate these intersections quickly, we can complete the task of finding $\text{POCA}(P_n)$ by a more detailed comparison of these candidates to e_i.

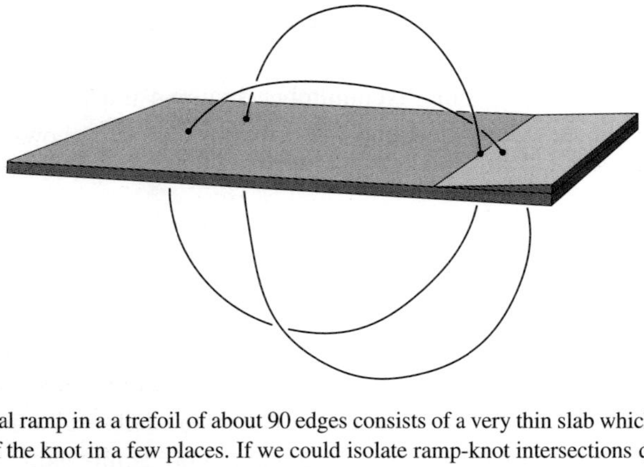

Fig. 3. A typical ramp in a a trefoil of about 90 edges consists of a very thin slab which only intersects the remainder of the knot in a few places. If we could isolate ramp-knot intersections quickly, it would reduce the number of edge-edge checks required to find the shortest POCA.

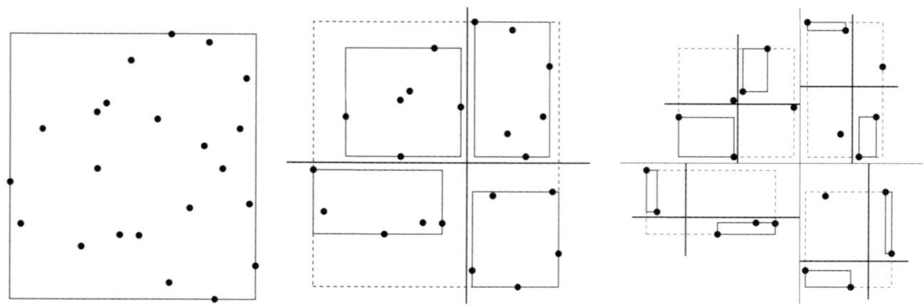

Fig. 4. From left to right, these pictures show three stages in the construction of a quadtree representation (the planar version of an octree representation) of a set of points. On the left, the bounding box of the entire point set is computed. This is the root of the tree. In the center, we see the points divided in two by x and y coordinates, and then grouped by quadrant into four subcollections, with bounding boxes as shown. On the right, we again divide the subcollections and group into subquadrants. The resulting 3-level tree is shown in Figure 5.

3. The Octree Data Structure

With the discussion in Section 2, we have reduced the problem of identifying edges e_j which may form locally minimal pairs with points on edge e_i to the problem of finding which edges of P_n intersect e_i's ramp. To do so efficiently, we will need a new data structure for P_n: the octree[11].

The octree representation of a collection of points in space is a tree where each node represents the bounding box of a subset of that collection. The eight daughter nodes of a parent represent the bounding boxes of subsets of the points in the parent box created by dividing that point set in two in each of the coordinate directions. The most detailed octree representation of a point set has leaf nodes which each contain a single point but it is common to stop subdividing when the point sets are smaller than some fixed number. Figure 4 illustrates three levels of this process for a set of points in the plane, while Figure 5 shows the resulting tree.

From the description above, one can observe that it is easy to build an octree using the recursive procedure of Algorithm 2.

For an n-point dataset, if one chooses each box partition so that no child box contains more than half the total number of points in the parent box, the number of levels in this tree is less than $\log_2 n$, and one expects this algorithm to run in $O(n \log n)$ time. However, this algorithm involves a nontrivial amount of overhead in keeping track of lists of points, and making procedure calls. Since we are very concerned with the final performance of our implementation, we now present a more insightful octree construction algorithm which has the same asymptotic

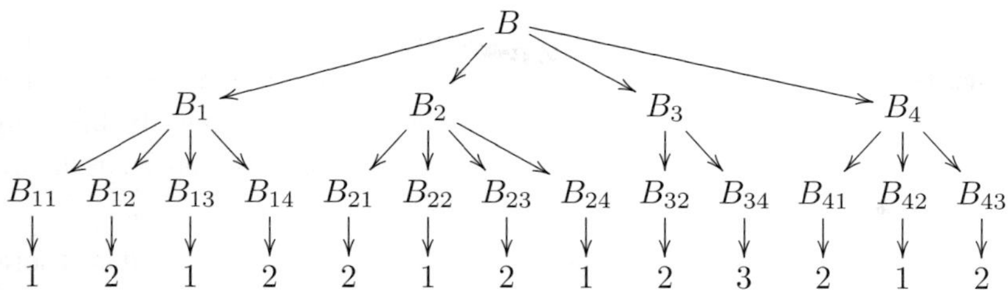

Fig. 5. This picture shows the quadtree constructed in Figure 4 in a more familiar form. The boxes are labelled according to the usual numbering convention for quadrants of the plane, where the first quadrant is on the top right and numbering proceedings counterclockwise. The final numbers show the number of points in each leaf box, and should be compared to the boxes shown in Figure 4 in the right-hand image.

Data : A node of the tree, a corresponding box B, the maximum number of points per box m and a list of points.
Result : An octree representation of this point list.
if *the list of points is no longer than* m **then**
  assign these points to this node;
  make this node a leaf of the tree;
  return;
else
  partition B into 8 child boxes;
  for *each child box* **do**
    create a sublist of points intersecting that box;
    recurse if this sublist is nonempty;
  end
end

Algorithm 2: One way to Build an Octree

time bound of $O(n \log n)$, but is much faster than Algorithm 2 in practice.

To describe the new algorithm, we start with a numbering scheme. As we mentioned before, it is conventional to denote the upper right quadrant of the plane by the number 1, and proceed counterclockwise to the fourth quadrant on the lower right. A more natural numbering scheme assigns each quadrant a 2-digit binary number, $d_x d_y$, where $d_x = 0$ has lower values of x (the left hand side) and $d_x = 1$ has higher values of x (the right hand side), while $d_y = 0$ denotes lower values of y (the bottom half), while $d_y = 1$ denotes higher values of y (the top half). For octants in 3-space, we could assign three digit binary numbers $d_x d_y d_z$

similarly.

Now consider the process of quadtree construction again. At the first subdivision, we divide the point set in two parts by x-coordinate and by y-coordinate. This gives us four groups of points, which we can number as above by the 2-digit binary numbers $d_x d_y$. These groups are the members of the 4 boxes in the next level of the tree, as we saw above.

But there is something else to notice here: If the collection of points is sorted by x and by y, the digits d_x and d_y for any particular point are the most significant binary digits of that point's *position* in the sorted array. Further, if we continue to subdivide the points into fourths by x and y, the next pair of binary digits associated to each point, $d_{x1} d_{y1}$ will be the next pair of binary digits in that point's position in the x and y arrays as well. Again, for octrees the situation is similar, but we sort by z as well, and create a sequence of 3-digit binary (or 1-digit octal) numbers.

Continuing this process, we see that each point in the collection has a unique *octal tag* generated by interleaving the binary digits of its position in the sorted x, y, and z arrays. This tag specifies its position in the octree. Further, if we made a least-first traversal of the octree (descending to octants in the order of their octal labels), the order in which we would encounter the points would be by increasing octal tags. These observations give rise to a new octree-building algorithm:

Data : A list of points in $\mathbb{R}^3$.
Result : An octree representation of point list.
Sort the points by x, y, and z coordinates;
Shuffle binary digits of array positions to create octal tags;
Sort again by octal tags;
Build tree from this traversal-ordered list;

Algorithm 3: A faster octree-building algorithm

The problem of building a tree from a traversal-ordered list of its contents is a standard one in computer science. Our particular solution is discussed in some detail in Section 5 below. We note that building the octree from the list also has time complexity $O(n \log n)$, since every node in the octree must be visited, but that this algorithm is still much faster than the previous method of octree construction (Algorithm 2). We are among many rediscoverers of this method of octree construction, which traces its roots to the "linear quadtree" construction of Gargantini[7].

4. The Core of the Algorithm

We can now describe our algorithm. Given any e_i, we must identify all edges e_j which might be part of a shortest POCA with e_i. Such edges must obey two conditions: they must intersect e_i's ramp, and they must be closer to e_i than the shortest POCA found so far. Since both conditions can be checked for sub-boxes of the octree, we can use them to eliminate groups of e_j from consideration before performing edge-edge checks.

In pseudo-code, this is a collection of n calls to the (recursive) Algorithm 4 (one for each e_i). We refer to the entire algorithm (minRad computation, octree construction by Algorithm 3, and calls to Algorithm 4 for each edge) as **Octrope**.

Data : An octree node, the current minimum POCA length s, the maximum number of edges per box m and a ramp from e_i.

Result : All minimum-length POCAs between e_i and edges in this subtree and (perhaps) a smaller value for the current minimum POCA length s.

```
if this box is within s of e_i then
    if this box intersects the ramp from e_i then
        if this box is a leaf then
            check the (at most m) edges against e_i;
            if POCAs of length ≤ s are found then
                update s;
                return list of minimum length POCAs;
            end
        else
            for each nonempty child node do
                recurse on the child node;
            end
        end
    end
end
```

Algorithm 4: Recursively identifying candidate e_j's.

Each call to this algorithm might require it to traverse the entire depth of the octree before reaching leaf nodes and performing the edge-edge checks. Yet this depth is bounded above by $\log_2 n$, so the expected running time for the algorithm is $O(\log n)$. In pathological cases, many or all of the boxes may intersect the ramp. If all the boxes intersect all the ramps, this algorithm may be asymptotically slower than the naive one: we are forced to visit $O(n \log n)$ tree nodes against each of n edges, for a total time complexity of $O(n^2 \log n)$. We have seen Algorithm 1 outperform Algorithm 4 only for a particularly bad class of examples: knots formed by connecting vertices chosen at random inside a fixed volume. (See Section 6 for details.)

5. Implementation Issues

While being able to replace an $O(n^2)$ algorithm with one which is $O(n \log n)$ will certainly save time *for large enough values of* n, there is no guarantee that this will help with problems of practical size. Indeed, Algorithm 4 threatens to consume a fair amount of overhead, while Algorithm 1 involves only edge-edge checks, which could be coded very efficiently. So in this section we turn our attention from the "$n \log n$" to its multiplier — from mathematics to program design. In this discussion, we'll refer to function names and prototypes from our publically available library version of **Octrope**, which is called `liboctrope`.

The depth of the octree. Since searching the octree involves some overhead, it is to be expected that we will not get the best performance from the deepest octree. Rather, we expect it to be more efficient to group some number of edges in each box and do simple checks between the current edge and the edges in an implicated leaf box.

We implement this by bounding the maximum number of levels in the tree by some ℓ and using that bound to calculate m, the maximum number of edges in any leaf box, by the formula $m = \left\lceil \frac{n}{2^{\ell-1}} \right\rceil$ (where $\lceil r \rceil$ is the least integer greater than or equal to r). The value of ℓ can be set by the user, using the `octree_set_levels` call or it will default to $\ell = \left\lceil \frac{3}{4} \log_2 n \right\rceil$, a formula at which we arrived empirically.

A concrete example. We will now trace through our implementation of the fast octree construction procedure of Algorithm 3 for a particular example: a Hopf link where each edge is given by a regular pentagon (see Figure 6 and Table 1). In this example, ℓ would default to 3 and m would then also equal 3.

Sorting edges. The algorithm begins by gathering all of the edges into a single n-element array which we call `by_oct`. To avoid double-checking edge pairs later on, we need each edge to "belong" to only one of the leaf boxes in our tree. So we identify each edge by its midpoint and store that, as well as the edge's length and its starting vertex in `by_oct`.

As we create `by_oct`, we also build `by_x`, `by_y` and `by_z`, three n-element arrays of pointers to the elements of `by_oct`. We then sort these by x, y, and z

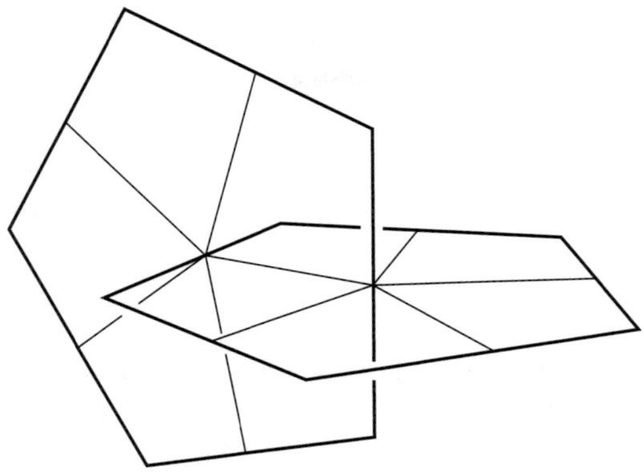

Fig. 6. Our example is a polygonal Hopf-link approximation composed of two regular pentagons. The lighter lines are the 9 minimum length POCAs for which the **Octrope** algorithm is searching. They extend from the midpoints of one side of each interlocked polygon to all the midpoints of the edges of the other.

Table 1. Hopf link vertices

Component	
1	2
$v_{00} = (14.5, 20, 0)$	$v_{10} = (0, 0, 14.5)$
$v_{01} = (23.5, -7.6, 0)$	$v_{11} = (0, 27.6, 23.5)$
$v_{02} = (0, -24.7, 0)$	$v_{12} = (0, 44.7, 0)$
$v_{03} = (-23.5, -7.6, 0)$	$v_{13} = (0, 27.6, -23.5)$
$v_{04} = (-14.5, 20, 0)$	$v_{14} = (0, 0, -14.5)$

The approximate vertices of the pentagonal Hopf link shown in Figure 6 are given here. We have rounded the numbers to the nearest tenth to simplify the table. This does not affect the octree-building procedure under discussion, but would change the picture shown in the figure above.

order. The result is shown in Table 2.

We divide `by_x`, `by_z`, and `by_z` into sections of m points each (shown in Table 2 by spacing) and walk through them, labeling the edges with the binary numbers of the sections in which they lie in the following unusual fashion: if $x = x_1 x_2 \cdots x_{\ell-1}$, $y = y_1 \cdots y_{\ell-1}$ and $z = z_1 \cdots z_{\ell-1}$ are the respective box numbers and their binary representations, we interleave those bits to produce a single octal number, $z_1 y_1 x_1 z_2 y_2 \cdots y_{\ell-1} x_{\ell-1}$. This is the *octal tag* of Section 3 above[a]. We then sort `by_oct` by that octal tag, as shown in Table 3.

[a]To construct octal tags, we take a single pass simultaneously through `by_x`, `by_y`, and `by_z`, starting with the second box, which has binary tag $00 \cdots 01$. As we walk the arrays, we spread the bits of the box number apart (e.g. $1101 \rightarrow 1001000001$) using a lookup table similar to that of Shaffer[18], shift them left 1 bit for y or 2 bits for z, and `OR` them with the tag constructed so far. The tags are thus built up over time and guaranteed to be correct only when we reach the end of the pass.

Table 2. The edges after sorting by x, y and z

by_x	by_y	by_z
$e_{03} : (-19, 6.2, 0)$	$e_{02} : (-11.75, -16.15, 0)$	$e_{13} : (0, 13.8, -19)$
$e_{02} : (-11.75, -16.15, 0)$	$e_{01} : (11.75, -16.15, 0)$	$e_{12} : (0, 36.15, -11.75)$
$e_{04} : (0, 20, 0)$	$e_{14} : (0, 0, 0)$	$e_{00} : (19, 6.2, 0)$
$e_{10} : (0, 13.8, 19)$	$e_{03} : (-19, 6.2, 0)$	$e_{01} : (11.75, -16.15, 0)$
$e_{11} : (0, 36.15, 11.75)$	$e_{00} : (19, 6.2, 0)$	$e_{02} : (-11.75, -16.15, 0)$
$e_{12} : (0, 36.15, -11.75)$	$e_{10} : (0, 13.8, 19)$	$e_{03} : (-19, 6.2, 0)$
$e_{13} : (0, 13.8, -19)$	$e_{13} : (0, 13.8, -19)$	$e_{04} : (0, 20, 0)$
$e_{14} : (0, 0, 0)$	$e_{04} : (0, 20, 0)$	$e_{14} : (0, 0, 0)$
$e_{01} : (11.75, -16.15, 0)$	$e_{11} : (0, 36.15, 11.75)$	$e_{11} : (0, 36.15, 11.75)$
$e_{00} : (19, 6.2, 0)$	$e_{12} : (0, 36.15, -11.75)$	$e_{10} : (0, 13.8, 19)$

In this table, we see the 10 edges of the pentagons in Table 1 sorted by x, y, and z. The edges are sorted by their midpoints, and numbered by the index of their first vertices. The spacing reminds us that since $m = 3$, we are grouping the midpoints by threes when constructing boxes.

Table 3. The edges with their octal tags.

edge	x-box	y-box	z-box	bits	octal	decimal
e_{02}	0	0	1	000100	04_8	4
e_{03}	0	1	1	000110	06_8	6
e_{00}	3	1	0	001011	13_8	11
e_{01}	2	0	1	001100	14_8	12
e_{12}	1	3	0	010011	23_8	19
e_{13}	2	2	0	011000	30_8	24
e_{10}	1	1	3	100111	47_8	39
e_{14}	2	0	2	101000	50_8	40
e_{04}	0	2	2	110000	60_8	48
e_{11}	1	2	2	110001	61_8	49

This table contains the edges with their box numbers in the x, y, and z directions, the binary numbers generated by interleaving the bits of these box numbers, and the corresponding octal tags (in octal and decimal). The data is sorted by octal tag, and so appears in the same order in which it appears in the `by_oct` array. In principle, as many as m edges can share an octal tag, which means that they occupy the same leaf node of the resulting octree, but this does not happen in our example.

As we discussed in Section 3, the sorted `by_oct` array is in the same order as that of a traversal of the full octree.

Building the tree. The actual building of the octree can be approached in various directions. We could simply use the `by_oct` array with no futher indexing, traversing it with binary searches (an approach which saves space at the expense of time). On the other hand, if we are to index it, we can build our index in a top-down fashion, establishing the root node and building out to the leaf nodes. We can build in a bottom-up fashion, partitioning off parts of `by_oct` as leaf nodes and collecting them together in groups until we reach a single top node. Or we can (and do) use a "sideways" or "limb-by-limb" approach. We take an array of ℓ box pointers and on them build the "left-hand limb" of the tree, all the way from the smallest numbered leaf box down to the root box. Each of the boxes knows its first edge in `by_oct` and how many edges it has (which are grouped together thanks to the octal sort).

Then we walk once through `by_oct`, watching the octal tags. As long as the tag is the same as the one before it, we simply increment the count of edges in that box. When it changes, we do a binary `XOR` with the previous tag to see how much they differ (that is, which of the octal digits changed). That tells how many of the boxes in this "limb" are complete. After some cleanup (which may include pruning the "limb") we leave those boxes and the create the new ones necessary to hold this edge. Figure 7 shows our example tree after this process is complete.

Searching the tree. We have now created the tree and can move into using it. We can check each edge and its ramp against the tree, looking for leaf boxes on which to run edge-edge checks. Since POCAs are symmetric, we do not ever want to compare the same pair of edges twice. To avoid this we do the edge-edge check only if the edge in question preceeds our chosen edge in `by_oct`. By so doing, we can eliminate entire boxes because their lowest edges are not in range. The improvement gained from technique has been significant.

6. Performance

We tested our algorithm using a 1.25 Ghz $G4$ Macintosh computer on high-resolution discretizations of trefoil knots and on (open) random walks. We compiled our code with `gcc 3.3` and used the `-O3` option. To make the tests, we compared the run times between `liboctrope` with the tree depth set to 1 and with the default tree depth of $\ell = \left\lceil \frac{3}{4} \log_2 n \right\rceil$. When the depth is 1, the octree consists of a single box and $\frac{n(n-3)}{2}$ edge-edge checks are performed during a run. This turns `liboctrope` into a fairly efficient implementation of Algorithm 1.

For both of these classes of knots, **Octrope** was much faster than our reference implementation of Algorithm 1. Figure 8 shows the relative performance of the two algorithms on trefoil knots given by $\gamma(\theta) = \left((1+\frac{2}{3}\cos 3\theta)\cos 2\theta, (1+\frac{2}{3}\cos 3\theta)\sin 2\theta, \frac{2}{3}\sin 3\theta\right)$. In fact, **Octrope** outperformed Algorithm 1 by an even greater margin on random walks.

For trefoils, Algorithm 1 was sometimes faster than `liboctrope` for very small numbers of edges. Figure 9 shows the performance of both algorithms near the crossover point in some detail.

To understand the effect of varying the number of levels in the octree, we also provide data for a 2499-edge random walk in Table 4. Here we see a trade-off between edge-edge checks and box/ramp checks as the octree resolution increases. Increasing the number of levels in the octree from 9 to 13 cuts the number of final edge-edge checks performed by a factor of 4, but doubles the number of box/ramp checks. Since the box/ramp checks are more computationally expensive, this is not a favorable ratio, and the overall execution time increases.

The data shows that we have been very effective at reducing the number of edge-edge checks. On average, **Octrope** compares each edge to less than 13 carefully chosen candidates when searching for minimum length POCAs.

It is worth noting that **Octrope** is not guaranteed to outperform the standard

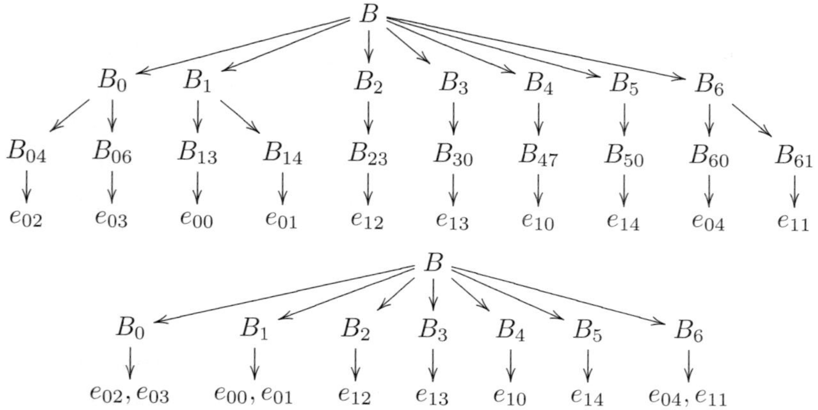

Fig. 7. These two trees show the octree as initially constructed (top) and after our pruning procedure (bottom). This has grouped some edges together in single nodes (such as e_{02} and e_{03}) and deleted some nodes with only one child (such as B_{23}). It is desirable to eliminate extra nodes of this kind, since even though we keep the octal tree as compact in memory as possible, each jump from node to node runs the risk of straying outside the memory cache of the processor and incurring a delay as more information is loaded from main memory.

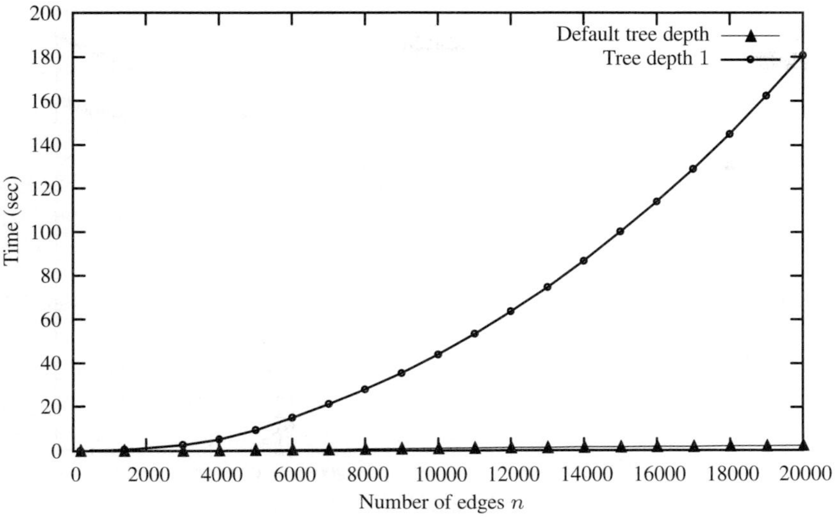

Fig. 8. This plot shows the time (in seconds) required to find the ropelength for trefoils as a function of the number of edges n in the polygonal knot. The timings were computed on a 1.25 Ghz Macintosh $G4$ computer, and represent averages over 10 runs (for times above one second), or 100 runs (for times below one second). The data marked with ▲ comes from `liboctrope` with the default tree depth of $\ell = \lceil \frac{3}{4} \log_2 n \rceil$, while the data marked with ∘ comes from `liboctrope` with the tree depth set to 1 to force $n(n-3)/2$ edge-edge checks.

Table 4. Comparison of algorithms for random walk

	Algorithm		
	Standard	**Octrope**	**Max depth**
Octree levels	1	9	13
Edge-edge checks	3,121,251	32,033	8189
Box/ramp checks	0	51,131	93,187
Time	1.9 sec	0.16 sec	0.22 sec

This table compares the performance of the `liboctrope` library on a 2500-edge random walk at three levels of tree depth: 1 (the standard $O(n^2)$ algorithm), $\lceil \frac{3}{4} \log_2 n \rceil$ (**Octrope**), and 13 (the maximum resolution).

algorithm, even for large numbers of edges. For instance, for random knots constructed by choosing vertices inside a fixed volume, neither ramp-checking nor distance checking eliminates a significant number of pairs from consideration. The edges are simply too long, and pass too close to one another to decide in advance which pairs are likely to control thickness. But even in this case, `liboctrope` was only a few times slower than the standard algorithm.

Currently, minimizing the ropelength of 850-edge knots by simulated anneal-

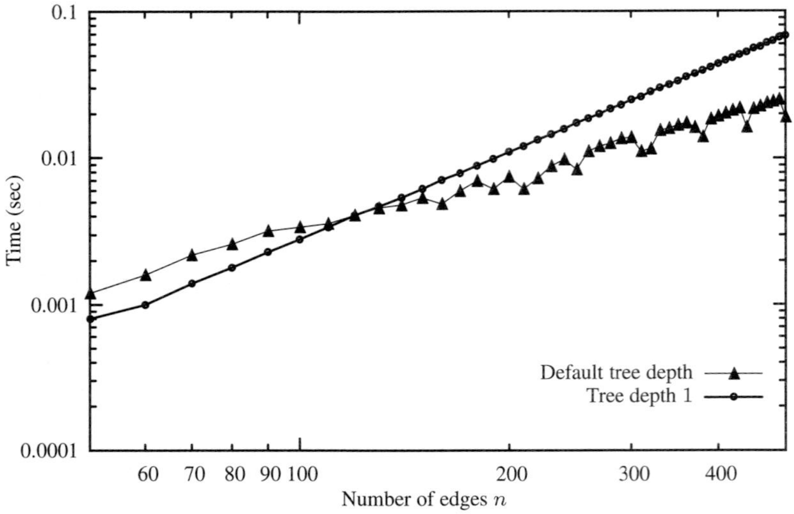

Fig. 9. This log-log plot shows the time (in seconds) required to find ropelength for trefoils as a function of the number of edges n in the polygonal knot near the crossover point where `liboctrope` becomes faster than Algorithm 1. As in Figure 8, the timings were computed on a 1.25 Ghz Macintosh $G4$ computer, and represent averages over 10 runs (for times above one second), or 100 runs (for times below one second). The data marked with ▲ comes from `liboctrope` with the default tree depth of $\ell = \lceil \frac{3}{4} \log_2 n \rceil$, while the data marked with ∘ comes from `liboctrope` with the tree depth set to 1 to force $n(n-3)/2$ edge-edge checks. The data shows that `liboctrope` is faster than our implementation of Algorithm 1 for trefoil knots with more than about 120 edges.

ing is a relatively taxing task, requiring a few weeks of computer time on a standard desktop machine. Our timings above show that the ropelength calculations involved in that process can be done 5 times faster using `liboctrope` or that calculating the ropelength of a 2000-edge knot with `liboctrope` would take the same time as finding that of the 850-edge knot does now.

7. Conclusions and Future Directions

We have given an outline of an improved algorithm for computing the ropelength of polygonal space curves in time $O(n \log n)$ and contrasted it to the previous standard algorithm which required time $O(n^2)$. We have implemented the algorithm efficiently in ANSI C, and given timings which show that our algorithm is also much faster in practice than previous methods used in the field.

The increase in speed from using our method should enable researchers to consider significantly more complicated knots, and to get much higher-resolution data for simpler knots. Both of these are valuable goals. It has always been a goal of the geometric knot theory community to apply our results to large biomolecules such as DNA and proteins. Since these curves may involve thousands of vertices, they have been out of the reach of tools based on Algorithm 1.

However, our methods do not entirely settle the problem of fast ropelength computation. As mentioned in the introduction, Cantarella, Fu, Kusner, Sullivan, and Wrinkle[4] have discovered tiny straight segments in a ropelength-critical simple clasp. These segments are a few one-thousands of one unit in length out of a total clasp length of about 6 units (a similar clasp has been constructed by Starostin[20]). To resolve these very small scale phenomena numerically will require ropelength-minimized configurations with tens of thousands of edges.

At about 3 seconds per ropelength computation on a standard desktop machine, it would simply be untenable to minimize the ropelength of a 20,000-edge knot using our library and simulated annealing on a desktop machine. However, **Octrope** parallelizes well, so one could bring supercomputing cluster machines to bear on the problem, reducing the time to evaluate a configuration to tenths or hundredths of a second. This might allow for a long enough cooling schedule to resolve some small-scale phenomena, but there is no guarantee.

We are hence considering two further approaches to the problem: the use of Edelsbrunner's "segment trees"[22] and an approach we call the **Multiresolution Ropelength Algorithm**. This algorithm is based on the idea that very high resolution knots can be well-approximated by subsampling the vertex set. If one keeps track of the distance between the subsampled knot and the original, one can again eliminate groups of edges from edge-edge checking. The potential advantages of this scheme are twofold: first, the construction of the corresponding tree is linear in time and second, the subsampled knots can be handled by **Octrope** itself. The disadvantage of the multiresolution algorithm is that it will not help with random walks or other very complicated knots, such as large protein backbones, as their subsamples will not be close to the original curve.

We would also like to observe that while the discussion above is phrased in terms of polygons, the general octree/ramp method is equally applicable to other discretization schemes for curves, such as biarcs. In that case, the relative speed advantage of this algorithm should be greater, since the "edge-edge check" for a pair of arcs or spline segments is much slower than the edge-edge check for polygonal edges described above.

In conclusion, we hope that our algorithm and implementation will become a standard software component in numerical investigations of the ropelength problem. If others can improve our code, we hope that they will do so, and invite them to contact us. We also hope that our public release of the library (the first that we know of in geometric knot theory since Brakke's Evolver[2]) will inspire others in the field to contribute from their personal and laboratory collections of code to the public domain. Those interested in obtaining `liboctrope` can turn to `http://ada.math.uga.edu/research/software/octrope/` for further information.

Acknowledgements

The authors are grateful to many colleagues, including Herbert Edelsbrunner, Mark Peletier, and John Sullivan, for discussions about ropelength and algorithms. The 2002-2003 VIGRE group in Geometric Knot Theory (in particular Xander Faber, Chad Mullikin, and Nancy Wrinkle) contributed to our understanding of the computational issues surrounding ropelength, and Monica Shaw and Allison Diana, members of the 2003 Summer Undergraduate Research Experience, worked on a prototype implementation of the octrope algorithm. Michael Piatek and Eric Rawdon served as the `liboctrope` beta-test team, as well as contributing insight on efficient code and library design. The authors would also like to acknowledge the support of the National Science Foundation through the University of Georgia VIGRE grant (DMS-00-89927), DMS-99-02397 (to Cantarella), and DMS-02-04826 (to Cantarella and Fu).

References

1. Leonard M. Blumenthal and Karl Menger. *Studies in geometry*. W. H. Freeman and Co., San Francisco, Calif., 1970. (The three-point curvature appears on page 320).
2. Kenneth A. Brakke. The surface evolver. *Experiment. Math.*, 1(2):141–165, 1992.
3. Gregory Buck and Jeremey Orloff. A simple energy function for knots. *Topology Appl.*, 61(3):205–214, 1995.
4. Jason Cantarella, Joseph H.G. Fu, Robert B. Kusner, John M. Sullivan, and Nancy Wrinkle. Criticality for the Gehring link problem. arXiv:math.DG/0402212, 2004.
5. Jason Cantarella, Robert B. Kusner, and John M. Sullivan. On the minimum ropelength of knots and links. *Invent. Math.*, 150(2):257–286, 2002.
6. Mathias Carlen, Ben Laurie, John H. Maddocks, and Jana Smutny. Biarcs, global radius of curvature, and the computation of ideal knot shapes. In *Physical and Numerical Models in Knot Theory and Their Application to the Life Sciences*. World Scientific, 2005.
7. Irene Gargantini. An effective way to represent Quadtrees. *Commun. ACM*, 25(12):905–910, 1982.
8. O. Gonzalez and R. de la Llave. Existence of ideal knots. *J. Knot Theory Ramifications*, 12(1):123–133, 2003.
9. O. Gonzalez, J. H. Maddocks, F. Schuricht, and H. von der Mosel. Global curvature

and self-contact of nonlinearly elastic curves and rods. *Calc. Var. Partial Differential Equations*, 14(1):29–68, 2002.

10. Oscar Gonzalez and John H. Maddocks. Global curvature, thickness, and the ideal shapes of knots. *Proc. Natl. Acad. Sci. USA*, 96(9):4769–4773 (electronic), 1999.
11. Chris L. Jackins and Steven L. Tanimoto. Oct-trees and their use in representing three-dimensional objects. *Comp. Graphics and Image Proc.*, 14:249–270, 1980.
12. Otto Krötenheerdt and Sigrid Veit. Zur Theorie massiver Knoten. *Wiss. Beitr. Martin-Luther-Univ. Halle-Wittenberg Reihe M Math.*, 7:61–74, 1976.
13. Kenneth C. Millett, Michael Piatek, and Eric Rawdon. Polygonal knot space near ropelength-minimized knots. Draft copy supplied by authors. As of this draft, the result in question is an unstated corollary of Theorem 3.6.
14. Alexander Nabutovsky. Non-recursive functions, knots "with thick ropes", and self-clenching "thick" hyperspheres. *Comm. Pure Appl. Math.*, 48(4):381–428, 1995.
15. Piotr Pierański. In search of ideal knots. In *Ideal knots*, volume 19 of *Ser. Knots Everything*, pages 20–41. World Sci. Publishing, River Edge, NJ, 1998.
16. Eric Rawdon. TOROS: Thickness or Ropelength Optimizing System. Personal Communication.
17. Eric Rawdon. *The Thickness of Polygonal Knots*. PhD thesis, The University of Iowa, 1997.
18. Clifford A. Shaffer. Bit interleaving for Quad- or Octrees. In Andrew S. Glassner, editor, *Graphics Gems*, pages 443–447. Morgan Kauffman, 1990.
19. Jana Smutny and John Maddocks. Approximation of space curves with biarcs. 2004. Preprint.
20. Eugene L. Starostin. A constructive approach to modelling the tight shapes of some linked structures. *Proc. Appl. Math. Mech.*, 3:479–480, 2003.
21. Heiko von der Mosel and Friedemann Schuricht. Characterization of ideal knots. *Calculus of Variations*, 2003. Online. DOI:10.1007/s00526-003-0216-y.
22. A. Zomorodian and H. Edelsbrunner. Fast algorithms for box intersections. *Internat. J. Comput. Geom. Appl.*, 12:143–172, 2002.

CHAPTER 18

TOPOLOGICAL ENTROPIC FORCE BETWEEN A PAIR OF RANDOM KNOTS FORMING A FIXED LINK

Tetsuo Deguchi

Department of Physics, Faculty of Science, Ochanomizu University
2-1-1 Ohtsuka, Bunkyo-ku, Tokyo 112-8610, Japan
E-mail: deguchi@phys.ocha.ac.jp

We discuss the linking probability of two random polygons, and evaluate the entropic force between them. We define the linking probability of a link L by the probability of a pair of random polygons of N nodes making the link L with their centers of mass separated by a distance R. Here we consider only such random polygons that have the trivial knot type. Through physical arguments we derive an analytic expression of the linking probability as a function of R, for the cases of the trivial and the Hopf links. Then we discuss the entropic force acting between such two random polygons forming the link L. Making use of the expression we evaluate the average size of random links, for the cases of a non-trivial link and the Hopf link. The average depends on the number of polygonal nodes, N, non-trivially.

1. Introduction

In statistical physics of polymers, topological constraints of polymer chains may play an important role [8,9]. For instance, in the experiment of ring polystyrenes [21] the osmotic pressure of ring polymers in solution does not vanish even at the θ condition of linear polymers. The anomalous osmotic pressure should be related to the topological constraint of ring polymers [12,25,2]. For polymer networks such as rubbers, their unique elastic properties should be closely related to entanglements among subchains in the networks [8,9,10]. However, the statistical mechanics of polymer networks is quite difficult, and still remains unsolved. As far as statistical properties are concerned, entanglement effects result in an entropic reduction to the system. In short, the entropy can become very small under topological constraints. Here we recall that the entropy of a system is given by the logarithm of the average number of possible microscopic configurations of the system.

If the number of configurations of a system decreases significantly due to some constraint, then the reduction of the entropy can lead to some physical effects, which we call *entropic effects*. In general, entropic effects can be expressed in terms of effective potentials or effective forces. We call such effective forces *entropic forces*. When we can explain an entropic effect in terms of an effective force arising from the constraint, then the physical meaning of the entropic effect should become rather clear. For an illustration, let us consider the pressure of an ideal gas system. The pressure corresponds to the entropic force, where the walls surrounding the gas gives the constraint on it. In fact, gas particles want to go away in order to enlarge the configuration space, however, the walls stop the particles going away so that the configuration space does not become large. The reaction of the pressure on the walls is thus a consequence of the reduction of the gas system's entropy.

Recently, making use of invariants of knots and links, several topological properties of ring polymers can be explicitly investigated through simulations of random polygons or self-avoiding polygons (SAPs) [1,2,11,20]. For instance, it has been shown in simulations [6,7,16,24] that random polygons with a fixed knot are larger in size than those of no topological constraint, when the number of polygonal nodes is large enough. Furthermore, it is also the case with the SAPs with small excluded volume [3,22,23]. (See also a separate contribution by M.K. Shimamura et al. in the present volume.) Hereafter we call random polygons or SAPs that have some fixed knot type random knots of the knot type. The effective swelling observed for random knots (in particular, for those of random polygons or SAPs with small excluded volume) is closely related to entropic repulsion among the segments of random knots induced by the topological constraint. So far we have only considered simulations of off-lattice polygons or SAPs. However, topological properties of SAPs on lattices have also been explicitly studied through numerical simulations [13,17,18,19,26].

In this paper we consider random links, i.e. pairs of random knots forming a given link. We discuss an entropic force acting between a pair of random trivial knots which are under the topological constraint of being equivalent to a fixed link. Here we denote by random trivial knots such random polygons that are topologically equivalent to the trivial knot. We define the linking probability of a link L by the probability of a pair of random trivial knots in a distance R making the link L. The entropic force is given by the logarithmic derivative of the linking probability with respect to the distance R. The entropic force is thus a physical consequence of the topological constraint on the two polygons. In order to evaluate the entropic force explicitly, we derive an analytic formula approximating the linking probability as a function of R, for the cases of the trivial link and the Hopf link. We also discuss the validity of the formulas, reviewing some simulation results of random links [2].

We present some useful physical assumptions on the linking probability. From the assumptions, we can systematically derive the analytic formula of the linking probability. The expression for the trivial link was introduced by Vologodskii et al. [27] in order to give fitting curves numerically. With the physical assumptions, however, we can show the physical meaning of the fitting formula. Furthermore, by using them, we can derive the analytic expression of the linking probability for the Hopf link, and consequently we obtain an expression for the entropic force between two random polygons making the Hopf link. The expressions for the Hopf link should be useful for statistical mechanical study on polymer networks. In fact, the expression for the trivial link has played a central role in the study on the elastic property of the model polymer network [10]. We note that the physical assumptions on the linking probability have been addressed in the previous research on random links [2].

The entropic forces discussed in the paper can be experimentally investigated in some physical systems. For instance, the entropic force for the trivial link should be related to the anomalous osmotic pressure observed in the experiment [21] of ring polystyrenes in solution. The entropic force associated with the probability of a link L corresponds to the entropic force acting on such a pair of ring polymers in the θ solution that have the topological constraint of forming the link L. The content of the paper consists of the following: In section 2, we briefly review the physical background of entropic forces in statistical mechanics. In section 3, we give some physical assumptions on the linking probability, and then derive its analytic expression as a function of the distance R between two random knots. In section 4, we derive expressions of topological entropic forces for the cases of the trivial link, nontrivial links and the Hopf link. In section 5, we evaluate the average size of random links for the cases of nontrivial links and the Hopf link. The average size depends nontrivially on the number of polygonal nodes of random knots. Interestingly, the average size of random links can be consistent with that of two randomly placed circles [5].

2. Statistical mechanical background of entropic forces

2.1. *Probability and the free energy of a macroscopic state*

We consider a system in thermal equilibrium with temperature T. In general, a macroscopic state j corresponds to a large number of microscopic states $\mu(j)$. For instance, a macroscopic state of a polymer chain is described by the distance x between the two ends of the linear chain. If two configurations of the polymer have the same distance x, then they belong to the same macroscopic state x. Here, microscopic states correspond to polymer configurations.

The probability $P_{\mu(j)}$ for the system at temperature T taking a microscopic state $\mu(j)$ follows the canonical distribution:

$$P_{\mu(j)} = \exp\left(-E_{\mu(j)}/k_B T\right) .$$

Here $E_{\mu(j)}$ denotes the energy of the microscopic state $\mu(j)$. Hereafter we denote it simply as E_j. In fact, if two microscopic states $\mu(j)$ and $\nu(j)$ belong to the same macroscopic state j, then they have the same energy: $E_{\mu(j)} = E_{\nu(j)}$.

If there are W_j microscopic configurations $\mu(j)$'s for the same macroscopic state j, then the probability P_j of the system having the macroscopic state j should be proportional to W_j: $P_j \propto W_j$. Thus we have

$$P_j = W_j \exp\left(-\frac{E_j}{k_B T}\right) = \exp\left(-\frac{E_j - k_B T \ln W_j}{k_B T}\right) .$$

Through the probability P_j, the free energy F of the system in the macroscopic state j is defined by $P_j = \exp\left(-F_j/k_B T\right)$ Therefore the free energy is given by

$$F_j = E_j - T S_j .$$

Here we recall that the entropy S of the system in the macroscopic state j is given by the following: $S_j = k_B \ln W_j$.

2.2. *Entropic forces*

In mechanics, the potential force is given by the derivative of the potential. For instance, the elastic force $f(x)$ of an elastic spring with displacement x is derived from the potential energy $U(x)$ as

$$f(x) = -\frac{d}{dx} U(x) .$$

In statistical mechanics, the potential energy is replaced by the free energy. Let us consider a polymer chain in solution of temperature T. A macroscopic state of the polymer chain is specified by the end-to-end distance x. The free energy $F(x)$ is given by $F(x) = E(x) - TS(x)$, where $E(x)$ and $S(x)$ denotes the internal energy and the entropy, respectively. The force acting on the ends of the polymer chain with end-to-end distance x is given by

$$f(x) = -\frac{d}{dx}F(x) = -\frac{d}{dx}E(x) + T\frac{d}{dx}S(x) \,.$$

We call the second term in the right hand side the *entropic force.* It is indeed derived from the entropy. For the polymer chain, the internal energy $E(x)$ does not depend on the end-to-end distance x significantly, and we may simply set $E(x) = 0$. Thus, we have

$$f_{entropic}(x) = T\frac{d}{dx}S(x) \,. \tag{1}$$

3. Random links

3.1. *Definition of random linking probabilities*

Let us first introduce random knotting probabilities. We define the probability of random knotting of a knot K by the probability of a random polygon or a self-avoiding polygon having the knot type K. We denote it by $P_K(N)$, where N denotes the step number of the polygon, i.e. the number of polygonal nodes. Here random polygons and self-avoiding polygons are defined by three-dimensional polygons with their vertices given randomly under a certain rule. Here the former has no excluded volume, while the latter has excluded volume.

We can construct random polygons (or self-avoiding polygons) having a given knot through the following procedure: we first construct a large number of random polygons of N vertices, and then select such random polygons out of them that have the fixed knot type. There are several rules for generating random polygons in three dimensions. For instance, in the simulation of Ref. [2], a large number of Gaussian random polygons have been constructed by using the conditional probability [4]. Furthermore, we have some knot invariants which are practically useful for detecting knot types of polygons generated in computer simulation [1,11,20].

Let us now introduce the probability of random linking. We denote by $P_L^{K_1,K_2}(R;N)$ the probability that a given pair of N-noded random polygons that have knot types K_1 and K_2, respectively, make a given link

type L, where their centers of mass are separated by distance R. For an illustration, we consider a few examples of the random linking probability.
(a) The probability of the trivial link: $P_{0_1^2}^{0,0}(R;N)$

The symbol $P_{0_1^2}^{0,0}(R;N)$ denotes the probability that a pair of N-noded random polygons with the trivial knot 0 makes the trivial link 0_1^2 when they are placed randomly as their centers of mass being separated by distance R. In short, it is the probability of a pair of random unknots in distance R making the unlink. We denote it also by $P_{triv}^{0,0}(R;N)$.
(b) The probability of being a nontrivial link: $P_{non}^{0,0}(R;N)$

Let us denote by $P_{non}^{0,0}(R;N)$ the probability of a pair of unknots in distance R making a nontrivial link. Clearly it is given by $P_{non}^{0,0}(R;N) = 1 - P_{triv}^{0,0}(R;N)$
(c) The probability of the Hopf link: $P_{2_1^2}^{0,0}(R;N)$

The symbol $P_{2_1^2}^{0,0}(R;N)$ denotes the probability that a pair of random trivial knots make the Hopf link 2_1^2 when their centers of mass have a given distance R. We denote it also by $P_{Hopf}^{0,0}(R;N)$.

We may consider a different version of the random linking probability. Let us denote by $P_L^{\infty\infty}(R;N)$ the random linking probability averaged over all knot types for two polygons. It is defined as follows

$$P_L^{\infty,\infty}(R;N) = \sum_{K_1,K_2} P_L^{K_1,K_2}(R;N)P_{K_1}(N)P_{K_2}(N)\,. \tag{2}$$

When the number of polygonal nodes N is small, the two versions of the linking probability, $P_L^{K_1,K_2}(R;N)$ and $P_L^{\infty,\infty}(R;N)$, give numerically similar values. However, they should be quite different when the step number N is very large or the link L is complicated enough. In the pioneering work [27] $P_0^{\infty,\infty}(R;N)$ and $P_{2_1}^{0,0}(R;N)$ have been evaluated by using the Alexander polynomial for $N \leq 80$.

We shall sometimes normalize the distance R between the centers of mass of two polygons as

$$r = \frac{R}{\sqrt{\langle R_{g,K}^2 \rangle}}\,, \tag{3}$$

where the symbol $\langle R_{g,K}^2 \rangle$ denotes the mean square radius of gyration for random polygons with knot type K. We shall give its definition shortly.

Let us denote by $\vec{x}_j$ the position vector of the jth node of a random polygon. Then, the position vector of the center of mass, $\vec{x}_{\rm CM}$, is given by $\vec{x}_{\rm CM} = \sum_{j=1}^{N} \vec{x}_j/N$. For a random polygon with N polygonal nodes we define the mean square radius of gyration $\langle R_g^2 \rangle$ by

$$\langle R_g^2 \rangle = \left\langle \frac{1}{N} \sum_{j=1}^{N} (\vec{x}_j - \vec{x}_{\rm CM})^2 \right\rangle .$$

Here the symbol $\langle \cdot \rangle$ denotes the average over all configurations of polygons. We define the mean square radius of gyration for random polygons that have knot type K

$$\langle R_{g,K}^2 \rangle = \left\langle \frac{1}{N} \sum_{j=1}^{N} (\vec{x}_j - \vec{x}_{\rm CM})^2 \right\rangle_K , \tag{4}$$

where the symbol $\langle \cdot \rangle_K$ now denotes the average over all configurations of such polygons that have the fixed knot K. Hereafter, we abbreviate $\langle R_g^2 \rangle$ and $\langle R_{g,K}^2 \rangle$ by R_g^2 and $R_{g,K}^2$, respectively. Furthermore, we denote by $R_{g,K}$ the square root of $\langle R_{g,K}^2 \rangle$, i.e. we have $R_{g,K} = \sqrt{\langle R_{g,K}^2 \rangle}$.

3.2. *On simulation results of random linking probability*

We briefly review some simulation results on random linking probabilities. The probabilities of the trivial and the Hopf links have been evaluated for $N \leq 80$ by using the Alexander polynomial [27]. In order to implement qualitatively a topological theory of rubber elasticity, the probability of forming a nontrivial link, $P_{non}(R;N)$, has been evaluated for self-avoiding polygons up to $N = 44$, where the topology is detected by using the Gaussian linking number [10]. The probabilities for the trivial and the Hopf links have been evaluated up to $N = 500$ through simulations using some knot and link invariants [2].

Let us discuss the simulation of Ref. [2]. The link types are detected by the Alexander polynomial evaluated at $t = -1$ ($|\Delta_L(-1,-1)|$) and the linking number ($|\Delta_L(1,1)|$), and the knot types of polygons are detected by the Vassiliev invariant of the second degree ($v_2(K)$) and the determinant of knot ($|\Delta_K(-1)|$). For an illustration, we show the linking probability of the Hopf link, $P_{2_1^2}^{0,0}(r;N)$ in the following (Fig. 12 of Ref. [2]).

3.3. *Analytic expressions of linking probabilities*

Through some physical arguments, we can derive an analytic formula which expresses the linking probability as a function of r, at least approximately. We review the argument given in Ref. [2]. Here we recall that the normalized

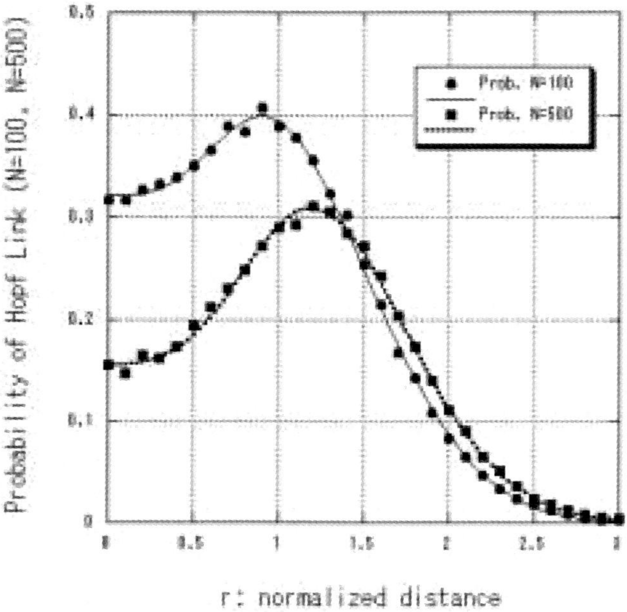

Fig. 1. Linking probability of the Hopf link, $P^{0,0}_{2^2_1}(r;N)$, for $N = 100$ and 500. 31 Data-points of Gaussian random polygons with fitting curves (16). (Fig. 12 of Ref. [2])

distance r is given by $r = R/R_{g,0}$, where $R_{g,0}$ is the square root of the mean square radius of gyration for random trivial knots of N nodes.

We consider the trivial link and the Hopf link cases. Hereafter we sometimes denote $P^{00}_{0^2_1}(N;r)$ by $P_0(r)$ for simplicity.

3.3.1. *Linking probability of the trivial link*

We first consider placing a pair of N-noded random polygons in three dimensions in such a way that their centers of mass have distance r. Then we move them away slightly so that the distance between the centers of mass becomes larger by dr, i.e. it becomes $r + dr$. In order to evaluate the difference in the probability, $dP_0 = P_0(r + dr) - P_0(r)$, we may have the following physical assumptions for the two random polygons:

(1) If they are *unlinked* in distance r, then they are unlinked even when placed again in distance $r + dr$,
(2) If they are *linked* in distance r, then they may become *unlinked* when placed in distance $r + dr$,
(3) The number of ways for moving two polygons from distance r to $r + dr$ should be proportional to the number of configurations of putting them in distance r, i.e. the volume of the spherical shell, $4\pi r^2 dr$.

Let us recall that $P_{non}(r)$ denotes the probability of being a nontrivial link, which is given by $1 - P_0(r)$. It follows from the assumptions 1 and 2 that the increase of the probability dP_0 should be proportional to $P_{non}(r)$. From the assumption 3, dP_0 should be also proportional to the volume of the thin spherical shell with width dr, i.e. $4\pi r^2 dr$. We thus have

$$dP_0(r) = C\,(1 - P_0(r))\,4\pi r^2 dr\,. \tag{5}$$

Here C is an unknown constant. By integrating (5), we have

$$P_0(N; r) = 1 - A\,\exp(-\alpha r^3)\,, \tag{6}$$

were $\alpha = 4\pi C/3$, and A denotes the integral constant.

Here we note that the three physical assumptions should be valid only approximately. For instance, the assumption 1 should be valid when r is rather large. However, they are quite useful for deriving the formula (6). The formula (6) as a function of distance r was first introduced by Vologodskii et al. as a fitting formula to the numerical data [27]. Furthermore, it has been improved later by replacing the term r^3 in the argument of the exponential by r^δ where δ takes real values [15]. In fact, the parameter δ improves the fitting curves. With the parameter δ we may assume that the constant C should depend on distance r as $C \propto r^{\delta-3}$.

Let us now discuss the N-dependence of the probability $P_0^{0,0}(r; N)$. The parameters A and α should depend on the number of polygonal nodes N. Through the simulation[2] we have

$$P_{0_1^2}^{0,0}(0; N) = (2.1 \pm 0.2)\,N^{-0.54\pm0.02}\,. \tag{7}$$

Thus, we have the following conjecture

$$A(N) = 1 - \frac{2}{\sqrt{N}}\,. \tag{8}$$

The N-dependence of α was also investigated in Ref. [2]. We have the following approximation (valid at least up to $N = 500$):

$$\alpha(N) = (0.55 \pm 0.03)\,N^{-0.16\pm0.01}\,. \tag{9}$$

Putting (8) and (9) into (6) we have the following analytic expression:

$$P_0^{0,0}(N;r) = 1 - \left(1 - \frac{2}{\sqrt{N}}\right) \exp(-\alpha(N)\, r^3)\,. \tag{10}$$

We remark that for the $\sqrt{N}$-dependence of $P_{0_1^2}^{0,0}(0;N)$ as shown in (8) there is a supporting argument based on a physical picture [14].

3.3.2. *Probability of being a nontrivial link*

Let us recall that $P_{non}^{0,0}(r;N)$ denotes the probability of a pair of random trivial knots of N nodes being a nontrivial link. It is given by

$$\begin{aligned} P_{non}^{0,0}(r;N) &= 1 - P_{0_1^2}^{0,0}(N;r) \\ &= A(N)\, \exp(-\alpha(N)\, r^3)\,. \end{aligned} \tag{11}$$

Here $A(N)$ and $\alpha(N)$ can be evaluated by the approximate formulas (8) and (9).

3.3.3. *Linking probability of the Hopf link*

Let us discuss the probability for two random trivial knots of N nodes making the Hopf link 2_1^2, when their centers of mass have the normalized distance r. Here we recall that it is denoted by $P_{2_1^2}^{0,0}(r;N)$. Here $K = 0$ denotes the trivial knot, and $L = 2_1^2$ the Hopf link. Hereafter we shall sometimes abbreviate $P_{2_1^2}^{0,0}(r;N)$ by $P_2(r)$, for simplicity.

Through similar physical arguments as the trivial link case, we can derive an analytic expression for the probability of the Hopf link $P_2(r)$ as a function of r. Let us put two polygons of N nodes with distance r between the centers of mass. The link type of the pair of polygons is given by the trivial link, the Hopf link, or a link that is neither the trivial nor the Hopf links.

(1) If it is trivial, it is again trivial even when the two polygons are placed in distance $r + dr$,
(2) If it is the Hopf link, it may become a different link when the pair is placed in distance $r + dr$ (with the coefficient, γ_1),
(3) If it is neither the trivial nor the Hopf links, it may become a Hopf link when the pair is placed in distance $r + dr$ (with the coefficient, γ_2).

In addition to the three assumptions, we may also assume that the number of ways for making the pair separate from r to $r+dr$ should be proportional to the volume of the spherical shell, i.e. $4\pi r^2 dr$. Thus we have

$$dP_2(r) = -\gamma_1 P_2 \, dv + \gamma_2 (1 - P_0 - P_2) \, dv \, . \tag{12}$$

Here γ_1 and γ_2 are unknown constants, and the spherical volume v is defined by

$$v = \frac{4}{3}\pi r^3 \, . \tag{13}$$

Integrating the inhomogeneous ordinary differential equation (12), we have

$$P_2(N; r) = \frac{\gamma_2 A}{\gamma_1 + \gamma_2 - \gamma_0} \exp\left(-\gamma_0 v\right) + C_2 \exp\left(-(\gamma_1 + \gamma_2) v\right) \, . \tag{14}$$

Here γ_0 is defined by $(4\pi/3)\gamma_0 = \alpha$. The parameters α and A have been defined for the trivial link in (6). The constant C_2 in (14) is an integral constant, which is found to be negative from the simulation result [2]. If C_2 is negative, the positivity of $P_2(r)$ in (14) leads to the following inequality

$$\gamma_0 < \gamma_1 + \gamma_2 \, . \tag{15}$$

It follows from the inequality that for large distance r, the first term of (14) becomes dominant in the probability of the Hopf link. In fact, the first term of (14) has the same r-dependence with the probability of being a nontrivial link, $P^{0,0}_{non}(r; N)$, as a function of r. Consequently, the entropic force of the Hopf link becomes equivalent to that of being a nontrivial link, when r is large. We shall see it explicitly in section 4.

N	B_1	B_2	β_1	β_2	χ^2
50	0.503 ± 0.008	0.107 ± 0.009	0.251 ± 0.004	3.81 ± 0.80	103
100	0.536 ± 0.012	0.215 ± 0.012	0.227 ± 0.004	1.85 ± 0.20	62
200	0.555 ± 0.016	0.309 ± 0.015	0.207 ± 0.004	1.29 ± 0.10	24
500	0.558 ± 0.023	0.402 ± 0.023	0.199 ± 0.005	0.887 ± 0.060	44

Table 1: Fitting parameters for the linking probability of the Hopf link, B_1, B_2, β_1 and β_2, in (16), and the χ^2 values for 31 data-points. (Table 7 of Ref. [2])

The approximate formula (14) has been applied to the simulation data[2], and the estimates of the fitting parameters and the χ^2 values are listed in the Table 1. As far as we see in Fig. 1, the fitting curves fit well to the data-points. However, the χ^2 values are not good for $N \leq 100$. Here we recall that there are 31 data-points for each curve.

The large χ^2-values for $N \leq 100$ suggest that the fitting formula (14) is not good enough for describing precisely the linking probability as a function of r. It seems that we can improve the χ^2 values by a generalized formula with five or six fitting parameters. However, the formula (14) should still be useful, since it is based on the physical arguments and is simply given by the solution to the differential equation (12). Furthermore, the formula (14) should give the minimal one with respect to the number of fitting parameters by which we can approximate the data-points at least roughly. It will be an interesting future problem to improve the formula (14) with respect to χ^2 values.

In Ref. [2] the formula (14) was expressed as

$$P_{2_1^2}^{0,0}(r;N) = B_1 \exp(-\beta_1 r^3) - B_2 \exp(-\beta_2 r^3)\,. \tag{16}$$

The fitting parameters of Ref. [2], B_1, B_2, β_1 and β, are related to γ_1, γ_2 and C_2 as follows

$$\beta_1 = \frac{4\pi}{3}\gamma_0\,, \qquad \beta_2 = \frac{4\pi}{3}(\gamma_1 + \gamma_2)\,,$$
$$B_1 = \frac{\gamma_2 A}{\gamma_1 + \gamma_2 - \gamma_0}\,, \qquad B_2 = -C_2\,. \tag{17}$$

Here we give two remarks. The estimate of β_1 should be equal to that of α, as we shall discuss in section 3.3.4. If we assume that B_2 is positive, we have the inequality: $B_1 > B_2$ in order to keep the linking probability positive at the origin, $P_{2_1^2}^{0,0}(0,N) > 0$.

3.3.4. *A consistency check*

The fitting parameter β_1 defined for the Hopf link should be equal to that of the parameter α for the trivial link, if the physical assumptions are valid. Comparing the estimates of $\beta_1(N)$ with the values of $\alpha(N)$ evaluated with the approximation (9), we see that they are numerically consistent. (See Table 2.) This gives a consistency check for the analytic expressions (6) and (14) for the linking probabilities of the trivial and the Hopf links.

4. Topological entropic forces

We now discuss an entropic force which is derived from the random linking probability of a given link type. It is an effective force associated with the topological constraint on a pair of random polygons being equivalent to the given link.

N	$\alpha(N)$	$\beta_1(N)$
50	0.29 ± 0.03	0.251 ± 0.004
100	0.26 ± 0.03	0.227 ± 0.004
200	0.24 ± 0.03	0.207 ± 0.004
500	0.20 ± 0.03	0.199 ± 0.005

Table 2: Consistency of $\alpha(N)$ and $\beta_1(N)$. The $\alpha(N)$ values are evaluated by (9).

Let us consider a pair of N-noded random polygons of the knot types K_1 and K_2. When the pair makes link L with distance r between the centers of mass of the two polygons, the entropy $S_L^{K_1,K_2}(R;N)$ of the pair is evaluated as follows

$$S_L^{K_1,K_2}(R;N) = k_B T \log P_L^{K_1,K_2}(R;N) + const. \tag{18}$$

The entropic force $f_L^{K_1,K_2}(R;N)$ acting between the pair of random polygons in distance r that makes the link L is given by the derivative of the entropy

$$f_L^{K_1,K_2}(R;N) = T\frac{d}{dR} S_L^{K_1,K_2}(R;N)\,. \tag{19}$$

Here we have followed the definition (1) of an entropic force.

For simplicity, we assume hereafter that the two polygons have the same knot type, $K_1 = K_2 = K$. In fact, we consider only the case of the trivial knot, $K = 0$. Furthermore, we employ the normalized distance $r = R/R_{g,K}$, where $R_{g,K}$ is defined in (4). We have

$$f_L^{K,K}(R;N) = T\frac{d}{dR} S_L^{K,K}(R;N) = \frac{k_B T}{R_{g,K}} \frac{d}{dr} \log P_L^{K,K}(r;N)\,. \tag{20}$$

4.1. *Entropic force for the trivial link*

Let us briefly denote by $S_{triv}(r)$ and $f_{triv}(r)$ the entropy and the entropic force associated with the probability $P_{triv}^{0,0}(r;N)$, respectively. The entropic force is an effective force acting between such a pair of random polygons of the trivial knot type that makes the trivial link, where their centers of mass have distance r. Here we recall that the polygons have N nodes and r denotes the normalized distance $R/R_{g,0}$, where $R_{g,0}$ is the average size of random polygons with the trivial knot type.

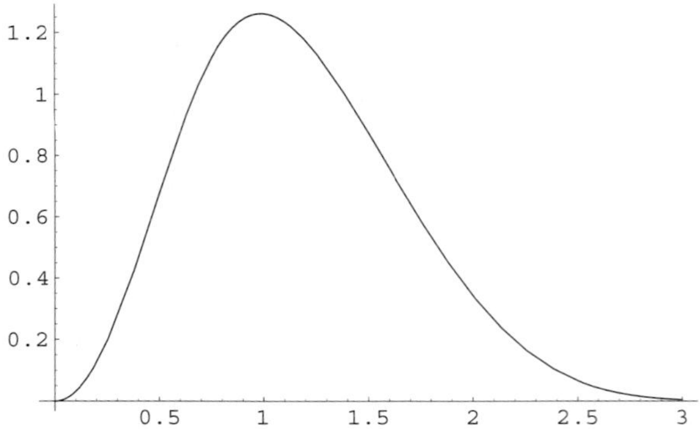

Fig. 2. Entropic force for the trivial link, $f_{triv}(r)$, for $N = 100$ as a function of r. Horizontal axis shows the normalized distance r, and the vertical axis denotes the ratio $f_{triv}(r)R_{g,0}/k_BT$.

Through the entropy $S_{triv}(r)$ together with the formula (6)

$$S_{triv}(r) = k_B \log P_{triv}^{0,0}(r) = k_B \log\left(1 - A\exp(-\alpha r^3)\right) , \tag{21}$$

we have the expression of the entropic force

$$f_{triv}(r) = \frac{3k_BT}{R_{g,0}} \frac{\alpha A r^2 \exp\left(-\alpha r^3\right)}{1 - A\exp\left(-\alpha r^3\right)} . \tag{22}$$

Here we recall that the parameters $A(N)$ and $\alpha(N)$ can be evaluated by (8) and (9), respectively.

The expression (22) leads to the following observations: (i) The entropic force between two unknots is always positive, i.e. it is repulsive; (ii) The entropic force vanishes at $r = 0$. This is because repulsive forces coming from different directions are canceled.

4.2. *Entropic force for the case of nontrivial links*

Let us denote by $f_{non}(r)$ the entropic force between a pair of random polygons forming a nontrivial link. Here the two polygons are supposed to be equivalent to the trivial knot. The entropy $S_{non}^{0,0}$ is given by

$$S_{non}^{0,0}(r) = k_B \log P_{non}^{0,0}(r; N) = k_B \log\left(A\exp(-\alpha r^3)\right) . \tag{23}$$

We thus have

$$f_{non}(r) = -\frac{3k_BT}{R_{g,0}} \alpha r^2 . \tag{24}$$

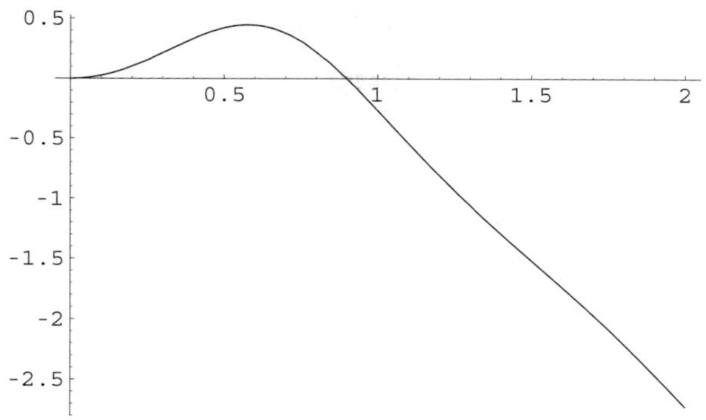

Fig. 3. Entropic force for the Hopf link, $f_{Hopf}(r)$, for $N = 100$ as a function of r. Horizontal axis shows the normalized distance r, and the vertical axis expresses the ratio $f_{Hopf}(r)R_{g,0}/k_BT$.

It is interesting to observe that the entropic force for nontrivial links is always negative, i.e. attractive. Furthermore, the expression (24) is rather simple.

4.3. *Entropic force for the Hopf link*

We denote by $S_{Hopf}(r)$ and $f_{Hopf}(r)$ the entropy and the entropic force associated with the linking probability $P^{0,0}_{Hopf}(r;N)$, respectively. Since the entropy $S_{Hopf}(r)$ is given by

$$S_{Hopf}(r) = k_B \log P^{0,0}_{Hopf}(r) + const. \tag{25}$$

the entropic force $f_{Hopf}(r)$ between linked polygons forming the Hopf link is given by

$$f_{Hopf}(r) = -\frac{3k_BT}{R_{g,0}}\beta_1 r^2 \frac{1 - \dfrac{\beta_2}{\beta_1}\dfrac{B_2}{B_1}\exp\left(-(\beta_2-\beta_1)r^3\right)}{1 - \dfrac{B_2}{B_1}\exp\left(-(\beta_2-\beta_1)r^3\right)}. \tag{26}$$

Here we recall that β_1 should be equal to α, and $B_2/B_1 < 1$ so that the probability $P^{0,0}_2(r;N)$ is positive. We also recall that $\beta_2 > \beta_1$, since $\beta_2 = 4\pi(\gamma_1+\gamma_2)/3 > 4\pi\gamma_0/3 = \beta_1$. We note that the entropic force for the Hopf link, $f_{Hopf}(r)$, vanishes at $r = 0$, similarly with the entropic force for the trivial link.

When the ratio $B_2\beta_2/B_1\beta_1$ is larger than 1, then the sign of the entropic force $f_{Hopf}(r)$ changes with respect to the value of r: it becomes repulsive near the origin such as $r \ll 1$, and attractive for large distances such as

$r \gg 1$. Let us denote by r_c the distance where the entropic force vanishes: $f_{Hopf}(r_c) = 0$. Then it is given by

$$r_c = \left(\frac{1}{\beta_2 - \beta_1} \log \left(\frac{B_2 \beta_2}{B_1 \beta_1} \right) \right) . \tag{27}$$

The value of r_c should be close to 1. From Fig. 1, we observe that the value of r_c becomes larger with respect to the step number, N: $r_c \approx 0.8$ for $N = 50$, while $r_c \approx 1.2$ for $N = 500$.

Let us compare the entropic force for the Hopf link, $f_{Hopf}(r)$, with that of nontrivial links, $f_{non}(r)$. When $r > r_c$, the expression (26) for the Hopf link can be approximated as follows

$$f_{Hopf}(r) \approx -\frac{3k_B T}{R_{g,0}} \beta_1 r^2 \left(1 - \frac{\beta_2 - \beta_1}{\beta_1} \frac{B_2}{B_1} \exp\left(-(\beta_2 - \beta_1) r^3 \right) \right) \quad (r > r_c) . \tag{28}$$

Thus, the entropic force for the Hopf link approaches that of nontrivial links asymptotically as r increases when $r \gg 1$.

5. Average size of random links

We now discuss the average size of random links. We can evaluate the mean square radius of gyration for such a pair of N-noded random polygons of knot K that makes a given link L. Here we make use of the analytic expression of the linking probability. We denote the mean square radius of gyration by the symbol $\langle R_g^2 \rangle_L^{K,K}$. The average size of random link L is given by the square root of it.

5.1. *The mean square radius of gyration for a random link consisting of two random knots*

Let us consider a pair of N-noded random polygons, A and B, that have the fixed knot type K. We assume that A and B makes a given link L. We assign to the nodes of polygon A the numbers from 1 to N, $\vec{x}_j$ for $j = 1, \ldots, N$, and to those of polygon B from $N+1$ to $2N$, $\vec{x}_j$ for $j = N+1, \ldots, 2N$. Then, the centers of mass of polygons A and B are given by $\vec{X}_A = \sum_{j=1}^{N} \vec{x}_j / N$ and $\vec{X}_B = \sum_{j=N+1}^{2N} \vec{x}_j / N$, respectively. The mean square radius of gyration for the random polygons A and B are given by $R_{g,K}^2$

$$\frac{1}{N} \sum_{j=1}^{N} \langle (\vec{x}_j - \vec{X}_A)^2 \rangle_K = \frac{1}{N} \sum_{j=N+1}^{2N} \langle (\vec{x}_j - \vec{X}_B)^2 \rangle_K = R_{g,K}^2 .$$

The mean square radius of gyration for random link L consisting of random trivial knots A and B is defined by

$$\langle R_G^2 \rangle_L^{K,K} = \frac{1}{2N} \sum_{j=1}^{2N} \langle (\vec{x}_j - \vec{X}_G)^2 \rangle_L^{K,K} \,, \tag{29}$$

where the center of mass $\vec{X}_G$ is given by

$$\vec{X}_G = \frac{1}{2N} \sum_{j=1}^{2N} \vec{x}_j \,.$$

It is easy to show

$$\langle R_G^2 \rangle_L^{K,K} = R_{g,K}^2 \left(1 + \frac{1}{4} \langle r^2 \rangle_L^{K,K} \right) , \tag{30}$$

where the mean square of the normalized distance r is given by

$$\langle r^2 \rangle_L^{K,K} = \langle (\vec{X}_A - \vec{X}_B)^2 \rangle_L^{K,K} / R_{g,K}^2 \,. \tag{31}$$

Here $\langle \cdot \rangle_L^{K,K}$ denotes the average over all configurations of two polygons A and B under the condition that they form link L and their knot types are fixed with K.

5.2. *Evaluation of the average size of random links*

We can evaluate $\langle r^2 \rangle_L^{K,K}$ by making use of the probability of random linking $P_L^{K,K}(r;N)$ as follows

$$\langle r^2 \rangle_L^{K,K} = \int_0^\infty r^2 \, P_L^{K,K}(r;N) \, 4\pi r^2 dr \Big/ \int_0^\infty P_L^{K,K}(r;N) \, 4\pi r^2 dr \,. \tag{32}$$

Applying the formula (32) we can explicitly evaluate the average size of random links for the two cases of nontrivial links and the Hopf link.

(a) The average size of nontrivial links

Let us discuss the case when a pair of random trivial knots forming a nontrivial link. We recall that $P_{non}^{0,0}(r;N) = 1 - P_0^{00}(r;N)$ denotes the probability of forming a nontrivial link. Substituting (11) into (32) we have

$$\langle r^2 \rangle_{non}^{0,0} = \frac{\Gamma(5/3)}{\alpha^{2/3}} \approx 1.34 \times N^{0.10} \,. \tag{33}$$

Here we note $\Gamma(5/3) \approx 0.902745$, and we have employed the approximation (9). The simulation result[2] shows that it is valid at least in the range: $50 < N < 500$. Thus, the ratio $\langle R_G^2 \rangle_{non}^{0,0} / R_{g,0}^2$ is given by

$$\langle R_G^2 \rangle_{non}^{0,0} / R_{g,0}^2 \approx 1 + 0.335 \times N^{0.10} \quad (50 < N < 500) \,. \tag{34}$$

Let us evaluate numerically the ratio $\langle R_G^2\rangle_{non}^{0,0}/R_{g,0}^2$. We have the following estimates: $0.335 \times N^{0.10} \approx 0.50$ for $N = 50$; $0.335 \times N^{0.10} \approx 0.62$ for $N = 500$. Thus, the average size of a nontrivial link is larger than that of the trivial knot. The ratio $\langle R_G^2\rangle_{non}^{0,0}/R_{g,0}^2$ is increasing with respect to the step number N. It is about 1.50 for $N = 50$, and then 1.62 for $N = 500$. However, the value at $N = 500$ is smaller than 2.0. Here we have $2N/N = 2$ for the ratio of the mean square size of random polygons of $2N$ nodes to that of N nodes when both of them have no topological constraint.

(b) The average size of the Hopf link

Let us discuss the mean square radius of gyration when a pair of random trivial knots makes the Hopf link. We recall that $P_{2_1^2}^{00}(r)$ gives the probability of forming the Hopf link. Substituting the expression (14) into (32) we have

$$\langle r^2\rangle_{Hopf}^{0,0} = \Gamma(5/3)\left(\frac{B_1}{\beta_1^{5/3}} - \frac{B_2}{\beta_2^{5/3}}\right) \Big/ \left(\frac{B_1}{\beta_1} - \frac{B_2}{\beta_2}\right) . \tag{35}$$

From the best estimates of the fitting parameters we have the following approximation:

$$\langle r^2\rangle_{Hopf}^{0,0} \approx 1.50 \times N^{0.11} . \tag{36}$$

Thus the ratio $\langle R_G^2\rangle_{Hopf}^{0,0}/R_{g,0}^2$ is evaluated as

$$\langle R_G^2\rangle_{Hopf}^{0,0}/R_{g,0}^2 \approx 1 + 0.375 \times N^{0.11} \quad (50 < N < 500) . \tag{37}$$

We note that the expression (37) of the ratio for the Hopf link is very lose to that of the case of nontrivial links, as shown in (34). We have thus similar values for the ratio $\langle R_G^2\rangle_{Hopf}^{0,0}/R_{g,0}^2$ with that of nontrivial links.

Finally, let us compare the average sizes of random links with that of two randomly placed circles. Suppose that we place randomly two circles with unit radius with distance r between their centers. Then, the probability $P(r)$ of the two circles being linked is rigorously given by the following [5]: $P(r) = 1 - r/2$ for $0 \le r < 2$, and $P(r) = 0$ for $r \ge 2$. Thus, we can evaluate the mean square of the distance between two centers of such linked circles that are randomly placed. We denote it by $\langle r^2\rangle_{linked}$. We have

$$\langle r^2\rangle_{linked} = \int_0^2 r^2 P(r) 4\pi r^2 dr \Big/ \int_0^2 P(r) 4\pi r^2 dr . \tag{38}$$

Evaluating the integrals, we have $\langle r^2 \rangle_{linked} = 8/5$. The result should be consistent with that of random links for some small values of N. In fact, from the evaluation (33) we have $\langle r^2 \rangle_{non}^{0,0} = 1.60$ for $N = 6$, and $\langle r^2 \rangle_{non}^{0,0} = 1.88$ for $N = 30$.

Acknowledgements

The author would like to thank the organizers: Jorge A. Calvo, Kenneth C. Millett and Eric J. Rawdon for their kind invitation to the AMS special session on "Numerical Methods, Calculations, and Simulations in Knot Theory and its Applications" at the Spring AMS Western Section Meeting (# 987) in San Francisco, CA, May 3-4, 2003. He is also thankful to the participants of the session for many useful discussions. The text of this paper is also related to the presentations at the conference "Knots, random walks and biomolecules" (co-organized by J.H. Maddocks and A. Stasiak and sponsored by the Bernoulli Centre of Swiss Federal Institute of Technology in Lausanne) held in Les Diablerets, Switzerland, July 14-17, 2003.

References

1. T. Deguchi and K. Tsurusaki, A new algorithm for numerical calculation of link invariants, Phys. Lett. A **174** (1993) 29–37.
2. T. Deguchi and K. Tsurusaki, Random knots and links and applications to polymer physics, in *Lectures at Knots '96*, edited by S. Suzuki, (World Scientific Publ. Co., Singapore, 1997) pp. 95–122.
3. T. Deguchi and M. K. Shimamura, Topological effects on the average size of random knots, Contemporary Math. **304** (2002) 93–114.
4. J. des Cloizeaux and M. L. Mehta, Topological constraints on polymer rings and critical indices, J. Physique **40** (1979) 665–670.
5. Y. Diao and E. J. Janse van Rensburg, Percolation of lined circles, in *Topology and Geometry of Polymer Science*, edited by S.G. Whittington, D. W. Sumners, and T. Lodge, (Springer-Verlag, New York, 1998) pp. 79–88.
6. A. Dobay, J. Dubochet, K. Millet, P.E. Sottas and A. Stasiak, Scaling behavior of random knots, Proc. Natl. Acad. Sci. USA **100** (2003) 5611–5615.
7. J. M. Deutsch, Equilibrium size of large ring molecules, Phys. Rev. E **59** (1999) R2539–R2541.
8. S. F. Edwards, *Statistical mechanics with topological constraints: I* Proc. Roy. Soc. **91** (1967) 513–519.
9. S. F. Edwards, *Statistical mechanics with topological constraints: II* J. Phys. A **1** (1968) 15–28.

10. R. Everaers and K. Kremer, Topological interaction in model polymer networks, Phys. Rev. E (1996) **53** R37–R40.
11. B. Ewing and K. C. Millett, A load balanced algorithm for the calculation of the polynomial knot and link invariants, in *The Mathematical Heritage of C.F. Gauss*, edited by G.M. Rassias, (World Scientific Publ. Co., Singapore, 1991) pp. 225–266.
12. K. Iwata, Macromolecules **18** (1985) 115.
13. E. J. Janse van Rensburg and S. G. Whittington, The dimensions of knotted polygons, J. Phys. A **24** (1991) 3935–3948.
14. A. Yu. Grosberg, Critical exponents for random knots, Phys. Rev. Lett. **85** (2000) 3858–3861.
15. K.V. Klenin, A.V. Vologodskii, V.V. Anshelevich, A.M. Dykhne and M.D. Frank-Kamenetskii, Effect of excluded volume on topological properties of circular DNA, J. Biomol. Struct. Dyn. **5** (1988) 1173–1185.
16. H. Matsuda, A. Yao, H. Tsukahara, T. Deguchi, K. Furuta and T. Inami, Average size of random polygons with fixed knot topology, Phys. Rev. E **68** (2003) 011102 (4 pages).
17. E. Orlandini, E. J. Janse van Rensburg, M. C. Tesi and S. G. Whittington, Random linking of lattice polygons, J. Phys. A **27** (1994) 335–345.
18. E. Orlandini, M. C. Tesi, E. J. Janse van Rensburg and S. G. Whittington, Entropic exponents of lattice polygons with specified knot type, J. Phys. A **29** (1996) L299–L303.
19. E. Orlandini, M. C. Tesi, E. J. Janse van Rensburg and S. G. Whittington, Asymptotics of knotted lattice polygons, J. Phys. A (1998) **31** 5953–5967.
20. M. Polyak and O. Viro, Gauss diagram formulas for Vassiliev invariants, Int. Math. Res. Not. No. 11 (1994) 445–453.
21. J. R. Roovers and P. M. Toporowski, Synthesis of high molecular weight ring polystyrenes, Macromolecules **16** (1983) 843–849.
22. M. K. Shimamura, and T. Deguchi, Gyration radius of a circular polymer under a topological constraint with excluded volume, Phys. Rev. E **64** (2001) 020801(R) (4 pages).
23. M. K. Shimamura and T. Deguchi, Finite-size and asymptotic behaviors of the gyration radius of cylindrical self-avoiding polygons, Phys. Rev. E **65** (2002) 051802 (9 pages).
24. M. K. Shimamura and T. Deguchi, Anomalous finite-size effects for the mean-squared gyration radius of Gaussian random knots, J. Phys. A **35** (2002) L241-L246 .
25. F. Tanaka, Osmotic pressure of ring-polymer solutions, J. Chem. Phys. **87** (1987) 4201–4206.
26. M. C. Tesi, E. J. Janse van Rensburg, E. Orlandini and S. G. Whittington, Knot probability for lattice polygons in confined geometries, J. Phys. A **27** (1994) 347–360.
27. A. V. Vologodskii, A. V. Lukashin and M. D. Frank-Kamenetskii, Topological interaction between polymer chains, Sov. Phys. JETP **40** (1975) 932–936.

CHAPTER 19

UNDER-KNOTTED AND OVER-KNOTTED POLYMERS: 1. UNRESTRICTED LOOPS

Nathan T. Moore, Rhonald C. Lua, Alexander Yu. Grosberg

Department of Physics, University of Minnesota,Minneapolis, MN 55455, USA

We present computer simulations to examine probability distributions of gyration radius for the no-thickness closed polymers of N straight segments of equal length. We are particularly interested in the *conditional* distributions when the topology of the loop is quenched to be a certain knot $\mathcal{K}$. The dependence of probability distribution on length, N, as well as topological state $\mathcal{K}$ are the primary parameters of interest. Our results confirm that the mean square average gyration radius for trivial knots scales with N in the same way as for self-avoiding walks, where the cross-over length to this "under-knotted" regime is the same as the characteristic length of random knotting, N_0. Probability distributions of gyration radii are somewhat more narrow for topologically restricted under-knotted loops compared to phantom loops, meaning knots are entropically more rigid than phantom polymers. We also found evidence that probability distributions approach a universal shape at $N > N_0$ for all simple knots.

1. Introduction

1.1. *The goal of this work*

Consider a random closed polygon of some N segments, all of equal length ℓ. What is the probability $w_{\mathrm{triv}}(N)$ that this polygon, considered as a closed curve embedded in $3D$, is topologically equivalent to a circle, that is, represents a trivial knot? What is the probability $w_{\mathcal{K}}(N)$ that it represents a knot of any other kind, $\mathcal{K}$? Such questions arose first in the context of DNA[1] and other polymers[2], and continue to attract significant attention to the present day. Although a large body of information has accumulated, mostly through computer simulations[3,4,5], final mathematical understanding of these questions remains elusive, despite their elementary formulation.

Meanwhile, a new set of questions came to the forefront in the last several years. For instance, what is the *conditional* probability density of the loop gyration radius given that its topology is fixed to be $\mathcal{K}$? As a first step, what is the average gyration radius of the loop with the given knot state $\mathcal{K}$? This latter question was first discussed by des Cloizeaux[6] and then re-visited theoretically[7,8] and computationally[9,10,11,12,13,14,15]. The excitement in the field is partially driven by the idea, first conjectured in the work[6], that topological constraints act effectively like self-avoidance, leading to the non-trivial scaling $\langle R_g^2 \rangle \sim N^{2\nu}$, where ν is the critical exponent known in the theory of self-avoiding walks, $\nu \approx 0.588 \approx 3/5$.

The distinction between the two groups of questions can be illuminated by the comparison with the concepts of annealed and quenched disorder, well known in the physics of disordered systems (see, for instance, book[16]). If the loop is phantom, i.e. if it can freely cross itself, then its topological state is annealed. In this case, we can ask what the probability is to observe a certain topological state $\mathcal{K}$. For the loop which is not phantom and cannot cross itself, the knot state is frozen, or quenched, and we can discuss physical properties of the loop, such as its size or entropy for every given knot state $\mathcal{K}$.

The main goal of this paper is to look more closely at the probability distributions of the gyration radius of the loops which are topologically constrained but not constrained otherwise. In section 2, we provide an overview of the previous results about the mean square averaged gyration radius as well as some related questions of method and simulation technique. We shall concentrate on the relatively simple knots, such as 0_1, 3_1, and 4_1, formed by rather long polymers, with N up to 3000. Using the terminology introduced in the recent work [15], we can say we shall be interested mostly in the under-knotted regime. This terminology makes simultaneous use of both annealed and quenched views of polymer topology. The idea is as follows. Consider real polymer loop with some quenched knot $\mathcal{K}$. It is considered over-knotted if upon topological annealing, allowing loop states to be sampled without topological constraints, the loop is likely to become topologically simpler than $\mathcal{K}$. Otherwise, the loop is considered under-knotted. Roughly, loop is under-knotted if it "wants" to have more knots, and it is over-knotted if it "wants" to have fewer knots. Whether a quenched loop is over- or under-knotted depends on the number of segments, N, and, in general, on some other conditions, such as solvent quality and the like. It is because the loop is under-knotted that it may swell, even if there is no excluded volume or self-avoidance. Here, however, terminology clarification is in order.

1.2. *Some terminology: non-phantom polymers and self-avoiding polymers are two different things*

We should first emphasize the difference between concepts of self-avoiding polymers and non-phantom polymers. These two concepts are quite frequently confused. The idea of self-avoidance always involves certain finite non-zero length scale, let say d, such that two pieces of a polymer cannot approach each other closer than d. For instance, if one thinks of a polymer as a little garden hose, then d is its diameter. Real polymers, of course, always have some excluded volume, or some thickness d. On the other hand, polymers which we call phantom are imagined to be able to switch from an under-pass to over-pass conformations, but, importantly, neither former nor later state violate the self-avoidance, or excluded volume, condition. Speaking about phantom polymers, we should intentionally close our eyes on the process - *how* the polymer passes from under- to over- state. This question is irrelevant when we address probabilities or equilibrium statistical mechanics. In some sense, the idea of a phantom polymer can be illustrated by the properties of a DNA double helix in the presence of topo-II enzymes[17]. Of course, this question of crossing mechanism becomes decisive if one wants to look at polymer dynamics without enzymes; for the studies of dynamics, the phantom model is meaningless, one should think in terms of reptation instead[18].

On a more quantitative level, it is known for the polymer with N segments of the length ℓ and diameter d that the excluded volume effect does not lead to appreciable swelling as long as $N \ll (\ell/d)^2$ [19]. For dsDNA at a reasonable ionic strength, this implies chain length up to about 2500 segments, or 75000 base pairs. In this sense, our testing of loops up to $N = 3000$, although dictated by our computational possibilities, is also meaningful for the important particular case of DNA. On this length scale, it is quite reasonable to neglect the self-avoidance condition, and at the same time to work with the polymer which is not phantom, because its topological state is quenched (unless enzymes are present).

2. Brief overview of our recent work[15]

Our most recent work has investigated the average size of knotted loops. The initial focus was on those loops with trivial knot topology, denoted 0_1, as their size has been addressed theoretically [6,8]. In collecting data through simulation we were able to gather statistically significant information about several other knots of low prime crossing number, specifically, 3_1, 4_1, 5_1 and

5_2 knots.

2.1. *Simulation methods*

Like others[13,10,14] our initial approach to the problem was to generate loops, compute the gyration radius for each of them

$$R_g^2 = \frac{1}{2N^2} \sum_{i=1}^{N} \sum_{j=1}^{N} r_{ij}^2 \ , \quad r_{ij} = |\vec{r}_i - \vec{r}_j| \ , \tag{1}$$

$\vec{r}_i$ being the position vector of the joint number i, and then analyze the generated conformations with several topological invariants.

Our loop generation routine is discussed in the Appendix A. Importantly, the loops are generated without any relation towards their topology. When a loop is generated, its knot type is assigned. Therefore, we can use the ensemble of all generated loops to address questions regarding the annealed topology, such as the population fractions of various knots. At the same time, we are able to determine average size, and more generally, the probability distribution of size for loops assigned any given knot type, which means, we can address the quenched topology questions. The number of loops and trivial knots used for analysis, as well as the associated CPU time is shown in table 1.

Table 1. Minimum number of loops used for statistics of each point

N	loops generated	trivial knots produced	CPU hours
15 to 480	10^6	10^6	0.02 to 288
510 to 990	7×10^5	10^5	41 to 1.7×10^3
1020 to 1701	7×10^5	10^4	230 to 8.2×10^3
1800 to 2301	10^6	10^3	3300 to 2×10^4
2400 to 2502	10^6	10^2	(3.6 to 7.7) $\times 10^3$
3000	10^6	9	3.5×10^3

To determine loop topology we compute several topological invariants[20]. For the loops with $N \leq 300$, we used Alexander invariant $\Delta(-1)$ and Vassiliev invariants of degree 2 and 3, v_2 and v_3. The loop was identified as a trivial knot when it yielded $|\Delta(-1)| = 1$, $v_2 = 0$, and $v_3 = 0$. For longer loops of $N > 300$, we were able to use only $\Delta(-1)$ and v_2 invariants, assigning trivial knot status to the loops with $|\Delta(-1)| = 1$, and $v_2 = 0$. The details of our computational implementation of these invariants are described elsewhere[21]. Of course, because of the incomplete nature of topological invariants, our knot assignment is only an approximation, and surely was sometimes in error.

2.2. *Knot population fractions*

We begin by addressing the annealed topology questions.

Theoretically, it is believed that the probability of a trivial knot is exponential in N:

$$w_{\rm triv}(N) = w_0 \exp\left(-N/N_0\right) \ , \tag{2}$$

at least, asymptotically when $N \gg 1$. Such exponential behavior was observed in a number of simulation works for a variety of models [4,5]. By now, it is already considered "obvious" by physicists in the field. It is indeed fairly obvious through the intuition gained by the study[22,23,24] (see also more recent work[25] and references therein) of exactly solvable model of winding around a point or a disc in $2D$. This model shows that typical Brownian trajectory (that is, polymer with $\ell \to 0$ and $N \to \infty$) tends to produce a diverging winding angle, that is, an infinite number of turns around the point-like obstacle. It does not seem to require a particularly great leap of imagination to conclude that at very large N some finite scale knots should be formed with a non-zero frequency everywhere along the polymer - and this leads exactly to a Poisson-like exponential formula (2).

With regard to the probabilities of other non-trivial knots, it may be argued that they should also be asymptotically exponential

$$w_{\mathcal{K}}(N) = w_0^{(\mathcal{K})} \exp\left(-N/N_{\mathcal{K}}\right) \ , \tag{3}$$

and, moreover, that the characteristic length should be the same as that for trivial knots: $N_{\mathcal{K}} = N_0$. This latter idea can be understood by saying that for every knot, the loop must eventually become strongly under-knotted if N increases without bound while knot is quenched. Formula (3) was also tested, albeit by a smaller number of simulations[5,26].

In the work[15], we fit formula (2) to our trivial knot data and found critical length, $N_0 = 241 \pm 0.6$ and $w_0 = 1.07 \pm 0.01$. This value of N_0 is consistent with the result on a rod-bead model[4,11] in the limit of excluded volume radius set to zero. In other works[4,5], somewhat larger values of N_0 were reported, closer to 300 or 330. We interpret this discrepancy as being due to the fact that we examined the model with all segments of the same length, while the works[4,5] dealt with Gaussian distributed segments.

We consider it an exciting challenge to understand why these two models exhibit differing values of characteristic knotting length.

Figure 1 shows our simulation data for trivial knots fraction, along with the data illustrating the relative frequency of other knot types. To the accuracy of our simulations, we do not see all non-trivial knot probabilities decaying with the same characteristic length N_0. However, we tried to determine N_K (see equation (3)) by fitting the data over a sliding window. For instance, Table 2 shows the fit parameters obtained on the interval $500 < N < 1150$, or on the interval on $1150 < N < 3000$. It is clearly seen that "apparent" characteristic length decreases. Although far from proof, this result is consistent with the theoretical argument behind formula (3) and allows one to hypothesize that the asymptotics is just very slowly achieved.

Table 2. Characteristic Lengths, N_K

knot type	N_K on $(500 < N < 1150)$	N_K on $(1150 < N < 3000)$
0_1	241	250
3_1	373	305
4_1	374	307
5_1	375	307
5_2	378	302

2.3. *Average size of different knots*

2.3.1. *Scaling of the trivial knot size*

When averaged over all loops, the mean square gyration radius, $\langle R_g^2 \rangle$, is equal to $N\ell^2/12$, which is two times smaller than the similar quantity for linear chains (see, for instance, book[27]; see also Appendix B). As regards $\langle R_g^2 \rangle$ averaged over only trivially knotted loops, the theorists[6,7,8] predicted, that trivial knots develop swelling behavior for $N \gg N_0$, in a way similar to objects which experience excluded volume forces:

$$\langle R_g^2 \rangle_{\text{triv}} = \begin{cases} \left(\ell^2/12\right) N & \text{if } N \ll N_0 \\ A\left(\ell^2/12\right) N^{2\nu} & \text{if } N \gg N_0 \end{cases}, \tag{4}$$

where scaling power is $\nu \approx 0.589$, and where N_0 is the same parameter introduced in formula (2).

We want to emphasize here that the first line of the prediction, formula (4), is not connected to any delicate and thus possibly unreliable theoretical

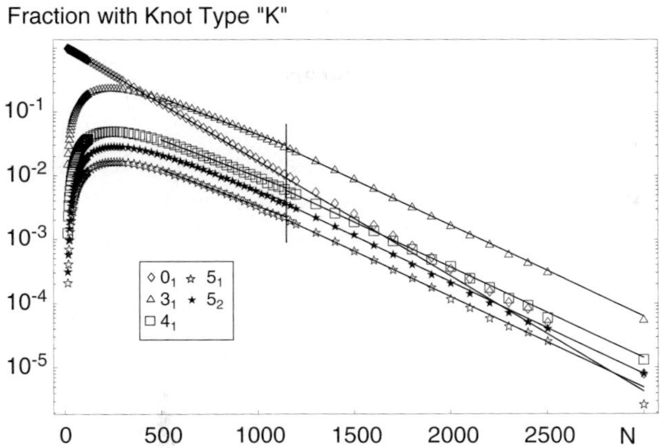

Fig. 1. The fraction of loops generated with trivially knotted topology followed the well known exponential form, equation (2), as a function of loop length N. Deviation from the fit line at large N is due to the incompleteness of topological invariants employed and reflects contamination of the supposedly trivial pool with some non-trivial knots. The fractional population curves for several different simple knot types are shown and labeled. Although their overall decay can be reasonably fit by exponents, the characteristic lengths $N_{\mathcal{K}}$ appear larger than N_0, which probably means that true asymptotics are very slowly achieved.

arguments, but rather comes out of almost pure common sense. Indeed, when $N \ll N_0$, according to formula (2), there is only marginal probability for a phantom loop to have any knot other than trivial. This means, the ensemble of trivially knotted loops at these N very nearly coincides with the ensemble of all loops, for which $\langle R_g^2 \rangle$ is certainly equal to $N\ell^2/12$.

Figure 2 demonstrates how our simulation results are consistent with formula (4). First of all, we see that indeed the Gaussian scaling $\langle R_g^2 \rangle = N\ell^2/12$ is recovered at N below N_0. Fitting the data over the interval $500 < N < 2500$, we found the parameters, $\nu \approx 0.58 \pm 0.02$ and $A \approx 0.44 \pm 0.03$. It is not only important that ν is consistent with expectations, it is also important that the value of pre-factor A provides for smooth cross-over between regimes at N very close to N_0, as expected (because A is very close to $N_0^{1-2\nu} \approx 0.42$).

2.3.2. *Corrections to scaling*

Can one pull the analogy between trivial knots and self-avoiding polymers further? The temptation in the field [10,12,13] has been to fit the trivial knot data with a more complex perturbation formula, motivated by the analogy with the excluded volume problem[14],

$$\langle R_g^2 \rangle = A \frac{\ell^2}{12} N^{2\nu} \left[1 + B \left(\frac{N_0}{N} \right)^{\Delta} + C \left(\frac{N_0}{N} \right)^{2\Delta} + \ldots \right] . \qquad (5)$$

To understand this formula, it is useful to recall its appearance in the better known context of the excluded volume problem (here, we re-phrase presentation in the book[19]). For the excluded volume (or self-avoiding) polymer, one first shows that the gyration radius can be written in the form $\langle R_g^2 \rangle = N\ell^2 f(x)$, where $f(x)$ is a universal function of the argument $x = (d/\ell)\sqrt{N}$ (where d and ℓ are segment thickness and length, respectively). For our purposes here, we denote $N_0^\star = (\ell/d)^2 > 1$ and then write $x = (N/N_0^\star)^{1/2}$. When x is small, $x \ll 1$, then $f(x)$ can be presented as an (asymptotic) perturbation series in integer powers of x. When x is large, $x \gg 1$, the leading term in $f(x)$ contains the non-trivial scaling power: $f(x) \sim x^{2\nu-1}$, and then the correction terms in this large x asymptotics involve negative powers of x, in most cases believed[14] to be integer negative powers: $f(x) \sim x^{2\nu-1} \left[1 + B/x + C/x^2 + \ldots \right]$. Using this formula to write $\langle R_g^2 \rangle$ in terms of $N_0^\star$, we obtain exactly the equation (5) (with $\Delta = 1/2$).

This consideration shows that for the excluded volume problem, formula (5) is only valid at $x \gg 1$, or $N \gg N_0^\star$. It is *not* an interpolation formula valid across the cross-over region $x \sim 1$; it does not connect two asymptotics smoothly. For the latter reason, it cannot be considered an interpolation for trivial knots. That is why we think it is not correct to fit the simulation data to this formula in the range of N other than $N \gg N_0$ (or at least $N > N_0$).

Unfortunately, our data do not allow for reasonable fit to this formula even in the range $N > N_0$. The reason is seen in the fact that our data represent a curve which seems to keep bending upwards as N increases, while formula (5) implies saturation of the log-log slope to that dictated by the power 2ν. A mechanical attempt to fit the formula to the data yields physically meaningless values for ν which are greater than unity.

Currently we do not know why data do not fit formula (5). One reason may be simply poor statistics and noisy character of data at large N. It might also be an indication of the knot pool contamination at large N because of the incompleteness of topological invariants. This is possible, but, in our opinion, not very likely given that trivial knot fraction does not deviate much from the exponential fit (see Figure 1). In the work[15], we attempted to address this question deeper, introducing the correction for the errors in knot assignment. It did not yield much change in terms of

$\langle R_g^2 \rangle$, making us a bit more confident that the problem might be somewhere else. For instance, it is possible that the formula (5) does not apply to trivial knots, indicating some restricted applicability of the very analogy between trivial knots and excluded volume polymers. Much work will be necessary to clarify this issue.

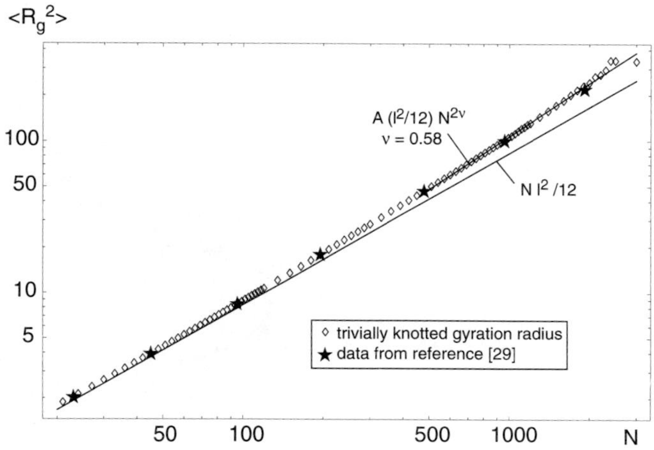

Fig. 2. Gyration radius averages over trivially knotted loops. The trivial knot average exhibits power law behavior at large N similar to that experienced by polymers which have excluded volume. The trivial knot data is systematically larger than the average over all loops, shown as the solid line in the figure. This topology driven swelling is seen to develop beyond the critical length about $N_0 = 241$. Independently collected data[29] is shown by stars ($\star$) and agrees with our results.

2.3.3. *Averaged sizes of non-trivial knots*

Our measurement of the swelling of non-trivial knots is shown in Figure 3. It is consistent with earlier work[10,13]. We find that the simple knots cross over from an over-knotted state, in which they are much smaller than the average sized loop to an under-knotted state in which they seem to approach the scaling of trivial knots in an asymptotic fashion. The inset in this image shows this asymptotic approach in the form of a small parameter of relative swelling, $\beta = 1 - \langle R_g^2 \rangle_K / \langle R_g^2 \rangle_0$ decaying with increasing N. Although hard to see in the data, we expect the asymptotic decay of β to approach, but never reach, $\beta = 0$. Consider the following argument. With the understanding gained from recent work addressing knot localization[33,34,35] we imagine the large N behavior of a simply knotted loop (eg a 3_1 or 4_1 knot) to look like the free coil formed by a slightly shorter trivially knotted loop. This comes because in the localization hypothesis, the knot on average occupies a very small portion of the loop. Supposing that N_s segments are taken up by the non-trivial knot, using scaling equation 4 predicts that $\beta = 1 - (1 - N_s/N)^{2\nu}$. So long as N_s is small in comparison to N, even as large as $N_s \sim N^{\omega}$, $\omega < 1$, β will exhibit a sort of power-law decay towards 0.

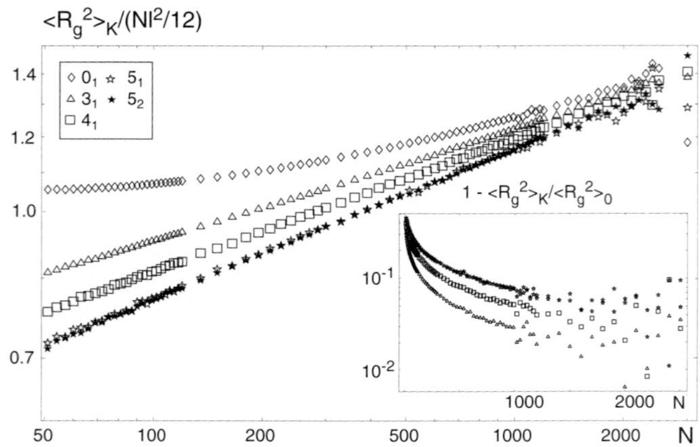

Fig. 3. Log-log plot of the mean square gyration radius, $\langle R_g^2 \rangle_{\mathcal{K}}$, of knot type $\mathcal{K}$, normalized by the topology blind average over all loops for several particular knot types. The inset, which shows the ratio of a particular knot gyration radius to the trivial knot gyration radius, $1 - \langle R_g^2 \rangle_{\mathcal{K}} / \langle R_g^2 \rangle_0$, demonstrates that all knots remain smaller than, but approach the size of, trivial knots.

3. Probability distributions of the loop sizes

Our data allow us to make one more step and to look not only at the averaged value of R_g^2 for trivial and some non-trivial knots, but also at the entire distributions of probability density of R_g^2, commonly referred to as "probability distributions" or "probability densities." We were able to generate and analyze histograms of quality (i.e. looking smooth when plotted, a minimum of 10^5 loops for each curve) for loops of size $N \leq 1200$. Predictably, the probability distributions are different for different topological classes, such as all loops versus loops of a certain knot type $\mathcal{K}$. Also predictably, the probability distributions of R_g^2 spread out as N increases. The latter observation suggests the idea of looking at the probability distributions of the re-scaled variable $\rho = R_g^2/\langle R_g^2 \rangle$, where the normalization factor $\langle R_g^2 \rangle$ is taken separately for each N and for each topological entity. Note that with this definition $P(\rho) = P(R_g^2)\langle R_g^2 \rangle$.

Our main findings are summarized in Figures 4, 5, and 6, where we present probability distributions $P(\rho)$ for the trivial knots 0_1 ($\diamond$), trefoils 3_1 (Δ), and 4_1 knots ($\square$). In the same figures we plot also for comparison the analytically computed probability distributions for linear chains and for all loops. For linear chains, the necessary distribution $P_{\text{chain}}(\rho)$ was found by Fixman a long time ago[30]; as described in Appendix B, we were able to derive a similar expression for the probability distribution over all loops, irrespective of topology. To avoid overloading the figures, we do not show the corresponding data points obtained for linear chains and for all loops, but they all sit essentially on top of the theoretical curves (confirming once again the ergodicity of our loop generation routine).

Comparing the shapes of probability distributions for all loops and those with identified quenched topology, we notice that the latter distributions are somewhat more narrow. We emphasize, that although the effect looks small for the eye, it is well above the error bars of our measurements. This means simple knots are less likely to swell much above their average size than other knots, and they are also less likely to shrink far below their average, again compared to other knots. Figures 4 and 5 show this stiffness in both large and small limits of ρ. In the region $\rho < 1.25$ the general notion that entropic stiffness goes with topological complexity seems to hold true, i.e. more complex knots are more difficult to stretch or compress than arbitrary loops of the same number of segments. That the opposite of this seems to be true in the large ρ region is a subtlety not yet fully understood. In any case, topology blind loops are by definition always more flexible than topology specific loops.

The small ρ limit is of particular interest given its relation to all problems involving collapsed polymers, such as proteins. A closer view of the small R_g region of the probability distribution is presented in Figure 5. There, the probability distributions are plotted in the semi-*log* scale against ρ and, in the inset, against $1/\rho$. This can be also understood as the plot of "confinement" entropy, which corresponds to the squeezing the polymer to within a certain (small) radius. The reason why we plot the data against $1/\rho$ is because both $P_{\text{chain}}(\rho)$ and $P_{\text{loops}}(\rho)$ at small ρ have asymptotics $\sim \exp(-\text{const}/\rho)$ (see formulae (B.12) and (B.13)), which corresponds to confinement entropy $\sim 1/\rho$, and which can be established by a simple scaling argument[31]. This $1/\rho$ behavior is seen clearly in Figure 5. Furthermore, we see indeed that compressing any specific knot, trivial or otherwise, is significantly more difficult than compressing a phantom loop. Analytical expression of entropy for knots is not known, thus far only the $R_g^{-3} \sim \rho^{-3/2}$ scaling at small ρ has been conjectured[32]. Although our data is qualitatively consistent with this prediction in terms of the direction of the trend, more data is needed for a quantitative conclusion.

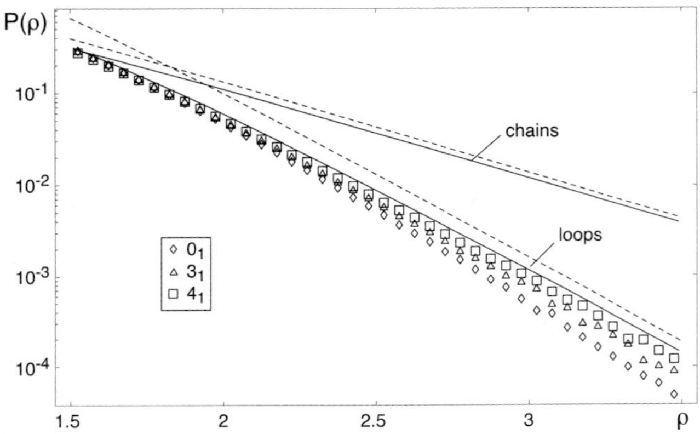

Fig. 4. The probability density plot for chains[30] (line), all loops (another line), and loops with certain knots (0_1 - ◇, 3_1 - Δ, 4_1 -□) in the range of large $\rho > 1$. Distributions are presented in terms of the scaling variable $\rho = R_g^2/\langle R_g^2 \rangle$. The asymptotics calculated in Appendix B, equations (B.12) and (B.13), are shown in the figure as dashed lines. Data for knots was generated from sets of $>$ 20 million loops each of 0_1, 3_1, and 4_1 topology, each loop $N = 90$ segments long.

More detailed comparison of probability distributions for different knots and different N are presented in the Figure 6. This figure shows a number of different probability curves under different conditions. The left column of this figure compares the topology of different objects while holding the length of the objects constant. The right column of the same figure shows comparisons of different lengths of the same topology. Significant in Figure 6 is the suggestion that probability distributions for different knots become very similar, if not identical with increasing N. Indeed in the left column in Figure 6 it is difficult to see the difference between the distributions for the three distinct topologies for $N = 1200$ or even for $N = 660$. One way to understand this effect is to consider the notion of knot localization[33,34,35]. The idea is that every strongly under-knotted loop at large N places its knot in some small fraction of its length, thus looking like a trivial knot,

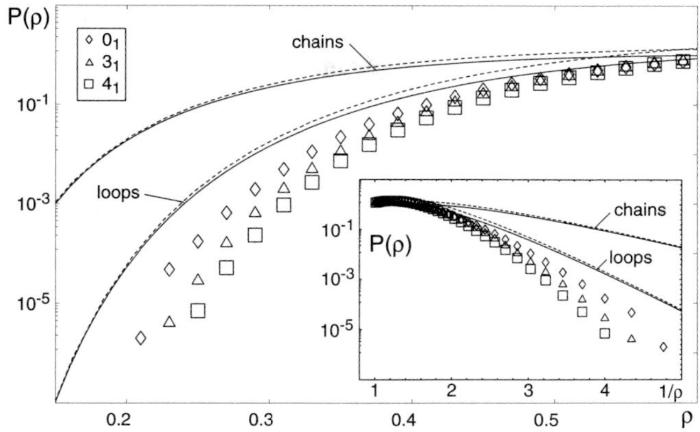

Fig. 5. The probability density plot for chains[30] (line), all loops (another line), and loops with certain knots (0_1 - ◇, 3_1 - Δ, 4_1 -□) in the range of small $\rho < 1$. Distributions are presented in terms of the scaling variable $\rho = R_g^2/\langle R_g^2\rangle$. The asymptotics calculated in Appendix B, equations (B.12) and (B.13), are shown in the figure as dashed lines. Data for knots was generated from sets of > 20 million loops each of 0_1, 3_1, and 4_1 topology, each loop $N = 90$ segments long. **Inset:** Semi-log probability density plot (or linear entropy plot) at small ρ against $1/\rho$.

with a small bump where the appropriate crossings reside. The collapse of $P_{\mathcal{K}}(\rho)$ for different, simple knot types, $\mathcal{K}$, to one curve at large N is consistent with this concept of localization. At the same time, Figure 6 suggests that probability distribution $P_{\mathcal{K}}(\rho)$ for each knot keeps evolving with N changing over the cross-over region at $N \sim N_0$.

4. Concluding remarks

To summarize, in this paper we presented computational results on knots in zero thickness loops of N rigid segments of equal length ℓ. To the accuracy of our measurements, our data are consistent with the idea that mean square gyration radius averaged over the loops which are topologically equivalent to trivial knots is larger than the similar quantity averaged over all loops irrespective of topology. The extent of this additional swelling appears similar to the swelling of self-avoiding walks compared to Gaussian random walks. Swelling is characteristic not only of trivial knots, but in general for under-knotted loops, in the sense that a topologically quenched loop swells if its knot state would have simplified upon annealing of its topology. We have examined not only averaged gyration radius, but also its probability distribution. We found that topologically under-knotted loops are relatively

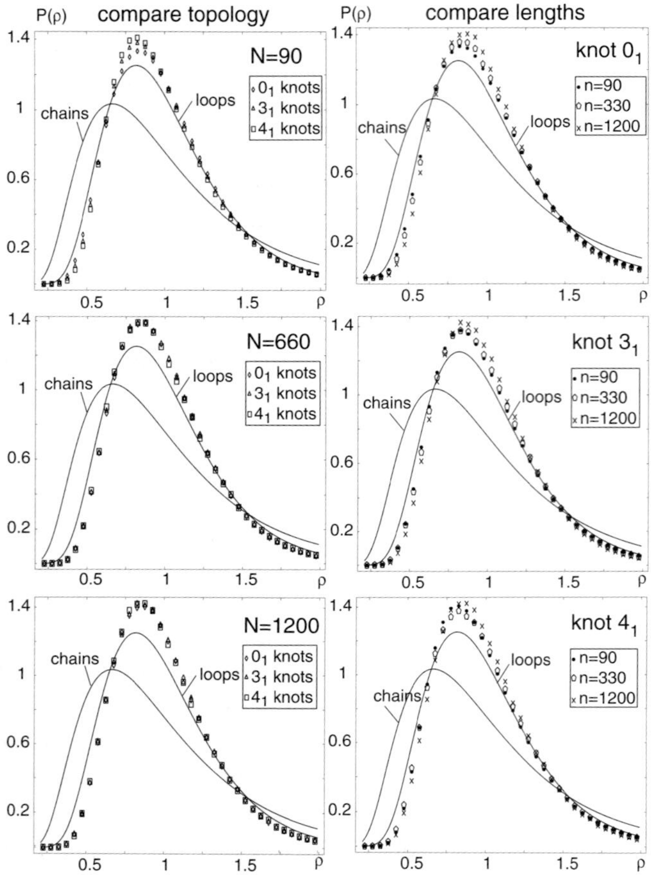

Fig. 6. The probability distributions, $P(\rho)$ for several different lengths, N. **Left Column:** Collapse of several different topologies to one curve at large N, (compare $N = 660$ or 1200 to $N = 90$), implies that one master curve for under-knotted loops exists, and that it is visible for 0_1, 3_1, and 4_1 knots at $N \geq 660$. **Right Column:** Curves for these simple topologies, as they differ in length, are certainly more similar to each other than they are to the average of all loops. Movement of the curves as N changes is not yet understood. Each knot curve was generated from simulation data in set of at least 10^5 loops.

unlikely to deviate far from their average sizes, either to smaller or to larger sizes. We have also found indication that the probability distribution of the gyration radius of simple knots becomes universal for all under-knotted loops when their length exceeds certain threshold. Importantly, our data confirm the existence of a cross-over at N of the order of N_0, the characteristic length of random knotting: it is only at $N > N_0$ that there is analogy between under-knotted loops and self-avoiding walks, at $N < N_0$ topological constraints have only a marginal effect on the trivial knots.

How far does the analogy go between self-avoiding polymers and topologically constrained ones? We were unable to confirm this analogy beyond simple scaling; it is unclear whether the $\langle R_g^2 \rangle$ dependence on N approaches its scaling form $N^{2\nu}$ in the same manner as it happens for self-avoiding walks. It is worth emphasizing that there is a field theoretic formulation for the self-avoiding walks[36], but there is nothing of this sort for knots. In our opinion, it remains an exciting challenge to find a solid understanding of the connection between fluctuation properties of the loop and its topology.

Acknowledgments

We thank A.Vologodskii for sharing with us his unpublished data[29] and T.Deguchi for useful discussions. Computations for this work were performed using Minnesota Supercomputer Institute facilities. This work was supported in part by the MRSEC Program of the National Science Foundation under Award Number DMR-0212302.

Appendix A. Loop generation

We generated loops of the length N divisible by 3 using the following method. To produce one loop, we generated $N/3$ randomly oriented equilateral triangles of perimeter 3ℓ. We consider each triangle a triplet of head-to-tail connected vectors. Collecting all N vectors from $N/3$ triangles, we re-shuffled them, and connected them all together, again in the head-to-tail manner, thus obtaining the desired closed loop.

A simpler method applicable for even N and re-shuffling vectors obtained from zero sum pairs often yields the loops with overlapping nodes. This happens when the re-shuffling results in the succession of some $2m < N$ vectors belonging to exactly m pairs and thus forming the closed sub-loop. The probability of such an event is of order unity, as the probability for the two vectors from the same pair to be next to each other after the re-shuffling scales as $1/N$, and there are $\sim N$ such pairs; more accurate calculation[28] shows that this probability approaches $1 - 1/e$ as $N \to \infty$.

For the triangles, the problem is not in any way as severe, because the probability for the three vectors of the triplet to be next to each other scales as $1/N^2$, while the number of triangles is still $\sim N$, so the overlapping loops are rare as $1/N$ (and the probability to have two, or, in general, m triplets

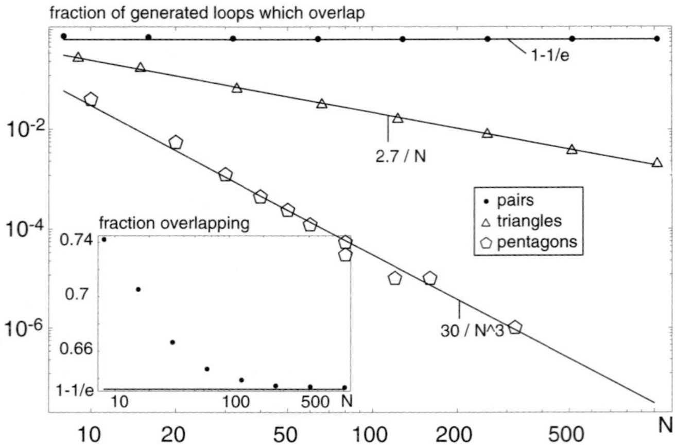

Fig. 7. The fraction of generated loops which overlapped within the resolution of computational accuracy. If a set of N/s polygons, where polygons have s sides, is used to generate a walk of length N, the fraction of generated loops, $\sim N^{2-s}$ will overlap exactly. This behavior is seen in the image.

to occupy completely the $3m$ stretch of the re-shuffled sequence does not change the $1/N$ estimate).

Our test measurements of the fraction of loops overlapping generated with pairs, triplets, and pentagons of vectors (squares are 2 pairs), shown in Figure 7, agree with this understanding. We see in this figure that the fraction overlapping at a certain N, when generated in polygons of s edges scales like N^{2-s}. We chose to generate with triplets to avoid the constant overlap implied by pairs, as well as avoiding the correlation implicit with larger sets of objects. Although generated with our method, these loops are not members of the set analyzed as they are not single stranded loops devoid of self-intersections, but rather a different physical class of objects with "petals." The simple example of $N = 9$, see Figure 8, illustrates this. Suppose that the three triangles generated have segment vectors, $(\vec{a_1}, \vec{a_2}, \vec{a_3})$, $(\vec{b_1}, \vec{b_2}, \vec{b_3})$ and $(\vec{c_1}, \vec{c_2}, \vec{c_3})$. By definition, each set of vectors within a triangle sums to zero, for example, $\vec{c_1} + \vec{c_2} + \vec{c_3} = 0$. A walk is then created by a random permutation of all of the segment vectors, for example, $(\vec{a_3}, \vec{c_2}, \vec{a_1}, \vec{b_1}, \vec{a_2}, \vec{b_3}, \vec{c_1}, \vec{b_2}, \vec{c_3})$. The problem of overlapping, described above, occurs whenever the elements of one or more complete triangles occur within a continuous section of the permutation vector. This subsection forms a complete loop, as does the rest of the chain and instead of a single loop, one has a diagram which looks something like a flower with multiple petals coming off of a center axis or set of axes.

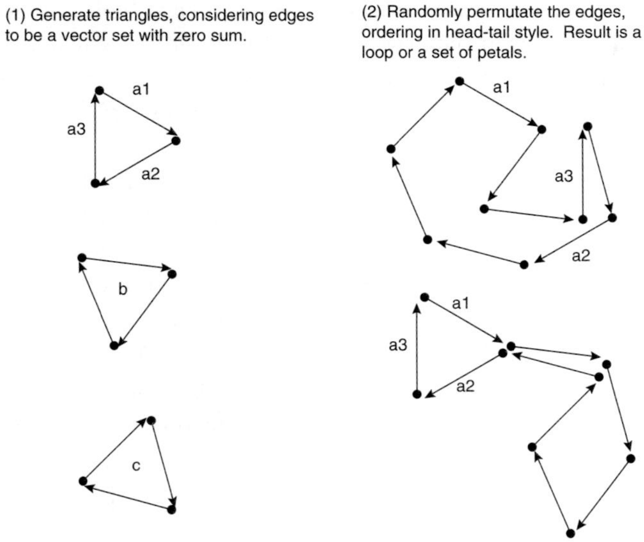

Fig. 8. Our generation routine can produce errant objects which are not loops.

In practice, there was also a totally different problem. At large N, our knot identification routine was sometimes failing because of the perceived triple crossing on the projection. A simple rotation by random Euler angles resolved this projection problem in all cases.

Appendix B. Probability distribution of all loops

In this Appendix we address the problem which, of its own, does not belong to the subject of knots. Namely, we consider a phantom loop, which can freely pass through itself, and determine the probability distribution of its gyration radius. In other, equivalent, words we consider the distribution of sizes over the ensemble of all possible loops of the given number of segments, N, irrespective of their topology. Our approach here closely follows that of the work[30] by Fixman, where he determined probability distribution for the gyration radius of the linear chains. To make our work self-contained, we reproduce below the main steps of Fixman derivation along with our results for phantom loops.

To begin with, we simplify the problem by transforming it from the gyration radius of a chain or a loop with N rigid segments of fixed length ℓ to the similar problem with a smaller number of Gaussian distributed segments. To achieve this, we group N segments in n blobs of N/n segments each. We denote as $b\vec{\eta}_k$ the end-to-end vector of each blob labeled k, where $b^2 = (N/n)\ell^2$. Note that mean squared gyration radius, which is well known for both chains and loops[19], can be expressed in terms of either N and ℓ or n and b: $\langle R_g^2 \rangle_{\text{chain}} = N\ell^2/6 = nb^2/6$ for chains and $\langle R_g^2 \rangle_{\text{loop}} = N\ell^2/12 = nb^2/12$ for loops.

If $N/n \gg 1$, then probability distribution for the unitless vector $\vec{\eta}$ is Gaussian, with zero mean and unit variance: $g(\vec{\eta}) = (3/2\pi)^{3/2} \exp\left(-3\eta^2/2\right)$. If, at the same time, $n \gg 1$, then computing the gyration radius (1) we can replace each blob with the concentrated mass N/n sitting, say, at the beginning segment of this blob. Then, formula (1) can be transformed to have just n (instead of N) points, where now $\vec{r}_{ij} = b\sum_{k=i}^{j} \vec{\eta}_k$. Accordingly, the gyration radius can be expressed as a quadratic form of the vectors $\vec{\eta}$. It is convenient to write it in the form

$$\rho \equiv \frac{R_g^2}{\langle R_g^2 \rangle} = A \sum_{k,m=1}^{n} G(k,m)\; \vec{\eta}_k \cdot \vec{\eta}_m \;, \tag{B.1}$$

where coefficient A is different for chains and loops and can be determined at the end to ensure the correct average ($\langle \rho \rangle = 1$), and where kernel $G(k,m)$ is as follows:

$$G(k,m) = \frac{k}{n^2} H(m-k) + \frac{m}{n^2} H(k-m) - \frac{km}{n^3} \;;\; H(x) = \begin{cases} 1 & \text{for } x > 0 \\ 1/2 & \text{for } x = 0 \\ 0 & \text{for } x < 0 \end{cases} . \tag{B.2}$$

We now note that the probability of the chain conformation specified by blob end-to-end vectors $b\vec{\eta}_1, b\vec{\eta}_2, \ldots, b\vec{\eta}_n$ is given by

$$Z_{\text{chain}}(\{\vec{\eta}\}) = \prod_{k=1}^{n-1} g(\vec{\eta}_k) \;. \tag{B.3}$$

Similar probability for the loop reads

$$Z_{\text{loop}}(\{\vec{\eta}\}) = \prod_{k=1}^{n} g(\vec{\eta}_k) \times \delta\left(\sum_{k=1}^{n} \vec{\eta}_k\right) \times \left(\frac{2\pi n}{3}\right)^{3/2} . \tag{B.4}$$

Compared to the distribution for the chains, we have here one more factor g, describing the connection between chain head and tail, making the loop; we have δ-function ensuring loop closing; and we have also the normalization factor.

Now, in order to compute probability distribution of ρ, we introduce the characteristic function

$$K(s) = \langle e^{\imath \rho s} \rangle = \int e^{\imath \rho s} Z\left(\{\vec{\eta}\}\right) d\{\vec{\eta}\} \ , \tag{B.5}$$

where Z is either Z_{chain} or Z_{loop}. Looking at the expressions for Z, (B.3) or (B.4), and for ρ, (B.1), we see that the three Cartesian components of vectors $\vec{\eta}$ decouple. Taking advantage of this decoupling, we can write $K(s) = [f(s)]^3$, where

$$f_{\text{chain}}(s) = \left(\frac{3}{2\pi}\right)^{(n-1)/2} \int \exp\left[-\frac{3}{2}\sum_{k=1}^{n-1} \eta_k^2 + \right.$$
$$\left. + \imath s A \sum_{k,m=1}^{n-1} G(k,m)\eta_k\eta_m \right] d\eta_1 d\eta_2 \dots d\eta_{n-1} \tag{B.6}$$

for chains, and

$$f_{\text{loop}}(s) = \left(\frac{3}{2\pi}\right)^{n/2} \left(\frac{n}{6\pi}\right)^{1/2} \int \exp\left[\imath p \sum_{k=1}^{n} \eta_k - \frac{3}{2}\sum_{k=1}^{n-1} \eta_k^2 + \right.$$
$$\left. + \imath s A \sum_{k,m=1}^{n-1} G(k,m)\eta_k\eta_m \right] d\eta_1 d\eta_2 \dots d\eta_n dp \tag{B.7}$$

for loops. In the later case, we have used the integral representation of the δ-function, thus the extra integration over p. These Gaussian integrals are easy to evaluate, because the matrix $G(k,m)$ is diagonalized, (we have omitted details[30]), by the unitary matrix $C(k,m) = \sqrt{2/n}\sin(\pi k m/n)$, revealing the eigenvalues of the G matrix, $1/k^2\pi^2$ with all integer k from 1 to n. Upon some algebra, we obtain for chains

$$f_{\text{chain}}(s) = \prod_{k=1}^{n-1} \left(1 - \imath \frac{2sA}{3k^2\pi^2}\right)^{-1/2} \simeq$$
$$\simeq \left[\underbrace{\prod_{k=1}^{\infty} \left(1 - \frac{z^2}{k^2\pi^2}\right)}_{\sin z/z}\right]^{-1/2} = \left(\frac{z}{\sin z}\right)^{1/2} , \tag{B.8}$$

where $z^2 = 2\imath As/3$. Similar manipulations for loops involve an extra integral over p:

$$f_{\text{loop}}(s) = \left(\frac{z}{\sin z}\right)^{1/2}\left(\frac{n}{6\pi}\right)^{1/2}\int_{-\infty}^{\infty}\exp\left[-p^2\frac{n}{3}\sum_{k=1}^{n}\frac{(1-(-1)^k)^2}{\pi^2k^2-z^2}\right]dp \simeq$$

$$\simeq \left(\frac{z}{\sin z}\right)^{1/2}\left[8\underbrace{\sum_{m=0}^{\infty}\frac{1}{\pi^2(2m+1)^2-z^2}}_{\tan(z/2)/4z}\right]^{-1/2} = \frac{z/2}{\sin(z/2)}\,, \quad \text{(B.9)}$$

where again $z^2 = 2\imath As/3$. Finally, we choose coefficient A based on the condition $\langle\rho\rangle = 1$, or $K'(s)_{s=0} = \imath$. This yields $A = 6$ for chains and $A = 12$ for loops. Therefore, we finally get

$$K_{\text{chain}}(s) = (\sin z/z)^{-3/2}\quad , \quad z^2 = 4\imath s\ , \quad \text{(B.10)}$$

(the result due to Fixman[30]), and

$$K_{\text{loops}}(s) = (2\sin(z/2)/z)^{-3}\quad , \quad z^2 = 8\imath s\ . \quad \text{(B.11)}$$

Knowing $K(s)$, finding the probability distribution $P(\rho)$ is the matter of inverse Fourier transform. Numerical inversion of Fourier transforms yield the curves presented in the figures 4, 5, and 6.

Analytically, asymptotic expressions can be found for both small and large ρ. For chains, the results of Fixman[30] (with small correction of the prefactor at large ρ by others[38,39]) read:

$$P_{\text{chain}}(\rho) \simeq \begin{cases} \frac{\pi^{5/2}}{\sqrt{2}}\rho^{1/2}e^{-\rho\pi^2/4} & \text{for } \rho \gg 1 \\ 9\sqrt{\frac{6}{\pi}}\rho^{-3}e^{-9/(4\rho)} & \text{for } \rho \ll 1 \end{cases}. \quad \text{(B.12)}$$

Similar expressions for loops read

$$P_{\text{loop}}(\rho) \simeq \begin{cases} \frac{\pi^6}{2}\rho^2e^{-\rho\pi^2/2} & \text{for } \rho \gg 1 \\ 324\sqrt{\frac{2}{\pi}}\rho^{-9/2}e^{-9/(2\rho)} & \text{for } \rho \ll 1 \end{cases}. \quad \text{(B.13)}$$

To obtain these results, it is convenient to re-write the inverse Fourier transform:

$$P_{\text{loop}} = \frac{1}{2\pi}\int K(s)e^{-\imath s\rho}ds = \frac{1}{2\pi\imath}\int_V \frac{\zeta^4}{\sin^3\zeta}e^{-\rho\zeta^2/2}d\zeta\ , \quad \text{(B.14)}$$

where in the latter integral $\zeta = z/2$ and integration contour V in complex ζ-plane is V-shaped, runs from infinity along the line with argument $3\pi/4$

to infinity along the line with argument $\pi/4$. In this form, it is conveniently seen that $P_{\text{loop}}(\rho) = 0$ at $\rho < 0$, as it must be, since ρ is a positive quantity. Furthermore, deforming the integration contour, we can establish that at $\rho \ll 1$ the integral is dominated by the saddle at $\zeta \simeq 3\imath/\rho$, while at $\rho \gg 1$ it is dominated by the residue at the third order pole in $\zeta = \pi$, yielding the results (B.13).

On a more physical note, it is important to realize that the exponential terms in equations (B.12) and (B.13) at small ρ are identical if written in terms of R_g, N and ℓ instead of ρ. Indeed, the leading term of the corresponding entropy (which is $-\ln P$) is equal to $9N\ell^2/24R_g^2$ for both chains and loops. Apart from the coefficient of 9/24, the scaling form of this result can be understood based on a simple argument considering confinement of either a chain or a loop in a cavity of the size $R \ll \ell\sqrt{N}$[37].

On the other hand, at large ρ chain entropy is $3(\pi R_g)^2/2N\ell^2$, while loop entropy is four times larger, it is $6(\pi R_g)^2/N\ell^2$. This can be understood as follows. For the chain, remembering that entropy of the state with end-to-end distance L is $3L^2/2N\ell^2$, Fixman noted[30] that large R_g conformations are dominated by the semi-circular shapes with $L = \pi R_g$. The loop obviously represents two such pieces, so loop entropy is twice entropy of the half-chain: $6(\pi R_g)^2/N\ell^2 = 2 \times \left(3(\pi R_g)^2/2(N/2)\ell^2\right)$.

References

1. M.Delbrueck, "Knotting Problems in Biology," *Mathematical Problems in the Biological Sciences, Proceedings of Symposium in Applied Mathematics* (American Mathematical Society, Providence, RI), v. **14**, pp 55-68, (1962).
2. H. L. Frisch., E. Wasserman, *J. A. Chem. Soc.* **83**, p3789, (1961).
3. A. Vologodskii, A. Lukashin, M. Frank-Kamenetskii *Sov. Phys. JETP*, v. **67**, p. 1875, 1974; A. Vologodskii, M. Frank-Kamenetskii *Sov. Phys. Uspekhi*, v. **134**, p. 641, 1981; A. Vologodskii *Topology and Physics of Circular DNA*, CRC Press, Boca Raton, 1992.
4. K. Koniaris, M. Muthukumar, *Phys. Rev. Lett.* **66**, 2211-2214 (1991).
5. T. Deguchi, K. Tsurusaki, *Phys. Rev. E.* **55**, 6245-6248 (1997).
6. J. des Cloizeaux, *J. Phys. Lett.* **42**, L433, (1981).
7. S. Quake, *Phys. Rev. Lett.* **73**, 3317, (1994).
8. A. Grosberg, *Phys. Rev. Lett.* **85**, 3858-3861 (2000).
9. J.M. Deutsch, *Phys. Rev. E.* **59**, 2539-2541 (1999).
10. A. Dobay, J. Dubochet, K. Millett, P. Sottas, A. Stasiak, *Proc. Natl. Acad. Sci. USA* **100**, 5611-5615 (2003).
11. T. Deguchi, M.K Shimamura, *Physical Knotting and Unknotting* eds. Calvo, J.A., Millett, K.C., Rawdon, E.J. Contemporary Mathematics book series, AMS (2002).

12. M.K. Shimamura, T. Deguchi, *Phys. Rev. E.*, **65**, 051802 (2002).
13. H. Matsuda, A. Yao, H. Tsukahara, T. Deguchi, K. Furuta, T. Inami, *Phys. Rev. E.* **68**, 011102 (2003).
14. E. Orlandini, M. Tesi, E.J. van Rensburg, S. Whittington, *J. Phys. A.* **31**, 5953-5967 (1998).
15. N.T. Moore, R. Lua, A. Grosberg, *Proc. Natl. Acad. Sci. USA* **101**, 13431 (2004).
16. J. Ziman, *Models of Disorder*, (Cambridge University Press, Cambridge, 1979).
17. J.-L. Sikorav, B. Duplantier, G. Jannink & Y. Timsit *J. Mol. Biol.*, v. **284**, p. 1279-1287, (1998).
18. M. Doi, S. F. Edwards *The Theory of Polymer Dynamics*, (Oxford University Press, Oxford 1986).
19. A.Y. Grosberg, A.R. Khokhlov, *Statistical Physics of Macromolecules*, p. 91 (AIP Press, NY, 1994)
20. C.C. Adams, *The Knot Book: an Elementary Introduction to theMathematical Theory of Knots*, (W.H. Freeman, 1994).
21. R. Lua, A.L. Borovinskiy, A.Y. Grosberg, *Polymer* **45**, p717-731 (2004).
22. F. Spitzer, *Trans. Amer. Math. Soc.* **87**, p187-197 (1958).
23. S. F. Edwards, *Proc. Phys. Soc. London* **91**, 513 (1967).
24. S. Prager, H. L. Frisch, *J. Chem. Phys.* **46**, n. 4, 1475-1483 (1967).
25. A. Grosberg and H. Frisch *J. Phys. A: Math. & Gen.* v. **36**, n. 34, p. 8955-8981 (2003).
26. T. Deguchi, K. Tsurusaki, *Lectures at Knots 96*, (World Scientific, Singapore, 1997) pp95-122.
27. Reference 19, page 64.
28. P. Flajolet, M. Noy, in Formal Power Series and Algebraic Combinatorics, eds. Mikhalev, A. V., Krob, D., Mikhalev, A. A. (Springer, 2000), pp. 191-201.
29. A. Vologodskii, private communication.
30. M. Fixman, *J. Chem. Phys.* **36**, 306-310 (1962).
31. Reference 19, page 42.
32. A.Y. Grosberg, S.K. Nechaev, E.I. Shakhnovich, *Le Journal de Physique (France)* **49**, 2095-2100 (1988).
33. V. Katrich, W.K. Olson, A. Vologodskii, J. Dubochet, A. Stasiak, *Phys. Rev. E.* **61**, 5545 (2000).
34. R. Metzler, A. Hanke, P.G. Dommesrnes, Y. Kantor, M. Kardar, *Phys. Rev. Lett.* **88**, 188101 (2002).
35. A. Hanke, R. Metzler, P.G. Dommesrnes, Y. Kantor, M. Kardar, *European Phys. Journal E* **12**, 347-354 (2003).
36. P.G. de Gennes, *Scaling Concepts in Polymer Physics*, (Cornell Univ. Press, Ithaca, NY, 1979).
37. Reference 19, formula 7.2.
38. W.C. Forsman, *J. Chem. Phys.* **42**, 2829-2835 (1965).
39. H. Fujita and T. Norisuye, *J. Chem. Phys.* **52**, 1115-1120 (1970).

CHAPTER 20

UNDER-KNOTTED AND OVER-KNOTTED POLYMERS: 2. COMPACT SELF-AVOIDING LOOPS

Rhonald C. Lua, Nathan T. Moore, Alexander Yu. Grosberg

Department of Physics, University of Minnesota
Minneapolis, MN 55455, USA
E-mail: grosbergphysics.umn.edu

We present a computer simulation study of the compact self-avoiding loops as regards their length and topological state. We use a Hamiltonian closed path on the cubic-shaped segment of a $3D$ cubic lattice as a model of a compact polymer. The importance of ergodic sampling of all loops is emphasized. We first look at the effect of global topological constraint on the local fractal geometry of a typical loop. We find that even short pieces of a compact trivial knot, or some other under-knotted loop, are somewhat crumpled compared to a topology-blind average over all loops. We further attempt to examine whether knots are localized or de-localized along the chain when the chain is compact. For this, we perform computational decimation and chain-coarsening, and look at the "renormalization trajectories" in the space of knots frequencies. Although not completely conclusive, our results are not inconsistent with the idea that knots become de-localized when the polymer is compact.

1. Introduction

1.1. *Goal and plan of this work*

In the preceding article[1], we have discussed the idea of two distinct scaling regimes for the closed polymer loop, we called them the over-knotted and under-knotted regimes. They are different in terms of the relation between a given loop with quenched topology and the imaginary loop with annealed topology, the latter being able to freely cross itself. Loosely speaking, for the over-knotted loop, topological constraints prevent the loop from untying its knots, from simplifying the knots or reducing their numbers. For the under-knotted loops, the effect is the opposite, for the topological constraints prevent the loop in this case from forming more knots. We argued that under-knotted loops tend to swell beyond the size of the phantom loop, while over-knotted loops tend to be more compact.

In the present article, we wish to extend this qualitative idea for another context, namely, for the polymers which are compressed by either external pressure or by some sort of poor solvent effect, the latter being equivalent to somewhat sticky segments. This problem is highly relevant, because most biopolymers are compact in their native conditions. DNA, for instance, is frequently so long that it would not fit inside the cell if allowed to fluctuate freely.

We shall argue first of all that the natural zeroth approximation for such a problem of compact loops can be formulated in terms of Hamiltonian paths on a cubic-shaped piece of a regular lattice in space. In the following sections we shall briefly describe our approach to computational generation of the maximally compact loops on a cubic lattice. This approach was formulated in the work[2]. Next we present probabilities of obtaining unknotted configurations, as well as probabilities of obtaining a few simple knots. We also present measurements of the average spatial extent of segments or sub-chains of compact conformations and analyze how these measurements depend on the topology or knot-type of the conformations.

Through the latter measurements, we address the question of how the global knot topology of a large loop affects the local fractal geometry of typical trajectories. An exciting general problem here, which we think is an important challenge for the future, is to which extent looking at local geometry of chains allows prediction of their global topology.

We should point out that compact loops very rapidly become under-knotted with increasing loop length with fixed knot state, because the ensemble of all loops (or, equivalently, the ensemble of states of a phantom loop) is dominated by very complex knots when the loop is compact. Thus, what we study is the local fractal geometry of under-knotted loops. This already suggests the result, which we shall explain in more depth later, that the pieces of under-knotted compact loop are locally somewhat crumpled compared to their phantom counterparts.

There is also an interesting question of whether the knots tend to be spread out or confined to a small portion of the polymer. Previous theoretical and computational work[3,4,5,6] have shown that knot-determining domains for non-compact loops are usually rather tight. For instance, the preferred size of the trefoil-determining portions of knotted polymer chains corresponds to just seven freely jointed segments[4]. In that work, the knotted domain is identified by looking for the minimal number of contiguous segments belonging to the circular chain such that, upon closure with an external planar loop, the new knot formed is of the same type as the original knot.

One possible interpretation would be that knot localization is the property of all under-knotted loops. From that point of view, it is interesting to look at knot localization in compact loops, as they are heavily under-knotted. So far, we are aware of only one numerical work[7] addressing this issue for prime *flat* knots in a model of self-attracting polymers with excluded volume. Here, when we say "flat" we mean that the polymer is strongly adsorbed onto a flat surface, or confined in a thin planar slit. It was found in[7] that these flat knots are localized in the high temperature swollen regime (consistent with theoretical prediction[6]), but become delocalized in the low temperature, collapsed globular phase. For three-dimensional polymers, the conjecture is that there is also a similar delocalization transition, i.e. the knots are delocalized for compact circular chains. We shall attempt to address this issue of knot (de)localization for compact circular chains.

1.2. *Why lattice model is natural for our purposes*

In order to address knots in densely compressed loops it is very important to realize the role of excluded volume, or self-avoidance. Physically, this is the short range repulsive forces which always exist and which prevent pieces of a polymer from penetrating each other. Mathematically, self-avoidance condition specifies the ensemble of allowed loop shapes. When the loop is not geometrically restricted, as it was examined in the preceding article[1], these excluded volume constraints are often irrelevant. That is why, for the purposes of the preceding paper[1], we argued it possible to view a polymer loop as just a closed continuous mathematical line, with no thickness. For such objects, the measure of trajectories with self-overlaps is exactly zero, and so the probabilities of all distinct knot topologies sum up to unity. For the compact loop, such a model becomes meaningless, because the limit of zero thickness is singular for loops restricted to within a certain volume. The meaningful model in this case implies that the density is maintained constant all across the allowed volume, which exactly corresponds to imposing the self-avoidance condition along with the volume restriction.

A simple model capturing the self-avoidance condition is a polymer represented as a path on the regular lattice in space, such as a cubic lattice. Knots in lattice polymers were examined in a number of works. In particular, it was proven in the works[8,9] that the probability to obtain a trivial knot upon random generation of the lattice polygon of N segments decays exponentially in N. We should emphasize that this is different from the problem addressed in our preceding article[1], because we looked at the loops with no volume exclusion, while every lattice model suitable for the study of knots and thus excluding self-intersections involves automatically some excluded volume. Unlike geometrically unrestricted loops, for the problem of compact loops excluded volume is what we must take into account. Therefore a lattice path is now a natural model to look at.

In order to make it compact, we now consider a segment of a cubic lattice, say, a cube of some size $m \times m \times m$, and consider a path of $N = m^3$ self-avoiding steps confined in this cube. Obviously, this is a Hamiltonian path. For the purposes of modelling the closed polymers, we shall consider here Hamiltonian closed paths, or Hamiltonian loops.

2. Brief overview of our recent results[2]

Fig. 1. A $14 \times 14 \times 14$ compact conformation on a lattice rendered by Piotr Pieranski. The three knot invariants calculated for this conformation correspond to a trivial knot. Using a knot-tightening algorithm called Shrink On No Overlaps[10] (SONO), Pieranski verified that the conformation is indeed unknotted by reducing the conformation into a small circular loop.

2.1. *Generation of compact loops*

The method we used to generate compact conformations on a lattice is described in Ref. 2. It is based on a combinatorial algorithm by Ramakrishnan *et al*[11]. The method essentially works by placing links randomly on the lattice, avoiding sub-cycles and dead ends, until a single loop fills the desired cubic lattice dimensions. For closed loops (Hamiltonian walks), conformation dimensions can only be even ($2 \times 2 \times 2$, $4 \times 4 \times 4$, etc.). Note that the method does not use a conventional process in which a single connected chain is grown to the desired size, because the rejection rate of any such process for compact chains rapidly becomes catastrophic even at modest chain length. Figure 1 illustrates an example of a $14 \times 14 \times 14$ conformation. Although our method is not free of biases and is not perfect, it is a significant improvement over the original algorithm by Ramakrishnan *et al.*

2.2. *Topology*

We identified the knot-type ($\mathcal{K}$) of a conformation by calculating the following three knot invariants: the Alexander polynomial evaluated at $t = -1$ ($\Delta(-1)_{\mathcal{K}}$), the Vassiliev invariant of degree two ($v_2(\mathcal{K})$) and the Vassiliev invariant of degree three ($v_3(\mathcal{K})$). For example, the knot invariants for an unknot or trivial knot ($\mathcal{K} \equiv 0_1$) are $|\Delta(-1)_{0_1}| = 1, v_2(0_1) = 0, v_3(0_1) = 0$. Although, it is possible for two distinct knots to have the same set of knot invariants, we expect the false identification of knots to be rare. For instance, the set of three knot invariants is distinct from those of (prime) knots with 10 crossings or fewer (249 knots in all) in their projection.

Using these knot invariants to classify the conformations, we collected data for the frequency of occurrence of the trivial knot and the first few simple knots (Figure 2). Computational data on trivial knot probability are customarily fit to an exponential, our last three data points giving $\sim \exp(-N/196)$.

An exponential fit should not be surprising, as the total number of conformations of length N grows exponentially with N. The estimate for the total number of compact conformations can be read out of the Flory[13] theory of polymer melts. On the cubic lattice, and in accord with simulations[14], the number of compact conformations is $\sim \exp(sN)$, where $s \approx 0.62$. From that point of view, the above mentioned result of trivial knot probability fitting to $\sim \exp(-N/196)$ implies that topologically restricting polymer to have the topology of a trivial knot only reduces entropy per segment by about $1/196 \approx 0.005$, which is a relatively insignificant amount compared to the entropy $s \approx 0.62$ itself. By contrast, to obtain an estimate from the other end, we can consider so-called crumpled conformations, similar to Peano curves. On the lattice, in the $2^k \times 2^k \times 2^k$ cube, they can be defined in the following way: the $2^k \times 2^k \times 2^k$ cube can be viewed as 8 smaller cubes $2^{k-1} \times 2^{k-1} \times 2^{k-1}$ each, and each smaller subcube can be further divided in a similar way, etc, down to the smallest $2 \times 2 \times 2$ cubes. We define the trajectory to be crumpled if it visits all the vertices within a given subcube before entering the next subcube of the same level. It is easy to prove that crumpled conformations are trivial knots. The exact recurrence relation can be written[15] for the total number of crumpled conformations for the polymer of $N = 8^k$ monomers; it yields the number of conformations of about $\sim \exp(s'N)$, with $s' \approx 0.30$. Thus, there remains a huge room for speculation regarding the asymptotic value of trivial compact knot probability at $N \to \infty$. Strictly speaking, we cannot claim it is 196 or close, we can only claim it is not larger than this quantity.

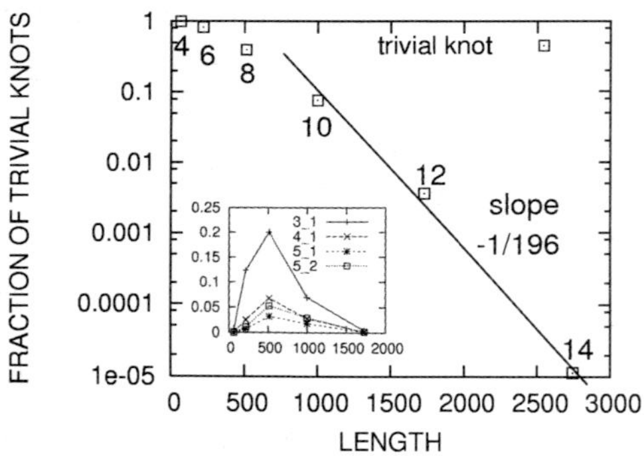

Fig. 2. Trivial knot probabilities for compact conformations of size $4 \times 4 \times 4$ to $14 \times 14 \times 14$. Inset shows the probabilities of the non-trivial knots 3_1 (trefoil), 4_1 (figure-eight), 5_1 (star), 5_2.

The inset in figure 2 shows the probabilities of some non-trivial knots. The existence of a maximum is easily explained qualitatively as follows. When N is small, the loop might be too short for a given knot, i.e. there is not enough "room" for the knot to exist. It is clear in the lattice model that there is a finite number of segments required to form any given knot, e.g. $N = 24$ for a trefoil[12]. At the other extreme, when N is large, there are many other knots "competing" for formation. The number of complex knots increases with N, yielding a decaying probability to locate any given knot.

We also addressed the question of the spatial extent of segments of compact conformations. In particular, we were interested in determining if the segments of knotted conformations are more stretched-out or more compact compared to segments of unknotted conformations. (Note that even though the entire conformation is maximally compact, a connected piece of it need not be.) To this end, we collected conformations of length N containing a particular knot and measured the mean-square end-to-end distances for segments or sub-chains of length up to $N^{2/3}$. We found that segments of knotted conformations are consistently more spread-out on average compared to segments of unknotted conformations. Figure 3 illustrates these results by plotting the ratios of mean-square end-to-end of segments of trefoil and figure-eight knotted conformations to unknotted conformations.

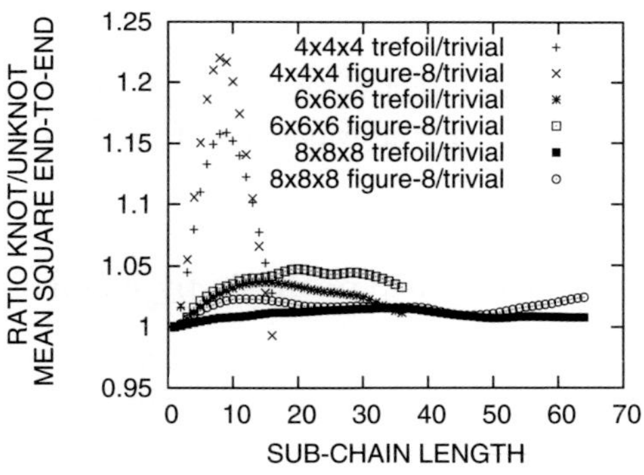

Fig. 3. Mean square end-to-end of segments (sub-chains) of trefoil and figure-eight knotted conformations relative to that of trivially knotted conformations ($4 \times 4 \times 4, 6 \times 6 \times 6, 8 \times 8 \times 8$). Pieces of knots are more extended compared to pieces of unknots.

In figure 4, instead of taking the ratio over trivial knots, we took the ratio over all conformations in the sample regardless of topology. Although figure 4 seems hardly different from figure 3, much of the points for the $8 \times 8 \times 8$ trefoils and figure-eights in figure 4 correspond to ratios less than unity. This means the trefoil and figure-eight knotted conformations begin to be more compact than typical conformations for size $8 \times 8 \times 8$, i.e. these knots cross over from their over-knotted to their under-knotted regimes. In fact, for $8 \times 8 \times 8$, the percentages of trivial, trefoil and figure-eight knots in a fairly generated sample of conformations are 40%, 20% and 6.7% respectively. The rest of the more complex knots pull the average mean

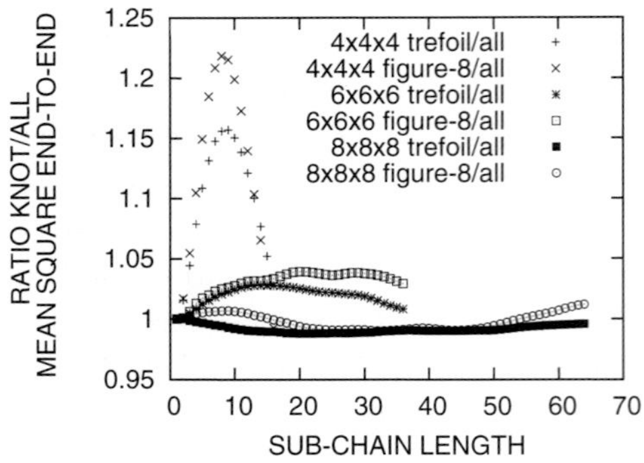

Fig. 4. Mean square end-to-end of segments (sub-chains) of trefoil and figure-eight knotted conformations relative to that of the entire sample ("All"). Pieces of conformations that are over-knotted are more extended; Pieces of conformations that are under-knotted are more compact.

square end-to-end to values larger compared to those of trivial, trefoil and figure-eight knots. For smaller conformations of size $4 \times 4 \times 4$ and $6 \times 6 \times 6$, the topology is dominated by trivial knots.

3. Testing knot localization hypothesis by renormalization

In this section we attempt to address the issue of knot localization for cicular chains using an idea inspired by field theory and polymer physics called *renormalization* or *decimation*[16]. The idea is to group the N segments of our original circular chain into N/g blocks or *blobs* of g units each. A new circular chain is formed by connecting the centers of mass of the blobs.

Our "renormalization" procedure works as follows. Starting from a batch of chains of length N with a given knot population distribution, we renormalize each chain to obtain a new batch of shorter chains of length $N_1 = N/g$. We then compute the probabilities to obtain various knots for the new batch of renormalized or decimated chains and compare that to the knot probabilities of the original batch. If the knots are localized, we expect a renormalization step using a "small" value of g to obliterate or "wash out" any memory of the original knot state. In other words, we expect a chain containing a localized knot and a chain containing no knots (or a localized knot of another type) to resemble each other after renormalization, with the value of g giving an idea of the size of the knotted domain. For example, a trefoil knot formed by six straight links (sticks) of equal length will get totally unknotted after grouping consecutive units into pairs ($g = 2$) to yield a chain containing just three (renormalized) units.

It is customary to present the results of a renormalization procedure by constructing a trajectory. In our case, the axes of the space in which the trajectory is plotted correspond to knot probabilities. Since the number of distinct knots is infinite, the dimension of this space is also infinite. To make the presentation more manageable, we plot the trajectory in a plane with one axis corresponding to the probability of trivial knots and the other axis corresponding to the probability of trefoil knots (Figure 5). In this plane, the important regions are marked. The point $(1, 0)$ at the lower right hand corner of figure 5 corresponds to all knots being trivial, while the point $(0, 1)$ at the upper left hand corner corresponds to pure trefoil knots. The line connecting these two points going diagonally corresponds to a sample containing both trivial and trefoil but no other knots. The region above this line, marked "FORBIDDEN", never gets visited by a trajectory by virtue of probabilities always summing to 1. The region below this line corresponds to a sample containing more complex knots as well as trivial and trefoil knots. By plotting the knot population in each step (labeled by g) of the renormalization procedure, one gets a picture of the evolution of knot complexity. After a sufficient number of iterations or for large enough g, all trajectories should terminate at the point $(1, 0)$ corresponding to completely unknotted chains.

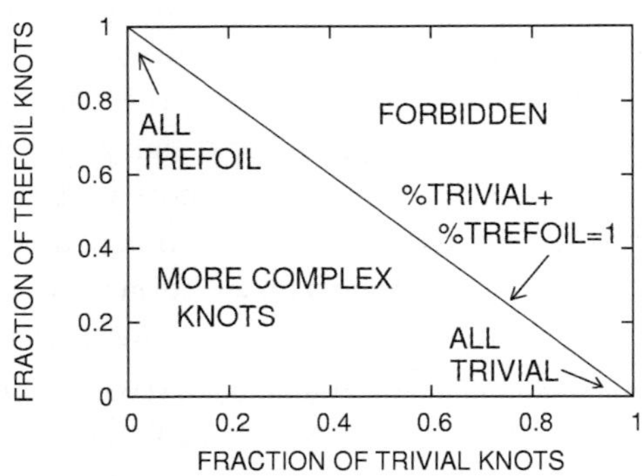

Fig. 5. The space in which the trajectories are plotted. One can also consider it as a two dimensional slice of the infinite dimensional space of knot probabilities.

We first tested this procedure by examining non-compact circular loops of length $N = 384$, using an algorithm[17] also described in a preceding

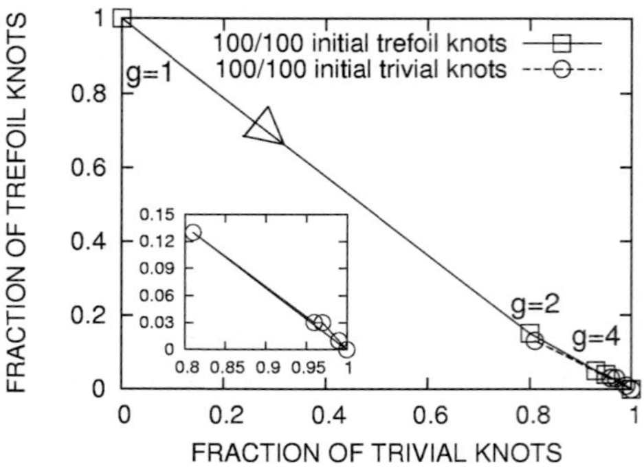

Fig. 6. Knot renormalization trajectories for initial non-compact loops of length $N =$ 384. Inset shows the trajectory only for initial unknotted loops. (Initial trefoil knots - squares, initial trivial knots - circles).

article. We generated a batch of 100 loops with trefoil knots and another batch of 100 unknotted loops or trivial knots. The loops in each batch were renormalized into blobs of size $g = 2, 4, 8, 16, 32, 64$ units and the fraction of trefoils and trivial knots were computed for each g. When the probabilities for these knots are plotted against each other, one obtains figure 6.

In figure 6, the trajectory for the 100 trefoils (squares) starts at the top of the vertical axis while the trajectory for the 100 trivial knots (circles) starts at the extreme right along the horizontal axis. The trajectory of the initial trefoils takes an almost straight path down towards the unknotted region (indicated by the arrow), implying that the tendency to unknot is overwhelming (although a few non-trivial and non-trefoil knots were produced). The trajectory for the initial trivial knots first takes a short step away from the starting point (circle, $g = 2$), meaning some knots (predominantly trefoil) are produced, then heads back towards the starting point.

The two trajectories first meet at $g = 2$, when the chains become roughly indistinguishable topologically. The fraction of trivial knots at this point is 80%, which is significantly larger than the probability of getting a trivial knot from a random sample of loops with number of segments given by $N_1 = N/g = 384/2 = 192$. For $N_0 = 241$, the empirical unknotting probability[17] is given by $w_{\rm triv} = \exp(-N_1/N_0) = 45\%$. These results seem to be consistent with the picture of a localized trefoil knot, using up just a few segments (6 or 7) of the entire chain.

The renormalization procedure outlined above was also applied to compact conformations on a cubic lattice of dimensions $8\times 8\times 8$ and $12\times 12\times 12$

(figures 7 and 8). Figure 7 illustrates the trajectories for initial $8 \times 8 \times 8$ compact conformations of 100 trefoils and 100 trivial knots. The trajectory of initial trefoils takes a downward path slightly skewed towards the origin. The trajectory of initial trivial knots makes a short excursion towards the origin then turns around and goes back to its starting point. The two trajectories seem to meet at about $g = 16$. The interpretation of a localized or delocalized knot is not straightforward, since on a cubic lattice, the minimum number of segments needed to form a trefoil is 24 instead of 6.

The trajectories deviate significantly from the diagonal, manifesting the fact that many more complex knots are formed as a result of the renormalization procedure. This result is not surprising, as the distance between renormalized segments may actually become smaller than the lattice constant of the original cubic lattice. In fact, renormalized segments may even overlap, after say $g = 4$, due to congruences arising from the integral coordinate positions of the original chain units.

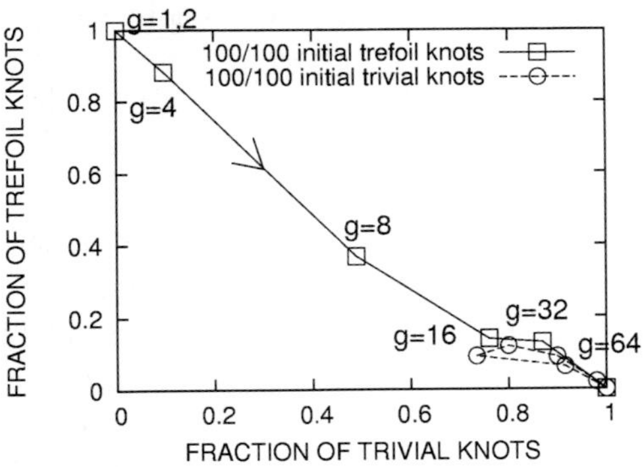

Fig. 7. Knot renormalization trajectories for initial $8 \times 8 \times 8$ compact conformations. (Initial trefoil knots - squares, initial trivial knots - circles).

Figure 8 illustrates the trajectories for initial $12 \times 12 \times 12$ compact conformations with $g = 4, 8, 16, 32, 64$. Aside from batches of 100 trefoils and 274 trivial knots, we also added a batch of 100 conformations (triangles) regardless of topology, representing typical compact conformations for this size. In this "all" knots batch, the probability for either a trivial knot or trefoil knot is about 2 in 500 (its starting point is thus practically located at the origin). In this regime, trivial knots and trefoil knots are overwhelmingly under-knotted. It can be seen from the trajectory of "all" knots that the

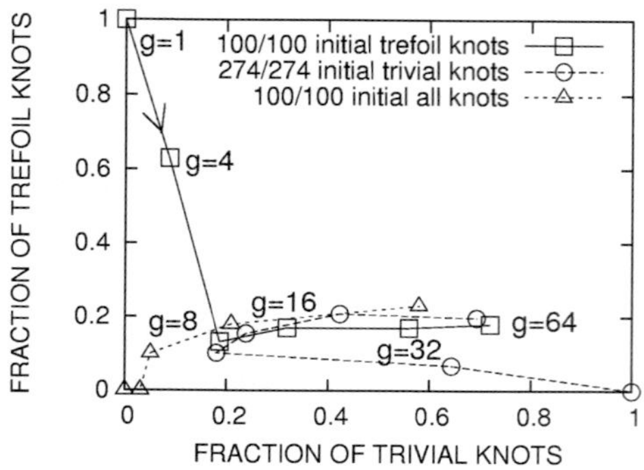

Fig. 8. Knot renormalization trajectories for initial $12 \times 12 \times 12$ compact conformations. (Initial trefoil knots - squares, initial trivial knots - circles, initial "all" knots - triangles).

memory of the initial knot state does persist longer the more complex the initial knotted state. This result is also complemented by the behaviour of the trajectories of initial trefoils and trivial knots, which come close to the origin at $g = 8$.

4. Conclusion

To conclude, we have examined the simple lattice model of compact polymer loops. To generate such loops computationally, we have employed what we believe is the least biased algorithm currently available[2]. We first confirmed that the probability to realize a trivial knot is dramatically suppressed in compact loops compared to their volume-unrestricted counterparts. We sampled compact loops of up to $N = 14^3 = 2744$ segments, and our data fit $\sim \exp(-N/N_0)$ with $N_0 \approx 196$. We emphasize that it remains unclear whether we have already reached the true $N \to \infty$ asymptotics, or in fact N_0 is smaller than our observed value of 196. When we looked at the end-to-end distance of the small pieces of polymer, with some g segments, $g \ll N^{2/3}$, we found that their end-to-end distances are smaller when the loop as a whole is a trivial knot, or, in general, when it is under-knotted, compared to the average over all loops or compared to complex knots or over-knotted loops. In principle, this allows for the possibility to make at least a probabilistic judgment of the global loop topology by observing its local fractal geometry. This is also consistent, at least qualitatively, with our findings in the preceding paper[1] that confinement entropy grows with decreasing size much sharper for the topologically constrained loops such as trivial knots compared to the topology-blind average over all loops. We presented also first attempts to address knot de-localization in the collapsed polymers by looking at the renormalization trajectories in the space of knots frequencies. We think that our results open prospects for deeper understanding of the interplay between knot topology and loop compaction, and that finding a more solid foundation for these studies is an acute challenge.

Acknowledgments

We thank P. Pieranski for his use of the SONO algorithm to confirm the trivial knot status of some of our biggest conformations, and also for producing the image presented in Figure 1. Computations were performed using resources from the Minnesota Supercomputer Institute and the Digital Technology Center at the University of Minnesota. This work was supported in part by the MRSEC Program of the National Science Foundation under Award Number DMR-0212302.

References

1. N.T. Moore, R.C. Lua, A.Yu. Grosberg *preceding article in this volume.*
2. R.C. Lua, A.L. Borovinskiy, A.Yu. Grosberg, *Polymer* **45**, 717 (2004).
3. A.Yu. Grosberg, A. Feigel, Y. Rabin, *Phys. Rev. E* **45**, 6618 (1996).
4. V. Katritch, W.K. Olson, A. Vologodskii, J. Dubochet, A. Stasiak, *Phys. Rev. E* **61**, 5545 (2000).
5. R. Metzler, A. Hanke, P.G. Dommesrnes, Y. Kantor, M. Kardar, *Phys. Rev. Lett.* **88**, 188101 (2002).
6. A. Hanke, R. Metzler, P.G. Dommesrnes, Y. Kantor, M. Kardar, *European Phys. Journal E* **12**, 347 (2003).
7. E. Orlandini, A.L. Stella, C. Vanderzande, *Phys. Rev. E* **68**, 031804 (2003).
8. D.W. Sumners, S.G. Whittington, *J. of Phys. A: Math.& Gen.* **21**, 1689 (1988).
9. N. Pippenger, *Disc. Appl. Math.* **25**, 273 (1989).
10. P. Pieranski, in *Ideal Knots*, Ed. A. Stasiak, V. Katritch, L.H. Kauffman, (World Scientific, Singapore, 1998), p. 20.
11. R. Ramakrishnan, J.F. Pekny, J.M. Caruthers, *J. Chem. Phys.* **103**, 7592 (1995).
12. Y. Diao, *J. Knot Theory Ramifications* **2**, 413 (1993).
13. P. Flory, *Principles of Polymer Chemistry* (Cornell University Press, Ithaca, 1953).
14. V.S. Pande, C. Joerg, A.Yu. Grosberg, T. Tanaka, *J. of Phys. A: Math & Gen.* **27**, 6231 (1994); Correction: *J. of Phys. A: Math & Gen.* **29**, 4753 (1996).

15. A.Yu. Grosberg, unpublished result.
16. P.G. De Gennes, *Scaling Concepts in Polymer Physics* (Cornell University Press, Ithaca, 1979).
17. N.T. Moore, R.C. Lua, A.Yu. Grosberg, *Proc. Natl. Acad. Sci. USA* **101**, 13431 (2004).

CHAPTER 21

ON THE MEAN GYRATION RADIUS AND THE RADIAL DISTRIBUTION FUNCTION OF RING POLYMERS WITH EXCLUDED VOLUME UNDER A TOPOLOGICAL CONSTRAINT

Miyuki K. Shimamura

Tetsuo Deguchi

Department of Physics, Faculty of Science, Ochanomizu University, 2-1-1 Ohtsuka, Bunkyo-ku, Tokyo 112-8610, Japan
E-mail: miyuki@degway.phys.ocha.ac.jp

We discuss the competition between the topological effect and the excluded volume effect on the average size of ring polymers with a fixed knot, reviewing some simulation results of Ref. [19]. Under a topological constraint, the average size of ring polymers can be much larger than that of no topological constraint. However, the effective expansion depends strongly on the excluded-volume parameter. We also discuss the radial distribution functions of segments of ring polymers with fixed knots. The numerical results of the radial distributions suggest that a topological constraint on ring polymers effectively leads to an entropic repulsion among polymer segments.

1. Introduction

1.1. *Motivation and contents of the paper*

A ring polymer is one of the simplest systems that exhibit the effect of a topological constraint. The topological state is given by a knot, and it is fixed after the ring polymer is formed. One of unsolved problems in statistical physics of macromolecules is how to formulate a topological constraint on ring polymers theoretically so that we can investigate topological effects on their statistical or dynamical properties. There are several simulation researches of ring polymers with fixed topology from the following aspects: fractions of various species of knotted DNAs [1,2,3,4]; the probability of making the trivial knot for Gaussian random polygons [5,6,7] and for self-avoiding polygons with excluded volume parameter [8,9]; knotting probability for lattice polygons [10,11,12]; knotting probabilities of nontrivial knots [13,14] and their critical exponents [15,16]; the average size of SAPs on lattice [17,18]; the average size of SAPs consisting of cylindrical segments [19,20,21]; some recent developments[22,23,24]. The first quantity of interest in the numerical studies is the *knotting probability*. We define the knotting probability $P_K(N)$ of a knot K as the probability that the topology of a random polygon of N nodes is given by the knot K. Under the topological constraint the available degrees of freedom in the configuration space of polygons may be severely restricted, and the entropy may be reduced significantly. The decrease of the polymer entropy is given by $\Delta S_K = -k_B \ln P_K(N)$, where $P_K(N)$ is the ratio of the volume of the configuration space under the topological constraint to that of no topological constraint.

It is an interesting question to what extent topologically different knots behave differently with respect to some physical quantities. For instance, we may consider the average size of a knotted ring polymer in solution, and its radial distribution function, i.e., the probability distribution of distances among the polymer segments. Both of them are fundamental in the physics of ring polymers. Recently it has been suggested [25,26,27] that the average size of a random polygon with the trivial knot should scale as N^ν, where the exponent ν is given by the exponent $\nu_{\rm SAW}$ of the self-avoiding walk (SAW). It is related to a conjecture that the topological effect on the average size of random polygons should play a similar role as the excluded-volume effect although they have no excluded-volume. In fact, we observe some numerical results that can be consistent with the interpretation that the topological constraint on ring polymers effectively leads to an entropic repulsion among polymer segments [28,29,30,31]. However, it is not clear how far topological constraints could produce the same effects as the excluded volume.

In this paper, we concentrate on knots of self-avoiding polygons (SAPs), i.e., knotted ring polymers made of real physical thickness in solution. We discuss how the excluded volume controls the topological effect on the average size of SAPs. We show numerically that the average size of ring polymers with a knot K can be much larger than that of no topological constraint. The effective expansion depends strongly on the excluded-volume parameter. We also discuss the radial distribution functions of segments of SAPs with fixed knots, and show that their numerical results are consistent with those of the average sizes of SAPs with fixed knots. Here the radial distribution function is defined by the probability distribution of the distance between a pair of segments in a ring polymer. The data of the radial distributions suggest that the topological constraint on ring polymers effectively leads to an entropic repulsion among polymer segments. The content of the paper is given in the following: In Sec. 2, we explain some points of simulation methods. We introduce SAPs consisting of cylindrical segments with radius r. In Sec. 3, making use of some knot invariants, we numerically evaluate the mean square radius of gyration of cylindrical SAPs with fixed knots, and calculate their radial distribution functions for the knots. In Sec. 4, we review numerical results on the asymptotic behaviors of the average size of ring polymers with fixed knots. In Sec.5, we show that the numerical data are consistent with Grosberg's argument using the $N - r$ diagram. The argument gives a good phenomenological explanation on the effective expansion under topological constraints. In the latter part of Sec.1, we discuss an analytic derivation of the effective expansion of SAPs with fixed knots. Thus, we review in the paper some important points of the average size of SAPs with fixed knots, and show their connection to the radial distribution functions.

In Sec. 3 and Sec. 4 we have selected the trefoil knot out of nontrivial knots, and discuss the mean square radius of gyration for SAPs with the trefoil knot. However, we expect that some similar behavior should hold also for other nontrivial knots. In fact, the expectation is consistent with the numerical data of SAPs with some nontrivial knots [32].

1.2. *Some theoretical results on the topological swelling of SAPs*

We can analytically derive the effective expansion of SAPs with fixed knots, making use of some results of random knotting together with some physical assumptions [21]. We first review some important results of the probability of random knotting.

Let us discuss the N-dependence of knotting probability. The knotting probability of the trivial knot, $P_{triv}(N)$, decays exponentially as N increases [33,34]. It is confirmed numerically for several models of random polygons and SAPs that we can approximate $P_{triv}(N)$ for $N > 50$ as [7,8]

$$P_{triv}(N) \approx C_{triv}\, N^{m(triv)} \exp(-N/N_c)\,, \tag{1}$$

where the exponent $m(triv)$ is given by zero. Here the estimate of the constant C_{triv} is close to 1.0 for several models of ring polymers. We call the parameter N_c the characteristic length of random knotting. For various nontrivial knots, it is numerically shown [13,16] that the probability $P_K(N)$ for $N > 50$ can be approximated by

$$P_K(N) \approx C_K\, (N/N_c)^{m(K)} \exp(-N/N_c)\,. \tag{2}$$

Here we note that the formula can be extended in order to explain some finite-size effects. The formula should be valid when N is large enough with respect to the complexity of a given knot K. If knot K is very complex, then the formula does not necessarily give a good approximation.

We define the mean square radius of gyration R^2 of a self-avoiding polygon by

$$R^2 = \frac{1}{2N^2} \sum_{n,m=1}^{N} \langle (\vec{R}_n - \vec{R}_m)^2 \rangle \,. \tag{3}$$

Here $\vec{R}_n$ is the position vector of the nth segment (or the nth node). The symbol $\langle \cdot \rangle$ denotes the ensemble average over all possible configurations of the polygon. We define the mean square radius of gyration R_K^2 for such a self-avoiding polygon that have a given knot type K by

$$R_K^2 = \frac{1}{2N^2} \sum_{n,m=1}^{N} \langle (\vec{R}_n - \vec{R}_m)^2 \rangle_K \,. \tag{4}$$

Here the symbol $\langle \cdot \rangle_K$ denotes the ensemble average over all possible configurations of the polygon with the fixed knot K. It is clear that R^2 and R_K^2's have the relation: $R^2 = \sum_K R_K^2 P_K(N)$.

We assume that when N is large enough, the R^2 of SAPs can be approximated as follows

$$R^2 = AN^{2\nu_{\rm SAW}} \left(1 + BN^{-\Delta} + O(1/N)\right) . \tag{5}$$

When N is large enough, we also assume that the R_K^2 of SAPs with knot K can be approximated by the expansion with respect to $1/N$. Here we introduce four parameters, A_K, ν_K, B_K and Δ_K corresponding to A, $\nu_{\rm SAW}$, B and Δ of (5), respectively. Furthermore, the amplitude ratio A_K/A does not depend on the knot type. We shall discuss some large-N behaviors of R_K^2 explicitly in Sec. 4 using the asymptotic expansion (10).

Let us now derive the effective expansion of SAPs with a fixed knot. We introduce parameter n by $n = N/N_c$. If N_c is very large and n is of the order of 1 with respect to N_c, then we have the following expansion:

$$R_K^2/R^2 = \alpha \left(1 + \Delta B_K n^{-\Delta} N_c^{-\Delta} + O(1/N_c)\right) . \tag{6}$$

Here α denotes the amplitude ratio A_K/A. In terms of n, the knotting probability (2) can be written as $P_K(N) = C_K n^{m(K)} e^{-n}$. Putting (6) into the relation $R^2 = \sum_K R_K^2 P_K(N)$, we have

$$\alpha = 1 + \Gamma(n) N_c^{-\Delta} + O(1/N_c) \tag{7}$$

where $\Gamma(n) = e^{-n} \sum_K (-\Delta B_K) C_K \, n^{m(K)-\Delta}$ and $\Delta B_K = B_K - B$. When n is $O(1)$, we can approximate the sum $\Gamma(n)$ over all knots with the first term, and we can show that $\Gamma > 0$, physically. (For details, see, p. 103 of Ref. [21]) Thus we should have $\alpha > 1$.

The approximation of the amplitude ratio α by the formula (7) should be effective when N_c is very large. Thus, for SAPs defined on lattices, we can show that there is no effective expansion of R_K^2 due to the topological constraint, since the characteristic length of random knotting, N_c, is very large such as $N_c \approx 10^5$ [10,11,12] . We should therefore have $R_K^2/R^2 = 1$. For several off-lattice self-avoiding polygons, however, N_c is not as large as that of SAPs on lattices. We have indeed $\alpha > 1$ for R_K^2 of cylindrical self-avoiding polygons with knot K, as we shall discuss in Sec. 3. Furthermore, the validity of the approximation (7) is not very clear for off-lattice polygons, since the expansion parameter $N_c^{-\Delta}$ may be not very small.

We remark that the approximation (7) of the amplitude ratio α should be valid for self-avoiding polygons, while it is not necessarily valid for random polygons. In fact, the exponent $\nu = 0.5$ of random polygons is different from that of self-avoiding walks, ν_{SAW}.

2. Cylindrical self-avoiding polygons

2.1. *Method for constructing ensembles of cylindrical SAPs*

Let us introduce a model of ring polymers in good solution. We consider self-avoiding polygons consisting of N rigid impenetrable cylinders of unit length and radius r: there is no overlap allowed for any non-adjacent pairs of cylindrical segments, while next-neighboring cylinders may overlap each other. We call them cylindrical self-avoiding polygons or cylindrical SAPs, for short. The cylinder radius r can be related to the stiffness of some stiff polymers such as DNAs [3,4,32].

In the simulations of the paper, we have constructed a large number of cylindrical SAPs by the cylindrical ring-dimerization method [9]. The method is based on the algorithm of ring-dimerization [6], and very useful for generating long self-avoiding polygons (for details, see Ref. [4]). In the cylindrical ring-dimerization method, a statistical weight is given to any self-avoiding polygon successfully concatenated. Thus, when we evaluate some quantity, we take the weighted average of it with respect to the statistical weight. Some details on the statistical weight of successful concatenation are given in Ref. [4].

2.2. *Numerical method of knot detection*

Let us describe the processes of our numerical experiments. First, we construct M samples of cylindrical SAPs with N nodes by the cylindrical ring-dimerization method. We put $M = 10^4$. Here we note that various knot types are included in the M random samples. Second, we make knot diagrams for the three-dimensional configurations of cylindrical SAPs, by projecting them onto a plain. Then, we calculate two knot invariants $\Delta_K(-1)$ and $v_2(K)$ for the knot diagrams [35]. Third, we select only such polygons that have the same set of values of the two knot invariants, and then evaluate physical quantities such as mean-squared gyration radius for the selected cylindrical SAPs.

The symbol $\Delta_K(-1)$ denotes the determinant of a knot K, which is given by the Alexander polynomial $\Delta(t)$ evaluated at $t = -1$. The symbol $v_2(K)$ is the Vassiliev invariant of the second degree. The two knot invariants are practically useful for computer simulation of random polygons with a large number of polygonal nodes [36].

2.3. *Characteristic length of random knotting N_c for cylindrical SAPs and rod-bead model*

Let us discuss the characteristic length of random knotting, $N_c(r)$, for cylindrical self-avoiding polygons with radius r of cylindrical segments. For a given knot K, $P_K(N, r)$ denotes the probability that the topology of an N-noded self-avoiding polygon with cylinder radius r is given by the knot type K. Let us assume that we have M_K self-avoiding polygons with a given knot type K among M samples of cylindrical SAPs with radius r. Then, the knotting probability $P_K(N, r)$ is given by $P_K(N, r) = M_K/M$.

For the cylindrical SAPs the knotting probability $P_{triv}(N, r)$ of the trivial knot, is given by

$$P_{triv}(N, r) \approx C_{triv} \exp(-N/N_c(r)) \,. \tag{8}$$

Here the estimate of the constant C_{triv} is close to 1.0 [9]. A semilogarithmic plot of the numerical estimates of the characteristic length $N_c(r)$ is given in Fig. 1.

For cylindrical SAPs, it is remarkable [9] that $N_c(r)$ can be approximated by an exponential function of r:

$$N_c(r) \approx N_c(0) \exp(\gamma r). \tag{9}$$

The best estimates of the two parameters $N_c(0)$ and γ are given by $N_c(0) = 292 \pm 5$ and $\gamma = 43.5 \pm 0.6$ [9].

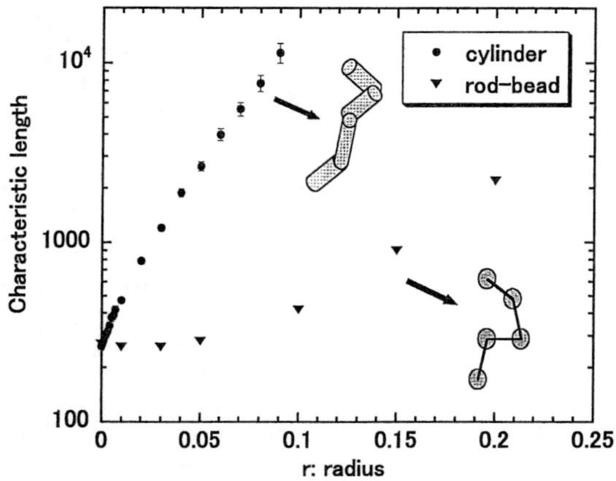

Fig. 1. Two series of estimates of the characteristic length $N_c(r)$ are shown by black circles for the cylindrical SAPs and black triangles [8,9,16,32] for the rod-bead model.

In the case of the rod-bead model, the estimates of the characteristic length $N_c(r)$ are shown in Fig. 1. For the rod-bead model, self-avoiding polygons are consisting of hard spherical beads with radius r whose centers are connected by bonds with unit length.[8,9,16,32]

The value of $N_c(r)$ is quite important for many properties of random knotting. At present, however, there is no theory for evaluating it. It might be related to the entanglement number N_e which plays a central role in the dynamics of dense polymer solutions. We note that $N_c(r)$ is of the order of 100, while N_e is of the order of 10. It should be quite a challenging problem to make some explicit relation between $N_c(r)$ and N_e.

3. Finite-size behaviors of R_K^2 for some knots

Let us now consider the ratio of R_K^2 to R^2 for a given knot K. If the ratio is larger (smaller) than 1.0, then the average size of SAPs with the knot K is relatively larger (smaller) than that of no topological constraint. We say that the SAPs with the knot K are effectively more (less) expanded. In Fig.2, the ratio R_K^2/R^2 versus the number N is plotted in a double-logarithmic scale for the trivial and trefoil knots. Here, we have depicted only the case of $r = 0.003$ among many sets of the cylindrical SAPs with 14 different values of cylindrical radius [19].

For the trivial knot, we see in Fig. 2 that the ratio R_{triv}^2/R^2 is greater than 1.0 when $N > 50$. Thus, the average size of the ring polymers with

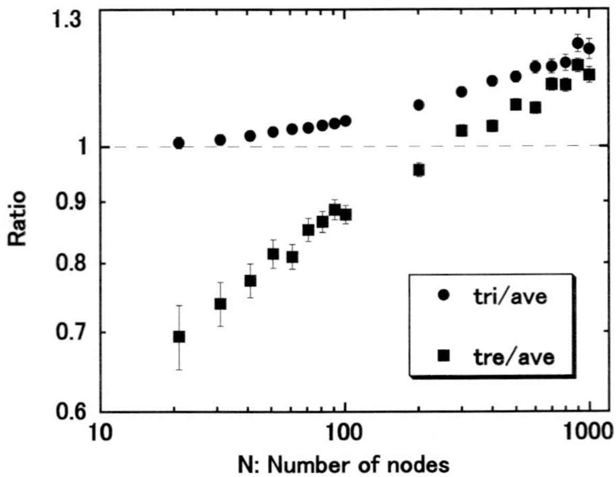

Fig. 2. Double-logarithmic plots of the ratio R_K^2/R^2 versus N for cylindrical SAPs with $r = 0.003$. R_{triv}^2/R^2 and R_{tre}^2/R^2 are shown by closed circles and squares, respectively.

the trivial knot enlarges under the topological constraint. It gives a typical example of effective expansion.

In Fig.2, the graph of the trivial knot is convex downwards: the ratio R_{triv}^2/R^2 is almost constant with respect to N for small N such as $N < 100$; for $N > 300$ the ratio R_{triv}^2/R^2 increases with respect to N with a larger gradient, and the graph can be approximated by a power law such as $R_{triv}^2/R^2 \propto N^{2\,\Delta\nu_{\rm eff}^{triv}}$. Here the symbol $\Delta\nu_{\rm eff}^{triv}$ denotes the effective exponent for the trivial knot. We note that the characteristic length $N_c(r)$ is approximately given by 300 for $r = 0.003$. Thus, we may say that the power law behavior is valid for $N > N_c(r)$.

For the trefoil knot, the graph can be approximated by a power of N such as $R_{tre}^2 \propto N^{2\,\Delta\nu_{\rm eff}^{tre}}$ through the range of $100 \leq N \leq 1001$. Here the symbol $\Delta\nu_{\rm eff}^{tre}$ denotes the effective exponent of the trefoil knot. In Fig. 2, we find that when $N < 100$ the ratio R_{tre}^2/R^2 is smaller than 1.0, while it is larger than 1.0 when $N > 300$. Thus, when N is small, the topological constraint of the trefoil knot gives effective shrinking to ring polymers, while it does not when N is large. For a nontrivial knot K, we expect that the ratio R_K^2/R^2 is less than 1.0 when N is small, while it can be larger than 1.0 when N is large.

3.1. *The radial distribution functions of random knots*

Let us discuss the effective expansion of SAPs with fixed knots through a different viewpoint. We introduce the probability distribution $G_K(d)$ that a given value d gives the distance between a pair of segments in a cylindrical SAP with knot K . We call it the radial distribution function of cylindrical SAPs with knot K. We note that dividing it by $4\pi d^2$, we have the pair correlation function. The probability distribution $G_K(d)$ should depend on the knot type K of cylindrical SAPs. We also consider the standard one, $G_{ave}(d)$, where we do not specify knot types of SAPs. For given random samples of M SAPs, we consider the two types, the standard probability distribution $G_{ave}(d)$ that is obtained by taking the average over all M polygons, and the probability distribution $G_K(d)$ with a knot K that is obtained through the average over M_K polygons with the knot K. Hereafter in the subsection, K is given by either the trivial or trefoil knot.

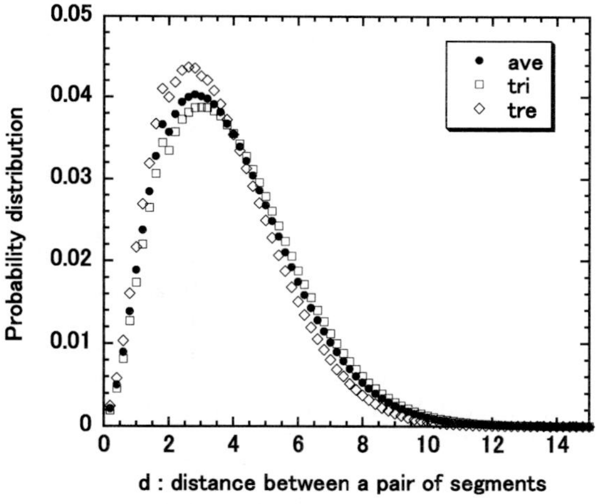

Fig. 3. Probability distribution for the distance d between a pair of segments in a cylindrical SAP for N =101 (r = 0.001). $G_{ave}(d)$, $G_{triv}(d)$ and $G_{tre}(d)$ are shown by closed circles, open squares and diamonds, respectively.

In Fig. 3, the numerical values of the probability distribution $G_K(d)$ for the three types ($G_{ave}(d)$, $G_{triv}(d)$, $G_{tre}(d)$) are plotted as a function of the distance d for N =101. We see in Fig. 3 that the probability distribution for the trefoil knot $G_{tre}(d)$ is larger than $G_{ave}(d)$ and $G_{triv}(d)$ when $d < 3$. However, when d is large, $G_{tre}(d)$ is smaller than $G_{triv}(d)$ and $G_{ave}(d)$.

It follows that the average distance between the segments of the SAPs of the trefoil knot is smaller than that of SAPs with no topological constraint and that of the trivial knot. Thus, the mean gyration radius of the trefoil knot, R_{tre}, should be smaller than that of the trivial knot, R_{triv}, or that of no topological constraint, R. The result is completely consistent with the observation of Fig. 2 that when $N = 100$, the ratio R_{tre}^2/R^2 is less than 1.0. We may consider that the polygons with the trefoil knot are more entangled than those of the trivial knot, so that they are smaller in size. In other words, the number of nodes, $N = 100$, should be rather small with respect to the complexity of the trefoil knot.

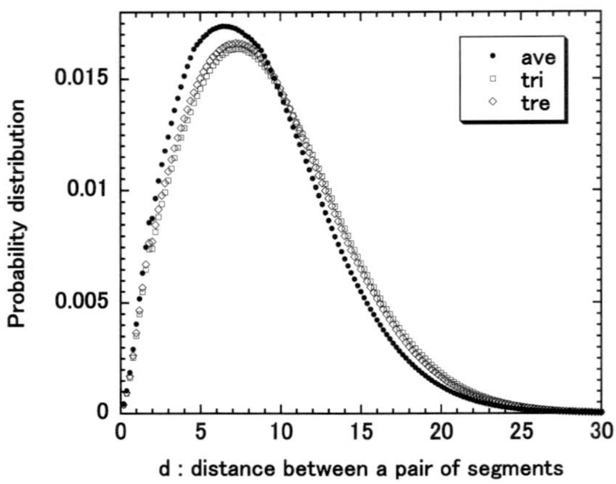

Fig. 4. Probability distribution for the distance d between a pair of segments in a cylindrical SAP for N =501 ($r = 0.001$). $G_{ave}(d)$, $G_{triv}(d)$ and $G_{tre}(d)$ are shown by closed circles, open squares and diamonds, respectively.

In Fig. 4, the numerical values of the probability distribution $G_K(d)$ for the three types are plotted as a function of d for N =501. We see that the probability distribution for the ring polymers without a topological constraint $G_{ave}(d)$ is larger than $G_{triv}(d)$ and $G_{tre}(d)$ when $d < 3$. However, when d is large, $G_{triv}(d)$ and $G_{tre}(d)$ are larger than $G_{ave}(d)$. It thus follows that the average distance between the segments of the SAPs of the trivial or trefoil knot is larger than that of SAPs with no topological constraint. The behavior of $G_K(d)$ is consistent with the fact that the average size of ring polymers with no topological constraint, R, is smaller than that of the trivial and trefoil knots, R_{triv} and R_{tre}, respectively. Thus, the numerical results of the probability distribution $G_K(d)$'s are consistent with the effective expansion of R_K^2.

The observation of Fig. 4 should be universal for any knot, when the number N of polygonal nodes is very large. The average distance between segments of SAPs of a fixed knot should be larger than that of SAPs with no topological constraint, if N is large enough with respect to the complexity of the knot. The enlargement of the average distance should be an evidence of the topological entropic repulsion among segments of ring polymers under a topological constraint.

Let us compare the peak values of graphs shown in Figs. 3 and 4. We find that the heights of the peaks of the probability distributions $G_K(d)$'s in Fig. 3 are higher than those of Fig. 4, respectively. Thus, the peak value of $G_K(d)$ of a knot K should decrease as the number of nodes N increases.

3.2. *Decrease of the topological effect under the increase of the excluded volume*

The effect of a topological constraint on the gyration radius decreases when the excluded volume increases. There are two examples: the decrease of ratio R_K^2/R^2 with respect to cylinder radius r while N being fixed, and the disappearance of the convexity or the crossover for the trivial knot, as we shall discuss shortly later.

Let us first discuss how the excluded volume can modify the effective expansion due to the topological constraint. As we clearly see in Fig. 5, the ratio R_K^2/R^2 decreases as cylinder radius r increases with N fixed, both for the trivial and trefoil knots. Thus, the effective expansion of SAPs under the topological constraint becomes smaller when the excluded volume becomes larger. It is quite nontrivial that the effective expansion given by the ratio R_K^2/R^2 decreases as cylinder radius r increases. Furthermore, one might expect that the effective expansion due to a topological constraint should also increase with respect to cylinder radius r, simply because the average size of ring polymers with larger excluded volume becomes larger. However, it is not the case for the ratio R_K^2/R^2.

Let us next discuss the convexity of the graph of the trivial knot, which we observe in Fig. 2. We consider how the convexity depends on the radius r. In Fig. 6, the graphs of the ratio R_{triv}^2/R^2 versus N are given in a double-logarithmic scale for four different values of cylinder radius r. Then, we see that the graph with $r = 0.05$ is less convex than that of $r = 0.003$. Thus, the convexity in the graphs of the effective expansion for the trivial knot

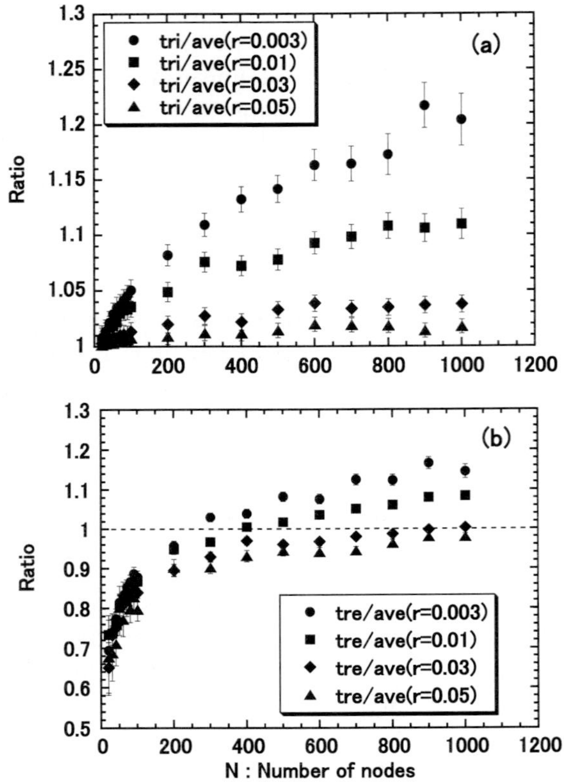

Fig. 5. Graphs of the ratio R_K^2/R^2 versus the number N in linear scales for $r = 0.003$, 0.01, 0.03 and 0.05 shown by closed circles, squares, diamonds and triangles, respectively: (a) for the trivial knot and (b) for the trefoil knot.

should be valid only when cylinder radius r is small.

Let us assume that the convexity of the graphs of R_{triv}^2/R^2 for the small r case should correspond to a crossover behavior of R_{triv}^2/R^2 with respect to N. Then, the crossover behavior could be related to that of random knots, which is recently discussed by Grosberg [27] for random polygons. Thus, we call the convexity of the trivial knot in Fig. 6 the crossover, hereafter in the paper.

It should be interesting to discuss the crossover behavior from the viewpoint of the competition between the topological and and the excluded-volume effects. Here we regard the crossover as a characteristic behavior derived from the topological constraint of being the trivial knot. As a work-

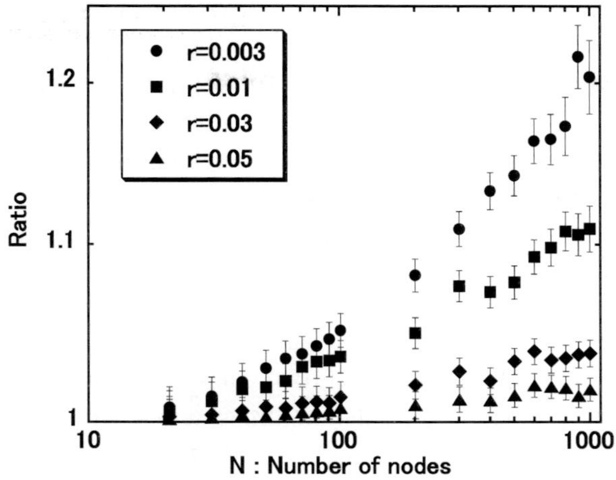

Fig. 6. Double-logarithmic plots of the ratio R^2_{triv}/R^2 versus N for r =0.003, 0.01, 0.03 and 0.05 shown by closed circles, squares, diamonds and triangles, respectively.

ing hypothesis, let us assume that the crossover should occur around at the characteristic length $N_c(r)$. Recall that $N_c(r)$ is larger than 1000 for $r = 0.03$ and 0.05, as we have estimated in the following: $N_c(r) \approx 1200$ for $r = 0.03$, and $N_c(r) \approx 2600$ for $r = 0.05$. If the above hypothesis would be valid, then the graphs for $r = 0.03$ and 0.05 should also be convex. In Fig. 6, however, there seems to be no change in the gradient of the double-logarithmic graph of R^2_{triv}/R^2 versus N for $r = 0.03$ or 0.05. The data of Fig. 6 for SAPs with $N < 1000$ do not strongly suggest that the ratio R^2_{triv}/R^2 for $r = 0.03$ increases significantly at $N = 2000$, although $N = 2000$ is larger than the $N_c(r)$ of $r = 0.03$. Here we note that the horizontal axis of Fig. 6 denotes the logarithm of the number of nodes N. We thus interpret from Fig. 6 that no assumed crossover of the trivial knot should occur for $r = 0.03$ or 0.05. The crossover as a topological effect may be diminished by the excluded-volume effect when $r \geq 0.03$.

3.3. *Interpretations through the characteristic length*

In terms of the characteristic length $N_c(r)$, we can explain some properties of the effective expansion of cylindrical SAPs under a topological constraint. The argument is consistent with the physical derivation of the effective expansion in the latter half of Sec. 1.

Let us first consider the case when the characteristic length $N_c(r)$ is very large. We show that the ratio R^2_{triv}/R^2 should be close to 1.0 for $N \ll N_c(r)$. Recall that the probability $P_{triv}(N)$ of the trivial knot decays exponentially with respect to the number N of polygonal nodes: $P_{triv}(N) \approx C_{triv} \exp(-N/N_c(r))$. If $N/N_c(r)$ is very small, the probability $P_{triv}(N)$ is close to 1.0, i.e., almost all SAPs have the trivial knot. Then, the mean-squared gyration radius with no topological constraint R^2 should be almost equal to that of the trivial knot R^2_{triv}. Consequently, the ratio R^2_{triv}/R^2 should be close to 1.0.

When $r \geq 0.05$, the characteristic length $N_c(r)$ is larger than 2600 for cylindrical SAPs. Then, the trivial knot is dominant among the possible knots generated in SAPs with $N < 1000$. Here we have constructed SAPs with $N \leq 1000$. Therefore R^2_{triv} should almost agree with R^2 for $N < 1000$, and there is no effective expansion under the topological constraint for SAPs with $N < 1000$: the R^2_{triv}/R^2 is close to 1.0. It is consistent with the data of Fig. 6.

Let us next consider the case when the characteristic length $N_c(r)$ is small or not large. Here we assume that the characteristic length is smaller than the length of SAPs we have investigated, i.e., $N_c(r) < 1000$. We then discuss that for $N > N_c(r)$ the mean square radius of gyration of SAPs with the trivial knot R^2_{triv} should be larger than that of no topological constraint R^2. In fact, various types of knots can appear in a given set of randomly generated SAPs of the cylinder radius r, since the probability of the trivial knot $P_{triv}(N)$ is exponentially small when $N > N_c(r)$. Here we note that the fraction of nontrivial knots is given by $1 - C_{triv} \exp(-N/N_c(r))$. We may thus expect that the ratio R^2_{triv}/R^2 should be indeed larger than 1.0 for $N > N_c(r)$. Here we consider the following points: when $N > N_c(r)$, the majority of SAPs generated randomly should have much more complex knots than the trivial knot; the mean square radius of gyration of N-noded SAPs with a very complex knot should be much smaller than that of the trivial knot.

4. Asymptotic behaviors of R^2_K

Let us discuss possible asymptotic behaviors of the mean square radius of gyration for ring polymers under a topological constraint. We may assume that when N is asymptotically large, R^2_K can be approximated by

$$R^2_K = A_K N^{2\nu_K} \left(1 + B_K N^{-\Delta_K} + O(1/N)\right) . \tag{10}$$

The expansion is consistent with renormalization group arguments. In order to analyze the numerical data systematically, however, we have introduced

ν_K as a fitting parameter. Thus, for the ratio R_K^2/R^2, we have the following expansion:

$$R_K^2/R^2 = (A_K/A)N^{2\Delta\nu_K}\left[1 + (B_K - B)N^{-\Delta} + O(1/N)\right]. \qquad (11)$$

Here we have put $\Delta\nu_K$ as a fitting parameter: $\Delta\nu_K = \nu_K - \nu_{\rm SAW}$. We have also introduced a technical assumption: $\Delta = \Delta_K$=0.5 in (10). As is discussed in §9.2.1 of Ref. [37], when we assume the form (11), there is no guarantee that the best curve of this form will reflect the true value of Δ_K, since we do not know the size of omitted correction terms. Rather, we may view the role of the parameter Δ_K in (10) as an aid to the extrapolation of a finite amount of data into the large-N asymptotic regime. Thus we may assume a fixed value for Δ_K, and we take the value $\Delta_K = 0.5$, following renormalization group arguments.

We have analyzed the data for the trivial and trefoil knots, applying the expansion (11) to the numerical data of R_K^2/R^2 for $N \geq 300$. Some of the best estimates of the three parameters are given in Tables 1 and 2 for the trivial and trefoil knots, respectively. From Tables 1 and 2, all the results of $\Delta\nu_K$ suggest that they should be given by 0.0, with respect to the confidence interval. It is rather clear from Tables 1 and 2 for a given cylinder radius r, the best estimates of $\Delta\nu_K$ are independent of the knot type.

Table 1. Fitting parameters A_K/A, $B_K - B$ and $\Delta\nu_K$ versus cylinder radius r: for the trivial knot.

r	A_K/A	$B_K - B$	$2\Delta\nu_K$	χ^2
0.003	1.213±1.065	-2.317±3.912	0.009±0.109	3
0.005	1.170±0.983	-1.684±3.985	0.007±0.104	4
0.01	1.106±0.836	-1.090±3.809	0.005±0.092	3
0.03	1.063±0.628	-0.518±3.166	-0.001±0.071	2
0.05	1.010±0.554	-0.143±3.039	0.002±0.066	1

Table 2. Fitting parameters A_K/A, $B_K - B$ and $\Delta\nu_K$ versus cylinder radius r: for the trefoil knot

r	A_K/A	$B_K - B$	$2\Delta\nu_K$	χ^2
0.003	1.202±0.905	-3.562±3.054	0.011±0.094	11
0.01	1.103±0.743	-3.220±2.870	0.013±0.084	1
0.03	1.058±0.790	-2.262±3.531	0.003±0.091	4
0.05	1.007±0.883	-2.422±4.119	0.007±0.107	4

There is another evidence supporting $\nu_K = \nu_{\rm SAW}$, i.e., $\Delta\nu_K = 0.0$, for trivial and trefoil knots. In fact, the plots of the ratio R_K^2/R^2 versus the number N in Fig. 5 are likely to approach some horizontal lines at some large N. The tendency of the graphs becoming flat for large N suggests that R_K^2 and R^2 should have the same exponent, i.e., $\nu_{\rm SAW}$. It is also the case with some other values of cylindrical radius r. From the two observations, we conclude that the difference of the exponents is given by 0.0: $\Delta\nu_K = 0.0$ for any value of r. There is thus no topological effect on the scaling exponent defined in the asymptotic expansion of R_K^2.

Let us give a remark on the fitting parameter B. In Tables 1 and 2 we observe that all the estimates of $B_K - B$ are negative. However, we also find that the errors are almost as large as the best estimates. Thus, it is not statistically clear whether the observation of negativity should be valid.

4.1. *The r-dependence of the amplitude ratio*

Let us discuss how the amplitude ratio A_K/A should depend on the radius r of cylindrical segments. Here we recall that the radius determines the excluded volume of the segments. Assuming $\nu_K = \nu_{\rm SAW}$ in the formula (11) we analyze the data of R_K^2/R^2 versus N again. We replace with α_K and β_K, A_K/A and $B_K - B$ in (11), respectively. We thus evaluate the amplitude ratio α_K, applying the following to the data of R_K^2/R^2 with $N \geq 300$:

$$R_K^2/R^2 = \alpha_K(1 + \beta_K N^{-\Delta} + O(1/N)) \,. \qquad (12)$$

Here we recall that $\Delta = \Delta_K$=0.5 in (11). The estimates of α_K versus r are shown in Fig. 7 in the double-logarithmic scale for the trivial, trefoil, and $3_1\sharp 3_1$ knots. To be precise, the values of α_K are a little larger than those of A_K/A given in Tables 1 and 2. The estimates of α_K are evaluated with the two-parameter fitting formula (12), while the estimates of A_K/A are derived through the three-parameter one (10).

We find in Fig. 7 that α_K should be independent of the knot type. In fact, the data points for the trivial, trefoil and the double-trefoil ($3_1\sharp 3_1$) knots overlap each other. Furthermore, the estimate of the parameter α_K becomes close to the value 1.0 when cylinder radius r is large enough. These two observations of Fig. 7 are also consistent with the simulation result of the self-avoiding polygons on the lattice [17,18].

Interestingly, we see in Fig. 7 that the ratio α_K decreases monotonically with respect to the cylinder radius r. For the data with $0.001 \leq r \leq 0.01$, we find that α_K is roughly approximated by a decreasing function of

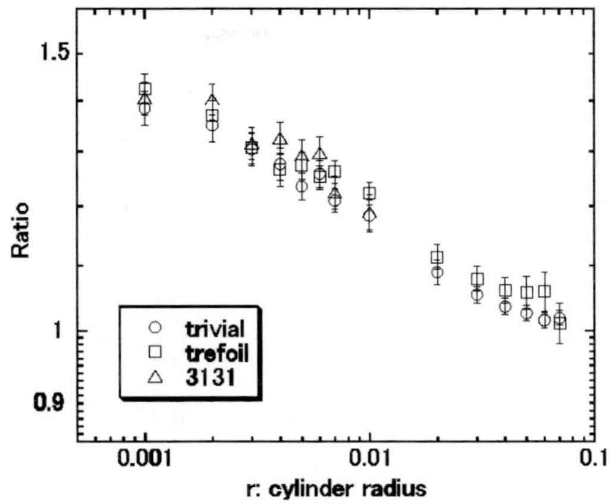

Fig. 7. Double-logarithmic plots of the amplitude ratio α_K versus cylinder radius r for the trivial, trefoil and double-trefoil ($3_1\#3_1$) knots shown by closed circles, squares and triangles, respectively. For the double-trefoil knot, the data points for $0.001 \le r \le 0.01$ are shown.

r such as $\alpha_K = \alpha_0 r^{\phi} \exp(-\psi r)$, with α_0=1.00±0.12, ϕ=-0.05±0.02 and ψ=5.78±4.79. The χ^2 value is given by 1.

5. N-r diagram

Let us discuss the finite-size behaviors of the ratio R_K^2/R^2 discussed in Sec. 3. In terms of graphs in the $N - r$ plane, we can illustrate the competition between the topological effect and the excluded-volume effect. We consider two lengths with respect to the number N of polygonal nodes: the characteristic length of random knotting, $N_c(r)$, and the "excluded-volume length", $N_{ex}(r)$. When $N > N_{ex}(r)$, the excluded-volume effect should be effective to N-noded SAPs with cylindrical radius r.

We now explain that $N_{ex}(r)$ is given by $N_{ex}(r) = 1/r^2$. For SAPs consisting of cylindrical segments with diameter d and length ℓ, the parameter z of the excluded volume is given by $z = \text{Const.} \sqrt{N} B/\ell^3 \propto N^{1/2}\, r$, where the second virial coefficient B of the polymer chain is given by $\ell^2 d$ [38]. Here we note that the ratio d/ℓ corresponds to the radius r of the cylindrical SAPs defined in Sec. 2. When $z \approx 1$, the excluded volume should be important. We may thus determine the number $N_{ex}(r)$ through the condition: $\sqrt{N_{ex}(r)}\, r = 1$.

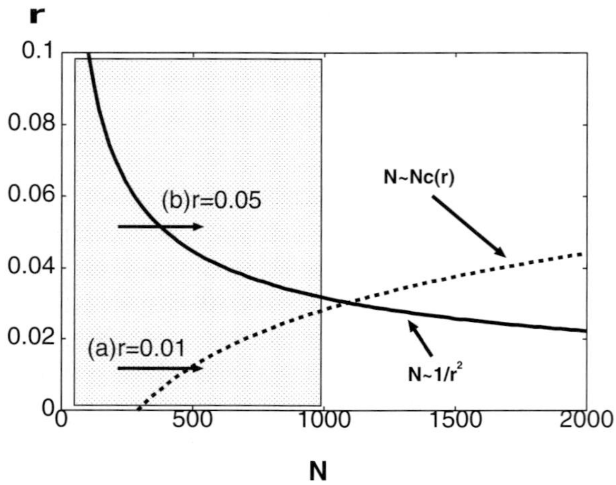

Fig. 8. $N - r$ diagram. The vertical line of the diagram expresses the r-axis and the horizontal one the N-axis. Graphs of $N = N_c(r)$ and $N = N_{ex}(r)$ are shown by two curved lines. The arrows (a) and (b) suggest the series of the data points of Fig. 5 for $r = 0.01$ and $r = 0.005$, respectively. All the data points in the paper are located in the shaded area.

Let us consider two curves in the $N - r$ plane: $N = N_{ex}(r)$ and $N = N_c(r)$. The graph $N_c(r) = N$ reaches the N axis at $N = N_c(0) \approx 300$. There is a crossing point (N^{*x}, r^*) for the two curved lines. The coordinates of the crossing point are approximately given by $N^* = 1300$ and $r^* = 0.03$. For a given simulation of the ratio R_K^2/R^2 with a fixed radius r, we have a series of data points located on a straight line parallel to the N axis.

Let us first consider the case of small values of r such as $r = 0.003$ and $r = 0.01$. From the simulation of Sec. 3, it is shown that the effective expansion due to the topological constraint is large. This is consistent with the following interpretation of the $N-r$ diagram: if we start from the region near the r axis and move in the direction of the N axis, then we cross the line $N = N_c(r)$ before reaching the other one $\sqrt{N}r = 1$; thus, we expect that the excluded-volume effect remains small when the topological effect becomes significant.

The above explanation should be consistent with the observation that the crossover of the trivial knot occurs near $N = N_c(r)$ for small values of r. Here we recall Figs. 2 and 6. When r is very small, then we cross the line of $N_c(r) = N$ almost at $N_c(0) \approx 300$.

When radius r is large such as $r = 0.03$ and 0.05, it is shown in Sec. 3 through simulation that the effective expansion is small: the ratio R_K^2/R^2 is close to 1.0. In the $N-r$ diagram, when we move rightwards from the region near the r axis with r fixed, we cross the line $\sqrt{N}\,r = 1$ before reaching the other line $N_c(r) = N$. Thus, the effective expansion as the topological effect should be small.

Finally, we should remark that some important properties of R_K^2 of cylindrical SAPs with radius r have been discussed systematically through scaling arguments with the blob picture by Grosberg [39]. In the note [39], the characteristic length $N_c(r)$ and the excluded-volume parameter z are explicitly discussed in the $N-r$ diagrams. It would thus be an interesting future problem to investigate how far the predicted properties of R_K^2 are consistent with simulation results.

Acknowledgements

We would like to thank the organizers: Jorge A. Calvo, Kenneth C. Millett and Eric J. Rawdon for their kind invitation to the AMS special session on "Numerical Methods, Calculations, and Simulations in Knot Theory and its Applications" at the Spring AMS Western Section Meeting (# 987) in San Francisco, CA, May 3-4, 2003. We are also thankful to the participants of the session for many useful discussions. The text of this paper is also related to the presentations at the conference "Knots, random walks and biomolecules" (co-organized by J.H. Maddocks and A. Stasiak and sponsored by the Bernoulli Centre of Swiss Federal Institute of Technology in Lausanne) held in Les Diablerets, Switzerland, July 14-17, 2003.

References

1. A.V. Vologodskii, A.V. Lukashin, M.D. Frank-Kamenetskii, and V.V. Anshelevich, Sov. Phys. JETP **39**,1059 (1974).
2. M. Le Bret, Biopolymers **19**, 619 (1980).
3. K.V. Klenin, A.V. Vologodskii, V.V. Anshelevich, A.M. Dykhne and M.D. Frank-Kamenetskii, J. Biomol. Struct. Dyn. **5**, 1173(1988).
4. M.K. Shimamura and T. Deguchi, J. Phys. Soc. Jpn. **70**, 1523 (2001).
5. J. des Cloizeaux and M.L. Mehta, J. Phys. (Paris) **40**, 665 (1979).
6. Y.D. Chen, J. Chem. Phys. **74**, 2034 (1981) ; J. Chem. Phys. **75**, 2447 (1981); J. Chem. Phys. **75**, 5160 (1981).
7. J.P.J. Michels and F.W. Wiegel, Phys. Lett. A **90**, 381 (1982).
8. K. Koniaris and M. Muthukumar, Phys. Rev. Lett. **66**, 2211 (1991).

9. M.K. Shimamura and T. Deguchi, Phys. Lett. A **274**, 184 (2000).
10. E.J. Janse van Rensburg and S.G. Whittington, J.Phys. A **23**, 3573 (1990).
11. S.G. Whittington, AMS Proc. Symp. Appl. Math. **45**, 73 (1992).
12. A. Yao, H. Matsuda, H. Tsukahara, M.K. Shimamura, and T. Deguchi, J. Phys. A: Math. Gen. **34**, 7563 (2001).
13. T. Deguchi and K. Tsurusaki, J. Knot Theory and Its Ramifications **3**, 321 (1994).
14. K. Tsurusaki and T. Deguchi, J. Phys. Soc. Jpn. **64**, 1506 (1995).
15. E. Orlandini, M.C. Tesi, E.J. Janse van Rensburg and S.G. Whittington, J. Phys. A: Math. Gen. **29**, L299 (1996).
16. T. Deguchi and K. Tsurusaki, Phys. Rev. E **55**, 6245 (1997).
17. E.J. Janse van Rensburg and S.G. Whittington, J.Phys. A **24**, 3935 (1991).
18. E. Orlandini, M.C. Tesi, E.J. Janse van Rensburg and S.G. Whittington, J. Phys. A: Math. Gen. **31**, 5953 (1998).
19. M.K. Shimamura and T. Deguchi, Phys. Rev. E **64**, R 020801 (2001).
20. M.K. Shimamura and T. Deguchi, Phys. Rev. E **65**, 051802 (2002).
21. T. Deguchi and M. K. Shimamura, Contemporary Math. **304**, 93-114 (2002) the proceedings of the AMS special session on "Physical Knotting and Unknotting" at the Spring Western Section Meeting, Las Vegas, NV, April 21-22, 2001, edited by J. A. Calvo, K. C. Millett and E. J. Rawdon
22. Y.-J. Sheng, P.-Y. Lai and H.-K. Tsao, Phys. Rev. E **58**, R1222 (1998).
23. V. Katritch, W. K. Olson, A. Vologodoskii, J. Dubochet and A. Stasiak, Phys. Rev. E **61**, 5545 (2000).
24. A. Dobay, P.-E. Sottas, J. Dubochet and A. Stasiak, Lett. Math. Phys. **55**, 239 (2001)
25. J. des Cloizeaux, J. Physique Letters (France) **42** , L433 (1981).
26. J. M. Deutsch, Phys. Rev. E **59**, R2539 (1999).
27. A. Yu. Grosberg, Phys. Rev. Lett. **85**, 3858 (2000).
28. M.K. Shimamura and T. Deguchi, J. Phys. A: Math. Gen. **35**, L241 (2002).
29. A. Dobay, J. Dubochet, K. Millett, P.E. Sottas and A. Stasiak, Proc. Natl. Acad. Sci. USA **100**, 5611 (2003)
30. H. Matsuda, A. Yao, H. Tsukahara, T. Deguchi, K. Furuta and T. Inami, Phys. Rev. E **68**, 011102 (2003) (4 pages).
31. A.Yu. Grosberg, talk at the conference "Knots, random walks and biomolecules" (co-organized by J.H. Maddocks and A. Stasiak and sponsored by the Bernoulli Centre of Swiss Federal Institute of Technology in Lausanne) held in Les Diablerets, Switzerland, July 14-17, 2003.
32. M.K. Shimamura, Thickness-Dependence of the Probability of Random Knotting and Application to Knotted DNA, Ph. D Thesis, Ochanomizu University, December 2000.
33. D.W. Sumners and S.G. Whittington, J. Phys. A: Math. Gen. **21**, 1689 (1988).
34. N. Pippenger, Discrete Applied Math. **25**, 273 (1989).
35. M. Polyak and O. Viro, Int. Math. Res. Not. No.11, 445 (1994).
36. T. Deguchi and K. Tsurusaki, Phys. Lett. A **174**, 29 (1993).
37. N. Madras and G. Slade, *The Self-Avoiding Walk*, Birkhäuser, Boston, 1993.

38. A.Yu. Grosberg and A.R. Khokhlov, *Statistical Physics of Macromolecules* (AIP Press, New York, 1994).
39. A.Yu. Grosberg, private communication (unpublished).

CHAPTER 22

THERMODYNAMICS AND TOPOLOGY OF DISORDERED KNOTS. CORRELATIONS IN TRIVIAL LATTICE KNOT DIAGRAMS

S.K. Nechaev[†,‡] and O.A. Vasilyev[‡]

[†] *LPTMS, Université Paris Sud,91405 Orsay Cedex, France*

[‡] *Landau Institute for Theoretical Physics, 117334 Moscow, Russia*

We investigate statistical properties of random lattice knots, the topology of which is determined by algebraic topological invariants. Representing the Jones-Kauffman polynomial invariant of a random knot diagram by a partition function of a Potts model with random coupling constants, we study the probability distribution of different topological types of random lattice knots. We are interested in the probability of the "knottedness" (the "knot complexity") of densely packed knots, which we measure via the degree of their Jones-Kauffman polynomial. In particular, we find the mean complexity of "daughter knots," obtained by cutting off a part of a trivial (i.e. unknotted) "parent" densely packed knot and closing up the "open tails" (the loose ends of the cut strands). We present arguments supporting the conjecture that the knot complexity n^* of a daughter knot of an unknotted parent one grows as $n^* \sim \sqrt{N}$, where N is the total number of vertices on the lattice. This result gives a strong support for the conjectured "crumpled globule" structure of collapsed unknotted closed polymer chains, in which the polymer forms a system of densely packed folds, mutually separated in a broad range of scales.

1. Thermodynamics of Random Knot Diagrams

New interesting problems are formulated, as a rule, in the boundary regions between traditional fields. This is clearly illustrated by the statistical physics of macromolecules, which arose due to the interpenetration of the solid state physics, statistical physics, and biophysics. Another example of a new, currently forming field is offered by the statistical topology born due to merging of the statistical physics, theory of integrable systems, algebraic topology, and group theory. The scope of the statistical topology includes, on the one hand, mathematical problems involved in the construction of topological invariants of knots and links based on some solvable models and, on the other hand, the physical problems related to determination of the entropy of random knots and links. In what follows, we pay attention predominantly to problems of the latter kind that can be rather conventionally separated into a subfield of "*probabilistic-topological*" problems.[1] Let us dwell on this class of problems in more detail.

Consider a lattice embedded into a three-dimensional space, for which Ω is the ensemble of all possible closed self-nonintersecting trajectories of N steps with a fixed point. Let ω denote some particular realization of such a trajectory. The task is to calculate the probability P for a knot of specific topological class for all trajectories $\omega \in \Omega$. This can be formally expressed as

$$P\{\text{Inv}\} = \frac{1}{\mathcal{N}(\Omega)} \sum_{\{\omega \in \Omega\}} \delta\Big\{\text{Inv}(\omega) - \text{Inv}\Big\} \tag{1}$$

where $\text{Inv}(\omega)$ is a functional representation of the invariant for a knot on trajectory ω, Inv is a topological invariant which characterizes the given topological type of the knot or link, $\mathcal{N}(\Omega)$ is the total number of trajectories, and δ is the delta function. The probability under consideration should obey the usual normalization condition $\sum_{\{\text{Inv}\}} P\{\text{Inv}\} = 1$. The entropy S of a knot of the given topological type is defined as

$$S\{\text{Inv}\} = \ln P\{\text{Inv}\} \tag{2}$$

Based on the above definitions, it is easy to note that the probabilistic-topological problems are similar to those encountered in the physics of disordered systems and sometimes, as demonstrated below, of the thermodynamics of spin glasses.[2] Indeed, the topological state plays the role of a "quenched disorder" and the functional $P\{\text{Inv}\}$ is averaged over the trajectory fluctuations at a fixed "quenched topological state," which is similar to calculation of a partition function for a spin system with "quenched disorder" in the coupling constant. In the context of this analogy, a question arises as to whether the concepts and methods developed over many years of investigation into the class of disordered statistical systems can be transferred to the class of probabilistic-topological problems, in particular, whether the concept of *self-averaging* is applicable to the knot entropy S.

The main difference between the systems with topological disorder and the standard spin systems with disorder in the coupling constant is a strongly nonlocal character of interactions in the first case: a topological state is determined only for the entire closed chain and is a "global" property of this chain. Therefore, we may speak of the topology of a part of some closed chain only in a very rough approximation. Nevertheless, below we consider a lattice model featuring a unique relationship between the topological disorder and the disorder in the local coupling constant for a certain disordered spin system on the lattice.

Every time when we deal with topological problems, there arises the task of classification of the topological states. A traditional topological invariant, known as the Gauss invariant, in inapplicable because this Abelian (commutative) characteristic takes into account only a cumulative effect of the entanglement, not reflecting the fact that the topological state depends significantly on the sequence in which a given entanglement was formed. For example, when some trial trajectory entangles with two (or more) obstacles, there may appear configurations linked with several obstacles simultaneously, while being not linked to any one of these obstacles separately. In this context, it is clear that most evident questions concerning calculation of the probability of a knot formation as a result of the random closing of the ends of a given trajectory cannot be solved in terms of the Gauss invariant because this characteristic is incomplete.

A very useful method of the classification of knots was offered by a polynomial invariant introduced by Alexander in 1928. A breakthrough in this field took place in 1975–1976, when it was suggested to use the Alexander invariants for classification of the topological state of a closed random trajectory computer-simulated by the Monte Carlo method. The results of these investigations showed that the Alexander polynomials, being a much stronger invariants compared to the Gauss integral, offer a convenient approach to the numerical investigation of the thermodynamic characteristics of random walks with topological constraints. The statistical-topological approach developed by Frank-Kamenetskii and Vologodskii[3] has proved to be very fruitful: the main results gained by now were obtained using this method with subsequent modifications.

An alternative polynomial invariant for knots and links was suggested by Jones.[4,5] This invariant arise from the investigation of the topological properties of braids.[6,7] V.F.R. Jones succeeded in finding a profound connection between the braid group relations and the Yang-Baxter equations representing a necessary condition for commutativity of the transfer matrix.[8] It should be noted that neither the Alexander, Jones, and HOMFLY invariants, nor their various generalizations are complete. Nevertheless, these invariants are successfully employed in solving statistical problems.

A clear geometric meaning of Alexander and Jones invariants was provided by the results of Kauffman, who demonstrated that Jones invariants are related to a partition function of the Potts spin model.[10] Later Kauffman and Saleur showed that the Alexander invariants can be represented by a partition function in the free fermion model.[11]

We employed the analogy between the Jones-Kauffman polynomial invariant and the partition function in the Potts model with ferro (f–) and antiferromagnetic (a–) links[12,13,14] to study the statistical properties of knots. In particular, the method of transfer matrices[15,16,17] was used to determine the probability $P\{f_K\}$ of finding a randomly generated knot K in a particular topological state characterized by the invariant f_K.

The idea of estimating $P\{f_K\}$ is briefly as follows: (i) the Jones-Kauffman invariant is represented as a partition function of the Potts model with disorder in the coupling constant; (ii) the thermodynamic characteristic of the ensemble of knots are calculated using the method of transfer matrices for the Potts spin system.

1.1. *Knots on lattices: model and definitions*

Consider an ensemble of randomly generated dense knots on a lattice in a three-dimensional space. A knot is called "dense" if the thread forming this knot entirely covers a rectangular lattice and has no freedom to fluctuate in the space. In this case, the knots of various topological types possess no configuration entropy and the probability of formation of a knot in a given topological type, is determined only by a local topology of the system. The application of this mode to real polymer systems will be discussed in the second part of this paper.

To determine topology of a thread in a three-dimensional space, we project the knot onto a plane in the general position (with no more than two knot segments intersecting at each point of the plane) and determining which segment passes above (overcrossing) or below (undercrossing) in accordance with the knot topology. Such a projection is called a knot diagram. In what follows we deal only with statistics of the knot diagrams. Evidently, this description implies an additional simplification of the model, but we believe that (in the phase of dense knots under consideration) the additional restrictions are not very significant since, as noted above, the knot entropy contains no contribution due to the string fluctuations.

Thus, let us consider a square lattice of the size $N = L \times L$ on a plane, which is rotated for convenience by the $\pi/4$ angle. The lattice is filled by a (dense) trajectory featuring intersections at each lattice site with clear indications of the threads passing above and below (see an example of the 3×3 lattice in Fig.1a). As can be readily verified, a dense trajectory on a lattice with an odd L is unique, that is, represents a knot with the topology unambiguously determined by the pattern of over- and undercrossing and the boundary conditions. In this case, the probability of realization of the knot belonging to a given topological type is determined by the distribution of over- and undercrossing in all lattice sites. The problem considered in this study is to describe the distribution of dense knots over topological classes for various kinds of the over- and undercrossing distributions.

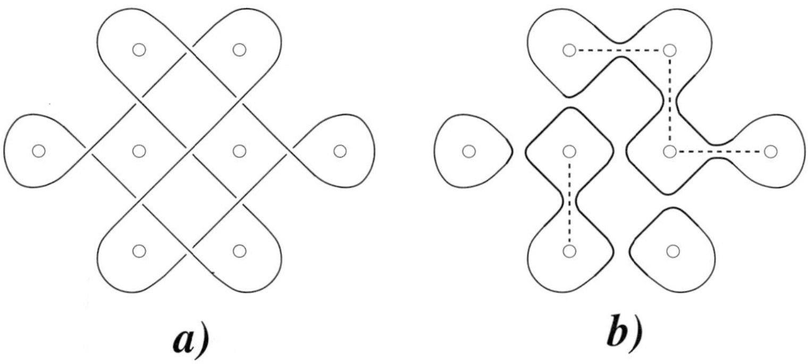

Fig. 1. a) A knot diagram on the $N = 3 \times 3$ lattice, b) A diagram splitting. Open circles indicate the spin positions in the Potts model; dashed lines show the graphs on the Potts lattice.

1.1.1. *Reidemeister moves and definition of the Kauffman invariant*

In solving any topological problem, the main point consists in comparing the knots. A knot diagram on the plane obey the following Reidemeister theorem:[18] *Two knots in the tree-dimensional space can be continuously transformed into each other if and only if the diagram of one knot can be transformed into that of the other knot by a sequence of local transformations (moves) of types I, II, and III* (see Fig.2).

As can be seen in Fig.2, the Reidemeister move I leads to the formation of a singularity on the plane during continuous fastening of the loop; this move is forbidden for smooth trajectories on the plane. Two knots are called regularly isotopic if their plane diagrams can be transformed into each other by means of the Reidemeister moves II and III. When the mutual

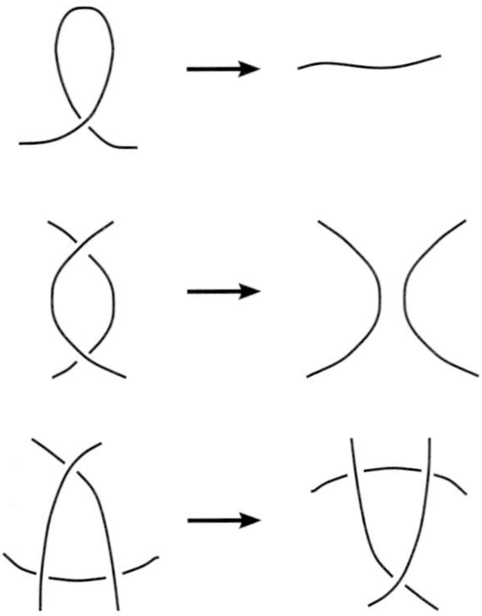

Fig. 2. The Reidemeister moves I, II, and II.

transformation of the knot diagrams requires using the Reidemeister moves of all three types, the knots are called ambiently isotopic.

Consider a two-dimensional knot diagram as a graph in which all intersection points (vertices) are characterized by the order of crossing (over- and undercrossing). Then each intersection point belongs to one of the two possible crossing modes. Let a kth point of the graph be characterized by the variable ϵ_k acquiring the values ± 1, depending on the mode of crossing.

Let us define the algebraic Kauffman invariant as a sum over all possible variants of splitting the diagram at the vertices. According to this, each splitting is ascribed a certain statistical weight by the following rule: a vertex with $\epsilon = +1$ is given the weight A or B corresponding to the horizontal or vertical splittings, respectively; for a vertex with $\epsilon = -1$, the weight B corresponds to the horizontal splitting and the weight A, to the vertical splitting. This definition can be illustrated by the following scheme:

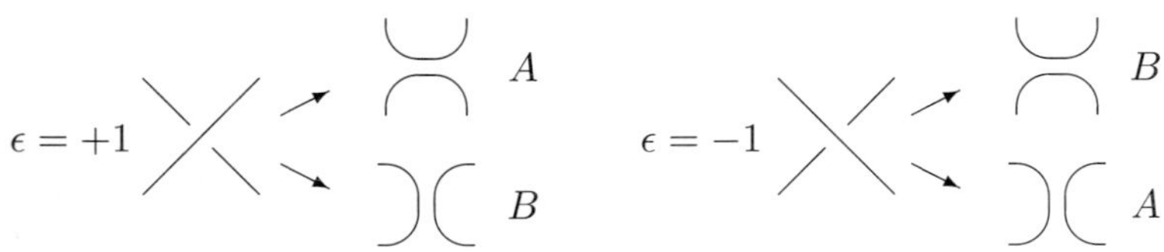

(3)

Thus, there are 2^N various microstates for the diagram of a knot possessing N vertices. Each state ω of the knot diagram represents a set of nonintersecting and non-selfintersecting cycles. The manifold of all microstates is denoted by $\{\omega\}$ (below, the braces $\{\dots\}$ will denote summing over these states). Let $S(\omega)$ be the number of cycles for the microstate ω. Consider a partition function

$$f_{KR} = \sum_{\{\omega\}} d^{S(\omega)-1} A^{l(\omega)} B^{N-l(\omega)} \tag{4}$$

where the sum is taken over all 2^N possible splittings of the diagram; $l(\omega)$ and $N - l(\omega)$ are the numbers of vertices with the weights A and B, respectively, for a given set of splittings of the microstate ω. Kauffman derived the following statement:[10] a polynomial f_{KR} of the variables A, B, and d representing a partition function (4) is a topological invariant of the regularly isotopic knots, if and only if the parameters A, B, and d obey the relationships: $AB = 1$ and $ABd + A^2 + B^2 = 0$. A proof of this statement[10] is based on verification of the invariance of the partition function f_{KR} with respect to the Reidemeister moves II and III. (The invariant for all tree Reidemeister moves is defined below.) The latter relationships pose the following restrictions on the parameters A, B, and d in Eq. (4):

$$\begin{aligned} B &= A^{-1}, \\ d &= -A^2 - A^{-2} \end{aligned} \tag{5}$$

which imply that invariant (4) is a polynomial of the single variable A.

1.1.2. *A partition function of the Potts model as a bichromatic polynomial*

Consider an arbitrary flat graph containing N vertices. Let ith vertex be characterized by a spin variable σ_i $(1 \le \sigma_i \le q)$ and each edge of the graph, connect ing the ith and jth spins $(1 \le \{i,j\} \le N)$, by the coupling constant $J_{i,j}$. The energy of the Potts model is defined[8] as

$$E = -\sum_{\{i,j\}} J_{i,j}\delta(\sigma_i, \sigma_j)$$

where the sum over $\{i,j\}$ is taken only for the adjacent spins connected by edges of the graph. Then, the partition function can be expressed as

$$Z = \sum_{\{\sigma\}} \exp\left(\sum_{\{i,j\}} \frac{J_{i,j}}{T}\delta(\sigma_i, \sigma_j)\right)$$

where $\{...\}$ denotes summation over all possible spin states, and the sum over $\{i,j\}$ is taken as indicated above. The last expression can be written in the following form:

$$Z = \sum_{\{\sigma\}} \prod_{\{i,j\}} (1 + v_{i,j}\delta(\sigma_i, \sigma_j)), \quad v_{i,j} = \exp\left(\frac{J_{i,j}}{T}\right) - 1 \tag{6}$$

A pair of adjacent spins i and j introduces into the product term a contribution equal to $\exp(\frac{J_{i,j}}{T})$ for $\sigma_i = \sigma_j$ and a unity contribution for $\sigma_i \neq \sigma_j$. Let us perform a procedure according to the following rules to a given spin configuration and the corresponding graph:

- an edge is removed from the graph if a contribution to the above product from the spins connected by this edge is unity;
- an edge is retained in the graph if the contribution from the spins σ_i and σ_j connected by this edge is $\exp(\frac{J_{i,j}}{T})$. This procedure ensures a one-to-one mapping between the spin configuration corresponding to a product term in the sum (6) and the related set of graph components.

Consider a graph G containing M edges and C connected components (an isolated vertex is considered as a separate component). Upon summing over all possible spin configurations and the corresponding subdivisions of the graph G, we may present the sum (6) in the following form:

$$Z = \sum_{\{G\}} q^C \prod_{\{i,j\}}^{M} v_{i,j} \tag{7}$$

where $\{G\}$ denotes summing over all graphs and $\prod_{\{i,j\}}^{M}$ is the product over all edges of the graph G. Note that expression (7) can be considered as an analytic continuation of the Potts spin system to non-integer and even complex q values.

For $v_{i,j} \equiv v$, expression (7) coincides with a well-known representation of the partition function of the Potts model in the form of a bichromatic polynomial.[8,19] The same expression is involved in the correspondence between the Potts model and the model of correlated percolation over sites and bonds suggested by Fortuin and Kastelleyn,[20] which serves a base for the Monte Carlo cluster algorithms developed by Swedsen and Wang[21] and Wolff[22] for the Potts model.

1.1.3. *Kauffman invariant represented as a partition function for the Potts model*

The Kauffman invariant of a given knot can be represented as a partition function for the Potts model on a graph corresponding to an arbitrary plane diagram of this knot, but we restrict the consideration below to an analysis of the knots on lattices.

Let us rewrite the Kauffman invariant in the form of a partition function for the Potts model determined in the preceding section, with M denoting the lattice knot diagram (see Fig.1a). The auxiliary variables $s_k = \pm 1$ describe the mode of the knot splitting in each kth lattice site, irrespective of the values of variables $\epsilon_k = \pm 1$ in the same vertices:

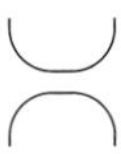
$s_k = +1$

$s_k = -1$

Let $\omega = \{s_1, s_2, \ldots, s_N\}$ be the set of all variables characterizing the mode of splitting in the lattice containing N intersections. The Kauffman invariant (4) can be written in the following form:

$$f_{KR}(\epsilon_1, ..., \epsilon_N) = \sum_{\{\omega\}} (-A^2 - A^{-2})^{S(\omega)-1} \exp\left(\ln A \sum_{k=1}^{N} \epsilon_k s_k \right) \quad (8)$$

Here, $\{\omega\}$ indicates summation over all values of the variables s_k (i.e. over all modes of splitting thy lattice diagram M) and the variables ϵ_k characterize a particular realization of "quenched" disorder in the system. Now we can demonstrate that configurations obtained as a result of splitting the diagram are in a one-to-one correspondence to configurations of the Potts model on a dual lattice.

Consider the Potts model lattice Λ corresponding to the lattice M (see Fig.1b), where open circles indicate positions of the Potts spins). The $b_{i,j}$ edge of lattice Λ corresponds to the kth site of the lattice diagram M. Therefore, disorder in the vertices of diagram M set by the variables ϵ_k coincides with the disorder in edges $b_{i,j}$ of the lattice Λ, that is, with the disorder in coupling constants. Let us define the disorder of $b_{i,j}$ for the lattice Λ via the coupling constants at the corresponding kth site of the lattice diagram M:

$$b_{i,j} = \begin{cases} -\epsilon_k, & \text{if the bond } (i,j) \text{ is vertical} \\ \epsilon_k, & \text{if the bond } (i,j) \text{ is horizontal} \end{cases} \quad (9)$$

It should be recalled that the definition of the Kauffman invariant (4) is based on splitting the lattice diagram M into polygons representing a system of closed densely packed nonintersecting contours (Fig.1b). For a given configuration of splitting the lattice diagram M and the corresponding dual lattice Λ, we take the following agreement: all edges of the lattice Λ not crossed by polygons of the lattice M are labelled. In Fig.1b, the labelled edges are indicated by dashed lines. All the remaining edges of Λ are unlabelled. In these terms, the exponent in the partition function in Eq. (8) can be written as follows:

$$\begin{aligned} &\sum_k s_k\epsilon_k = \sum_{\text{mark}}^{\text{hor.}} s_k\epsilon_k + \sum_{\text{mark}}^{\text{ver.}} s_k\epsilon_k + \sum_{\text{nonmark}}^{\text{hor.}} s_k\epsilon_k + \sum_{\text{nonmark}}^{\text{ver.}} s_k\epsilon_k = \\ &\sum_{\text{mark}}^{\text{hor.}} b_{i,j} + \sum_{\text{mark}}^{\text{ver.}} b_{i,j} - \sum_{\text{nonmark}}^{\text{hor.}} b_{i,j} - \sum_{\text{nonmark}}^{\text{ver.}} b_{i,j} = -\sum_{\text{all}} b_{i,j} + 2\sum_{\text{mark}} b_{i,j} \end{aligned} \tag{10}$$

where

$$\sum_{\text{mark}} b_{i,j} + \sum_{\text{nonmark}} b_{i,j} = \sum_{\text{all}} b_{i,j}$$

Let m_ω be the number of labelled edges and C_ω be the number of connected components in the labelled graph ω with N_p vertices (each vertex corresponding to a Potts spin). The Euler relationship for this graph is

$$S(\omega) = 2C_\omega + m_\omega - N_p$$

Now we can readily transform Eq. (8) to

$$\begin{aligned} f_{KR}(A, \{b_{i,j}\}) = (-A^2 - A^{-2})^{-(N_p+1)} \prod_{\text{all}}^{N} \left(A^{-b_{i,j}}\right) \times \\ \sum_{\{G\}} \left(-A^2 - A^{-2}\right)^{2C_\omega} \prod_{\text{mark}}^{m_\omega} \left(A^{2b_{k,l}} \left(-A^2 - A^{-2}\right)\right) \end{aligned} \tag{11}$$

by using (10) and the fact that N is odd. Comparing expressions (11) and (6), we obtain the equality

$$\sum_{\{G\}} \left(A^2 + A^{-2}\right)^{2C_\omega} \prod_{\text{mark}}^{m_\omega} \left(A^{2b_{i,j}} \left(-A^2 - A^{-2}\right)\right) \equiv \sum_{\sigma} \prod_{\{i,j\}} \left(1 + v_{i,j}\delta(\sigma_i, \sigma_j)\right) \tag{12}$$

in which the right-hand part coincides with a partition function of the Potts model represented in the form of a bichromatic polynomial. From this we obtain

$$v_{i,j} = A^{2b_{i,j}} \left(-A^2 - A^{-2}\right) = -1 - A^{4b_{i,j}}; \qquad q = (A^2 + A^{-2})^2$$

Since the disorder constants may acquire only discrete values $b_{i,j} = \pm 1$, we may write the following expression for the coupling constants $J_{i,j}$:

$$J_{i,j} = T \ln\left(1 - \left(A^2 + A^{-2}\right) A^{2b_{i,j}}\right) = T \ln\left(-A^{4b_{i,j}}\right) \tag{13}$$

Thus, we arrive at the following statement. The topological Kauffman invariant $f_{KR}(A)$ of the regularly isotopic knots on the lattice M can be represented in the form of a partition function for a two-dimensional Potts model on the corresponding dual lattice Λ:

$$f_{KR}(A, \{b_{i,j}\}) = K\left(A, \{b_{i,j}\}\right) Z\left(q(A), \{J_{i,j}(b_{i,j}, A)\}\right) \tag{14}$$

where

$$K\left(A, \{b_{i,j}\}\right) = \left(A^2 + A^{-2}\right)^{-(N_p+1)} \exp\left(-\ln A \sum_{\{i,j\}} b_{i,j}\right) \tag{15}$$

is a trivial factor independent of the Potts spins. Here, the partition function of the Potts model is

$$Z\left(q(A), \{J_{i,j}(b_{i,j}, A)\}\right) = \sum_{\{\sigma\}} \exp\left(\sum_{\{i,j\}} \frac{J_{i,j}(b_{i,j}, A)}{T} \delta(\sigma_i, \sigma_j)\right) \tag{16}$$

with the coupling constant $J_{i,j}$ and the number of states q given by the formulas

$$\frac{J_{i,j}}{T} = \ln(-A^{4b_{i,j}}), \qquad q = (A^2 + A^{-2})^2 \tag{17}$$

and the variable $b_{i,j}$ expressing disorder on edges of the lattice Λ corresponding to the lattice M. A relationship between $b_{i,j}$ and ϵ_k is defined in Eq. (9).

A specific feature of the partition function (16) is the existence of a relationship between the temperature T and the number of spin states q. For this reason, T and q cannot be considered as independent variables. Once a positive q value is fixed, the variable A can formally acquire the complex values according to Eqs. (17), where logarithm can be taken of a complex argument. The appearance of complex quantities in the partition function can be interpreted in two ways. On the one hand, this implies expansion of the domain of the partition function to the complex plane. On the other hand, the parameters T and $J_{i,j}$ do not enter explicitly into an expression for the Kauffman invariant and, hence, their complex values do not require any special consideration. Below we will be interested mostly in the probability distribution of the degree in the polynomial invariant and, therefore, can digress from particular values of the variables A, T, and $J_{i,j}$.

Thus, we have determined an invariant f_{KR} for the regularly isotopic knots, the diagrams of which are invariant with respect to Reidemeister moves II and II. In order to obtain an invariant f_{KR} for the ambiently isotopic knots with oriented diagram, the corresponding partition function has to be invariant with respect to the Reidemeister moves of all types. Let us characterize each oriented intersection by a variable $c_k = \pm 1$ according to the following rule

(a)  $c_k = -1$ (b)  $c_k = +1$

In addition, we define the knot twisting $Tw(\omega)$ as a sum of the $c_k = b_k$ values taken over all intersections: $Tw(\omega) = \sum\limits_k c_k$ The invariant $f_{KI}(\omega)$ of the ambiently isotopic knots can be expressed as follows:[10]

$$f_{KI}(\omega) = f_{KR}(\omega)(-A)^{3Tw(\omega)} \tag{18}$$

It should be noted that our boundary conditions imply that $c_k = b_k$. As is known, the Kauffman polynomial invariant $f_{KI}(\omega; A)$ of the ambiently isotopic knots as a function of variable A is equivalent to the Jones polynomial invariant $f_J(\omega; x)$ of the variable $x = A^4$. Now we can use formulas (14) and (15) to express the Jones invariant through a partition function of the Potts model. Let a partition function of the Potts model have the form $Z(t; q) = \sum\limits_E H(E, q)t^{-E}$ where $t = e^E$; E denotes the energy levels over which the sum is taken, and $H(E, q)$ is the degree of degeneracy of an energy level E for a given q value. Taking into account that $x = A^4 = -t$ and using formulas (14), (15) and (18), we obtain the following expression for the Jones polynomial invariant:

$$f_J = \sum_E H(E; q = 2 + x + x^{-1})(-x)^{-E}(1 + x)^{-N_p - 1}(-\sqrt{x})^{N_p + 1 + \sum\limits_{\{i,j\}} b_{i,j}} \tag{19}$$

By the degree of a polynomial invariant we imply the degree of a variable x in the Jones polynomial invariant f_J. It will be born in mind that the degree of the Kauffman polynomial invariant f_{KI} of the ambiently isotopic knots is obtained from the corresponding degree of the Jones polynomial invariant through simply multiplying by a factor of four.

As indicated above, out task is to calculate the probability $P\{f_J\}$ of finding a knot on a lattice in the topological state with a preset Jones-Kauffman invariant $f_J(x, \{\epsilon_k\})$ among all the 2^N possible disorder realizations $\{\epsilon_k\}$, $k = 1, \ldots, N$. This probability can formally be written as

$$P_N\{f_J\} = \frac{1}{2^N} \sum_{\{\epsilon_k\}} \delta\Big(f_J(x, \{\epsilon_1, \ldots, \epsilon_N\}) - f_J\Big)$$

Thus, the topological disorder determined by a random independent selection of intersections of the $\epsilon = +1$ and $\epsilon = -1$ types represents a random quenched external field. When we use the Jones invariant, each topological class (homotopic type) is characterized by a polynomial. In this case, precise identification of the homotopic type for a knot on a lattice containing N intersections will require N variables. Since the number of various homotopic types increases as 2^N , it is very difficult to study the probability of each separate homotopic type characterized by N variables. For this reason, we introduce a simplified characteristic of a knot on a lattice — the degree n of the polynomial invariant $f_J(x)$:

$$n = \lim_{|x| \to \infty} \frac{\ln f_J(x)}{\ln(x)} \tag{20}$$

For a trivial knot $n = 0$ and, as the knot complexity grows, the degree increases (not exceeding N). Thus, ensemble of all 2^N realizations of knots on a lattice can be divided into subclasses characterized by the degree n $(0 \leq n \leq N)$ of the Jones polynomial invariant. In these terms, we will study the probability that a randomly selected knot belongs to one of these subclasses (and is characterized by the degree n of the Jones invariant). Random knots can be generated by two methods:

(1) place a fixed number of intersections with $\epsilon = -1$ on the knot diagram with N vertices (accordingly, the other vertices belong to intersections of the $\epsilon = +1$ type);
(2) place intersections of the $\epsilon = -1$ and $\epsilon_k = -1$ types in each vertex with the probabilities p and $1 - p$, respectively.

As can be readily seen, a trivial knot having all intersections of the $\epsilon = +1$ type corresponds to a partition function with the ferro (f–) and antiferromagnetic (a–) bonds distributed in accordance with the rule (9). An impurity (corresponding to $\epsilon_k = -1$ of the knot diagram) will be considered as a change in the sign of $b_{i,j}$ relative to the values characterizing trivial knots. Note that we must differentiate between the notion of impurity (a change in the sign of $b_{i,j}$) from the a-bond with $b_{i,j} = -1$. For a trivial knot with all $\epsilon_k = 1$, the lattice contains no impurities while containing the a-bonds.

1.2. *Auxiliary constructions and numerical methods*

1.2.1. *The form of a lattice for the Potts model and the positions of ferro- and antiferromagnetic bonds*

Now we describe the geometry of a lattice for the Potts model corresponding to a knot diagram of the $N = L \times L$ size. This study is restricted to square lattices, although all considerations remain valid for rectangular lattices as well. Figure 3a shows an example of the trivial knot on a lattice with $N = 5 \times 5$. Positions of the Potts spins corresponding to this lattice diagram are indicated by circles. The Potts spin lattice corresponding to this knot is depicted in Fig.3b, where the f-bonds (that would be horizontal in Fig.1) are indicated by solid lines and the a-bonds (that would be vertical in Fig.1) are indicated by dashed lines. In Fig.3c, the same lattice is transformed into a rectangular one of the $L_h \times L_v$ type, where $L_h = L+1$ and $L_v = (L+1)/2$ (in this particular case, $L_v = 3$, $L_h = 6$. The partition function of the Potts model is studied below for a rectangular lattice of this type.

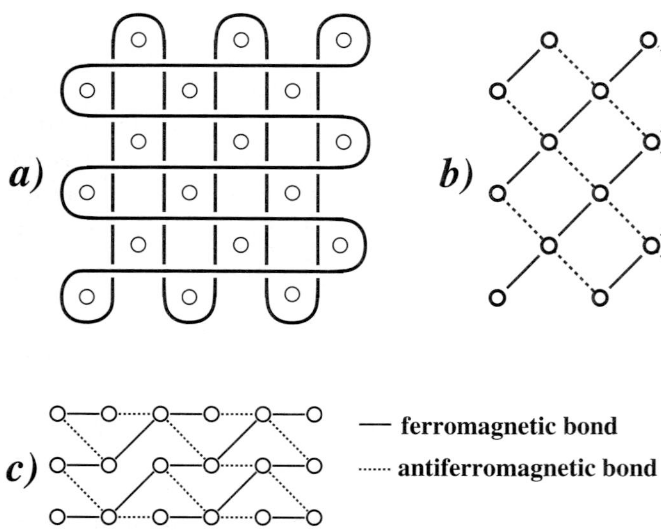

Fig. 3. A knot on $N = 5 \times 5$ lattice; **b)** Potts spin configuration corresponding to this lattice, and **c)** the same spin configuration reduced to a rectangular lattice $L_h \times L_v = 3 \times 6$.

The fact that the height of the lattice for the Potts model is half that of the lattice knot diagram (see Figs.3a–c) is very convenient in the case of using the method of transfer matrices, where the computational time expenditure exponentially depends on the lattice height. Let $N = L \times L$ be the total number of bonds in the lattice, with $N_+ = \sum_{\{i,j\}} \delta(b_{i,j}, 1)$ representing the number of f-bonds and $N_- = \sum_{\{i,j\}} \delta(b_{i,j}, -1)$ the number of a-bonds. An impurity, corresponding to $\epsilon_k = -1$ in the initial lattice, is described by a change in sign of the corresponding constant $b_{i,j}$ for the Potts model lattice.

It should be emphasized once again that by impurity we imply a change in the type of intersection from $\epsilon_k = +1$ to $\epsilon_k = -1$, which corresponds to a change in sign of the coupling constant $b_{i,j}$ for this intersection, rather than to the a-bond (such bonds are present in the Potts model lattice of the trivial knot). A trivial knot is characterized by the absence of impurities on the Potts model lattice $N_+ = \frac{N+1}{2}$, $N_- = \frac{N-1}{2}$. Note that the Kauffman invariants for a knot and its mirror image are different. This is due to the following property of the Jones invariant: the invariant of the mirror image of a knot is obtained by substituting $t \to t^{-1}$ (and $A \to A^{-1}$), after which the distribution of powers in the polynomial is asymmetric with respect to the substitution $p \to 1 - p$.

1.2.2. *The method of transfer matrix*

It should be recalled that we are interested in determining the probability distribution of the degree n (averaged over various types of intersections in the knot diagram) of the Jones-Kauffman invariant. According to definition (20), the coefficient at the term of the degree n in the Jones polynomial of a randomly generated knot is insignificant. Let us fix a certain value n_0, generate an ensemble of random knot diagrams, and determine the fraction of knots the polynomials of which contain the degree $n = n_0$.

The traditional approach to numerical analysis of a q-component Potts spin system assumes that each possible state of the column L_v spins corresponds to an eigenvector of the q^{L_v}-dimensional transfer matrix. With computers of the P-II-300 type and a reasonable computational time, the maximum possible size of the transfer matrix is 100–200. This corresponds to a spin strip width $L_v = 8$ for $q = 2$ and $L_v = 4$ for $q = 3$. Obviously, this method cannot be used for investigation of the Potts model with large q values.

A method of transfer matrices applicable to the case of arbitrary q was developed by Blöte and Nightingale.[15] The basic idea of this method is that each eigenvector corresponds to a subdivision of the column into clusters of "coupled" spins. By definition, the spins in each cluster are parallel. Therefore, the number of "colors" q can be considered simply as a parameter taking any values (including non-integer and even complex). A detailed description of this method can be also found in Jacobsen and Cardy[16] and in Salas and Sokal,[17] where this approach was used to study the Potts model with impurities on the bonds and to determine the roots of a partition function for the Potts model in the complex plane.

Below we will briefly describe the principles of determining a basis set in the space of coupled spins and constructing transfer matrices in this basis. This description follows the results reported in Salas and Sokal.[17]

A basis set corresponds to a set of various subdivisions of the spin column into clusters of coupled spins. We must take into account that the subdivision should admit realization in the form of a flat graph on a half-plane. Now we can describe a recursive procedure for determining the basis set. Let a basis set be available for a column with the height L_v. Upon adding one more spin from below, we obtain a column with the height $\tilde{L}_v$. Then we take, for example, a subdivision $v_1(L_v)$ corresponding to the first basis vector L_v and begin to generate subdivisions corresponding to the basis vectors for the column $\tilde{L}_v$ by attaching the added spin to those existing in $v_2(L_v)$. In each step, we check for the possibility of attaching this spin within the framework of the flat graph on a half-plane. after the attempt at attaching the added spin to all existing clusters, we generate a subdivision in which the added cluster is separate. Then we take the next subdivision (for example, $v_2(L_v)$) for the L_v column and repeat the generation procedure.

The procedure begins with a single spin as depicted in Fig.4. This spin is assigned the index of unity as belonging to the first cluster. Then another spin is added from below to the same column. Accordingly, the strip of two spins may occur in one of the two states: $v_1(L_v = 2)$, in which the spins are coupled and belong to cluster 1, or $v_2(L_v = 2)$, in which the spins are not coupled and belong to clusters 1 and 2 (Fig.4). A basis set for the strip of three spins is obtained by adding another spin from below and coupling this spin to all clusters. The clusters are enumerated by integers top to bottom, while vectors in the basis set are enumerated in the order of generation. As can be readily seen, the first basis vector corresponds to a state in which all spins are coupled to each other and belong to the same cluster, while the last basis vector corresponds to the state where all spins are uncoupled and the number of clusters equals to the number of spins in the column. The total number of vectors in the basis set for a strip of

m spins is determined as the Catalan number $C_m = \frac{1}{m+1}\binom{2m}{m}$ with the corresponding index number.[17]

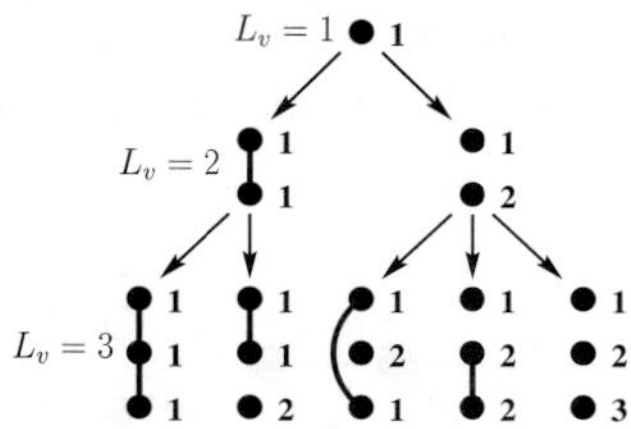

Fig. 4. A schematic diagram illustrating the step-by-step generation of subdivisions into clusters for a column of spins corresponding to the knot state vectors.

Let us denote by $v_i(\backslash k)$ a subdivision obtained upon separating the kth spin from the subdivision v_i to form a separate cluster; $v_i(\{k,l\})$ will denote a subdivision in which the clusters containing kth and lth spins are combined to form a common cluster. Following a method described in Salas and Sokal,[17] let us form the matrix

$$D_{i,j}(k) = \delta(v_j(\backslash k), v_i) + q\delta(v_j(\backslash k), v_j)$$

As can be seen, an element of the matrix $D_{i,j}(k)$ is non-zero if the kth spin is not coupled to any other spin in the subdivision v_i. When the same spin is not coupled to any other in the jth subdivision as well, the matrix element is q, otherwise it is taken equal to unity. We will also introduce the matrix $C(k,l)$ defined as

$$C_{i,j}(k,l) = \delta(v_j(\{k,l\}), v_i)$$

the elements of which are equal to unity if and only if the subdivision v_i is obtained from the subdivision v_j by combining clusters containing the kth and lth spins.

Now let us consider a lattice structure with the first two rows such as depicted in Fig.5d (for an even index $j+1$ in the second column) and Fig.5e (for an odd index $j+1$ in the second column). The spin columns are enumerated by index j: $1 \le j \le L_h$, while the spin position in a column is determined by index i: $1 \le i \le L_v$.

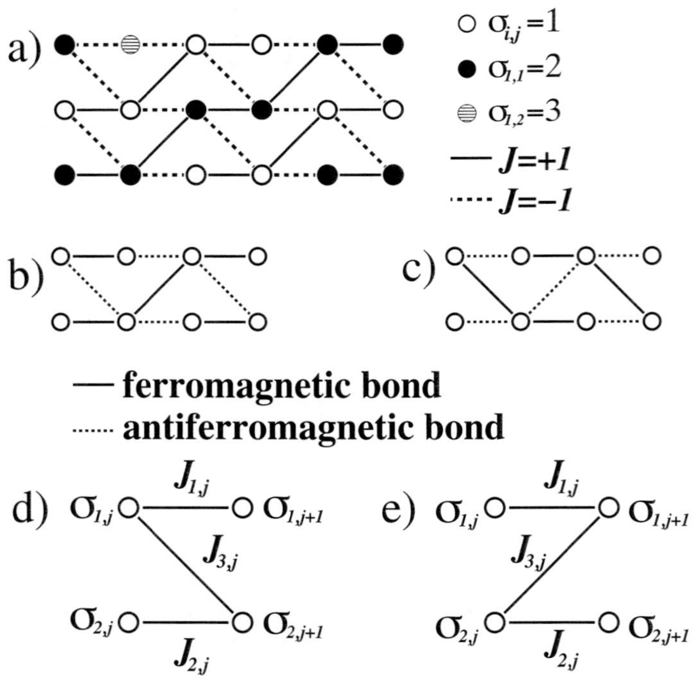

Fig. 5. a) An example of the configuration of bonds in which a minimum energy of the Potts model is not reached for $q = 2$; b), c) the arrangement of f- and a-bonds on the Potts lattice with $N = 3 \times 3$ and the impurity concentrations $p = 0$ and $p = 1$, respectively; (c, d) the arrangement of bonds between col- umns with even and odd numbers, respectively.

According to these constructions, the transfer matrices of the type

$T^{even}(j+1=2k)$, which corresponds to adding an even column $j+1=2k$ after an odd one $j=2k-1$ (we add the columns from the right-hand side), contain bonds between the spins $\sigma_{i,j=2k-1}$ and $\sigma_{i+1,j=2k}$. The transfer matrices of the type $T^{odd}(2j+1)$, which corresponds to adding an odd column $j+1=2k+1$ after an even one $j=2k$, contain bonds between the spins $\sigma_{i,j=2k}$ and $\sigma_{i-1,j+1=2k+1}$.

Let the contribution to a statistical weight corresponding to the f-bond $b(\sigma_{i,j},\sigma_{i',j'})=1$ be expressed as

$$t = \exp(\beta J)$$

where $\beta=\frac{1}{T}$. Then the a-impurity $b(\sigma_{i,j},\sigma_{i',j'})=-1$ corresponds to the weight

$$t^{-1} = \exp(-\beta J)$$

and the total contribution of a given configuration to the statistical weight of the system is $t^{b(\sigma_{i,j},\sigma_{i',j'})}$. Let us define a matrix for the horizontal bond $(\sigma_{i,j},\sigma_{i,j+1})$ as

$$P(i,j) = I(t^{b(\sigma_{i,j},\sigma_{i,j+1})}-1)+D(i,j)$$

and the matrices for "sloped" bonds as

$$\begin{aligned} R^{even}(i,j) &= I + C(i,j)(t^{b(\sigma_{i,j},\sigma_{i+1,j+1})}-1) \\ R^{odd}(i,j) &= I + C(i,j)(t^{b(\sigma_{i,j},\sigma_{i-1,j+1})}-1) \end{aligned}$$

where I is the unit matrix.

Then the transfer matrices can be expressed as follows:

$$\begin{aligned} T^{even}(j) &= P(L_v,j)R^{even}(L_v-1,j)P(L_v-1,j)\ldots R^{even}(1,j)P(1,j) \\ T^{odd}(j) &= P(1,j)R^{odd}(2,j)P(2,j)\ldots R^{odd}(L_v,j)P(L_v,j) \end{aligned}$$

and the partition function is

$$Z(t) = u^T \prod_{k=1}^{L_h/2-1} T^{even}(L_h)\left(T^{odd}(2k+1)T^{even}(2k)\right)v_{max}$$

where v_{max} is a basis vector with the maximum number corresponding in our representation to the state in which all spins belong to different clusters and $u^\top=\{q^{N_{CL}(v_1)},q^{N_{CL}(v_2)},...,q^{N_{CL}(v_{max})}\}$ so that $u^\top v_i = q^{N_{CL}(v_i)}$, where $N_{CL}(v_i)$ is the number of various clusters in the subdivision v_i. For the first basis vector, $N_{CL}(v_1)=1$ there is a single cluster, and for the last vector, $N_{CL}(v_{max})=L_v$, there are L_v different clusters.

The general algorithm of the calculation is as follows.

(1) Generate basis set vectors for a spin strip of the required width;
(2) Use this basis set to generate the matrices $R(i,j)$, $P(i,j)$ in cases of f- and a-bonds (σ_i, σ_j)
(3) Generate a distribution of impurities on the lattice using a random number generator;
(4) Generate the transfer matrices for the Potts spin model and the polynomial invariant;
(5) Calculate a partition function for the Potts model, the polynomial invariant, the minimum energy, and the degree of the polynomial.
(6) Repeat points 3–5 and perform averaging over various realizations of the distribution of impurities on the lattice.

For each realization of the impurity distribution, the program simultaneously determines both the polynomial invariant and the partition function of the Potts model for an arbitrary (but fixed) q value. This method allows us to study correlations between the degree of the polynomial and the minimum energy within the framework of the Potts model on the corresponding lattice.

1.3. *Results of the transfer matrix computations*

1.3.1. *Correlations between the degree of Jones polynomial of the lattice knot and the minimum energy of the Potts model*

Let us consider dependence of the degree of the Jones polynomial on the type of the partition function for the Potts model. Formula (19) explicitly relating the Jones-Kauffman invariant to the Potts model shows that, if the variable x were not entering into an expression for the degeneracy of the energy level $H(E; q = 2 + x + x^{-1})$, the degree of the polynomial invariant would always correspond to a term of the partition function with the minimum energy.

Taking into account dependence of the degree of degeneracy on the variable x, we can see that contributions to the coefficient at this power in some cases mutually cancel each other and this coefficient turns zero. A simple example is offered by a system free of impurities, in which case the minimum energy is $E_{min} = -(N+1)/2$ and the Jones polynomial is identically unity (with the degree being zero).

Nevertheless, there is a strong correlation between the degree n of the Jones polynomial and the minimum energy E_{min} of the corresponding Potts model. Since the minimum energy E_{min} of the Potts model is always sign-definite and cannot be positive, we use below a positive quantity representing the absolute value of the minimum energy $|E_{min}| = -E_{min}$. Figure 6a shows a joint probability distribution $P\left(\frac{n}{N}, \frac{|E_{min}|}{N}\right)$ of the normalized degree $\frac{n}{N}$ and the normalized minimum energy $\frac{|E_{min}|}{N}$ obtained for a lattice with $N = 49$ and an impurity concentration of $p = 0.5$ by averaging over $N_L = 10^5$ impurity configurations. Here and below we use only the normalized quantities determined on the $[0, 1]$ segment for the energy and $[-1, 1]$ for the degree of the polynomial (the latter value can be negative). This allows us to plot the curves for various lattice dimensions on the same figure. In Fig.6a, the probability distribution is described by the level curves, the spacing between which corresponds to a probability difference of 0.001. As can be seen, there is a strong correlation between the degree of the polynomial invariant and the minimum energy of the corresponding Potts model. This relationship can be quantitatively characterized by the the coefficient of correlation.

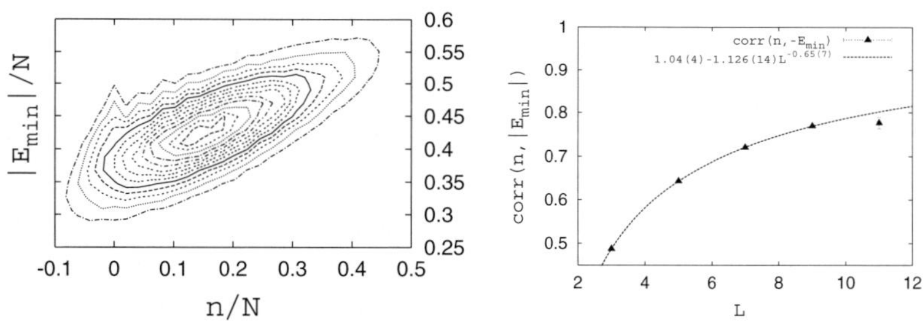

Fig. 6. a) A joint probability distribution $P\left(\frac{n}{N}, \frac{E_{min}}{N}\right)$ of the normalized maximum polynomial degree $\frac{n}{N}$ and the normalized minimum energy $\frac{E_{min}}{N}$ described by the level curves with a step of 0.001 for a lattice with $N = 49$ and an impurity concentration of $p = 0.5$; b) The coefficient of correlation between the degree of the polynomial invariant and the minimum energy as a function of the lattice size L; dashed curve shows the approximation of these data by a power function of L.

It should be recalled that the coefficient of correlation between random quantities x_1 and x_1 with mathematical expectations $\langle x_1 \rangle$ and $\langle x_2 \rangle$ and dispersions $\Delta x_1 = \langle x_1^2 \rangle - \langle x_1 \rangle^2$ and $\Delta x_2 = \langle x_2^2 \rangle - \langle x_2 \rangle^2$, (where $\langle \dots \rangle$ denotes averaging) is determined by the formula

$$\mathrm{corr}(x_1, x_2) = \frac{\langle x_1 x_2 \rangle - \langle x_1 \rangle \langle x_2 \rangle}{\sqrt{\Delta x_1 \Delta x_2}}$$

The correlation coefficient equal to ± 1 corresponds to a linear relationship between x_1 and x_2.

The values of the coefficient of correlation between the degree n_{max} of the polynomial invariant and the minimum energy $|E_{min}|$ for the impurity concentration $p = 0.5$ on lattices with various the linear dimensions $L = 3, 5, 7, 9$, and 11 are presented in Table 1. These values were obtained by averaging over $N_L = 10^5$ impurity realizations for $L = 3 \div 7$, $N_L = 2 \times 10^4$ for $L = 9$, and $N_L = 10^3$ for $L = 11$. As is seen, the correlation increases with the lattice size.

Table 1. Mean value of the correlation coefficient $\text{corr}\,(n_{max}, |E_{min}|)$ and the corresponding statistical error for several lattice sizes.

L	$\text{corr}\,(n, \|E_{min}\|)$	$\Delta\text{corr}\,(n, \|E_{min}\|)$
3	0.4871	0.0021
5	0.6435	0.0022
7	0.7205	0.0007
9	0.7692	0.0013
11	0.7767	0.0129

Figure 6b shows the approximation of these data by a power function $\text{corr}\left(\frac{n}{N}, \frac{|E_{min}|}{N}\right) = 1.04(4) - 1.126(14)L^{-0.65(7)}$. Obviously, the correlation coefficient cannot be greater than unity. The last expression is the result of ignoring the higher terms of expansion in powers of $\frac{1}{L}$. The approximation shows that there is a relationship between the degree of the polynomial invariant and the minimum energy of the Potts model, the degree of correlation increasing with the lattice size. In what follows, the results for the degree of the polynomial invariant will be accompanied by data for the minimum energy.

The presence of a correlation between the degree n of the Kauffman polynomial invariant and the minimum energy E_{min} of the corresponding Potts model for each particular impurity realization on the lattice is of interest from theoretical viewpoint and can be used in the numerical experiments. At present, the limiting lattice size N_L used in the method of transfer matrix does not exceed $N_L = 11$, which is determined by a large volume of necessary computations. At the same time, the minimum energy for the Potts model can be calculated within the framework of the standard Monte Carlo technique, which poses a less strong requirements to the computational facilities and, hence, can be readily applied to the lattices of significantly greater size.

1.3.2. *The probability distribution of the degree of the polynomial invariant and the minimum energy of the corresponding Potts model*

Here we present the results of determining the probability distribution $P(n)$ of the maximum polynomial degree n. Figure 7a shows the $P(n)$ curves for $L = 7$ and the impurity concentrations $p = 0.1$ (depicted by crosses), 0.2 (squares), and 0.5 (circles). The data were obtained by averaging over $N_L = 10^5$ impurity realizations. The statistical errors are smaller than the size of symbols.

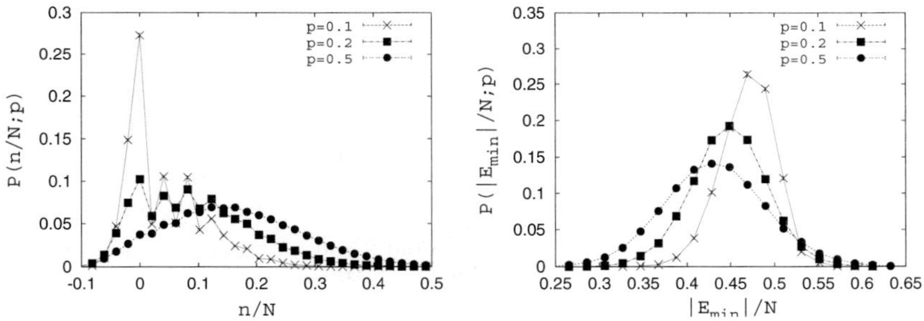

Fig. 7. a) The probability $P(\frac{n}{N};p)$ of the normalized maximum polynomial degree $\frac{n}{N}$ and b) The probability $P\left(\frac{|E_{min}|}{N};p\right)$ of the normalized minimum energy modulus E_{min} for a 7×7 lattice and various impurity concentrations $p = 0.1, 0.2$ and 0.5.

As can be seen, the $P\left(\frac{n}{N}\right)$ curve for small impurity concentrations is nonmonotonic. As the p value increases, the probability distribution function becomes monotonic and approaches in shape to the Gauss function. Figure 7b shows the corresponding probability distribution of the normalized minimum energy $\frac{|E_{min}|}{N}$ obtained for the same lattice size $(L = 7)$ and impurity concentrations (denoted by the same symbols). This function appears as more monotonic and approaches the Gauss function already at small impurity concentrations $(p \simeq 0.2)$.

The shapes of the probability distribution observed for a fixed impurity concentration $(p = 0.5)$ and various lattice dimensions are shown in 8a (for the maximum polynomial degree) and Fig.8b (for the minimum energy). Data for the lattice size $L = 3$ (squares), 5(*squares*), and 7 (circles), and 9 (triangles) were averaged over $N_L = 10^5$ $(L = 3, 5,$ and 7) and data for $L = 9$, over $N_L = 5 \times 10^3$ impurity realizations. The probability was normalized to unity: $\sum_n P(n) = 1$ For this reason, an increase in the lattice size is accompanied by growing number of the values that can be adopted by the normalized degree of $P(\frac{n}{N})$, while the value of the probability distribution decreases approximately as $\frac{1}{N}$. As is seen, the probability distribution $P\left(\frac{n}{N}\right)$ for lattices of smaller size ($L = 3$ and 5) is nonmonotonic; as the L value increases, the distribution becomes a smooth Gaussian-like function.

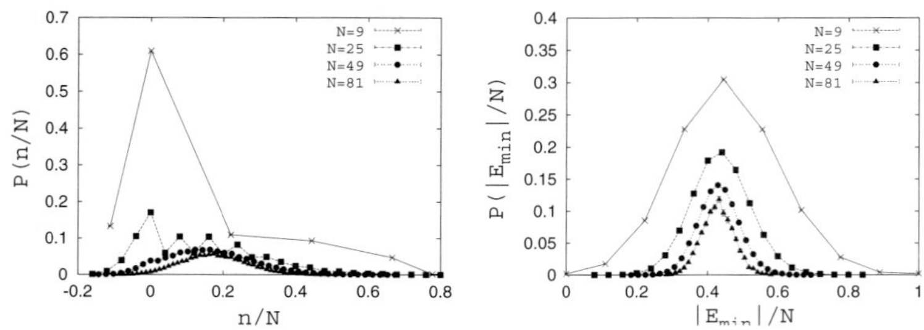

Fig. 8. a) The probability of the mean normalized degree $\left\langle \frac{n}{N} \right\rangle$ of the polynomial invariant and b) The probability of the normalized mean minimum energy modulus $\left\langle \frac{|E_{min}|}{N} \right\rangle$ for the square lattices with $L = \sqrt{N} = 3, 5, 7$, and 9 and an impurity concentration of $p = 0.5$.

Thus, we may conclude that the probability distribution $P\left(\frac{n}{N}\right)$ of the degree of the polynomial invariant for small-size lattices is nonmonotonic as a result of the boundary effects even for a considerable impurity concentration. As the lattice size increases, the nonmonotonic character disappears and the probability distribution becomes a smooth function. Apparently, we may ascertain that, whatever small is the impurity concentration p, there is a lattice size N ($N \gg \frac{1}{p}$) such that the corresponding probability distribution is smooth and Gaussian-like. Thus, we may suggest that the probability distribution $P(n)$ on the lattices of large size is determined by the Gauss function, the main parameters of which are the mathematical expectation (mean value) of the maximum polynomial degree and the dispersion. Plots of the mean value of the maximum polynomial degree $\left\langle \frac{n}{N} \right\rangle$ and the corresponding dispersion W^2_{knot} as functions of the impurity concentration p are presented in Figs. 9a and Fig.9, respectively, for the lattices size of $L = 3$ (crosses), 5 (squares), and 7 (circles). The data for each point were obtained by averaging over $N_L = 10^5$ impurity realizations for $L = 3$ and 5 and $N_L = 5 \times 10^4$ for $L = 7$.

As expected, both the mean value and the dispersion of the maximum polynomial power turn zero for $p = 0$ and 1 (trivial knot) and reach maxi-

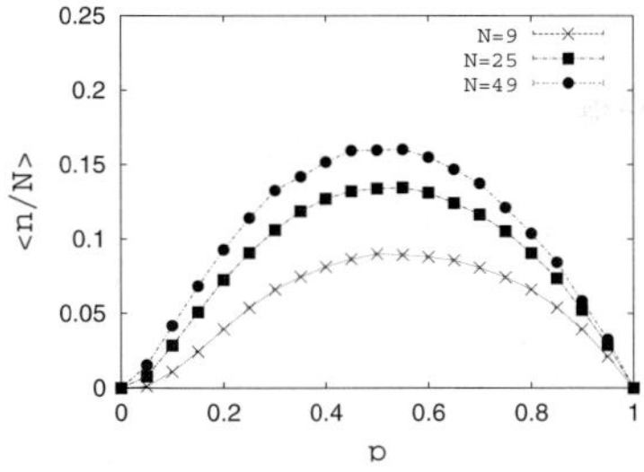

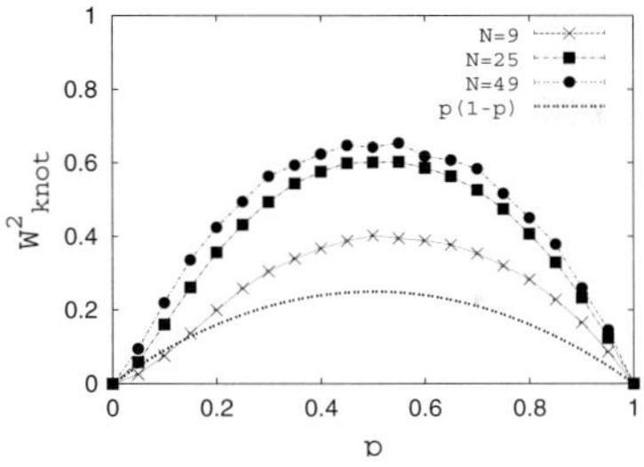

Fig. 9. Plots of a) the mean normalized maximum polynomial degree $\langle \frac{n}{N} \rangle$, and b) the dispersion $W^2_{knot}(p)$ of the $\frac{n}{N}$ value distribution versus the impurity concentration p in the range from 0 to 10.

mum at $p \simeq 0.5$. Note that these functions are not symmetric with respect to the transformation $p \to 1-p$, since the Jones polynomial invariant of a mirror knot (with all overcrossing changed for undercrossing and vice versa) is obtained by changing variables $x \to x^{-1}$, whereby the maximum polynomial power of the mirror knot corresponds to the minimum degree of the original polynomial, taken with the minus sign. However, this asymmetry disappears with increasing lattice size N as a result of increase in the amount of impurities and in the number of possible impurity realizations employed in the averaging. For the comparison, Fig.9b shows the function $\frac{1}{2}p(1-p)$ representing the dispersion of the distribution function in the hypothetical case when the maximum polynomial degree n is a linear function of the number of impurities M. Thus, a difference between W^2_{knot} and $\frac{1}{2}p(1-p)$ characterizes dispersion of the distribution of the maximum polynomial degree at a fixed number of impurities M (it should be recalled that the impurity occupies each lattice site with a probability p and the total number of impurities M fluctuates).

Figures 10a and 10b (with the parameters and notations analogous to those in Fig.9a and Fig.9b) shows data for the absolute value of the mean minimum energy $\left\langle \frac{|E_{min}|}{N} \right\rangle$ and the corresponding dispersion W^2_{Potts}.

The asymmetry of the mean minimum energy plot is related to the fact that the number of f–bonds is greater than that of the a–bonds by one at $p = 0$ and is smaller by one, at $p = 1$. The probability distribution of the minimum energy is treated in more detail in[32].

We have studied dependence of the mean normalized maximum polynomial degree on the lattice size L for $p = 0.5$ (Fig.8a). The results were averaged over $N_L = 10^5$ impurity realizations for $L = 3, 5$, and 7 and over $N_L = 1.5 \times 10^4$ realizations for $L = 9$. As can be seen, the $\langle \frac{n}{N} \rangle$ value tends

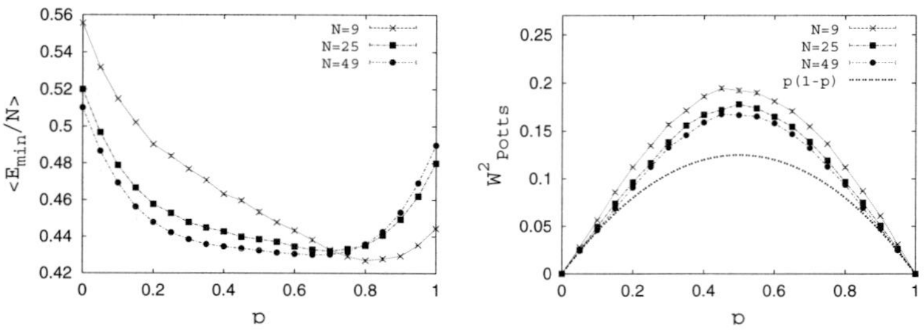

Fig. 10. Plots of a) the mean normalized minimum energy modulus $\left\langle \frac{|E_{min}|}{N} \right\rangle$ and b) the dispersion $W^2_{Potts}(p)$ of the $\frac{|E_{min}|}{N}$ value distribution versus the impurity concentration p in the range from 0 to 10.

to a certain limit with increasing $L \to \infty$. We approximated the results by a power function to obtain

$$\left\langle \frac{n}{N} \right\rangle \simeq 0.334(8) - 0.41(2)L^{-0.48(5)} \tag{21}$$

Analogous data for the normalized minimum energy in the Potts model are depicted in Fig.8b. The mean absolute value of the minimum energy decreases with increasing lattice size and can be approximated by a power function of the type

$$\left\langle \frac{|E_{min}|}{N} \right\rangle \simeq 0.4185(7) + 0.119(3)L^{-1.11(4)}$$

Some features of the minimum energy distribution in the Potts model with random ferro– and antiferromagnetic bonds are discussed in the Appendix for $q \geq 4$.

2. Physical Applications

In this part the methods described above are used to argue for the nontrivial fractal structure of an unknotted polymer "parent" ring in a compact state. We analyze the typical topological complexity of "quasi-knots," coinciding with some parts of unknotted "parent" ring up to closure of open tails. We show that the condition for the whole parent knot to be trivial implies that each quasi-knot is "almost" trivial (all definitions are given below).

It is known that the non-phantomness of polymer chains causes two types of interactions: a) volume interactions, vanishing for infinitely thin chains, and b) topological interactions, present even for chains of zero thickness. For sufficiently high temperatures, the polymer molecule is a strongly fluctuating system without a reliable thermodynamic state. However for temperatures below some critical value, θ, the macromolecule exhibits a dense drop-like weakly fluctuating globular structure.[23,24] In classical works[23,24] devoted to the investigation of the coil-to-globule phase transition without topological constraints, it has been shown that for $T < \theta$ the globular state can be described in the virial approximation by using only two- and three-body interaction constants: $B = b\frac{T-\theta}{\theta} < 0$ and $C = \text{const} > 0$. See Grosberg and Khokhlov[25] and Cloizeaux and Jannink.[26] The approach developed by Lifshitz[23,24] is regarded as the basis of the modern statistical theory of collapsed state of polymer systems.

In the globular phase of an unknotted macromolecule topological constraints play the role of an additional repulsion. One might suppose that the parts of the unknotted chain deeply penetrate each other by loops as is shown in Fig.11a. However, this is not the case. It has been argued, mainly in the framework of scaling theory, that the absence of knots in a densely packed polymer ring causes a very unusual fractal structure of the chain trajectory, strongly affecting all thermodynamic properties of the macromolecule in the globular phase.[27] The corresponding structure of a collapsed unknotted polymer ring was called a *crumpled* fractal globule. The chain trajectory in the crumpled globule densely fills the volume that all parts of the chain become segregated from each other in a broad region of scales (see Fig.11b). So, the line path in a crumpled globule resembles the well known Peano curve[28] schematically shown in Fig.11c.

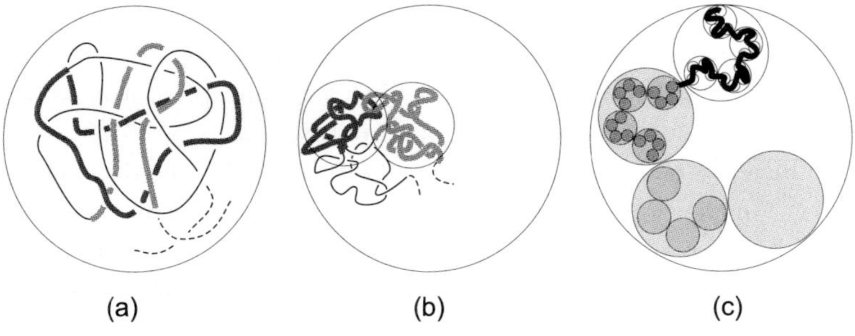

Fig. 11. a) The globular (collapsed) state of a polymer chain with an arbitrary loop penetrating the entire globule; b) The crumpled globular state with a test loop segregated "in itself." This structure resembles the Peano curve schematically shown in c).

Experimental examination of the fractal structure of an unknotted polymer ring is a difficult problem. Some measurements can be interpreted as an indirect verification of the crumpled globule structure: the two-stage dynamics of the collapse of a linear macromolecule after an abrupt change of the solvent quality,[29] and the effect of compatibility enhancement in a melt of linear and ring macromolecules.[30] Recently, new results of a numerical investigation of correlations of parts of unknotted selfavoiding rings completely filling the cubic lattice (the so-called Hamiltonian walks) of size up to $22 \times 22 \times 22$ show a tendency to the formation of a crumpled structure.[31]

It is shown below that the investigation of the distribution of random knots over the topological classes and the analysis of topological correlations in trivial knots helps one to understand the structure of the phase space of an unknotted polymer in the globular phase and to validate the crumpled globule conjecture "from first principles."

3. The model of densely packed knots on a lattice and the concept of "quasi-knots"

Below we present arguments in support of the crumpled globular structure based upon the direct determination of the topological state of a *part* of a long polymer chain under the condition that the chain, as a whole, forms an unknotted loop. In our work we use methods given by topological knot invariants,[5,10,19] which have been successfully applied to give solutions of specific statistical problems.[13,14,32]

First of all, one has to define the topological state of a part of a ring polymer chain. Of course, a mathematically rigorous definition of the topological state exists only for closed or infinite paths. Nevertheless, the everyday experience tells us that an open, but sufficiently long, rope can be knotted. Hence, it is desirable to introduce the concept of a *quasi-knot* to provide a topological description of open paths.

The idea of quasi-knot was first used by I.M. Lifshits and A.Yu. Grosberg[33] in the context of polymers. They suggested that the topological state of a linear polymer chain in collapsed (globular) state is defined much better than the topological state of a random loop. Actually, the distance between the ends of the chain in a globule is of order $R \sim aN^{1/3}$, where a is the size of a monomer and N is a number of monomers in a chain. Taking into account that R is significantly smaller than the contour length N and that the density fluctuations are negligible, we may define the topological state of a path in a globule as the topological state of the composition of the chain itself with an added segment connecting its ends. This composite structure can be regarded as a quasi-knot for an open chain in the collapsed state. The influence of boundary conditions (closure of open tails) on the topological state of a quasi-knot will be discussed later.

The role of topological constraints in polymer statistics, namely the random knotting probability in confined geometries, was numerically considered by Tesi et al;[34] their work includes a paper devoted to the determination of the equilibrium entanglement complexity of polymer chains in melts.[35] In the present work, we make a crucial further step, considering the topological state of a part of an unknotted ring polymer in a confined geometry, where the compact configuration of the chain is represented by a densely packed lattice knot.

Let us describe the model under consideration. The crumpled globule is modelled by a "densely packed" knot. We call a knot *densely packed* if its projection on the plane completely fills a rectangular lattice M of size $L_v \times L_h$ — see Fig.12.

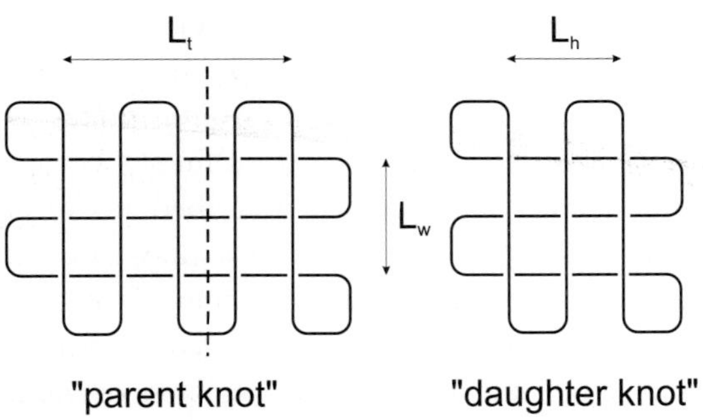

Fig. 12. The trivial knot 3×5 and the knot 3×3 obtained by cutting a knot 3×5 and closure of open tails.

The lattice M is entirely covered by a single string, which can cross itself at each vertex of the lattice in two different ways: "above" or "below." The topology of a knot is defined by these overpasses and underpasses and the prescribed boundary conditions. The "woven carpet" shown in Fig.12 corresponds to the trivial knot. Let us enumerate the vertices of the lattice by using the index k and attribute to each vertex the variable ϵ_k according to the rule:

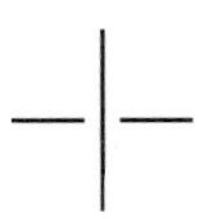 $\epsilon_k = -1;$ 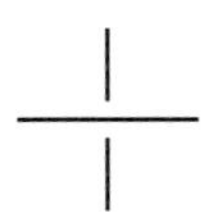 $\epsilon_k = +1$

Our task can now be rephrased as follows. We have an ensemble of randomly generated crossings $\{\epsilon_k\}$

$$\begin{cases} \epsilon_k = -1 \text{ with probability } p \\ \epsilon_k = +1 \text{ with probability } 1-p \end{cases} \tag{22}$$

created independently at each vertex of the lattice. The collection of crossings $\{\epsilon_k\}$ defines the *parent* knot. Let us suppose that the parent knot is trivial. For example, it is clear that if $\epsilon_k = +1$ for all $k = [1, N]$, then the parent knot is certainly trivial. By changing the sign of some ϵ_k from "+" to "−" we can create a nontrivial topological state. Later on we will call such a sign change an *impurity*. Hence, p in eq.(22) is the concentration of impurities.

Now we cut off a part of the trivial parent knot and close the open tails of the strings[a] as shown in Fig.12. This way we obtain well defined *daughter* knot. Below we study the typical topological state of such daughter knots under the condition that the parent knot is trivial.

To avoid any possible confusion, we apply our model to the polymer loop located in a thin slit between two horizontal plates as it is shown in Fig.13. It is evident that a closed chain in a thin slit becomes a quasi two-dimensional system. Our lattice model described above is oversimplified (even for polymer chains in a slit) because it does not take into account spatial fluctuations of the knotted polymer chain. However, we may expect that our model properly describes the condensed (globular) structure of a polymer loop. Such a belief is based on the weakness of chain fluctuations in a globule.[33] The weakness of fluctuations leads to a reliable thermodynamic structure of a globular polymer with constant density.

3.1. *Comparison of different definitions of the "knot complexity"*

In our work we characterize the topological state of knots by the degree of their Jones-Kauffman polynomial invariants.[5,10] Let us recall that the Kauffman invariant in the variable A is related to the Jones polynomial in the variable x by the equality $x = A^4$.[10] In brief, the construction of algebraic knot invariant is as follows. Let us take the "Potts lattice" Λ dual to the "knot lattice" M — see Fig.14b.

[a]We always cut along an even vertical line (cutting along an odd one may yield a link, rather than a knot.

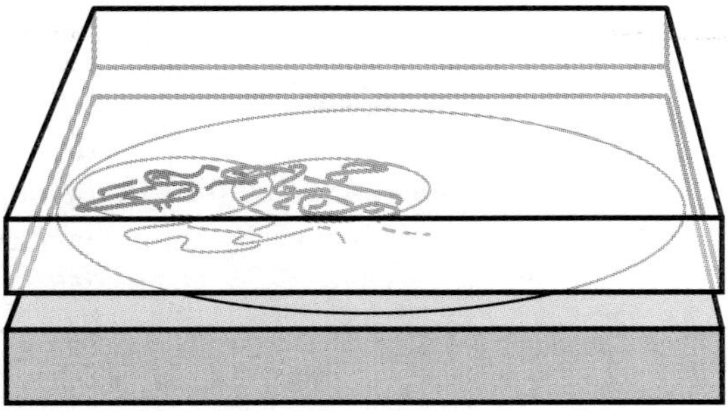

Fig. 13. The condensed polymer ring in a thin slit.

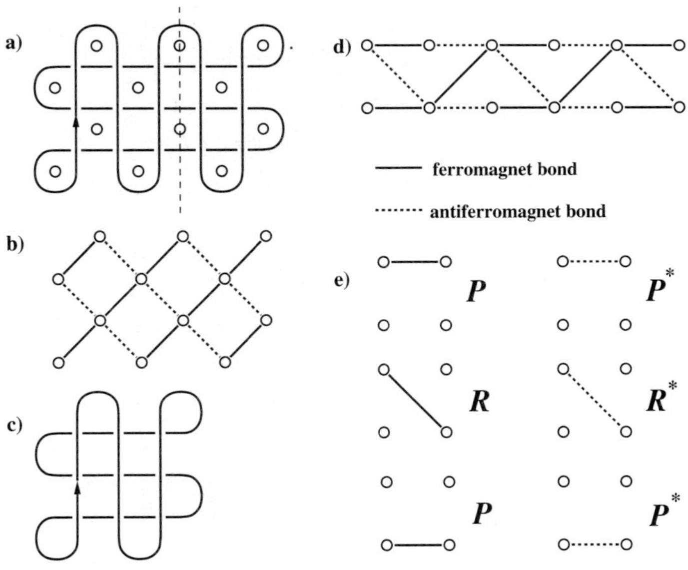

Fig. 14. a) The trivial knot 3 × 5; b) The lattice of Potts spins, corresponding to the trivial knot 3 × 5; c) The same lattice of Potts spins transformed to the 2 × 6—strip; d) The transfer matrix, corresponding to ferro- and antiferro-magnetic bonds on the Potts lattice.

In Fig.14a positions of Potts spins are depicted by circles. We redefine the set of variables b_{ij} on the bonds of the Potts lattice (23) as follows:

$$b_{i,j} = \begin{cases} -\epsilon_k, & \text{for right-down bond } (ij) \\ \epsilon_k, & \text{for right-up bond } (ij) \end{cases} \tag{23}$$

Let us also introduce the auxiliary variables c_k according to the following rule

 $c_k = +1;$ 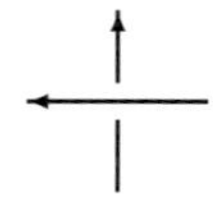 $c_k = -1$

Once the variables c_k are defined and the direction along the thread is chosen, we fix uniquely the knot "twisting" by the following equation

$$Tw = \sum_{k=1}^{N} c_k \tag{24}$$

In Fig.14 the bonds having the variables $b_{ij} = 1$ and $b_{ij} = -1$ are shown by solid and dashed lines respectively.

It was explicitly shown in the first part of the paper and elsewhere[13,32] that the Jones knot invariant $f_K(x)$ of ambiently isotopic knots on the lattice M can be expressed as the partition function of the Potts model on the dual lattice Λ:

$$f_K\Big(x, \{b_{ij}\}\Big) = K\Big(x, \{b_{ij}\}\Big)\, Z\Big(q(x), \{J_{ij}(b_{ij}, x)\}\Big) \tag{25}$$

where

$$K\Big(x, \{b_{ij}\}\Big) = \left(-\sqrt{x} - \frac{1}{\sqrt{x}}\right)^{-(N_p+1)} \left(-x^{1/4}\right)^{-\sum\limits_{\{ij\}} b_{ij} + 3\sum\limits_{\{k\}} c_k} \tag{26}$$

is a coefficient, which does not depend of the Potts spins and $N_p = \frac{1}{2}(L_v + 1)(L_h + 1)$ is the total number of Potts spins. The Potts partition function $Z\Big(q(x), \{J_{ij}(b_{ij}, x)\}\Big)$ appearing in (25) reads:

$$Z\Big(q(x), \{J_{ij}(b_{ij}, x)\}\Big) = \sum_{\{\sigma\}} \exp\left(\sum_{\{ij\}} \frac{J_{ij}(b_{ij}, x)}{T} \delta(\sigma_i, \sigma_j)\right) \tag{27}$$

where the coupling constant J_{ij} and the number of spin states q are related to the variable x as follows:

$$-x^{b_{ij}} = e^{\frac{J_{ij}}{T}}, \quad q = x + 2 + x^{-1} \tag{28}$$

It is worth noting that there is a relation between the temperature T and the number of Potts spins states, q, in the partition function (27). Thus, the variables T and q cannot be considered as independent quantities. Later on we shall be interested in the degree of the Jones polynomial and the physical meaning of the parameters q, T, and J_{ij} is not important for us.

Let us pay attention to the fact that each sign change of the variable ϵ_k (and hence, of b_{ij}) causes a sign change of the variable c_k. Hence, the resulting polynomial contains only integer degrees of x. Later on we write

$b_{ij} = 1$ and $b_{ij} = -1$ for "ferromagnetic" and "antiferromagnetic" bonds. In Fig.14b they are shown by solid and dashed lines respectively. Let us stress that even for the trivial knot there is a set of antiferromagnetic bonds on the Potts lattice, see Fig.14a and Fig.14b,c. Let us recall that in our terminology "impurity" stands for the *sign change* of a bond to the trivial knot, but not for antiferromagnetic bond itself. The expression (26) leads to some simplifications useful in numerical computations. As it was mentioned already, $b_{ij} = c_k$ (where the vertex k belongs to the bond (ij)) for the trivial knot. So, the signs "$+$" and "$-$" of c_k form a chessboard. The creation of an "impurity" causes simultaneous sign changes in b_{ij} and c_k. Hence the identity $b_{ij} = c_k$ holds for any knot configuration with the prescribed direction of strings. Taking into account that

$$-\sum_{\{ij\}} b_{ij} + 3\sum_{\{k\}} c_k = 2\sum_{\{ij\}} b_{ij}$$

we may rewrite (26) as follows

$$K\Big(x, \{b_{ij}\}\Big) = \left(-\sqrt{x} - \frac{1}{\sqrt{x}}\right)^{-(N_p+1)} (\sqrt{x})^{\sum\limits_{\{ij\}} b_{ij}} \tag{29}$$

As it has been pointed out in the first part of the paper, each topological type of a knot is characterized by a polynomial invariant $f_K(x)$, however it is sufficient and much more convenient for our particular goals to discriminate knots by more rough characteristics – by the degree n of the polynomial $f_K(x)$: $n = \lim_{|x|\to\infty} \frac{\ln f_K(x)}{\ln x}$. For trivial knots $n = 0$; as knots become more complex, n grows up to its maximal value $n \sim N$. Such a definition of knot complexity was proposed for the first time in a paper from 1992.[14]

Certainly, knot complexity n is rather rough topological characteristic which does not allow to distinguish "topologically similar" knots. However, the discrimination of topological states of knots (or links) by the degree n of the corresponding polynomial invariant enables us to introduce a "metric" in the phase space of topological states and, hence, to compare knots.

There are many definitions of knot complexity. Some authors use the concept of *minimal number of crossings*.[36,37,38,39,40,41] In other works,[38,42,43] knot complexity is associated with a properly normalized logarithm of a kind of knot torsion, $\log|\Delta_K(-1)|$, where $\Delta_K(t)$ is the Alexander polynomial of the knot K. Another approach deals with the fashionable concept of *knot inflation*.[44] The topological invariant considered in this book is defined as the quotient μ of the contour length of the knot made from an elastic tube to its diameter in the maximal uniformly inflated configuration. Such an approach was introduced and exploited by Grosberg et al.[44,45] An estimate of knot complexity using the *knot energy* has been discussed in classical works.[46,47]

Knot invariants like the minimal number of crossings as well those built on the basis of the knot inflation concept, are close to the invariants defined as the degree of the polynomial used in our works.[14,32] Despite the fact that these invariants may differ in the recognition of particular knots, they lead to the same statistical results. This belief has the following grounds. All invariants such as the *minimal number* of crossings, the *quotient* μ of the maximal inflated knot configuration, and the *degree* of the algebraic knot polynomial have one common ancestor – the so-called *primitive path.* The concept of primitive path appeared in the physical literature in the 1970ies in the works on entanglements in polymer systems. Introduced by P. de Gennes,[48] the primitive path was successfully used for an investigation of topological effects in the dynamics of individual chains in concentrated polymer solutions. Later on, the same concept was successfully used in the computation of equilibrium properties of polymer chains of specific topology in lattices of topological obstacles.[b,49,50]

The notion of primitive path might be elucidated as follows. Consider a closed path of fixed length entangled with the lattice of obstacles (see Fig.15a). Performing an affine extension (inflation) of the lattice of obstacles (preserving the length of the path), one arrives at the unfolded "fully stretched configuration," see fig.15c. Just this configuration is called the *primitive path* and it characterizes the topological state of a path with respect to the lattice of obstacles. Let us associate the properly normalized length of the chain and the spacing between obstacles, with the length of an elastic tube and its diameter. In that way, the relationship between the primitive path, the minimal number of crossings and the quotient μ in the maximal inflated tube configuration becomes clear by construction. In more details this relationship was discussed in Grosberg et al.[45] In Fig.16 we schematically show the connections between different definitions of knot complexity known in mathematical and physical literature.

In 1991 it was recognized that the concept of primitive path has a straightforward interpretation in terms of a geodesic line in a space of constant negative curvature.[51] In the forthcoming section, we show how the geodesic length, in turn, may be related via its matrix representation to the degree of the polynomial invariant. Despite the fact that our con-

[b]One can imagine the obstacles as small vertical rods (at the nodes of the lattice) through which the polymer cannot pass.

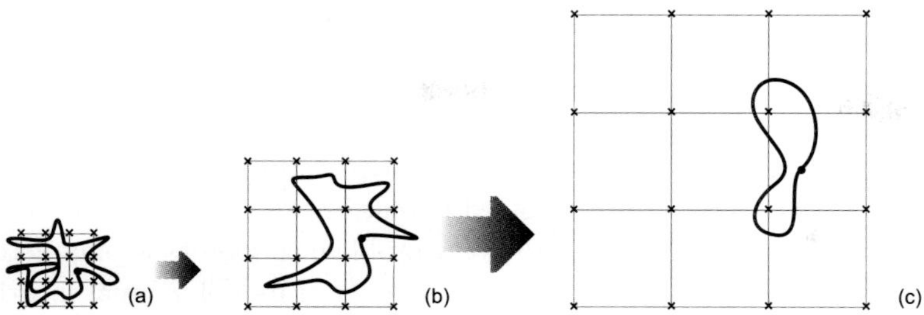

Fig. 15. Affine inflation of a lattice of obstacles (the length of an entangled path and its topology with respect to the lattice are fixed).

struction is restricted to the particular case of knots on narrow strips, this idea can be used to attack more general models. The relationship between primitive paths and the maximal degree of the corresponding polynomial invariants completes the set of relations between various definitions of knot complexities. The discussed connections are shown in Fig.16.

Let us note that the relationship of other knot complexities, such as $\log|\Delta_K(-1)|$ or the knot energy, with the concept of the primitive path are not clear to us at present.

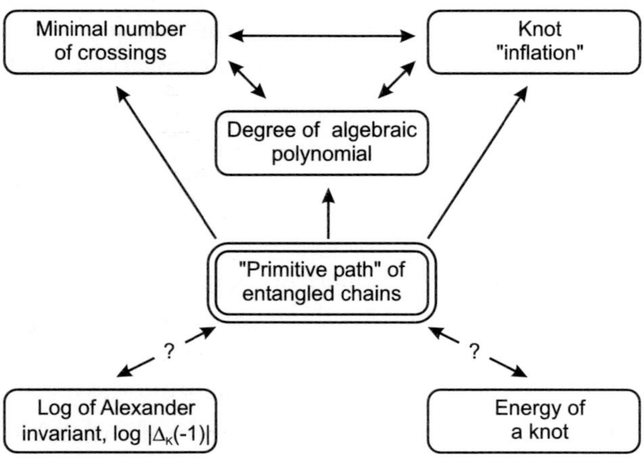

Fig. 16. Various definitions of the "knot complexity."

We believe that our choice of knot complexity as the degree of the polynomial invariant leads to a purely analytical description. That seems to be the single advantage of our definition as compared to the other ones, presently accessible only in numerical simulations.

3.2. *Topological correlations of (quasi)knots on a strip*

In this section we investigate by numerical and analytical methods the typical topological state of quasi-knots under the condition that the parent knot is trivial. It might be useful to repeat, that by definition, quasi-knot coincides with a part of an unknotted parent loop after the closure of open tails. In fact, we slightly modify the very problem as soon as we characterize the topological state of a knot by the degree n of its polynomial invariant. Namely, we study the topological correlations in all knots characterized by zero degree of the knot polynomial.

In our numerical investigations, we restrict ourselves to the topological correlations in knots on strips of sizes $L_v \times L_h$, where L_v ($L_v = 3,\ 5$) and L_h are the width and the length of the strip. In most cases we consider an essentially asymmetric distribution of crossings, namely with probabilities $p = 0.05$ and $p = 0.1$ to increase the fraction of trivial knots (let us recall that for $p = 0$ and $p = 1$ the corresponding configurations of crossings represent the trivial knot).

3.2.1. *Unconditional distributions*

Here we study numerically the typical knot complexity n and the (unconditional) probability for a random knot to have the complexity $n = 0$ on strips $L_h \gg L_v$. According to (20), the Jones-Kauffman invariant can be rewritten as the Potts partition function normalized by the coefficient $K\Big(x, \{b_{ij}\}\Big)$ (see Eq.(15)) independent of the configuration of the Potts spins. Hence, the degree n of the Jones-Kauffman invariant is proportional to the free energy of the Potts spin system.

Mean knot complexity. Our numerical computations show that for long strips ($L_h \gg L_v$) one can neglect the boundary effects produced by shorter side L_v. In this limit we get $n \sim L_h$. At the same time, for short strips $L_h \sim L_v$ the boundary effects cause a deviation from the linear behavior of n as a function of L_h. The corresponding data for strips of widths $L_v = 3$ and $L_v = 5$ are shown in Table 2. We have averaged the results over 30 000 realizations of topological disorder for each L_h. To estimate the numerical error, we have split the ensemble of realizations onto 10 subsets of 3000 configurations each.

Unconditional probability of a trivial knot. The dependence of the unconditional trivial knot formation probability $P(L_h)$ on L_h on the strips of fixed widths $L_v = 3, 5$ for the concentration of impurities $p = 0.05$ and

Table 2.

$L_v = 3$	
p	$n(L_h)$
0.05	$0.0697 \times L_h - 0.483$
0.1	$0.138 \times L_h - 0.51$

$L_v = 5$	
p	$n(L_h)$
0.05	$0.159 \times L_h - 1.053$
0.1	$0.31 \times L_h - 1.094$

$p = 0.1$ are shown in Table 3. The averaging is performed over 10×3000 samples. We see that the probability of the formation of a trivial knot exponentially decays with the strip length L_h.

Table 3.

$L_v = 3$	
p	$P(L_h)$
0.05	$1.049 \exp(-0.037 \times L_h)$
0.1	$0.957 \exp(-0.067 \times L_h)$

$L_v = 5$	
p	$P(L_h)$
0.05	$1.156 \exp(-0.079 \times L_h)$
0.1	$1.231 \exp(-0.157 \times L_h)$

In Fig.17 the unconditional probability $P(p)$ of the formation of a trivial knot is plotted as a function of concentration of impurities p on the strips $L_v \times L_h$ for $L_v = 3$ and $L_h = 3, 5, 7$. The averaging is performed over 10×10000 samples. We can see from Fig.17 that the probability of the formation of a trivial knot does not depend on p, but only on L_h for intermediate values of $p \sim 0.5$ and large $L_v \geq 9$.

3.2.2. *Conditional distributions ("Brownian Bridges")*

Now we turn to the investigation of topological properties of quasi-knots. Consider a trivial densely packed parent knot, situated in a strip of width L_v and length L_t. Cut a part of the strip of length L_h ($0 < L_h < L_t$) and close the remaining open tails. In this way we get a new (daughter) knot of the same width L_v but of shorter length L_h. In Fig.12 the parent and a daughter knots are shown. All lattice dimensions L_v, L_t, L_h are taken to be odd.

The main attention is paid to the following question. How is the triviality of the parent knot $L_v \times L_t$ reflected in the topological properties of the daughter (quasi)knot $L_v \times L_h$? We consider two cases: $L_h = \frac{1}{3} L_t$, and $L_h = \frac{1}{2}(L_t - 1)$. We determine the conditional probability for the daughter

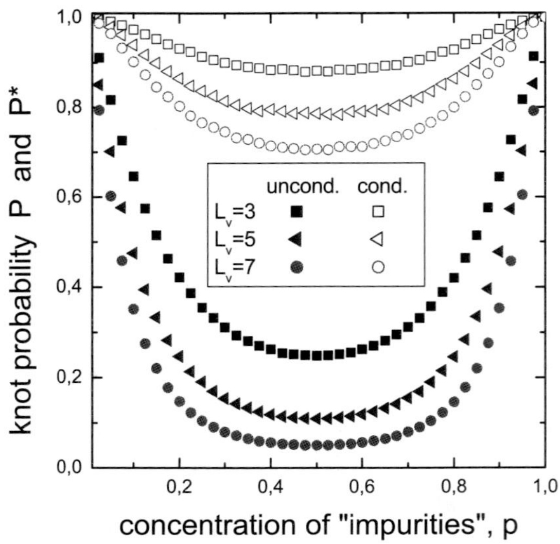

Fig. 17. Probability of a trivial knot P and conditional probability of trivial daughter (quasi)knot P^* on a strip $L_v \times L_h$ for $L_v = 3$ and $L_v = 3, 5, 7$ as function of impurity concentration p.

(quasi)knot to be trivial and find its mean complexity under the condition that the parent knot is trivial by definition.

The problem under consideration is typical for the theory of Markov chains, where the conditional probability, i.e., the so called *Brownian Bridge* (BB) has been repeatedly studied. The investigation of statistics of BB supposes, first of all, the determination of the probability $P(\mathbf{x}, t|\mathbf{0}, T)$ for a random walk to start at the point $\mathbf{x} = \mathbf{0}$, to visit the point $\mathbf{x}$ at some intermediate moment $0 < t < T$, and to return to the initial point $\mathbf{x} = \mathbf{0}$ at the moment T. The same question has been investigated for BB on the graphs of free groups, on Riemann surfaces[52,53] and for products of random matrices of groups $PSL(2, \mathbb{R})$ and $PSL(2, \mathbb{Z})$.[51,54]

It is easy to understand that our topological problem allows natural interpretation in terms of BB. As it has already been mentioned,[13,32] the degree n of the Jones-Kauffman polynomial defines a scale in the space of topological states of knots. That allows us to compare knots and to talk about their respective complexity or simplicity. Consider the phase space Ω of all topological states of densely packed knots on the lattice. Select from Ω

the subset $\Omega_{n=0}$ of knots with $n = 0$. Cut a part (say, one half, or one third) of each knot in the subset $\Omega_{n=0}$, close the open tails and investigate the topological properties of the resulting knots. Just such situation has been qualitatively studied and the *crumpled globule* (CG) concept was formulated mainly on the basis of heuristic scaling arguments.[27] The CG-hypothesis states the following: if the whole densely packed lattice knot is trivial, then the topological state of each of its daughter knot is "almost" trivial. Further on we formulate this statement rigorously and confirm it numerically.

Mean complexity of a daughter knots. We consider strips of width $L_v = 3$. The concentration of impurities is set to $p = 0.05$. The averaging is performed over 3×10^4 preselected realizations of trivial parent knots. The mean complexity $n^*(L_h)$ of daughter knots on strips $L_v \times L_h$ $(L_v = 3)$ under the condition that the parent knot on the strips $L_t = 3L_h$ (triangles) and $L_t = 2L_h + 1$ (crosses) are trivial, are shown on Fig.18b. We have compared the complexity of daughter (D) knots (triangles, crosses) with the unconditional (U) complexity of random knots (squares) on the same strips. It is easy to see, that the complexity of D-knots is sufficiently smaller than the complexity of U-knots. The functional dependence of n^* on L_h for D-knots shall be discussed later.
Probability of the daughter knot to be trivial. Here we investigate the probability $P^*(p)$ to create a trivial daughter knot on a strip $L_v \times L_h$ under the condition that it is a part of a trivial parent knot on the strip $L_v \times L_t$, where $L_t = 2L_h + 1$. Results for $L_v = 3$ and $L_h = 3, 5, 7$ are shown in Fig.17 by crosses, squares and triangles respectively. As expected, the probability of a daughter knot to be trivial becomes substantially higher if we impose the condition that this daughter knot is a part of a trivial knot.

3.2.3. *Matrix representation of Kauffman knot invariants on the strips*

We are interested in the functional dependence $n(L_h)$ of the daughter knot for a fixed strip width L_v. The direct investigation of this question involves some technical difficulties. To avoid them, we use some properties of products of random matrices. For a knot on the strip of width $L_v = 3$ we

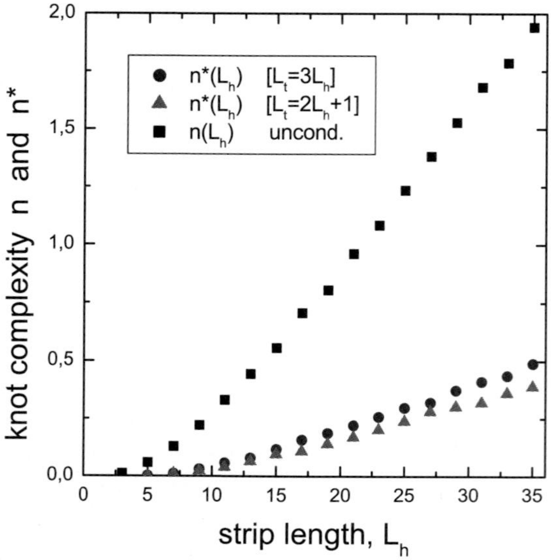

Fig. 18. Complexity $n^*(L_h)$ of a daughter (quasi)knots on a strip of width $L_v = 3$ and length L_h for concentration of impurities $p = 0.05$ under condition that the parent knot on strips of lengths $L_t = 3L_h$ (circles) and $L_t = 2L_h + 1$ (triangles) is trivial. For comparison, the complexity of random unconditional knots $n(L_h)$ on a strip of length $\times L_h$ is plotted by squares.

introduce the matrices

$$\begin{aligned} P &= \tfrac{1+x}{\sqrt{x}} \begin{pmatrix} -\sqrt{x} & 0 \\ \frac{\sqrt{x}}{1+x} & \frac{1}{\sqrt{x}} \end{pmatrix}; \quad R = \sqrt{x} \begin{pmatrix} -\sqrt{x} & -\frac{x+1}{\sqrt{x}} \\ 0 & \frac{1}{\sqrt{x}} \end{pmatrix} \\ P^* &= \tfrac{1+x}{\sqrt{x}} \begin{pmatrix} -\frac{1}{\sqrt{x}} & 0 \\ \frac{\sqrt{x}}{1+x} & \sqrt{x} \end{pmatrix}; \; R^* = \sqrt{x} \begin{pmatrix} -\frac{1}{\sqrt{x}} & -\frac{\sqrt{x}}{x+1} \\ 0 & \sqrt{x} \end{pmatrix} \end{aligned} \tag{30}$$

corresponding to ferro- and antiferromagnetic bonds as shown in Fig.14d.

The polynomial knot invariants can be constructed in terms of the product of these matrices. For example, the resulting transfer matrix T of the knot shown in Fig.14a and corresponding to the bond arrangement in Fig.14b,c is:

$$M \equiv \begin{pmatrix} M_{11} & M_{12} \\ M_{21} & M_{22} \end{pmatrix} = PR^*PP^*RP^*PR^*PPR^*PP^*RP^* \tag{31}$$

In the particular case of the 3-string braid, the Jones-Kauffman knot polynomial can be written as follows:[32]

$$f_K(x) = (-1-x)^{-(N_p+1)} \left(\sqrt{x}\right)^{N_p+N_b+1} \times \\ \Big((x+x^{-1}+2)M_{12} + (x+x^{-1}+2)^2 M_{22}\Big) \tag{32}$$

where $N_p = \frac{1}{2}(L_v+1)(L_h+1)$ is the number of Potts spin components; $N_b = \sum_{ij} b_{ij}$ is the sum over all b_{ij} and T_{12} and T_{22} are the elements of the last column of the transfer matrix T.

Let us recall that the expression of the knot invariant contains the coefficient $d = (-1-x)^{-(N_p+1)} \left(\sqrt{x}\right)^{N_p+N_b+1}$ (see (32)). So it is convenient to add $\log d$ to the degree n_λ and introduce the modified value $\tilde{n}_\lambda = n_\lambda - \frac{1}{2}(N_p+1-N_b)$. For a strip of width $L_v = 3$ and concentration of impurities $p = 0.05$, the data for $\tilde{n}_\lambda$ in the interval $L_h \in [20, 60]$ is well approximated by the linear function

$$\tilde{n}_\lambda(L_h) = -1.16(2) + 0.074 \times L_h \tag{33}$$

This result confirms the linear dependence of the knot complexity on L_h.

To extract the information about the dependence $n^*(L_h)$ for daughter knots on the 3-string strip in is useful to split the problem into two parts.

1. Relationship between the conditional knot complexity n^* and the conditional logarithm of the modified highest eigenvalue $\tilde{n}^*_\lambda$ of the transfer matrix M. Define $\tilde{n}^*_\lambda$, the mean logarithm of the modified highest eigenvalue of the product of the first L_h matrices of the transfer matrix T under the condition that the whole transfer matrix T has its modified highest eigenvalue equal to zero. This construction is schematically shown below:

$$\tilde{n}_\lambda\Big(M\Big) \equiv \tilde{n}_\lambda\Big(\underbrace{\overbrace{PR^*PP^*RP^*PR^*PP}^{\tilde{n}^*_\lambda(L_h)=?} R^*PP^*R \ldots R^*PP^*R}_{L_t}\Big) = 0$$

In Fig.19b we analyze the correlation coefficient $\mathrm{corr}(n^*, \tilde{n}^*_\lambda)$ of the knot complexity n^* and the conditional logarithm of the modified highest eigenvalue $\tilde{n}^*_\lambda$

$$\mathrm{corr}(n^*, \tilde{n}^*_\lambda) = \frac{\langle n^*\, \tilde{n}^*_\lambda\rangle - \langle n^*\rangle \langle \tilde{n}^*_\lambda\rangle}{\sqrt{(\langle (n^*)^2\rangle - \langle n^*\rangle^2)(\langle (\tilde{n}^*_\lambda)^2\rangle - \langle \tilde{n}^*_\lambda\rangle^2)}} \tag{34}$$

The difference $1 - \mathrm{corr}(n^*, \tilde{n}^*_\lambda)$ is plotted in Fig.19a (the function $1.57\, L^{-0.4}$ is added for comparison). We see, that the correlation between $\tilde{n}^*_\lambda$ and n^* approaches 1 as L_h tends to infinity. That behavior allows to conclude that the functional dependencies $n^*(L_h)$ and $\tilde{n}^*_\lambda(L_h)$ are identical.

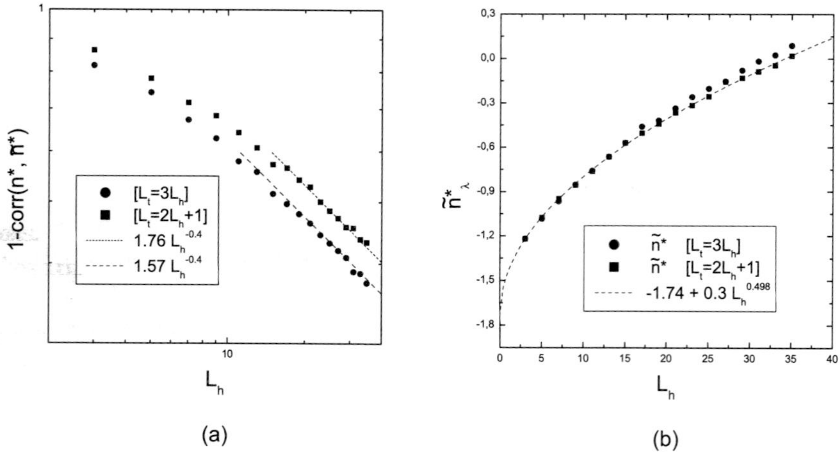

Fig. 19. a) The correlation function $\mathrm{corr}(n, \tilde{n}_\lambda)$ as a function of L_h for fixed $L_v = 3$ and $L_t = 3L_h$ (circles) and $L_t = 2L_h + 1$ (squares); b) Conditional mean logarithm of the modified highest eigenvalue $\tilde{n}^*_\lambda$ of the highest eigenvalue of the product of transfer matrices for D-knots $L_t = 3L_h$ (circles) and $L_t = 2L_h + 1$ (squares).

2. Functional dependence $\tilde{n}_\lambda(L_h)$ of the conditional mean logarithm of the modified highest eigenvalue on the strip of length L_h. Once the correlations between n and $\tilde{n}_\lambda$ are established, we may work with the value $\tilde{n}_\lambda$ only and analyze the conditional functional dependence $\tilde{n}^*_\lambda(L_h)$. The reason for this auxiliary construction is the fact that the numerical investigation of the dependence $\tilde{n}^*_\lambda(L_h)$ is a much simpler numerical task than the direct analysis of the dependence $n^*(L_h)$.

Computing numerically the conditional mean logarithm of the modified highest eigenvalue $\tilde{n}^*_\lambda(L_h)$ for $L_t = 3L_h$ and $L_t = 2L_h + 1$ we obtain the results shown in Fig.19b. As expected (see the discussions in the conclusion), the obtained data are well approximated by the square-root function shown in Fig.19b by the dotted line:

$$\tilde{n}^*_\lambda(L_h) = -1.74 + 0.30 \times L_h^{0.498} \tag{35}$$

Thus we have strong arguments in support of the conjecture that the average knot complexity n^* of a daughter knot on a N-vertex lattice (under the condition that the parent knot is trivial), has the following functional dependence

$$n^* \sim \sqrt{N}$$

A discussion supporting this conjecture is presented at the end of next section.

4. General conclusions

1. An analysis of the relationship between the partition function of the Potts model and the Jones-Kauffman polynomial invariant, in combination with the results of numerical calculations for the Potts model, allowed us to study the probability distribution $P(\frac{n}{N}; p, N)$ of the degree n for various concentrations (probabilities) p of an "impurity" representing intersections of the $\epsilon = -1$ type in the sites of a square lattice with $N = L \times L$ for $L = 3 - 11$.

2. For the lattices of small size with small impurity concentrations p, the probability distribution $P(\frac{n}{N}; p, N)$ of the normalized maximum polynomial invariant degree n is not smooth, showing an alternation of more and less probable states. This behavior indicates that the knots of certain topological types are difficult to realize for $p \ll 1$ on a square lattice as a result of geometric limitations.

3. As the impurity concentration (probability) p increases, the probability distribution $P(\frac{n}{N}; p, N)$ of the normalized maximum polynomial invariant degree n becomes smooth, with the shape approaching that of the Gauss function. A typical value of the "knot complexity" η for $p = 0.5$ and sufficiently large lattices ($N \gg 1$) can be obtained by extrapolating the expression (21):

$$\eta = \lim_{N\to\infty} \left\langle \frac{n}{N} \right\rangle \approx 0.334$$

4. There is a correlation between the degree of the polynomial invariant of a knot and the minimum energy of the corresponding Potts lattice model for $q \geq 4$.

5. An analytical expression was obtained for the probability distribution function $P(\frac{E_{min}}{N}; p, N)$ at relatively small $p \sim \frac{1}{N}$ impurity concentrations.

These results have been obtained for square lattice diagram. However, all computations can be straightforwardly generalized to the case of rectangular lattices. In particular, it would be of interest to check whether the distribution of knots over the topological types depends only on the number of intersections on the lattice diagram, or it depends on the diagram shape as well.

We believe that the proposed combination of analytical and numerical methods for the investigation of topological problems using the models of statistical physics offers both a promising means of solving such topological problems and a new approach to the standard methods of investigation of disordered systems. This can be illustrated by the following fact. One of the main concepts in the statistical physics is the principle of additivity of the free energy of a system, that is, proportionality of the free energy to the system volume N. By interpreting the free energy as a topological characteristic of the "complexity" of a knot, we may conclude that the complexity of a typical knot increases linearly with the system volume N. This property is, in turn, a fundamental manifestation of the non-Abelian (non-commutative) character of the phase space of knots.

6. The correlation between n and $\tilde{n}_\lambda$ established in the previous section permits us to consider the distribution of Lyapunov exponents for products of transfer matrices instead of the distribution of degrees of a polynomial invariant. By definition, the Lyapunov exponent δ of the matrix T is the logarithm of the highest eigenvalue of T.

Qualitatively, the behavior (35) can be understood by considering the limiting distribution of Lyapunov exponents of the first m matrices in the product of N random unimodular 2×2 matrices under the condition that the product (as a whole) is equal to the unit matrix. In an explicit analysis of the problem,[51] the distribution of the Lyapunov exponent δ of the first $m = cN$ $(0 < c < 1)$ matrices in the product of N non-commuting $PSL(2, \mathbb{R})$-matrices whose elements are independently distributed in some finite interval under the Brownian Bridge condition (i.e., under the condition that the whole product is equal to the unit matrix) was studied. For $N \gg 1$ the exponent δ has the following asymptotic behavior:

$$\delta \sim \sqrt{N} \tag{36}$$

(compare to (35)). Without the Brownian Bridge condition (i.e., for "open" chains of matrices) the standard Fürstenberg result[55] was recovered:

$$\delta \sim N. \tag{37}$$

In order to clarify the behavior (36) we mainly restrict ourselves to qualitative considerations. The Brownian Bridge problem for Markov chain of N identically distributed noncommutative random matrices may be reformulated in terms of a random walk in the space of constant negative curvature. Now it is well known[52,53,54] that the BB condition "kills" the influence of the curvature. The limiting distribution turns out to be Gaussian with zero mean. This fact can be illustrated by the following simple computation.

Consider a random walk in the space of the constant negative curvature $\gamma = -1$ (the hyperbolic space) with the metric $ds^2 = d\mu^2 + \sinh^2 \mu \, d\Phi^2$, where $d\Phi^2$ is the square of the increment in the space of angles. The probability of a path to start from the point $\mu = 0$ and to end at a time moment t in a *particular* point located at distance μ from the origin in the Lobachevsky space is well known:

$$P(\mu, t) = \frac{e^{-t}}{8\sqrt{(\pi t)^3}} \frac{\mu}{\sinh \mu} \, e^{-\frac{\mu^2}{4t}} \tag{38}$$

(the diffusion coefficient D is set to 1). Eq.(38) was first obtained in a 1959 paper[56]. Correspondingly, the probability to find a walker at the time moment t at *some* point at the distance μ from the origin is:

$$\mathcal{P}(\mu, t) = P(\mu, t)\mathcal{N}_s(\mu) \tag{39}$$

where $\mathcal{N}_s(\mu) = \sinh^2 \mu$ is the surface area of the sphere of radius μ in the hyperbolic space.

The difference between P and $\mathcal{P}$ becomes crucial in non-Euclidean geometry. By using the definition of the Brownian Bridge we, can easily compute the conditional probability that the N-step random walk begins and ends at the origin ($\mu = 0$) and after t first steps reaches some point at the distance μ from the origin in the hyperbolic space. This probability is:

$$\mathcal{P}(\mu, t|0, N) = \frac{P(\mu, t)P(\mu, N-t)}{P(0, N)} \sinh^2 \mu = \frac{N^{3/2}\, \mu^2}{8\pi t^{3/2}(N-t)^{3/2}} \, e^{-\frac{\mu^2}{4}\left(\frac{1}{t} + \frac{1}{N-t}\right)} \tag{40}$$

Thus, we arrive at the Gaussian distribution with zero mean.

The behavior (40), found for random walks on Riemann surfaces of constant negative curvature, has a straightforward relationship to the conditional distribution of Lyapunov exponents of products of noncommutative random matrices. We can identify hyperbolic space H with the noncommutative group $SL(2, \mathbb{C})/SO(3)$. Let us consider the Brownian Bridge for $\mathcal{H} = SL(2, \mathbb{C})/SO(3)$. Namely, consider the product of N random matrices $\mathcal{M}_k \in \mathcal{H}$ $(0 \le k \le N)$ under the condition that the product $M = \prod_{k=1}^{N} \mathcal{M}_k$ is identical to the unit matrix. We are interested in the distribution of Lyapunov exponents δ for the first m matrices in the product M. The stochastic recursion relation for the vector $\mathbf{W}_k$ reads:

$$\mathbf{W}_{k+1} = \mathcal{M}_k \mathbf{W}_k; \qquad \mathbf{W}_0 = \text{const} \tag{41}$$

where $\mathcal{M}_k \in \mathcal{H}$ for all $k \in [0, N]$. The Brownian Bridge condition means that

$$\mathbf{W}_N = \mathbf{W}_0 \tag{42}$$

For simplicity consider the case in which each matrix in the product is close to the unit one:

$$\mathcal{M}_k = 1 + M_k; \qquad \text{norm}[M_k] \ll 1 \tag{43}$$

The discrete dynamical equation (41) under the condition (43) may be replaced by a differential equation. Its stationary distribution is defined by an appropriate Fokker-Plank equation for random walks in hyperbolic space. The distribution of the Lyapunov exponent δ for the conditional product of random matrices is given by (40). So, under the conditions (42) and (43) for δ, we have recovered the usual Gaussian distribution. Hence the mean value $\langle\delta_m\rangle$ of the Lyapunov exponent of any first $m = cN$ ($c =$ const, $0 < c < 1$) matrices has a square root dependence on N ($N \gg 1$):

$$\langle\delta_m\rangle \sim \sqrt{N}$$

This statement can be reformulated in topological terms. As it was shown above, the correlation between the knot complexity and the Lyapunov exponent tends to 1 as $L_h \to \infty$. Thus, we can conclude that the typical complexity n of a Jones-Kauffman invariant of a daughter knot, obtained by cutting a part $N = L_v \times L_h$ of a trivial parent knot of size $L_v \times L_t$ ($L_h = cL_t$, $0 < c < 1$), grows as

$$n \sim \sqrt{N} \sim \sqrt{L_h}$$

for $N \gg 1$. On the contrary, for a random knot of size $N = L_v \times L_h$ without the Brownian Bridge condition, the complexity grows as[32]

$$n \sim N \sim L_h$$

in agreement with the Fürstenberg theorem.[55]

Therefore the relative complexity n/N of a daughter knot, which is part of a trivial knot, tends to 0:

$$\lim_{N\to\infty} \frac{n}{N} = \lim_{N\to\infty} \frac{\text{const}\sqrt{N}}{N} = 0$$

Figures 17 and 18 confirm our statement. Actually, the mean complexity of the daughter knot is smaller than the typical complexity of a random knot of the same size without the BB condition. The conditional probability for a part of a trivial knot to be also trivial, is sufficiently higher than the corresponding unconditional probability for a random knot (of the same size) to be trivial, see Fig.17. Thus we directly validate the conjecture that the topological state of any daughter knot which is part of a trivial parent knot is close to the trivial one. Hence, the parts of a polymer chain in the crumpled globule are weakly knotted in a broad range of scales.

7. The said above serves as an indirect verification of a previous conjecture[57] concerning the possibility to formulate some topological problems of collapsed polymer chains in terms of path integrals over trajectories with prescribed fractal dimension and without any topological ingredients. Namely, in an ensemble of collapsed polymer chains the essential fraction of trajectories has the fractal (Hausdorff) dimension of the dense packing state $D_f = d$ (d is the space dimension). Conversely, there are reasons to assume that almost all paths in an ensemble of trajectories with fractal dimension $D_f = d$ (where $d \leq 3$) are topologically isomorphic to knots, close to trivial ones.

The partition function of trivial polymer rings can be expressed as a path integral over the set Ω of closed paths with fixed topological invariant:

$$Z = \int_{\Omega} D\{r\}e^{-H} = \int \dots \int D\{r\}e^{-H}\Delta[n], \tag{44}$$

where $D\{r\}e^{-H}$ is the integration over the Wiener measure and $\Delta[n]$ extracts paths with algebraic topological invariant of degree $n = 0$.

If our assumption is correct, integration over Ω in (44) can be replaced by integration over all paths without any topological restriction, but with special fractal measure $D_f\{r\}e^{-H_f}$:

$$Z = \int \dots \int D_f\{r\}e^{-H_f} \tag{45}$$

The usual Wiener measure $D\{r\}e^{-H}$ is concentrated on the trajectories with fractal dimension $D_f = 2$. We believe, that for the description of the statistics of densely packed trivial polymer loops, the measure $D_f\{r\}e^{-H_f}$ with fractal dimension $D_f = d$ ($d \leq 3$) should be used.

Our conjecture can be rephrased in the following way. *The fraction of* ***trivial knots*** *in an ensemble of trajectories with Hausdorff dimension* $D_f = 3$ *in 3D space is substantially higher than the fraction of trivial knots in an ensemble of trajectories with Hausdorff dimension* $D_f = 2$ *(i.e., ofordinaryrandomwalks) in 3D space.* We believe that it would be interesting to check this conjecture in numerical simulations.

Acknowledgments

This study is a logical development of the ideas formulated by one of the authors (S.K.N) in collaboration with A.Yu. Grosberg in 1992-1993 .[13,14] The authors are grateful to A.Yu. Grosberg for fruitful comments and to A.B. Sossinsky for constructive criticism and helpful discussions. We appreciate the useful comments of J.-L. Jacobsen concerning the realization of numerical algorithm used in our work. O.A.V thanks the laboratory LPTMS (Université Paris Sud, Orsay) for hospitality. We appreciate also the help of the Supercomputer Center (Russian Academy of Sciences) for available computational resources. This work is partially supported by ACI grant "Knots and braids" (Nouvelles Interfaces des Mathematiques, France).

Appendix: Minimum energy distribution in the Potts model with random ferro- and antiferromagnetic bonds

The minimum energy of a spin system is independent of q for $q \geq 4$. In this study of the topological invariants of random lattice knots, we are interested in determining behavior of the system for $x \to \infty$(see Eq. (20)). With an allowance for the relationship $q = 2 + x + x^{-1}$, this implies $q \to \infty$ for the corresponding Potts model. Some features of the dependence of the free energy of the Potts spin system on the number of states q can be established based on simple considerations. In cases when the lattice contains no impurities and the a-bonds are arranged as depicted in Figs.3b and 3c, the minimum energy is independent of q for $q \geq 2$. Under these conditions, two spin values $\sigma = \{1, 2\}$ are sufficient to "create" a configuration corresponding to the minimum energy. In the presence of impurities, the minimum energy depends on q for $q \geq 2$. This is illustrated in Fig.5a for a 5×5 lattice with a single impurity on the bond between $\sigma_{1,1}$ and $\sigma_{1,2}$ spins. For the spin configuration energy to reach a minimum value of $E_{min} = -12$ for the given distribution of bonds, it is necessary that all spin variables in the lattice sites connected by f-bonds $J = 1$ (in Fig.5, these sites are indicated by open and filled circles) would acquire the same values, while the spin variables in the sites $\sigma_{1,1}$ and $\sigma_{1,2}$ connected by a-bonds $J = -1$ (in Fig.5a, these sites are indicated by filled and hatched circles) would acquire different values (e.g., $\sigma_{1,1} = 2$, $\sigma_{1,2} = 3$). However, it is impossible to assign the values of spin variables in a system with $q = 2$ so that the spins in clusters $\sigma_{1,1}$, $\sigma_{1,2}$ (as well as in the adjacent cluster) indicated by unlike symbols (black versus open or hatched) were different. For this reason, a minimum energy of the spin state for $q = 2$ is $E_{min} = -11$ (instead of -12).

As was demonstrated, a minimum energy of the spin configuration in the Potts model corresponding to a given distribution of a–bonds may depend on the number of spin states q. This fact can be represented as follows:

to reach the state with minimum energy, it is necessary that the spin variables in the lattice sites connected by f–bonds would acquire for the most part the same values (so as to form clusters), while the spin variables in the sites connected by a–bonds would be possibly different. Thus, a given distribution of the a–bonds corresponds to a subdivision of the lattice into independent clusters of spins. Spins belonging to the same cluster are connected predominantly by the f–bonds, while spins of different clusters are connected by a–bonds.

The energy reaches minimum if the adjacent clusters in a given subdivision possess different values of the spin variable. We may bring each value of the spin variable into correspondence with a certain color. Then a minimum energy corresponds to the lattice subdivision into clusters painted so that all spins in one cluster are of the same color, whereas adjacent clusters have different colors. Figure 5a shows an example of the configuration which cannot be painted in this way using two colors, while three colors allow reaching the goal. In mathematics, there is a theorem concerning the task of "painting maps," according to which any configuration on a surface possessing a a topology of a sphere can be painted using four (or more) colors. In other words, any subdivision of a lattice into clusters can be painted using four (or more) colors so that the adjacent clusters would possess different colors. If each color corresponds to a certain value of the spin variable $q = \{1, 2, 3, 4, ...\}$, we may assign the q values (for $q \geq 4$) so that spins in the adjacent clusters would possess different values (colors). There is no impurity configuration (and the corresponding lattice subdivision into clusters) such that four values of the spin variable would be insufficient to reach the state of minimum energy.

The above considerations allow us to formulate the following statement: For an arbitrary configuration of a-impurities on a Potts model lattice, a minimum energy of the spin system is independent of the number of spin states q for $q \geq 4$.

As is known, the Potts model exhibits a first-order phase transition at $q \geq 4$. Note also that the parameter in the Kauffman invariant becomes real just for $q \geq 4$.

References

1. S.K. Nechaev, Usp.Fiz.Nauk **168**, 369 (1998) [Sov.Phys. Uspekhi, **41**, 313 (1998)].
2. M. Mezard, G. Parisi, M. Virosoro, *Spin glass theory and beyond*, (WCPC: Singapore, 987)
3. M.D. Frank-Kamenetskii and A.V. Vologodskii, Usp. Fiz. Nauk **134**, 641 (1981) [Sov.Phys. Uspekhi **24**, 679 (1981)]; A. V. Vologodskii et al., Zh.Eksp.Teor.Fiz. **66**, 2153 (1974) [Sov.Phys. JETP **39**, 1059 (1974)]; A.V. Vologodskii, A.V. Lukashin, and M.D. Frank-Kamenetskii, Zh.Eksp.Teor.Fiz. **67**, 1875 (1974) [Sov.Phys. JETP **40**, 932 (1975)]; M.D. Frank-Kamenetskii, A.V. Lukashin, and A.V. Vologodskii, Nature (London) **258**, 398 (1975).
4. V.F.R. Jones, Ann.Math., **126**, 335 (1987)
5. V.F.R. Jones, Bull.Am.Math.Soc. **12**, 103 (1985)
6. J. Birman, *Knots, Links and Mapping Class Groups*, Ann.Math.Studies, **82** (Princeton Univ. Press, 1976)
7. A.B. Sossinski and V.V. Prasolov, *Knots, Links, Braids and Three-dimensional Manifolds* (Nauka, Moscow, 1998).
8. R.J. Baxter, *Exactly Solved Models in Statistical Mechanics* (Academic, New York, 1982; Mir, Moscow, 1985).
9. W.B.R. Likorish, Bull.London Math.Soc., **20**, 558 (1988); M. Wadati, T.K. Deguchi, Y. Akutso, Phys.Rep., **180**, 247 (1989)
10. L.H. Kauffman, Topology, **26**, 395 (1987)
11. L.H. Kauffman, H. Saleur, Comm. Math. Phys., **141**, 293 (1991)
12. F.Y. Wu, J. Knot Theory Ramific., **1** 47 (1992)
13. A.Yu. Grosberg, S. Nechaev, J. Phys. (A): Math. Gen., **25** 4659 (1992)
14. A.Yu. Grosberg, S. Nechaev, Europhys. Lett., **20** 603 (1992)
15. H.W. Blote and M.P. Nightingale, Physica A, **112**, 405 (1982)
16. J.L. Jacobsen and J. Cardy, cond-mat/9711279; Nucl.Phys. B, **515** [FS], 701 (1998)
17. J. Salas and A.D. Sokal, cond-mat/0004330; J.Stat.Phys., **104**, 609 (2001)
18. K. Reidemeister, *Knotentheorie* (Berlin: Springer, 1932)
19. F.Y. Wu, Rev.Mod.Phys. **54**, 235 (1982)
20. C.M. Fortuin and P.M. Kastelleyn, Physica **57**, 536 (1972)
21. R.H. Swedsen and J.S. Wang, Phys.Rev.Lett., **58**, 86 (1987)
22. U. Wolff, Phys.Rev.Lett., **62**, 361 (1988)
23. I.M. Lifshitz, JETP, **55** 2408 (1968)
24. I.M. Lifshits, A.Yu. Grosberg, A.R. Khokhlov, Rev. Mod. Phys., **50** 683 (1978)
25. A.Yu. Grosberg, A.R. Khokhlov, *Statistical physics of macromolecules* (New York: AIP Press, 1994)
26. J.des Cloizeaux and G. Jannink, *Polymers in Solution: Their Modelling and Structure* (Oxford Univ. Press: New York, 1990)
27. A.Yu. Grosberg, S.K. Nechaev, E.I. Shakhnovich, J.Phys.(Paris), **49** 2095 (1988)
28. B.B. Mandelbrot *The Fractal Geometry of Nature*, (San Francisco: Freeeman,

1982)
29. B. Chu, Q. Ying, A. Grosberg, Macromolecules, **28** 180 (1995)
30. W.L. Nachlis, R.P. Kambour, W.J. McKnight, Polymer, **39** 3643 (1994)
31. P.Lua, A.Borovinskiy, A.Grosberg, cond-mat/0305406; talk at *Special Session on Numerical Methods, Calculationsand Simulations in Knot Theory and its Application*, 987th AMS Meeting, San Francisco, CA, May 3-4, 2003
32. O.A. Vasilyev, S.K. Nechaev, JETP, **93** 1119 (2001)
33. I.M. Lifshitz, A.Yu. Grosberg, JETP, **65** 2399 (1973)
34. M. Tesi, E. Janse van Rensburg, E. Orlandini, S. Whittington, J. Phys. (A): Math. Gen., **27** 347 (1994)
35. E. Orlandini, M. Tesi, S. Whittington, J. Phys. (A): Math. Gen., **33** L181 (2000)
36. P. Tait, Trans. Royal Soc. Edinburgh **28** 145 (1877)
37. P.Bangert, M.Berger, R. Prandi, J. Phys. (A): Math. Gen., **35** 43 (2002)
38. E. Janse van Rensburg, D. Sumners, E. Wasserman, S. Whittington, J. Phys. (A): Math. Gen, **25** 6557 (1992)
39. S.R. Quake, Phys. Rev. Lett., **73** 3317 (1994)
40. Y. Sheng, P. Lai, Phys. Rev. E, **63** 021506 (2001)
41. M.Shimamura, T.Deguchi, `cond-mat/0207282`, `cond-mat/0211504`
42. C. Soteros, D. Sumners, S. Whittington, Math. Proc. Camb. Phil. Soc., **111**, 75 (1992)
43. E.J. Janse van Rensburg, The probability of knotting in lattice polygons (preprint)
44. *Ideal Knots*, Eds. A. Stasiak, V. Katrich, L.H. Kauffman, Series of Knots and Everything, **19** (WSPC: Singapore, 1998)
45. A.Yu.Grosberg, A.Feigel, Y.Rabin, Phys.Rev.E, **54**, 6618 (1996)
46. M. Freedman, Z. He, Z. Wang, Ann. Math. **139** 1 (1994)
47. A. Kholodenko, D. Rolfsen, J. Phys. (A): **29** 5677 (1996)
48. P.G. de Gennes, *Scaling Concepts in Polymer Physics* (Cornell Univ. Press: Ithaca, 1979)
49. E. Helfand, D. Pearson, J. Chem. Phys., **79** 2054 (1983); M. Rubinstein, E. Helfand, J. Chem. Phys., **82** 2477 (1985)
50. A. Khokhlov, S. Nechaev, Phys. Lett. (A) **112** 156 (1985)
51. S. Nechaev, Ya.G. Sinai, Bol. Soc. Bras. Mat., **21** 121 (1991)
52. L.B. Koralov, S.K. Nechaev, Ya.G. Sinai, Prob. Theory Appl., **38** 331 (1993)
53. P. Bougerol, Probab. Theory and Rel. Fields **78** 193 (1988)
54. A.V. Letchikov, Russ. Math. Surv., **51** 49 (1996)
55. H. Fürstenberg, Trans. Amer. Math. Soc., **198** 377 (1963); V.N. Tatubalin, Prob. Theory Appl., **10** 15 (1965), Prob. Theory Appl., **13** 65 (1968)
56. M.E. Gerzenshtein, V.B. Vasilyev, Prob. Theory Appl., **4** 424 (1959); F.I. Karpelevich, V.N. Tatubalin, M.G. Shur, Prob. Theory Appl., **4** 432 (1959)
57. A.Yu. Grosberg, S.K. Nechaev, Polymer topology, Adv. Polym. Sci., **106** 1, in *Polymer Characteristics*, (Springer: Berlin, 1993); S.K. Nechaev, *Statistics of Knots and Entangled Random walks*, (WSPC: Singapore, 1996)

CHAPTER 23

GENERATING LARGE RANDOM KNOT PROJECTIONS

Yuanan Diao[†], Claus Ernst[‡] and Uta Ziegler[‡]
[†] *Department of Mathematics*
University of North Carolina at Charlotte
Charlotte, NC 28223
[‡] *Department of Mathematics*
Western Kentucky University
Bowling Green, KY 42101

In this paper, the problem of generating large random knot projections is explored. Regular knot projections without the usual over and under information at each crossing the can be viewed as 4-regular planar graphs. Two methods are introduced that generate 4-regular planar graphs that can be viewed as projections of prime knots. Various questions that arise along the way are explored.

1. Introduction

It is well known in knot theory that tabulating knots and links is a very difficult task. Currently, the existing knot tabulations are all based on the crossing numbers of the knots. The number of knots and links with crossing number n increases exponentially and it quickly becomes impractical to tabulate all knots and links of the given crossing number. Consequently, not much is known about large knots. Let $\mathcal{L}_n$ be the set of all links with crossing number n and let $\mathcal{AL}_n$ be the set of all alternating links; then most members of $\mathcal{L}_n$ are non-alternating composite links. Let $\mathcal{K}_n$ ($\mathcal{AK}_n$) be the set of all (alternating) prime knots with crossing number n. A typical member of $\mathcal{K}_n$ is a non-alternating prime knot, that is probably hyperbolic. If n is large, then the sizes of the sets $\mathcal{K}_n$, $\mathcal{AK}_n$, $\mathcal{L}_n$ and $\mathcal{AL}_n$ are very large. For example, the size of $|\mathcal{AL}_n|$ can be approximated as follows:[13]

$$\frac{a_{n-1}}{8(2n-3)} \leq |\mathcal{AL}_n| \leq \frac{a_{n-1}}{2},$$

where $a_n \approx \frac{11.5}{4\sqrt{\pi}} n^{-5/2} 6.148^{n-3/2}$. Furthermore it is known that the ratios of $\frac{|\mathcal{AK}_n|}{|\mathcal{K}_n|}$ and $\frac{|\mathcal{AL}_n|}{|\mathcal{L}_n|}$ approach zero as n grows to infinity.[15]

While there are many different measures of knot complexity (including the genus, the unknotting number, the bridge number, the ropelength, and various knot energies), the crossing number of the knot is probably the most often used. In this paper, a "large" knot or link, means a knot or a link with a large crossing number. It is a fundamental problem in knot theory to find the relations between the crossing number and the other knot complexity measures. The authors are particularly interested in the relations between the crossing number and the ropelength of a knot and this is a the main motivation of this paper. While this is a very hard question in general, much has been achieved despite the difficulties. It has been shown[4] that the ropelength $L(K)$ of a knot K is bounded below by

$$L(K) \geq \frac{1}{2}\left(17.334 + \sqrt{17.334^2 + 64\pi Cr(K)}\right). \tag{1}$$

For relatively small knots, the above formula produces the best known theoretical lower bounds of the ropelengths of these knots. The lower bounds of the ropelengths of knots with crossing number up to 20 produced by this formula are listed below.

$Cr(K) =$	3	4	5	6	7	8	9	10	11
$L(K)$ $>$	23.70	25.29	26.74	28.08	29.33	30.51	31.64	32.70	33.73

$Cr(K) =$	12	13	14	15	16	17	18	19	20
$L(K)$ $>$	34.71	35.66	36.58	37.46	38.32	39.16	39.97	40.76	41.54

For large knots, the above lower bound is inferior to the following bound obtained by Buck and Simon:[1]

$$L(K) \geq b\,(Cr(K))^{\frac{3}{4}}, \tag{2}$$

where b is some positive constant. The constant b is estimated to be at least 1.105 by a result in Buck and Simon[1] and it is improved to at least 2.135 in Rawdon and Simon.[11] This three fourths power is also shown to be achievable for some knot families.[3,5] That is, there exists an infinite family $\{K_n\}$ of knots and a constant $a_0 > 0$ such that $Cr(K_n) \to \infty$ as $n \to \infty$ and $L(K_n) \leq a_0 \cdot (Cr(K_n))^{3/4}$. On the other hand, it is also known that $L(K)$ is bounded above by $c(Cr(K))^{3/2}$ (where $c > 0$ is a constant).[8] It is widely suspected that this $3/2$ power can be reduced and it is possible that the ropelength of any knot K may be bounded above by $O(Cr(K))$ or

$O(Cr(K)\ln^p Cr(K))$ for some positive power p. It has been shown recently that the power 1 can be realized by a family of knots[7] and any power between 3/4 and 1 can be realized by some family of knots.[6] However, what happens between the powers 1 and 3/2 remains a challenging problem.

In general, relating different measures of knot complexity to the crossing number is difficult since in order to obtain information on how these measures of knot complexity grow with the crossing number n, one is presented with a host of seemingly intractable problems. Foremost, determining the variable n proves to be impossible in practice. Unless a knot diagram is alternating, deciding the minimum crossing number n for a given large non-alternating knot K cannot be done unless one knows that the knot belongs to the few knot families for which the crossing number is known. Determining the value of any knot complexity measure dependent on n seems even more daunting. So it is natural for one to turn to numerical methods in an effort to gain insight into the issue. However, in the case of relating the crossing number to other knot complexity measures, this approach has not gained much momentum. One of the reasons is the lack of an effective way of generating large random prime knots with a fixed crossing number.

With the long-term goal of determining how the measures of complexity grow with n for members of $\mathcal{K}_n$, $\mathcal{AK}_n$, $\mathcal{L}_n$ and $\mathcal{AL}_n$, this paper addresses the short-term goal of generating members of these sets with a computer. More specifically, since the exact crossing number for non-alternating knots can not determined, we propose two methods here that generate regular projections of knots. These regular projections are viewed as 4-regular planar graphs G with n vertices (that is the traditional over/under information at each crossing is ignored). Such graphs are called regular projection graphs or RP-graphs. An RP-graph G can be transformed easily into a knot or link L by simply adding the overpass/underpass information to each vertex. Assigning the overpass/underpass information at each vertex of G in a random way is likely to result in a knot or link L whose crossing number is smaller than the number of vertices in the RP-graph G. However, if the overpass/underpass assignment to each vertex of G is made in an alternating way and the RP-graph G is 4 edge-connected, then the obtained alternating knot is a prime knot and thus a member of $\mathcal{AK}_n$ and the diagram would represent a minimum projection of a certain knot in $\mathcal{AK}_n$. In this paper, we are concerned with the problem of how to generate the random knot projections, not with the problem how to realize a knot with the given crossing number from such a projection diagram.

The first method to generate an RP-graph presented here is based on the principle of a *blossom tree*.[13] With a runtime of $O(n)$ a 4-regular planar graph with n vertices is randomly generated. If this graph is the regular projection of a link with k components then the components can be determined in $O(n)$ steps and in $O(n^3)$ time this graph can be modified into a graph that is the projection of a knot. The final product of the method is an RP graph that is a regular projection of an alternating prime knot. The number of vertices in this RP-graph is less than or equal to n.

A recent result says that for any knot K with crossing number n, there exists a knot K' that is topologically equivalent to K such that K' has a regular projection that is Hamiltonian and at most $4n$ crossings in this regular projection.[8] The algorithm used to produce the $n^{3/2}$ power upper bound for $L(K)$ is based on the existence of a Hamilton cycle in an RP-graph of K'. The problem of finding a Hamilton cycle in a planar graph is known to be NP-complete.[10] Even for the more restricted class of 4-regular planar graphs finding a Hamilton cycle is still a difficult problem. Because of this, the second method introduced here may have a special advantage since it generates 4-regular planar graphs G based on an existing Hamilton cycle in it: the Hamilton cycle could be used in further computations (such as computing an upper bound on the ropelength). Although this method has a larger average run time than the first method, it may be well worth the effort.

In the next section, some basic concepts are introduced. We will then describe the two methods in Section 3 and Section 4. Section 5 contains some numerical results comparing the two methods.

2. Basic Concepts

A *graph* G consists of a set $V(G)$, a set $E(G)$, and an incidence relation which says that every element of $E(G)$ is incident with two elements of $V(G)$. The elements of $V(G)$ are called the *vertices* of G, and $V(G)$ is called the *vertex set* of G. The elements of $E(G)$ are called the *edges* of G, and $E(G)$ is called the *edge set* of G.

Let G be a graph. The *degree* of a vertex v of G is the number of edges of G incident with v. The graph G is called *k-regular* if every vertex of G

is of degree k. Let e be an edge of G incident with vertices u and v. We say that e *connects* u and v, u and v are the *ends* of e, and u and v are *adjacent*. We also use uv or vu to denote e when there is only one edge connecting u and v. If $u = v$ then e is called a *loop edge*.

A graph H is a *subgraph* of a graph G if $V(H) \subset V(G)$ and $E(H) \subset E(G)$. In the case that $V(H) = V(G)$, H is called a *spanning subgraph* of G. Two graphs G and H are said to be *isomorphic* if there exist bijections $f : V(G) \longrightarrow V(H)$ and $g : E(G) \longrightarrow E(H)$ such that $e \in E(G)$ is incident with $x, y \in V(G)$ if, and only if, $g(e) \in E(H)$ is incident with $f(x), f(y) \in V(H)$.

A *path* P of length $k - 1$, where $k \geq 2$, is a graph which is isomorphic to the graph with vertex set $\{v_1, v_2, ..., v_k\}$ and edge set $\{v_i v_{i+1} : i = 1, \ldots, k-1\}$. We say that v_1 and v_k are the *ends* of P, P is *from* v_1 *to* v_k, and P is *between* v_1 *and* v_k. A *cycle* C of length k, where $k \geq 3$, is a graph which is isomorphic to the graph with vertex set $\{v_1, v_2, ..., v_k\}$ and edge set $\{v_i v_{i+1} : i = 1, \ldots, k-1\} \cup \{v_k v_1\}$. A *Hamilton cycle* in a graph G is a spanning subgraph that is a cycle. A graph with a Hamilton cycle is said to be *Hamiltonian*. A graph with no cycle is a *tree*. A tree is *rooted* if there is a single vertex in the tree labeled as root.

A *geometric realization* of a graph G in R^2 or R^3 is such that the vertices of G are represented by distinct points in the space, every edge of G is represented by a simple curve in R^2 or R^3 connecting the two points representing its ends, and these simple curves representing the edges do not intersect each other except possibly at their ends. A graph G is *planar* if it has a geometric realization in a plane. Such a geometric realization is called a *plane graph*. A planar *map* consists of a planar graph and a cyclic order of the edges around each vertex.

4-regular plane graphs are related to knots and links as follows: A projection p of a knot or link L in R^3 is a continuous function $p : R^3 \to R^2$. The image $p(K)$ is called the projection of the knot or link K into a plane. $p(K)$ is a closed curve (or collection of closed curves if K is a link) in R^2 that may contain self-intersecting points. A self-intersecting point is also called a *crossing* of the projection. The *multiplicity* of a crossing in the projection is the number of strands that pass through that point. We say that a projection is a *regular projection* if there are only finitely many crossings in $p(K)$ and all crossings are of multiplicity 2, that is there are exactly two points on L that map to any of the crossings. It is a well-known result in knot theory that for any knot in R^3 most projection directions give rise to regular projections.[2] Furthermore, a regular projection $p(K)$ of K is called a *minimum projection* of K if it is a regular projection with $Cr(K)$ crossings.

Let K be a knot or link and let G be the image of a regular projection of K. If we treat the crossings in G as vertices and the arcs of G joining these crossings as edges, then G can be viewed as a 4-regular plane graph. Thus, from now on, we may view a regular projection G as a 4-regular plane graph G called an *RP-graph* of K, where "RP" serves as a reminder that the graph is obtained as a regular projection of K. If G arises from a minimum projection of K, we then call it a *minimum RP-graph* of K. Note that any 4-regular plane graph is an RP-graph of some knot or link. Moreover a RP-graph has a specified cyclic order of the edges around each vertex. Thus in the context of this paper, a 4-regular planar RP-graph is equivalent to a 4-regular planar map and the terms are used interchangeably.

For any $X \subset V(G)$, let $G - X$ denote the subgraph of G obtained from G by deleting vertices of G in X and edges of G with at least one end in X. Similarly, for any $Y \subset E(G)$, we use $G - Y$ to denote the subgraph of G obtained from G by deleting the edges in Y (but keeping all vertices of G). We say that G is *k-connected*, where k is a positive integer, if $|V(G)| \geq k+1$ and, for every subset $X \subset V(G)$ with $|X| < k$, $G - X$ is connected. We say that G is *k-edge-connected* if, for every $Y \subset E(G)$ with $|Y| < k$, $G - Y$ is connected. The *connectivity* (respectively, *edge-connectivity*) of G is the largest integer k such that G is k-connected. It is easy to see that a 4-regular plane graph is either 2-edge connected or 4-edge connected. If G is a 2-edge connected 4-regular plane graph that arises as a regular projection from some knot or link K, then either G is not a minimum RP-graph of K or K is not a prime knot or link. This motivates the following definition: A 4-regular plane graph is *diagrammatically prime* if it is 4-edge connected.

Let G be a 4-regular RP-graph, let v be a vertex of G, and let e_1, e_2, e_3, e_4 be the edges of G incident with v. Suppose that e_1, e_2, e_3, and e_4 occur around v in this cyclic order as shown in Figure 1. Then we say that e_i is *opposite* to e_j if $|j - i| = 2$, and e_i and e_j are *adjacent* otherwise.

One can travel along the edges of a 4-regular plane graph G and follow the convention that if one travels an edge e leading to a vertex v then one has to continue along the edge opposite to e. In this way G can be viewed as the union of several closed curves. G has k *components* if G is the union of k closed curves. G is a *knot graph* if the number of components in G is one. Of course, the number of components of G is the same as the number

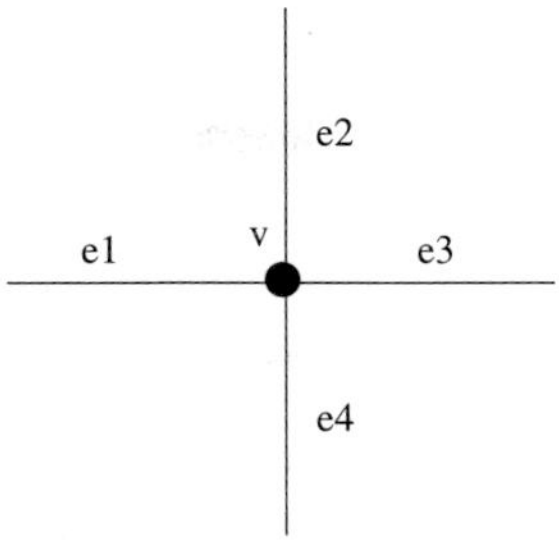

Fig. 1. Pairs of opposite edges: $\{e_1, e_3\}$, $\{e_2, e_4\}$.

of components in any link K which has G as an RP-graph.

In the following two sections we give two methods that can be used to computer generate large diagrammatically prime RP-graphs G (or planar maps) of knots in a random manner. However in the generation process the number of vertices in the graphs varies from one graph to the next. If the user wants a specific interval into which the number of vertices of a generated graph should fall then one needs to run the process several times until a graph with the desired number of vertices is generated.

3. Algorithm One; Generating large, diagrammatically prime RP-graphs of knots

This method first generates a 4-regular planar map with n vertices (Step 1), combines the components of this RP-graph into a knot graph (Step 2) and decomposes the knot graph into diagrammatically prime knot graphs (Step 3).

Step 1. Creating a 4-regular planar map. Here an algorithm introduced by Schaeffer and Zinn-Justin[14,12] is used to generate 4-regular planar RP-graphs with n vertices in $O(n)$ time. A short outline of the algorithm is given below. In the following definition, two labels are used to mark the degree one vertices of a graph: one is called a *bud* and the other is called a *leaf.* In Figure 2 below, a bud is marked by an arrow and a leaf is marked as a solid dot. A vertex with degree more than 1 is marked by a circle.

Definition 1: A *blossom tree* is a rooted plane tree such that:
(i) vertices of degree one are marked as either buds or leaves;

(ii) every inner vertex has degree four and is incident to exactly one bud;

(iii) the root is a leaf.

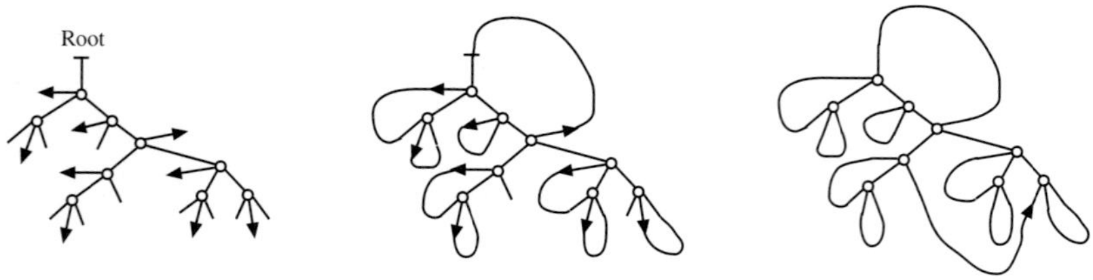

Fig. 2. On the left is a blossom tree with 9 vertices, 11 leaves and 9 buds. One leaf is marked as the root. The figure in the center shows how the blossom tree can be changed into a 4 regular plane graph by connecting each bud to the nearest leaf in counterclockwise direction. This operation leaves two leaves remaining, which are connected by an oriented edge. On the right the final rooted 4-regular map is shown.

To generate a blossom tree, at first a random binary plane tree (a tree such that all internal vertices have degree three) is generated with n internal vertices and $n+2$ leaves (one of which is the root of the tree). Now at each internal vertex one adds an edge whose end is called a *bud*, randomly in one of three different possible ways, see Figure 2. In the figure the buds are the arrows at the end of some of the edges. It has been shown that there are exactly $\frac{3^n}{n+1}\binom{2n}{n}$ rooted blossom trees with n internal vertices.[14,12] Furthermore, each blossom tree can be generated with equal probability. The *closure* of a blossom tree is created by connecting the n buds with the nearest unmatched leaf in a counterclockwise direction. (This procedure uses a counterclockwise order of the leaves and buds of the tree around the infinite face the tree defined in the plane.) The arrows on each bud will be deleted after the bud has been connected to a leaf. After all buds have been matched to leaves, there are two unmatched leaves A and B in the infinite face and all edges are now unoriented. Finally, one connects the two unmatched leaves by an edge which will get an orientation in counterclockwise direction. (There are two such edges possible, one edge connecting leaf A to leaf B, the other connecting leaf B to leaf A. Only one of these is shown in Figure 2.) Such a 4-regular planar map with an oriented edge is called a *rooted planar map*. (Note that the "rooted" in the rooted planar map is not related to the root of the original binary tree used to create the map.) The number of such maps is $\frac{2}{n+2}\frac{3^n}{n+1}\binom{2n}{n}$ and such a map can be constructed in $O(n)$ time.[14,12] Moreover the algorithm generates such a map with a uniform probability distribution on the space of rooted 4-regular planar maps with n vertices. In our case, the orientation of the single edge in the rooted map is deleted since we are not interested in rooted 4-regular planar maps but just in RP-graphs (or planar maps) which have no orientation on any edge. Deleting the orientation from this single edge results in a non-uniform distribution on the space of PR-graphs. Thus the 4-regular planar RP-graphs are generated with a non-uniform distribution.

Step 2. Creating a knot graph: The 4-regular RP-graphs so generated are usually not diagrammatically prime, and not a knot projection graph either, since they usually have many components. Numerical experiments show that a 4-regular planar map with 2^{10} vertices generated this way has an average number of about 51 components, and a 4-regular planar map with 2^{12} vertices so generated has an average of about 199 components.[14] In fact, it is conjectured that the number of components in such a 4-regular planar map is proportional to the number of vertices of the graph.[14] Since the algorithm should generate diagrammatically prime RP-graphs of knots, the RP-graphs need to be modified into RP knot graphs.

Given a 4-regular planar map G with n vertices and k components. The idea is to combine components in a manner which does not modify the number of vertices while ensuring that the modified RP-graph remains a 4-regular planar map. This can be done if a face f of G is bound by edges of at least two different components of G. Suppose that walking around f in one direction, the edges v_1v_2 and w_1w_2 from two different components are found, where, for both edges, the subscript 1 indicates the vertex first encountered. If the two edges are disjoint then the two edges are replaced by two new edges v_1w_2 and w_1v_2 as shown in the left half of Figure 3. This edge replacement reduces the number of components by one, but does not reduce the number of vertices. If no such disjoint edges can be found on any face of G and there is still more than one component in G, then there are two edges from two different components that intersect at either one or both of their endpoints. In this case an edge replacement move is still possible and results in the modifications shown in the right half of Figure 3. However such a move introduces one or two loop edges. Continuing this process one obtains a 4-regular graph with n vertices and one single component.

To combine the components randomly, a vertex v in G is picked at random where two different components of G cross at v. The faces of G containing v are processed in a random order. For each face, starting at v

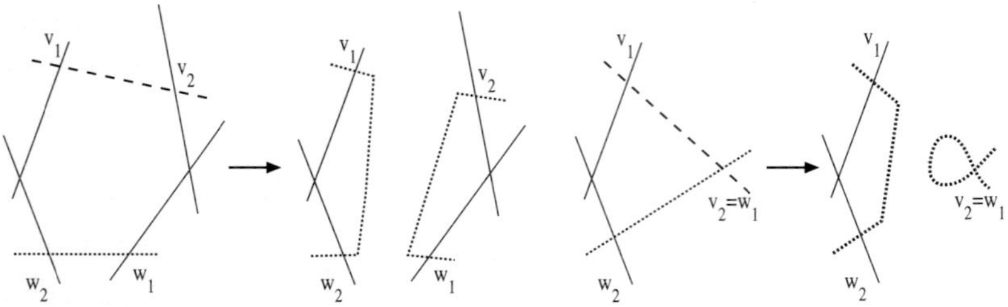

Fig. 3. Two cases of an edge replacement move. On the left using two disjoint edges v_1v_2 and w_1w_2, on the right using two edges v_1v_2 and w_1w_2 which have one vertex in common. In this case a replacement move introduces one loop edge.

the algorithm searches through the edges of the face in a clockwise direction. The algorithm search for an edge which belongs to a different component and is disjoint from the first edge of the face (incident to v). If the algorithm finds such an edge replacement move of the above type is carried out. If an edge from a different component is found that is adjacent to the first edge, then this is stored and the search for a replacement move continues. once a replacement move is carried out a new vertex v is picked at random. If none of the faces of G containing v can be used for a replacement move using two disjoint edges then a vertex different from v is picked at random and the procedure is repeated. If no edge replacement move using two disjoint edges is available, a replacement move using two adjacent edges is carried out. Such a move leads to a loop edges as shown on the right in Figure 3. Thus the algorithm avoids the creation of loop edges whenever possible. (It should be noted that it is possible that an already existing loop edge will be eliminated by later edge replacement moves.) This process of combining the components through edge replacement moves can be carried out in $O(n^3)$ steps.

The exact number of 4-regular planar maps G of n vertices with one component is unknown, and no claim is made that the above algorithm produces such graphs with a uniform distribution for a given vertex number n. Clearly any 4-regular planar one-component map with n vertices and without loop edges can be obtained by the algorithm (with a certain probability). Some graphs containing loop edges may also be produced, however. Therefore, no claim is made that this procedure will always generate a 4-regular planar one-component map with n vertices and without loop edges. Notice that in the algorithm, a loop edge is avoided whenever possible. This is no drawback since any minimal RP-graph of a knot K does not contain loop edges and generally knot RP-graphs are preferred whose numbers of vertices are close to (if not equal to) the crossing numbers of the corresponding knots.

Step 3. The knot projection graphs obtained up to this point are not necessarily diagrammatically prime. To obtain diagrammatically prime 4-regular graphs a given 4-regular planar knot graph is split into 4-edge connected components using an algorithm by Dowker and Thistlethwaite,[9] which can be carried out in $O(n^2)$ steps. In this last step, the control over the number of vertices in the final diagrammatically prime knot graphs is lost in that a graph with a fixed number of n vertices now falls apart into several graphs of varying sizes. See Section 5 for further discussions.

4. Algorithm Two; Generating large, diagrammatically prime Hamiltonian RP-graphs of knots

In this section, the second method that generates diagrammatically prime knot graphs G with a built-in Hamilton cycle is discussed. This method relies on the fact that every knot K has a regular projection of at most $4Cr(K)$ crossings with a Hamilton cycle.[8] The main idea is to start with a single cycle C of n vertices (Step 1) which will become the Hamilton cycle of the graph G under construction. Next, edges are added to the cycle C in a random manner to create a 4-regular graph which is non-planar with very high probability (Step 2). The resulting non-planar graph is modified into a 4-regular planar graph (Step 3), whose components are combined into an RP-graph of a knot (Step 4) and finally the RP-graph is decomposed into diagrammatically prime RP-graphs (Step 5). Although it is far from clear how the RP-graphs of knots generated in such a manner are distributed in the space of all RP-graphs with up to n vertices (as with any other method at this point), all possible RP-graphs of knots can be generated using this algorithm. Figure 4 shows a typical RP-graph of a knot generated with method two.

Step 1. Start with a cycle C with n vertices $v_1, v_2, \ldots, v_n$ such that the order of the vertices matches the order of $v_1, v_2, \ldots, v_n$ on the cycle C. Note that the cycle C will become the Hamilton cycle of the graph G under construction.

Step 2. Notice that two edges have to be added at each vertex. Since each edge added is incident to two vertices, a total of n edges need to

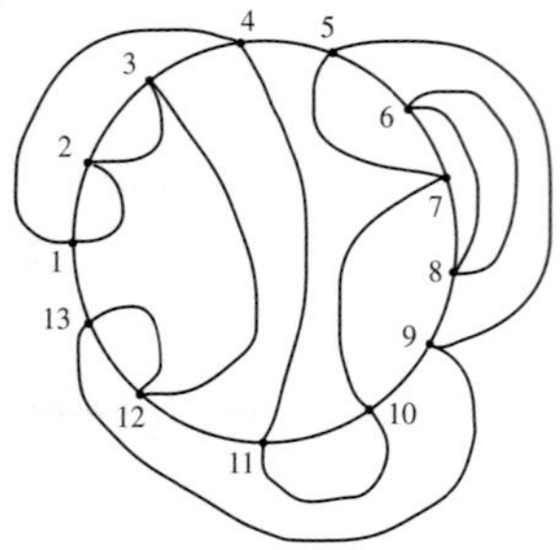

Fig. 4. A small diagrammatically prime RP-graph of a knot with 13 crossing generated using method 2. The Hamilton cycle is the circle and the vertices are numbered along the cycle.

be added. These n edges are divided into two groups: those in the region bounded by the Hamilton cycle (called B-edges) and those in the unbounded region (called U-edges). Each such edge e can be represented by a triple of integers of the form (i, j, k). Here, the indices i and j imply that e is incident to vertices v_i and v_j, $k = 0$ if e is a U-edge and $k = 1$ if e is a B-edge. For example, $(2, 5, 0)$ means an outside edge incident to vertices v_2 and v_5 and $(1, 8, 1)$ means the inside edge incident to vertices v_1 and v_8.

The following procedure generates random triples for each of these n edges:

Sub-step 1: Generate a random permutation of the vector $(1, 1, 2, 2, ..., n, n)$. The result is a vector of $2n$ entries: $(j_1, j_2, ..., j_{2n})$. The n edges not on the Hamilton cycle are to be chosen as the pairs (j_1, j_2), (j_3, j_4), ..., (j_{2n-1}, j_{2n}).

Sub-step 2: Generate a random vector $(k_1, k_2, ..., k_n)$ where each k_i is 0 or 1 with equal probability.

Sub-step 3: Merge the two vectors to create the n triples of integers representing the edges which need to be added: (j_1, j_2, k_1), (j_3, j_4, k_2), ...,(j_{2n-1}, j_{2n}, k_n).

Steps 1 and 2 in this algorithm can be carried out using $O(n)$ steps. Up to this point the algorithm generates a uniform distribution among all possible configurations of the n edges not on the cycle C. Furthermore, the numbers can always be arranged so that $j_1 \leq j_2, j_3 \leq j_4$ and so on, since this does not change the definition of the edge. Notice that at this stage of the algorithm there are several problems. The graph generated so far may have several components. In addition, it may be impossible to embed in R^2 the n edges given by the list (j_1, j_2, k_1), (j_3, j_4, k_2), ...,(j_{2n-1}, j_{2n}, k_n) without many self intersections of these edges. Thus the graph defined by these edges will most likely not be a planar graph. In fact there may be many non equivalent ways to use the triples (j_1, j_2, k_1), (j_3, j_4, k_2), ...,(j_{2n-1}, j_{2n}, k_n) to create nonequivalent (up to isotopy) non-planar 4-regular graphs. In the next step we change the list (j_1, j_2, k_1), (j_3, j_4, k_2), ...,(j_{2n-1}, j_{2n}, k_n) to a new list of n triples whose edges can be embedded into the plane in a unique way without any self-intersections of these edges (except at the vertices on the Hamilton cycle C, of course).

Step 3. It can be easily shown that two edges generated by the triples $e = (i_1, j_1, k_1)$ and $f = (i_2, j_2, k_2)$ must intersect each other in their interior if and only if

a. $k_1 = k_2$ (e and f are both U-edges or both B-edges) and

b. either

$$i_1 < i_2 < j_1 < j_2 \text{ or } i_2 < i_1 < j_2 < j_1. \tag{3}$$

An edge e is randomly chosen and it is checked if the above conditions are satisfied for any of the other edges. If the conditions are never satisfied for any of the other edges, e remains unchanged. None of the other edges intersects and e is marked. A new edge e that is not already marked is randomly selected. If two triples $e = (i_1, j_1, k)$ and $f = (i_2, j_2, k)$ satisfy (3), then with equal probability, the two triples e and f are replaced by either (i_1, i_2, k), (j_1, j_2, k) or (i_1, j_2, k), (i_2, j_1, k).

Lemma 2: *The above procedure reduces the number of pairs of triples which satisfy the above conditions by at least one.*

Proof: Assume that $e = (i_1, j_1, k)$ and $f = (i_2, j_2, k)$ satisfy (3). Clearly the conflict between e and f is removed in that none of the two potential new pairs of triples (i_1, i_2, k), (j_1, j_2, k) or (i_1, j_2, k), (i_2, j_1, k) satisfies (3).

It suffices to show two conditions:

(i) If there is an edge $h = (i_3, j_3, k)$ such that neither h and e nor h and f satisfy (3), then none of the 4 potential new triples (i_1, i_2, k), (j_1, j_2, k) or (i_1, j_2, k), (i_2, j_1, k) together with h satisfy (3).

(ii) If there is an edge $h = (i_3, j_3, k)$ such that either h and e or h and f satisfy (3), then at most one of each pair of the 4 potential new triples (i_1, i_2, k), (j_1, j_2, k) or (i_1, j_2, k), (i_2, j_1, k) together with h satisfy (3).

In case (i), there are two subcases $i_1 < i_2 < j_1 < j_2$ and $i_2 < i_1 < j_2 < j_1$. Since they are similar, only the subcase $i_1 < i_2 < j_1 < j_2$ is proven here. The situation is shown in Figure 5 (on the left) assuming that e, f and h are B-edges. (The situation is similar if all three are U-edges.) Since h does not satisfy (3) when paired with either e or f, it must be true that i_3 and j_3 are between either i_1 and i_2 or i_2 and j_1 or j_1 and j_2 or j_2 and i_1 when the vertices are viewed along the Hamilton cycle. Thus h and any of the four potential new triples does not satisfy (3).

The proof of case (ii) involves a similar analysis as in case (i) above and is left to the reader. □

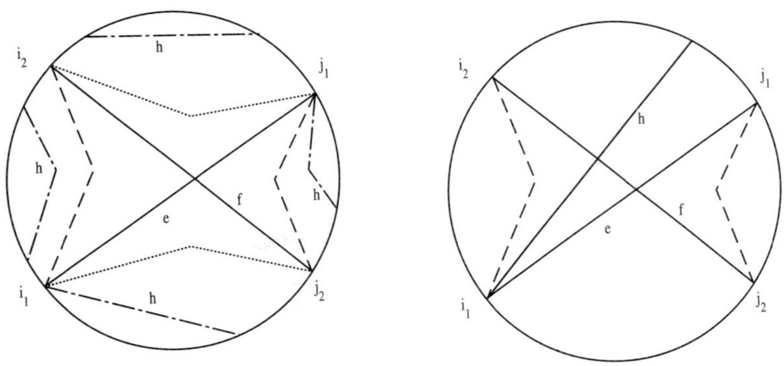

Fig. 5. On the left: The edges e and f satisfying the condition (3) are shown together with the four new triples (i_1, i_2, k), (j_1, j_2, k) (dashed line segments) or (i_1, j_2, k), (i_2, j_1, k) (dotted line segments). In addition 4 potential edges h are shown which do not satisfy (3) with any of the other shown edges. On the right: replacing the edges e and f with the two dashed edges also eliminates the intersection between the edges e and h.

Figure 5 (on the right) shows that one such step could reduce the pairs of triples which satisfy the above condition(3) by more than one.

Once two triples satisfying the above condition have been replaced by two different triples, the process is repeated all over again. That is an unmarked edge e is randomly selected and it is checked whether the above conditions are satisfied for any of the other edges. This process is continued until all edges are marked. This results in a set of n triples which contain no pair satisfying (3). Such a set of n triples defines up to isotopy a 4-regular plane graph G with n vertices and a Hamilton cycle.

Since there are at most $O(n^2)$ pairs of triples satisfying the above condition (3), this step of the algorithm requires a polynomial time of the order $O(n^3)$ (With up to $O(n)$ steps one either finds a pair of triples satisfying the condition (3) or one can mark an edge, resulting in an order $O(n^3)$ in the worst case).

Step 4. Creation of an RP-graph of a knot from G. The situation now is similar to Step 2 in the first method and a similar procedure as in method one is carried out. The difference is that edges that are on the Hamilton cycle can not be used in any edge replacement move since this destroys the Hamilton cycle. Thus on a face of the graph G one needs to find two edges that belong to different components of G and that are not on the Hamilton cycle before carrying out an edge replacement move. Otherwise the algorithm remains unchanged. In the Lemma below we show that this algorithm almost always works.

Lemma 3: *The above algorithm leads to an RP-graph G of a knot with an Hamilton cycle C in all but one case.*

Proof: If in Step 2 a graph G is created in which the cycle v_1, v_2, ..., v_n forms a component (that is, there is exactly one U and one B edge attached at each v_i), then the algorithm fails. This happens since we only allow edge replacement moves with edges that are not on the Hamilton cycle. Moreover, if exactly one U and one B edge are attached at each v_i, then this property remains unchanged by an edge replacement move. Thus the Hamilton cycle remains one component. In such a case the algorithm produces a two-component graph with one component equal to the Hamilton cycle C and the other component containing all the edges which are not on the Hamilton cycle C.

It needs to be shown that the algorithm works in all other cases. Notice that if all the U-edges do not belong to one component, then G has a face in the U-region that contains two edges not on the Hamilton cycle of different components. This implies that an edge replacement move is possible. So if the algorithm is carried out and there is no further edge replacement left, then all the U-edges must belong to one component. Similarly all B-edges must belong to one component of G. Under these assumptions, it can be shown that all the edges on the Hamilton cycle belong to one component. Given that $v_{i-1}v_i$ is an edge in one component then the edge v_iv_{i+1} belongs to the same component. This can be seen as follows: If at v_i there is a B-edge and a U-edge attached then this is obvious. If there are two B-edges (or two U-edges) at v_i then both edges of the Hamilton cycle must be of the same component due to the fact that both B-edges (or U-edges) belong to the same component.

Now consider the component D of G containing the Hamilton cycle C. If there is a single vertex on the Hamilton cycle which is attached to two B-edges (or two U-edges), then D contains the Hamilton cycle and all the B-edges (or all U-edges). It is now easy to see that D must indeed be equal to the whole graph G. □

Notice that the special case when one U- and one B-edge is attached at each vertex v_i happens only if the vector $(k_1, k_2, ..., k_n)$ generated in sub-step 2 of step 2 has an equal number of ones and zeros. Moreover after the merging of the vector $(k_1, k_2, ..., k_n))$ with the pairs (j_1, j_2), (j_3, j_4), ..., (j_{2n-1}, j_{2n}) to the vector (j_1, j_2, k_1), (j_3, j_4, k_2), ..., (j_{2n-1}, j_{2n}, k_n) each vertex v_i appears exactly once in a triple $(j_i, j_{i+1}, 0)$ and once in a triple $(j_k, j_{k+1}, 1)$ for some indices i and k. It can be shown that the probability for such an event to happen is exponentially small. For the large number of vertices (see section 5) in which we are interested, the algorithm never encountered this situation in our simulations. If such a rare event does happen, the algorithm simply jumps back to step 1 and generates an entirely new graph.

Step 5. The planar graphs obtained up to this point are not necessarily diagrammatically prime. To obtain diagrammatically prime 4-regular RP-graphs the same algorithm as in Step 3, method one can now be carried out to decompose the graph into parts that are diagrammatically prime. Note that there are some differences to method one however. First in method one it is possible that the graph G obtained at this stage is 1-connected. Here, a graph that is Hamiltonian can still have loop edges but it must be 2-connected. Furthermore the Hamilton cycle naturally breaks apart into smaller Hamilton cycles for each of the diagrammatically prime pieces. Thus the diagrammatically prime graphs obtained are all Hamiltonian.

5. Data and Comparisons

Using 10,000 runs for varied initial values of n, large samples of diagrammatically prime RP-graphs were generated. In both methods one cannot control the exact number of crossings in the diagrammatically prime RP-graphs since a reduction procedure is required for the originally constructed planar maps with n vertices. Figure 6 shows the frequency of diagrammatically prime RP-graphs generated using method one, while Figure 7 shows the same information using method two. Figure 7 does not contain the graph for $n = 10000$ since the runtime of method two is quite large, see also Figure 11. One would expect that the larger diagrammatically prime RP-graphs are less frequent than smaller diagrammatically prime RP-graphs. However for method one the data as shown in Figure 6 for $n = 1000$, 3000 and 7000 indicates that this ratio (of the larger diagrammatically prime RP-graphs and the total number of graphs generated) does not behave linearly: there is a pronounced maximum for $n = 1000$ and $n = 3000$. These maxima may occur because the vertex numbers of the diagrammatically prime RP-graphs for a given n are dependent on each other. It seems that these maxima become less pronounced as n becomes larger. Most of the diagrammatically prime RP-graphs are quite small. For example, in method one, the $10,000$ runs for $n = 3000$ generated a total of $1,009,728$ diagrammatically prime RP-graphs; and of those only $27,820$ (or 2.76 percent) have more than 30 vertices. For $n = 500$ about 7.6 percent of the diagrammatically prime RP-graphs generated have more than 30 vertices. This indicates that as n grows, the percentage of large diagrammatically prime RP-graphs declines. However the overall number of large diagrammatically prime RP-graphs increases also rapidly since many more diagrammatically prime RP-graphs of knots are generated in one run of the algorithm for larger values of n, see Figure 8.

The behavior of the Hamiltonian RP-graphs generated using method two as shown in Figure 7 is similar to the behavior for the RP-graphs generated using method one in that the large majority of diagrammatically prime RP-graphs is very small. For $n = 500$ (or $n = 3000$) about 8.2 (or 4.0) percent of the diagrammatically prime RP-graphs generated have more than 30 vertices. Again, this indicates that as n grows, the percentage of large diagrammatically prime RP-graphs declines. As in the case for $n = 1000$ and $n = 3000$ for method one there is a pronounced local maxima for each curve. One can speculate that these maxima would start to disappear (as happens in method one) with larger values of n, however currently no

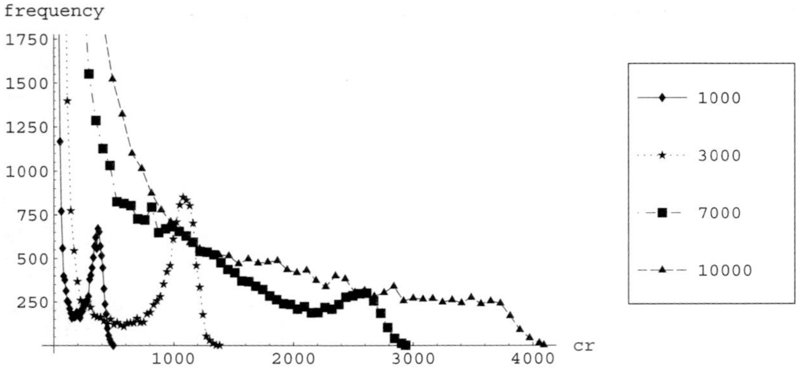

Fig. 6. The frequency of diagrammatically prime RP-graphs generated using method one and $n = 1000$, 3000, 7000 and $10,000$ for the initial number of vertices. Each curve is based on $10,000$ runs.

numeric evidence for this exists. The large runtime required by method two (Figure 11) would make this difficult to check.

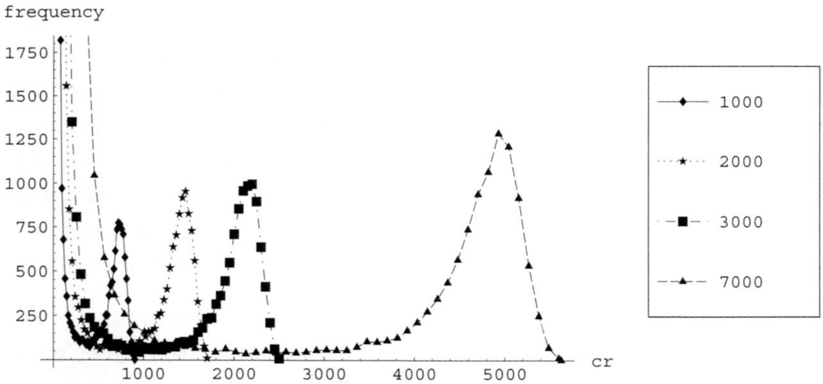

Fig. 7. The frequency of diagrammatically prime RP-graphs generated using method two and $n = 1000$, 2000, 3000 and 7000 for the initial number of vertices. Each curve is based on $10,000$ runs.

In method two there are more diagrammatically prime RP-graphs generated in each run, see Figure 8, and for a given n the largest of these RP-graphs have a larger vertex number than any of the RP-graphs generated with method one. Moreover for both methods it appears that the number of diagrammatically prime RP-graphs in a given one component RP-graph generated by method one or two (before the last step in the algorithms) grows at the order $O(n)$.

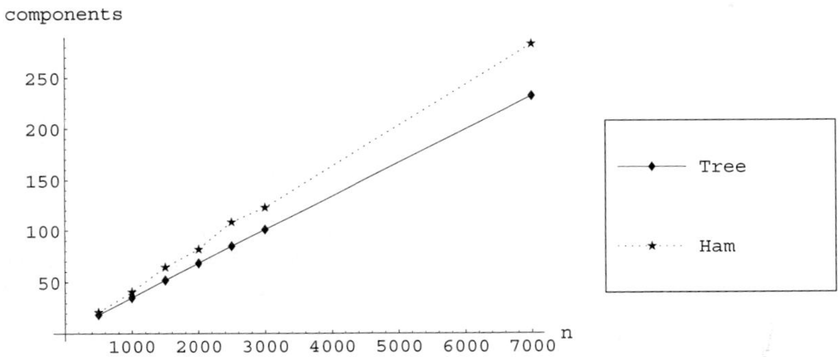

Fig. 8. The average number of diagrammatically prime RP-graphs generated with one attempt for both methods and different values of n. Each data point is based on 10,000 runs.

Figure 9 shows that method two generates larger diagrammatically prime RP-graphs. In fact the vertex number of the largest components generated with method two were almost twice the vertex number of the largest component generated using method one. Of course, these numbers are also a function of the number of trials and one would expect that as the number of runs tends to infinity, the size of the largest diagrammatically prime RP-graph would be close to n itself for both methods.

Figure 10 shows the average size of a diagrammatically prime RP-graphs generated using both methods. It appears that either this average size grows extremely slowly or for any large n the average size is bounded above by a different constant for each method. The average size for $n = 7000$ is 16.15 for method one and 24.77 for method two.

Figure 11 shows a comparison of the average run-time of both methods. One can see that method two has a much longer average run-time. Elementary data fitting suggests that the runtime of method one grows as $3.6 \times 10^{-8} n^2$, while method two has a run-time that grows as $6.8 \times 10^{-9} n^3$.

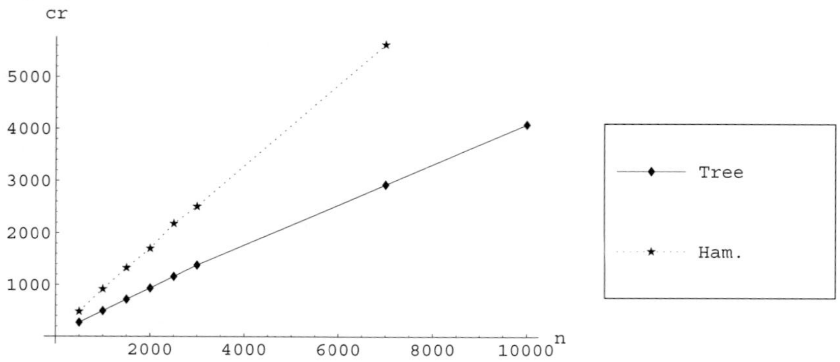

Fig. 9. The crossing number of the largest diagrammatically prime one-component RP-graphs generated using 10000 runs. Method two produces RP-graphs which are almost twice as large as the RP-graphs produced with method one.

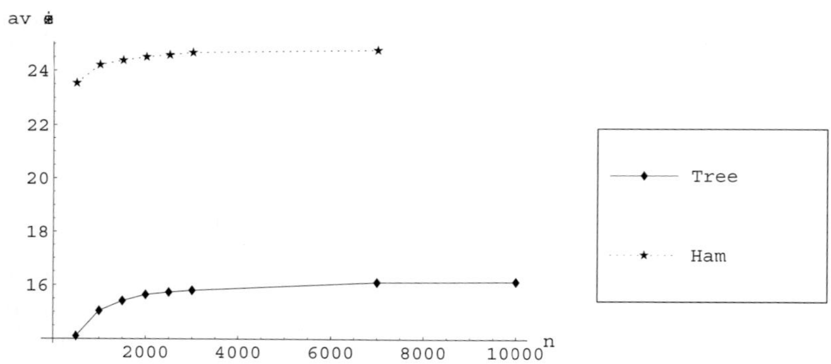

Fig. 10. The average crossing number of the diagrammatically prime one component maps generated using 10000 runs. Method two produces maps which are on average about 50 percent larger than the maps generated with method one.

6. Conclusion

We have introduced two methods that generate large random diagrammatically prime RP-graphs of knots effectively. To obtain actual knots from these diagrams one assigns an overpass/underpass at each crossing of the diagrams. If this does not lead to an alternating knot diagram then one can ask what is the relation between the actual crossing number of the resulting knot and the vertex number of the RP-graph of the knot? This

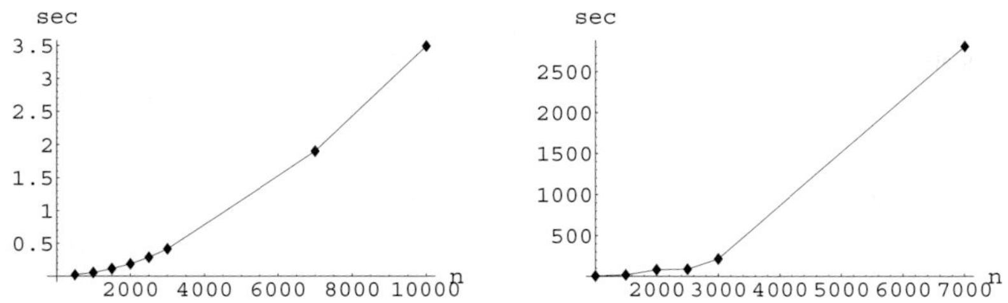

Fig. 11. The average run time of the two methods in seconds. Method one on the left, method two on the right.

is a difficult question since there is no known algorithm to determine the crossing number of a generic non-alternating knot.

There are many other unresolved issues here. For example, what is the distribution of the diagrammatically prime RP-graphs so generated? Is it uniform? Or is it more likely to have certain RP-graph than others? Since the exact number of diagrammatically prime RP-graphs of knots with n vertices is unknown one can only expect partial answers. It seems clear that for a given n the two methods produce different distributions of diagrammatically prime RP-graphs of knots, see Figure 9 or Figure 10. One has to expect that method one generates knots with a more uniform probability than method 2. Step 1 in method one generates very little bias, however the following steps (2 and 3 of method one) do introduce non-uniform probability (as they will also do in method two). In method two a lot of bias is introduced in steps 1 through 4. One can speculate that 4-regular planar graphs that have a large number of different Hamiltonian cycle are a lot more likely to be generated by method two than graphs that only have a small number of different Hamiltonian cycles. The sole advantage of method two is that the Hamiltonian cycle in the knot diagram is already known. Applying the algorithm mentioned above[8] for generating an embedding of the diagram on the cubic lattice requires finding a Hamiltonian cycle in the knot diagram. This will be easier for diagrams generated by method two than it will be for diagrams generated by method one.

To summarize: the two methods guarantee only the following at this point:

1. Any minimum RP-graph of a prime knot can be generated using method one.

2. Any Hamiltonian projection diagram of a prime knot can be gener-

ated by method two. It was recently shown that any knot has a Hamiltonian projection diagram with at most four times as many crossings as the crossing number of the knot.[8] Thus method two can also generate a RP-graph of any knot, however the graphs generated may not be minimal.

3. The data shows that it is possible to obtain fair size samples of large random knot diagrams within a reasonable time frame using these methods.

Acknowledgments

The authors are partially supported by NSF Grant #DMS-0310562.

References

1. G. Buck and J. Simon, *Thickness and Crossing Number of Knots*, Topology Appl. **91**(3) (1999), pp. 245–257.
2. G. Burde and H. Zieschang, *Knots*, De Gruyter (1985).
3. J. Cantarella, R. B. Kusner and J. M. Sullivan, *Tight Knot Values Deviate from Linear Relations*, Nature, **392** (1998), pp. 237-238.
4. Y. Diao, *The Lower Bounds of the Lengths of Thick Knots*, J. Knot Theory Ramifications **12** (2003), no. 1, 1–16.
5. Y. Diao and C. Ernst, *The Complexity of Lattice Knots*, Topology Appl. **90**, no. 1-3 (1998), pp. 1-9. **12** (2003), no. 1, pp. 1–16.
6. Y. Diao and C. Ernst, *Realizable powers of ropelengths by non-trivial knot families*, JP J. Geom. Topol. 4 (2004), no. 2, 197–208.
7. Y. Diao, C. Ernst and M. Thistlethwaite, *The Linear Growth in Rope Length of a Family of Knots*, J. Knot Theory Ramifications **12** (2003), no. 5, 709–715.
8. Y. Diao, C. Ernst and X. Yu, *Hamiltonian Knot Projections and Lengths of Thick Knots*, Topology Appl. 136 (2004), no. 1-3, 7–36.
9. C. H. Dowker and M. Thistlethwaite, *Classification of knot projections*, Topol. Appl. **16** (1983), pp. 19–31.
10. R. M. Karp, *Reducibility among combinatorial problems*, In R. Miller and J. Thatcher, editors, *Complexity of Computer Computations*, Plenum Press, 1972. pp. 85–103.
11. E. Rawdon and J. Simon, *The Moebius energy of thick knots*, Topology Appl. **125** (2002), no. 1, pp. 97–109.
12. P. Schaeffer, *Bijective census and random generation of Eulerian planar maps with prescribed vertex degrees*, Electron. J. Combin. 4 (1997), no. 1, Research Paper 20, 14 pp. (electronic).
13. C. Sundberg, and M. B. Thistlethwaite, *The rate of growth of the number of prime alternating links and tangles*, Pacific J. Math., Vol.182, No.2 (1998), 329–358.
14. P. Schaeffer, and P. Zinn-Justin, *On the asymptotic number of plane curves and alternating knots*, Experiment. Math. 13 (2004), no. 4, 483–493.
15. M. B. Thistlethwaite, *On the structure and scarcity of alternating links and tangles*, J. Knot Theory Ramifications, Vol.7, No.7 (1998), 981–1004.

CHAPTER 24

MINIMAL FLAT KNOTTED RIBBONS

Louis H. Kauffman

Department of Mathematics, Statistics and Computer Science (m/c 249)
851 South Morgan Street
University of Illinois at Chicago
Chicago, Illinois 60607-7045
E-mail: kauffman@uic.edu

This paper gives mathematical models for flat knotted ribbons, and makes specific conjectures for the least length of ribbon (for a given width) needed to tie the trefoil knot and the figure eight knot.

1. Introduction

We study framed knots, configured as knotted ribbons, where a ribbon is a space homeomorphic to a rectangle. Think of that rectangle as a long strip of paper, and imagine tying a knot in the strip and then pulling it gently tight so that it becomes a flat folded knotted strip. See Figure 3 for an illustration of this process for the trefoil knot. The folds in the strip should all be of the type shown in Figure 2. The question about such knots is this: What is the least length of strip (for a given width) needed to make a flattened version of the knot?

This question, about length to width ratio for flat knots, is analogous to a question about length to radial width (so-called *thickness*) for knotted tubes embedded in three dimensional space. The question for knotted tubes has a good-sized recent literature [1,2,5,7], but to this date there is no specific conjecture for the exact thickness of even the trefoil knot. In the present paper, we make specific conjectures for minimal length to width ratio for the trefoil and figure eight as knotted flat ribbons. It is clear from this initial exploration of knotted ribbons that there is much to think about and very interesting relationships between geometry and topology in this domain.

The paper is organized as follows. Section 2 gives our mathematical

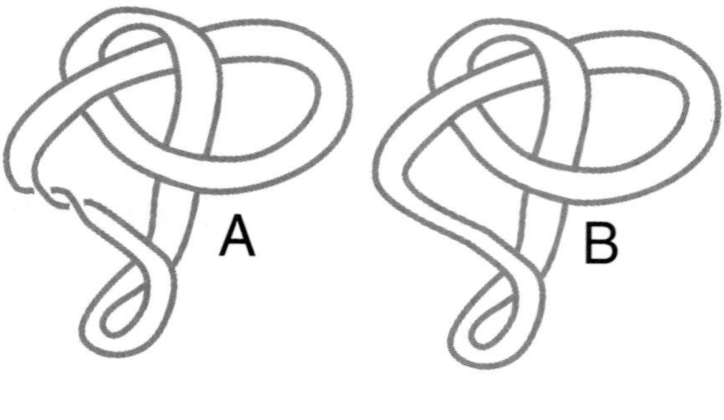

Fig. 1. Knotted Ribbons

model for flat knotted ribbons, in terms of immersed piecewise linear ribbons in the plane. We emphasize two approaches to the length to width ratio, one for closed knots, and one for truncated knots. In Section 3 we analyze the geometry of the folded trefoil and conjecture its minimal ratio for the truncated trefoil knot. In Section 4, we analyze the geometry of the folded figure eight knot and conjecture its minimal ratio. Section 5 is a discusssion of possible relationships with fully three dimensional issues.

2. A Mathematical Model for Flat Knotted Ribbons

In Figure 1 we depict two knotted ribbons (A and B) embedded in three dimensional space. The projection of ribbon A to the plane is *not* the sort of flattened knot that we are considering here. The twist in the embedding flattens to a singular projection that we wish to avoid. The projection of ribbon B to the plane is suitable for our purposes.

In Figure 2 we provide a mathematical model for the sort of folded strip that we have in mind. In the mathematical model, there is a piecewise linear curve in the plane with transverse self-intersections. This curve is the projection of the center-line of the strip to the plane after all folds have been made. Each fold is modeled as follows: Consider an angle in the piecewise linear embedding. Place a *mirror segment* at each angle so that it makes equal angles to each of the parts of the embedding that form the angle. The two parts of the embedding at the angle now have the appearance of rays being reflected by the mirror segment. Then the image of the folded ribbon in the plane is *by definition* a bundle of parallel rays reflected by the mirror segments. See Figure 2 for an illustration of this geometry. It is assumed that separate straight line bundles intersect either along mirrors, or transversely in the pattern of the self-crossings of the core diagram. A *flat*

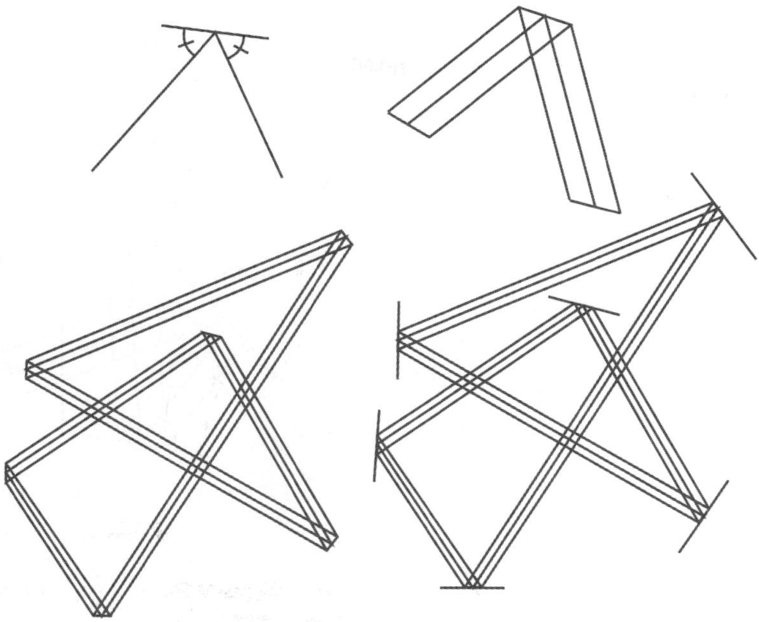

Fig. 2. Mirrors and Folds

ribbon immersion is such an immersion of a ribbon into the plane. A *flat knotted ribbon* is a choice of weaving that overlies a flat ribbon immersion. In such a weaving, the width of any straight interval of parallel rays must be chosen so that *no bundle of parallel lines can overlap a fold line when it is intended to weave that interval with the intervals meeting at the fold.* Such an overlap would be the same as a situation where the bundle of lines overlapped one of the mirrors, and would have to be either entirely over, or entirely under the mirror fold. Such a choice is obtained by first selecting an over or under crossing at each self-intersection of the center-line, and then extending these choices compatibly across the families of reflected rays. In this way, any piecewise linear knot diagram gives rise to a collection of flat knotted ribbons of varying width. For a given diagram, there will be a unique largest width knotted ribbon. We would like, within the topological type of a given knot, to determine the least length of ribbon needed for a given choice of width.

In order to discuss the issue of the length of a given knotted ribbon (for a given width), there are two choices. We can assume that the knotted ribbon is in the traditional form of a closed piecewise linear loop, and take

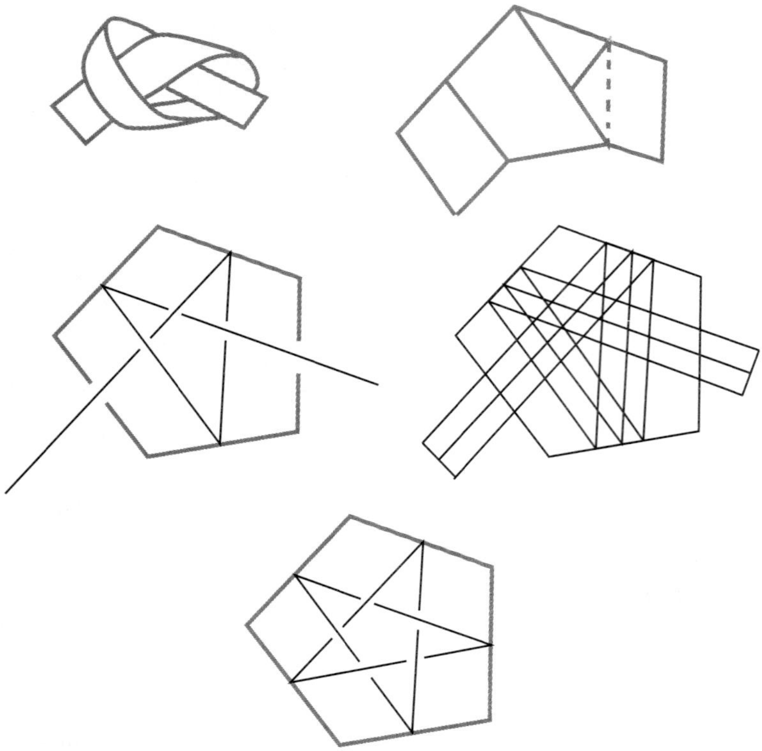

Fig. 3. A Pentagonal Mirror

the length of the core curve. This is the *closed length* of the knotted ribbon. Alternatively, we can choose to cut and truncate the knotted ribbon and measure the length of the unfolded ribbon that results from the cut. In some cases, there is a clear way to make the truncation. For example, in Figure 3 we illustrate how a trefoil knot occurs in the form of a woven pentagon. In this form it is natural to truncate the ribbons flush with the sides of the pentagon. When there is such a choice, we shall call the resulting length the *truncation length* of the knotted ribbon.

In Figure 3, we illustrate the process of forming a trefoil knot and pulling it to make a pentagonal flat knotted ribbon. We also show the internal mirror structure of this knot. The edges of a regular pentagon are the mirror segments. A ray of light enters the pentagon and is reflected three times before exiting. In the diagram, the over and undercrossings of the trefoil knot have been indicated. Note that the pentagonally folded trefoil knot is seen here as a limiting case of flat banded knots whose mirrors are segments on the sides of a regular pentagon. There is no way, using this particular geometry, to extend the lengths of the mirrors beyond the lengths of the pentagonal sides and still maintain the possiblity of an embedding.

We also show in Figure 3 how a totally internally self-reflecting pentagon would behave. In this case the alternating over and undercrossing pattern is actually not realizable by straight lines in three space, but could be effected by a weaving process, as we have described.

3. The Trefoil Knot, the Pentagon, and the Golden Ratio

We begin with an experiment that one can perform with the pattern of the trefoil knot and a strip of paper. Tie the strip into a trefoil and pull it gently tight and fold it so that you obtain a flat knot. As you pull it tight, a regular pentagon will appear.

It is intuitively clear that this pentagonal form of the trefoil knot uses the least length of paper for a given width of paper (to make a flattened trefoil). At this writing, we do not have a proof of this statement. It is the purpose of this section to analyze the pentagonal geometry and to make a quantitative version of this conjecture.

We can determine the length to width ratio of the strip of paper obtained from this pentagonal construction, by first cutting the knot exactly along the edges of the pentagon, and then unfolding this strip. See Figure 4.

The lengths l and d that label the strip in Figure 4 are the edge length l and the chord length d of the pentagon. To see this, contemplate Figure 4, or better yet, perform the experiment of making the knot, cutting it flush to the pentagon and unfolding the strip. The length of the whole strip is $L = 2(l + d)$.

In Figure 4 there is a blown-up picture of the pentagon with the edge-length l, the chord d and the width of the strip W labeled. You can see that the width is indeed the length of a particular perpendicular dropped to the chord. It is also apparent from this diagram that

$$W^2 + (d - l)^2/4 = l^2.$$

Thus we can solve for the width W of the strip in terms of l and d. The result is

$$W = \frac{\sqrt{3l^2 - d^2 + 2ld}}{2}.$$

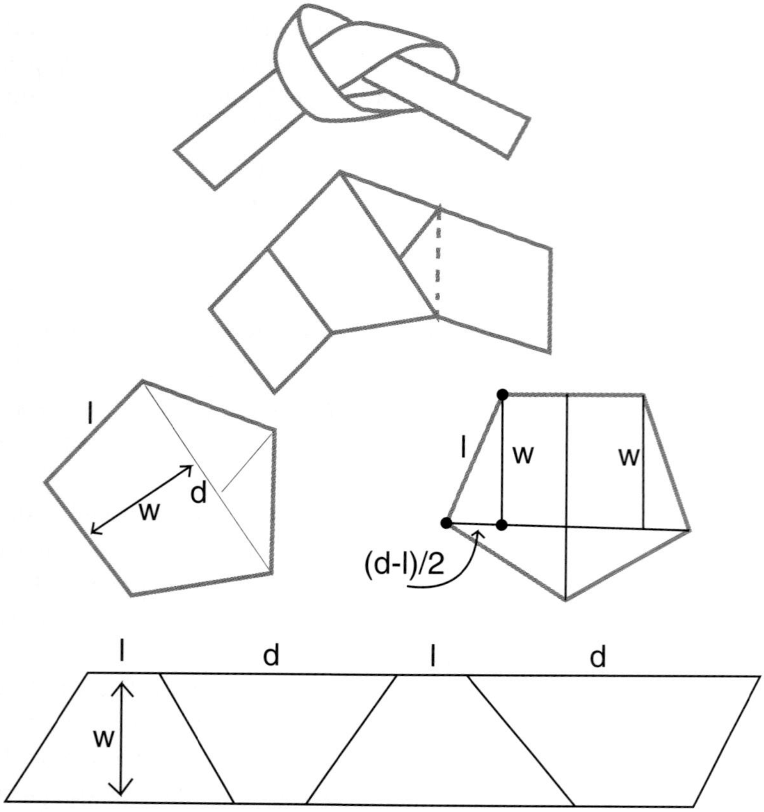

Fig. 4. Pull and Unfold Trefoil

Since the length of the strip is $L = 2(d+l)$, the ratio of length to width is

$$\begin{aligned} L/W &= \frac{4(d+l)}{\sqrt{3l^2 - d^2 + 2ld}} \\ &= \frac{4((d/l)+1)}{\sqrt{3-(d/l)^2+2(d/l)}}. \end{aligned}$$

Let $\phi = d/l$. Then we have

$$L/W = \frac{4(\phi+1)}{\sqrt{3-\phi^2+2\phi}}.$$

This is a formula for the length to width ratio of the flattened (truncated) pentagonal trefoil knot. Now we use a classical fact about the pentagon.

Fact. The ratio $\phi = d/l$ of the chord of a regular pentagon to its edge length is equal to the golden ratio. That is, $\phi = (1+\sqrt{5})/2$ and $\phi^2 = \phi + 1$. To see this fact, contemplate Figure 5.

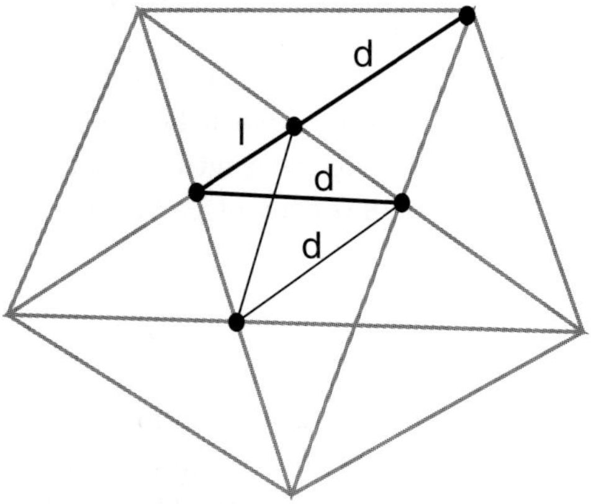

Fig. 5. The Pentagon

The small pentagon has its edge l and chord d labeled. We embed the small pentagon in the larger one and observe via a parallelogram and by similar triangles that $d/l = (d+l)/d$. Thus, with $\phi = d/l$, we have $d/l = 1 + l/d$ whence $\phi = 1 + 1/\phi$. This is sufficient to show that ϕ is the golden ratio. Note how the golden ratio appears here through the way that a pentagon embeds in a pentagon, a pentagonal self-reference.

We conclude that

$$L/W = \frac{4(\phi+1)}{\sqrt{3-\phi^2+2\phi}} = \frac{4(\phi+1)}{\sqrt{2+\phi}}.$$

Thus

$$L/W = \frac{4(\phi+1)}{\sqrt{2+\phi}} = \frac{4}{\sqrt{7-4\phi}}.$$

(We thank Robert Gray[3] for observing this last equality.) This shows how the golden ratio appears crucially in this fundamental parameter associated with the trefoil knot.

Numerically, we find that

$$L/W = 5.5055276818846941528288383276435507\cdots.$$

We conjecture that

$$L/W == 4/\sqrt{7-4\phi} = 4/\sqrt{5-2\sqrt{5}}$$

is indeed the minimum length to width ratio for a flattened truncated trefoil knot.

Remark. It is well-known in recreational mathematics that the trefoil ribbon produces a pentagon [6]. To our knowledge, this is the first time this construction has been analyzed for its length to width ratio, and it is the first time that the construction has been placed in the context of the search for minimal ribbon length.

4. Figure Eight Knot and a Hexagon

In this section we consider the analogous question for the figure eight knot. Figure 6 illustrates the experiment for the figure eight knot that leads to a flat folded hexagon. As the figure shows, the hexagon is not regular. It has two opposite parallel sides of the same length, and four sides that are equal in length to one another.

Consider Figure 7, which shows the hexagon formed by the tightened ribbon and the center line of the ribbon (appearing as five line segments with four folds), together with a sixth line segment connecting the starting point (A) and ending point (H) of this folded center line. Place the figure on a coordinate plane so that the point $O = (0,0)$, $A = (a,b)$, and $B =$

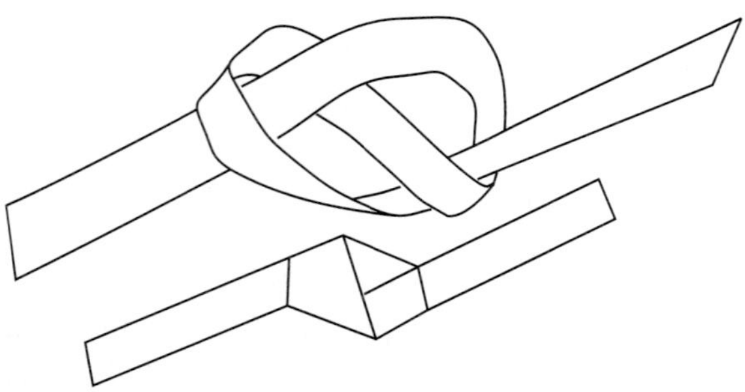

Fig. 6. Figure Eight Knot and its Hexagon

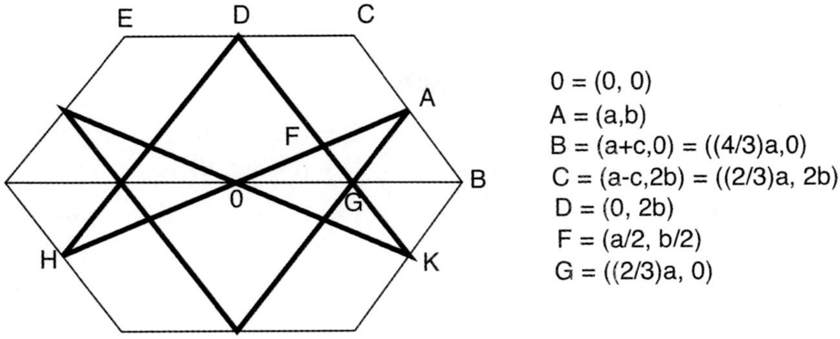

Fig. 7. Geometry of Folded Figure Eight Hexagon

$(a+c,0)$. Since A is the midpoint of BC, we must have $C=(a-c,2b)$ and $D=(0,2b)$. Finally, since lines BC and DG are parallel,

$$G = B + (D - C) = ((a+c) - (a-c), (2b) - (2b)) = (2c, 0).$$

On the other hand, $\angle AGB = \angle ABG$ because AG is parallel to DH, so $G = (a-c, 0)$ implying that $a = 3c$. Thus $B = (\frac{4}{3}a, 0)$, $C = (\frac{2}{3}a, 2b)$, and $G = (\frac{2}{3}a, 0)$.

Now the angle of incidence at A equals the angle of reflection, so $\angle BAG = \angle CAF$. Furthermore, FG and AB are parallel, so $\angle BAG = \angle AGF$ and $\angle CAF = \angle AFG$. Hence $\angle AFG = \angle AGF$. Thus $\triangle GAF$ is isosceles, so that $|AF| = |AG|$.

Since F is the intersection point of the lines OA and DG, it satisfies the equations

$$y = \frac{b}{a}x \qquad \text{and} \qquad y = \frac{-3b}{a}x + 2b.$$

Hence $F = (\frac{a}{2}, \frac{b}{2})$. Then $|AF|^2 = |AG|^2$ implies

$$\frac{1}{4}a^2 + \frac{1}{4}b^2 = \frac{1}{9}a^2 + b^2,$$

so that

$$b = \sqrt{\frac{5}{27}}a.$$

The center line of the ribbon starts at (a,b) and proceeds to $(0,-2b), (-a,b), (a,-b)$, and $(0,2b)$, before finishing at $(-a,-b)$. Thus the

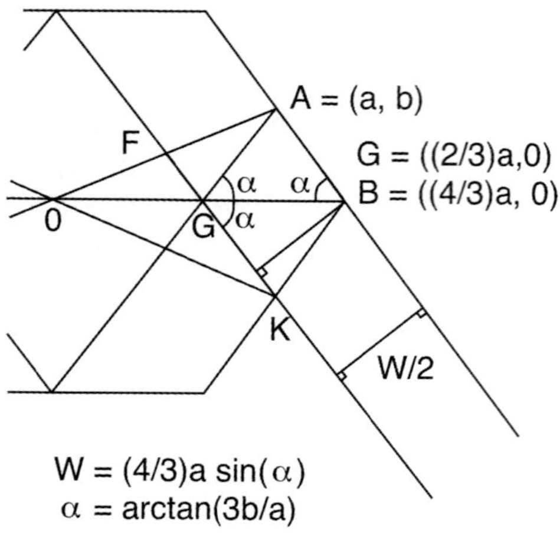

Fig. 8. Finding the width W.

length of the ribbon is

$$L = 4\sqrt{a^2 + 9b^2} + \sqrt{4a^2 + 4b^2} = \frac{32}{3}\sqrt{\frac{2}{3}}\, a$$

In order to find the width W of the ribbon, view Figure 8. Here we observe that $W/2$ is the distance between the parallel lines through AB and FG. It follows from our coordinate calculations that $|AB| = |AG|$ since

$$|AB| = |(a, b) - (4a/3, 0)| = |((a, b) - (2a/3, 0)| = |AG|.$$

From this, and the equality of alternate interior angles between parallel lines, it follows that the three angles indicated by α in Figure 8 are indeed equal to one another. That is, $\angle AGB = \angle ABG = \angle BGK$. It then follows at once that $W = (4/3)a\sin(\alpha)$ and that $\alpha = \arctan(3b/a) = \arctan(\sqrt{5/3})$. Hence

$$W = (4/3)a\sin(\arctan(\sqrt{5/3})) = (4/3)a\sqrt{5/3}/\sqrt{1 + 5/3} = a\sqrt{10}/3.$$

and so

$$L/W = \frac{32}{3}\sqrt{\frac{2}{3}}\frac{3}{\sqrt{10}} = \sqrt{\frac{2}{3}}\frac{32}{\sqrt{10}} = \frac{32}{\sqrt{15}} \approx 8.26236447190916\cdots.$$

As expected, the figure eight knot is longer than the trefoil knot. We conjecture that $\frac{32}{\sqrt{15}}$ is the minimal (truncation) length to width ratio for the figure eight knot.

Fig. 9. An alternate, longer, flat figure eight knot

Even at the level of the figure eight knot there are other ways to make the folds. But experiments so far reveal that the L/W ratio is larger for these other constructions. An example of an alternative fold is shown in Figure 9. We leave it to the reader to verify that this folding geometry leads to more length than the semi-regular hexagon described above.

5. Discussion

Some workers have thought about the relationship between thickness for flat knots and thickness for knotted tubes. For example Kusner [4] informed the author of this paper that one can get an upper bound of the form

$$Ropelength(K) \leq c_1\, Ribbonlength(K) + c_2\, Crossing(K)$$

He remarks that it seems that the best one can expect for ribbonlength is a linear relation

$$c_3\, Crossing(K) \leq Ribbonlength(K) \leq c_4\, Crossing(K)$$

with minimum crossing number. Here *Ropelength* and *Ribbonlength* refer respectively to minimal length to width ratios for tubular and flat knots, while *Crossing* refers to the minimal crossing number.

It is clear that much more work remains to be done in this field. The immediate appearance of geomerty, in relation to flattened knotted ribbons, is encouraging. The problems of minimization discussed in this paper can be explored both theoretically and via computer models. There should be important and interesting relationships between the structure of ribbons, the structure of ropes, and geometry in three dimensional space. Values such as the minimal ropelength and minimal ribbonlength for knots are fundamental topological quanta associated with the knots. It is to be expected that such numbers will be related to the geometry, topology and physics of these entities.

Acknowledgments

Most of this effort was sponsored by the Defense Advanced Research Projects Agency (DARPA) and Air Force Research Laboratory, Air Force Materiel Command, USAF, under agreement F30602-01-2-05022. The U.S. Government is authorized to reproduce and distribute reprints for Government purposes notwithstanding any copyright annotations thereon. The views and conclusions contained herein are those of the authors and should not be interpreted as necessarily representing the official policies or endorsements, either expressed or implied, of the Defense Advanced Research Projects Agency, the Air Force Research Laboratory, or the U.S. Government. (Copyright 2004.) It gives the author great pleasure to acknowledge support from NSF Grant DMS-0245588, to give thanks to the University of Waterloo and the Perimeter Institute in Waterloo, Canada for their hospitality during the preparation of this research, and to thank Jorge Alberto Calvo for a crucial correction in the computation of the figure eight knot ribbon.

References

1. G. Buck and J. Simon, Thickness and crossing number of knots, *Topology and its Appl.* **91** (1999), 245-257.
2. J. Cantarella, R. Kusner, J. Sullivan, On the minimum ropelength of knots and links, 2001, arXiv:math.GT/0103224.
3. R. Gray, (private e-mail)
4. R. Kusner, (private e-mail)
5. E. Rawdon, Approximating the thickness of a knot. In A. Stasiak, V. Katritch, and L. Kauffman, editors, "Ideal Knots", pp. 143-150. World Sci. (1998).
6. M. Schneider, "A Beginner's Guide to Constructing the Universe", Harper Collins Pub. (1994), p. 103, p. 230.
7. A. Stasiak, V. Katritch and L. Kauffman (editors), "Ideal Knots", World Sci. Pub. (1998), Vol 19 Series on Knots and Everything, (414 pages).

CHAPTER 25

QUADRISECANTS OF KNOTS WITH SMALL CROSSING NUMBER

Gyo Taek Jin

Department of Mathematics
Korea Advanced Institute of Science and Technology
Daejeon 305-701 Korea
E-mail: trefoil@kaist.ac.kr

It is known that every nontrivial knot has at least two quadrisecants. In this paper we illustrate quadrisecants for several configurations of the knots of crossing number up to five. Given a knot, we mark each intersection point of each of its quadrisecants; replacing the subarcs between marked points by straight segments gives a modified knot called the quadrisecant approximation. We conjecture that this always has the same knot type as the original.

1. Introduction

A *knot* is a locally flat simple closed curve in the Euclidean space $\mathbb{R}^3$. A *quadrisecant* of a knot K is a straight line L such that $K \cap L$ is a set of four points.[3,5,6] The *knottedness* of a knot K is the minimal number of singular points on the boundary of locally flat singular spanning disks of K.[2,6]

Theorem 1: A generic polygonal knot with knottedness k has at least $k^2/2$ quadrisecants.[6]

Corollary 2: *Every nontrivial generic polygonal knot has at least two quadrisecants.*

A polygonal knot is said to be in *general position* if no quadruple of vertices are coplanar, no triple of edges are linearly dependent, and no quadruple of edges are on a single hyperboloid of one sheet.

Proposition 3: *Every polygonal knot in general position has only finitely many quadrisecants.*

Budney, Conant, Scannell and Sinha have the same result.[1] Since every smooth knot is arbitrarily close to a polygonal knot, we expect that smooth knots in general position have finitely many quadrisecants. Quadrisecants of knots in general position are studied by Kuperberg.[3]

Lemma 4: *If four edges are mutually disjoint, then there are at most two quadrisecants to them.*

Proof: Three of the four edges belong to the same family of generators of a unique hyperboloid of one sheet. Then the fourth edge cannot be any generator of the hyperboloid. Therefore it intersects the hyperboloid in at most two points. A generator of the hyperboloid which passes through the intersection point and belongs to the other family is a quadrisecant only if it meets each of the first three edges. □

Proof of Proposition 3 : For any quadrisecant, the four secant points must belong to four distinct edges. If the four edges are mutually disjoint, then they have at most two quadrisecants, by Lemma 4. If two edges are adjacent, they determine a plane. By the general position condition, the extension of each of the other two edges intersects the plane. The line determined by the two intersection points is a quadrisecant if it meets all the four edges in distinct points. □

In this work we have located quadrisecants for some knots of crossing number not greater than five. We used Maple 8.00 for all the computations and graphics. For smooth knots, all solutions are numerical. For polygonal knots whose vertices have rational coordinates, the coefficients of the equations of the quadrisecants are rational numbers if any of the two edges involved are adjacent and roots of integral quadratic equations if all four edges are mutually disjoint. Therefore solutions to polygonal knots are exact if the vertices have rational coordinates. All of our examples meet the conjecture of Morton and Mond that there are at least $n(n-1)/2$ quadrisecants where n is the minimal crossing number.[5]

Let K be a knot which has finitely many quadrisecants. Then they cut K into finitely many subarcs. Straightening each of the subarcs with the end points fixed, we obtain a polygonal knot $\widehat{K}$ which may have self-intersections. We call $\widehat{K}$ the *quadrisecant approximation* of K. It is interesting that our examples show that taking quadrisecant approximation does not change the knot type.

Conjecture 1: *If K has finitely many quadrisecants, then the quadrisecant approximation $\widehat{K}$ has the knot type of K. Furthermore K and $\widehat{K}$ have the* *same set of quadrisecants.*

2. The knot 3_1 — the trefoil knot

We first used the following parameterization of the trefoil knot T_1.

$$x(t) = (2 + \sin 3t) \cos 2t$$
$$y(t) = (2 + \sin 3t) \sin 2t$$
$$z(t) = \cos 3t$$

As this knot has the $2\pi/3$-rotational symmetry, it has a multiple of three quadrisecants. It has only three quadrisecants. The left figure below is the projection of T_1 into a plane perpendicular to the quadrisecant shown in the right figure below.

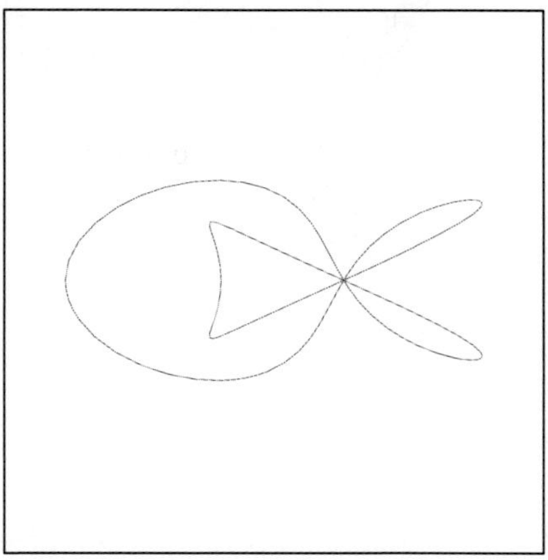

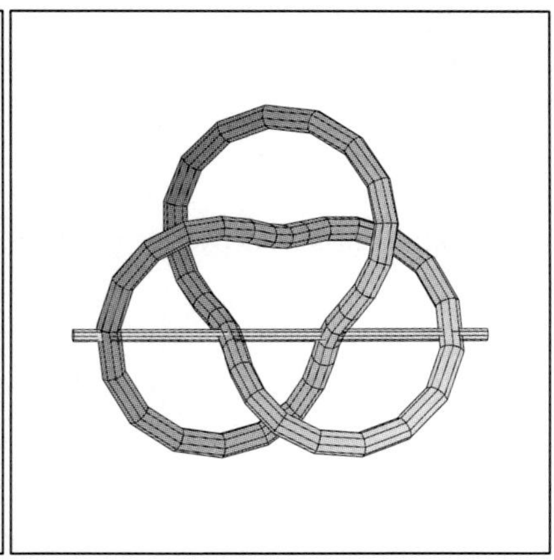

The left figure below is T_1 with all of its three quadrisecants. The right figure below is the quadrisecant approximation $\widehat{T}_1$.

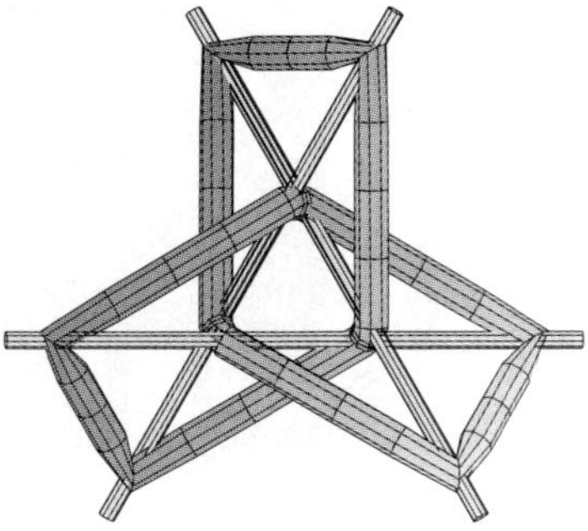

The polygonal knot T_2 with vertices cyclically located at $(-6, 23, -9)$, $(-6, -23, 9)$, $(23, -6, -9)$, $(-17, 17, 9)$, $(-17, -17, -9)$, $(23, 6, 9)$ is also a trefoil knot. It has only three quadrisecants. All the secant points have rational coordinates. The left figure below is T_2 with all of its quadrisecants. The right figure below is the quadrisecant approximation $\widehat{T}_2$. There are no other quadrisecants of $\widehat{T}_2$ than those three of T_2.

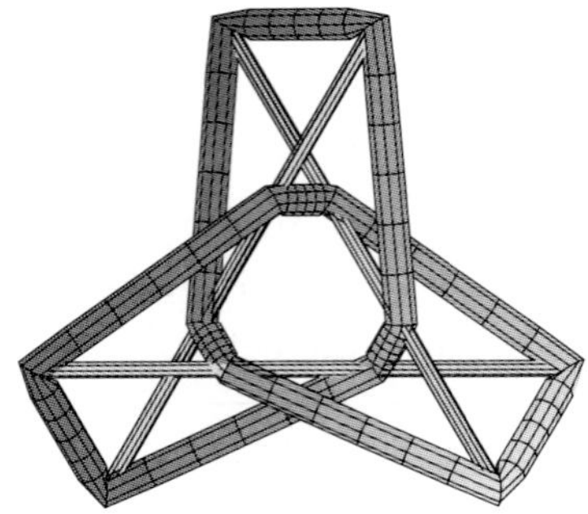

The knot T_3 given by the parameterization

$$\begin{aligned} x(t) &= (2 + \sin 2t) \cos 3t \\ y(t) &= (2 + \sin 2t) \sin 3t \\ z(t) &= \cos 2t \end{aligned}$$

is also a trefoil knot and has a π-rotational symmetry. It has four quadrisecants as shown below left. On the right is the quadrisecant approximation $\widehat{T}_3$.

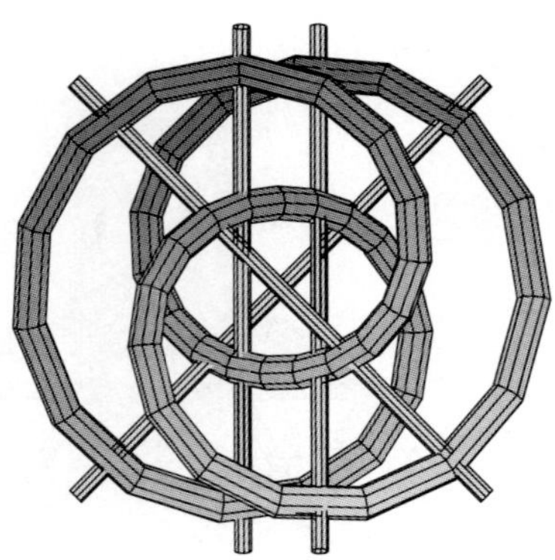

3. The knot 4_1 — the figure eight knot

The knot F_1 with the following parameterization is a figure eight knot.

$$x(t) = 32\cos t - 51\sin t - 104\cos 2t - 34\sin 2t + 104\cos 3t - 91\sin 3t$$
$$y(t) = 94\cos t + 41\sin t + 113\cos 2t - 68\cos 3t - 124\sin 3t$$
$$z(t) = 16\cos t + 73\sin t - 211\cos 2t - 39\sin 2t - 99\cos 3t - 21\sin 3t$$

It has the property that the coordinate functions are polynomials in $\sin t$ and $\cos t$ of degree three which is minimal for all figure eight knots. This minimal degree is an invariant called the *harmonic index*.[8] F_1 has six quadrisecants. In each row of the figures below, the left figure is the projection of F_1 into a plane perpendicular to the quadrisecant shown in the right figure.

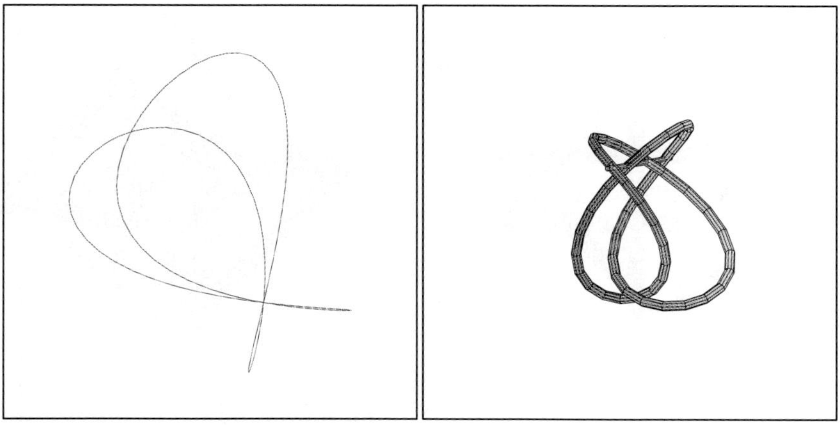

The left figure below is F_1 together with all of its six quadrisecants. The right figure below is the quadrisecant approximation $\widehat{F}_1$.

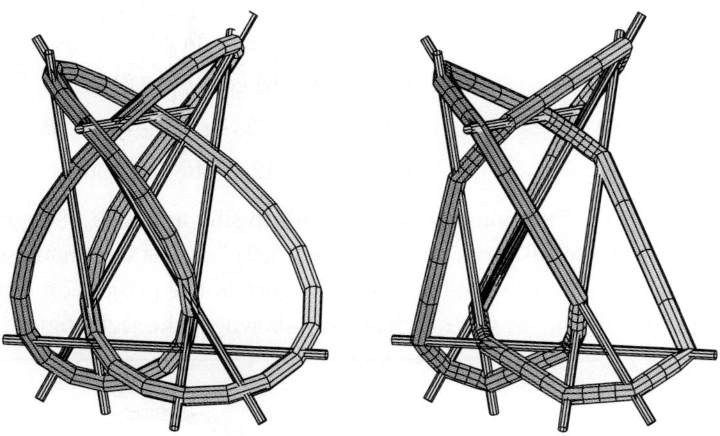

The polygonal knot F_2 whose vertices are cyclically located at $(10, 10, 2)$, $(-3, -10, 2)$, $(-10, 2, -10)$, $(3, 2, 10)$, $(10, -10, -2)$, $(-3, 10, -2)$, $(-10, -2, 10)$, $(3, -2, -10)$ is also a figure eight knot. It has six quadrisecants. All the secant points have rational coordinates. The left figure below is F_2 together with the six quadrisecants. The right figure below is the quadrisecant approximation $\widehat{F}_2$. There are no other quadrisecants of $\widehat{F}_2$ than those six of F_2. Each of F_2 and $\widehat{F}_2$ has a π-rotational symmetry and the $\pi/2$-rotation about the axis of symmetry gives the mirror image.[4]

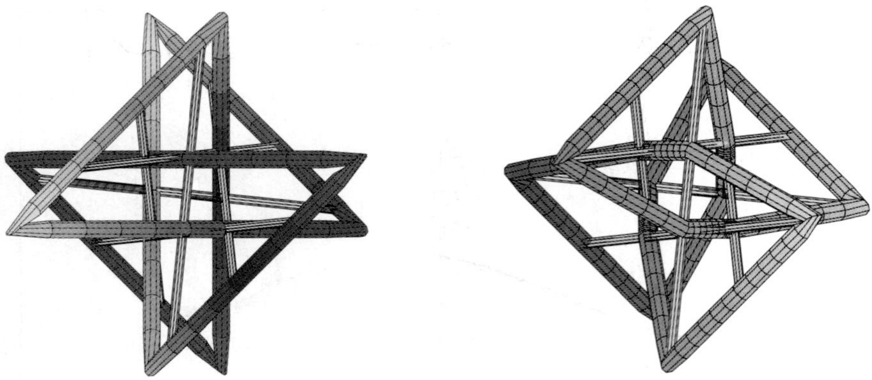

4. The knot 5_1 — the torus knot of type (5,2)

The knot C with the following parameterization is a type $(5,2)$ torus knot.

$$\begin{aligned} x(t) &= 88\cos t + 115\sin t - 475\cos 2t - 127\sin 2t - 87\cos 3t \\ &\qquad + 36\sin 3t + 11\cos 4t - 19\sin 4t \\ y(t) &= 89\cos t - 32\sin t - 172\cos 2t + 294\sin 2t + 76\cos 3t \\ &\qquad + 102\sin 3t - 61\cos 4t + 113\sin 4t \\ z(t) &= 44\cos t - 69\sin t + 34\cos 2t + 223\sin 2t + 16\cos 3t \\ &\qquad + 120\sin 3t + 42\cos 4t - 125\sin 4t \end{aligned}$$

The coordinate functions are polynomials in $\sin t$ and $\cos t$ of degree four which is minimal for all torus knots of type $(5,2)$.[8] C has ten quadrisecants. In each row of the figures below, the left figure is the projection of C into a plane perpendicular to the quadrisecant shown in the right figure.

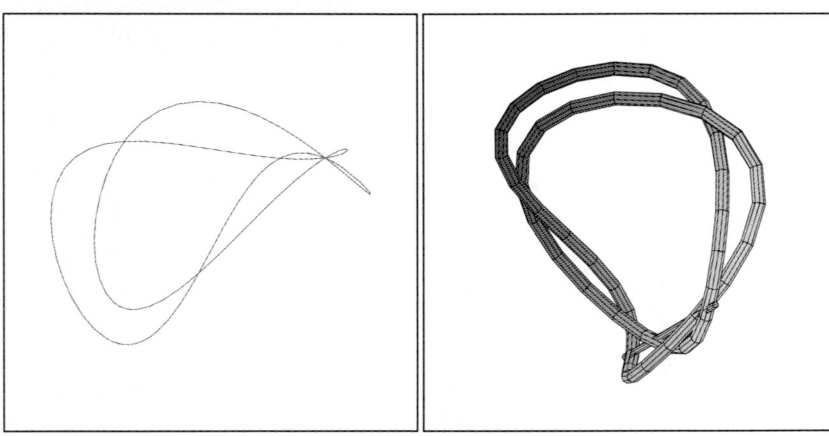

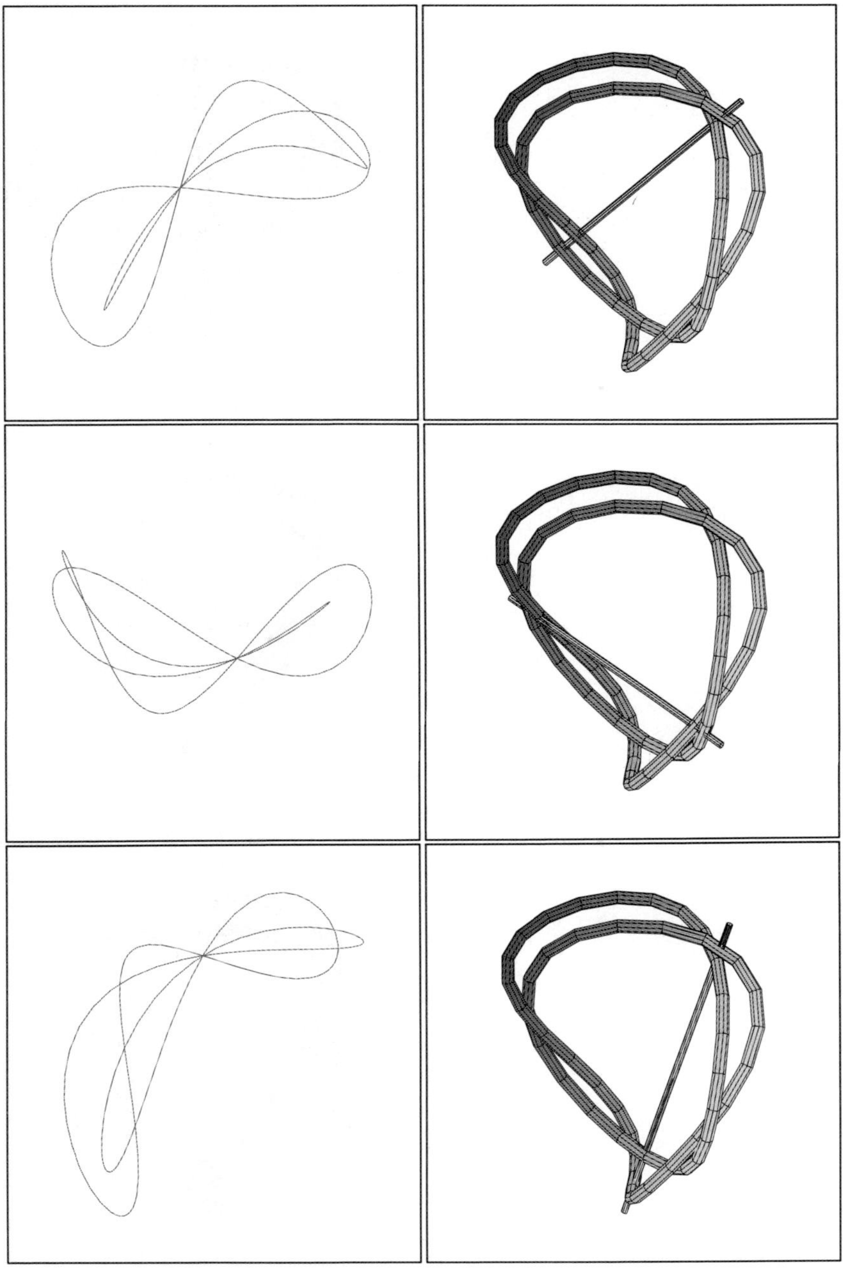

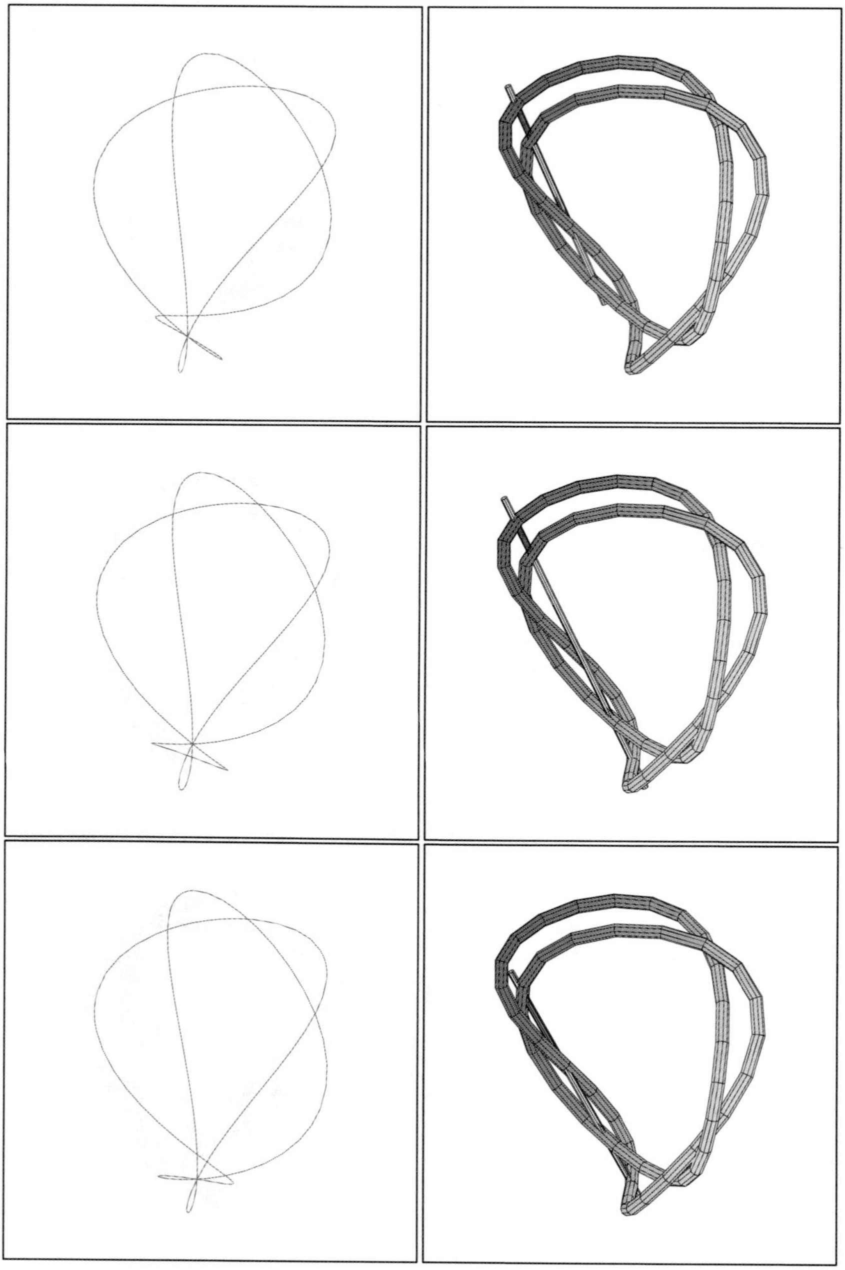

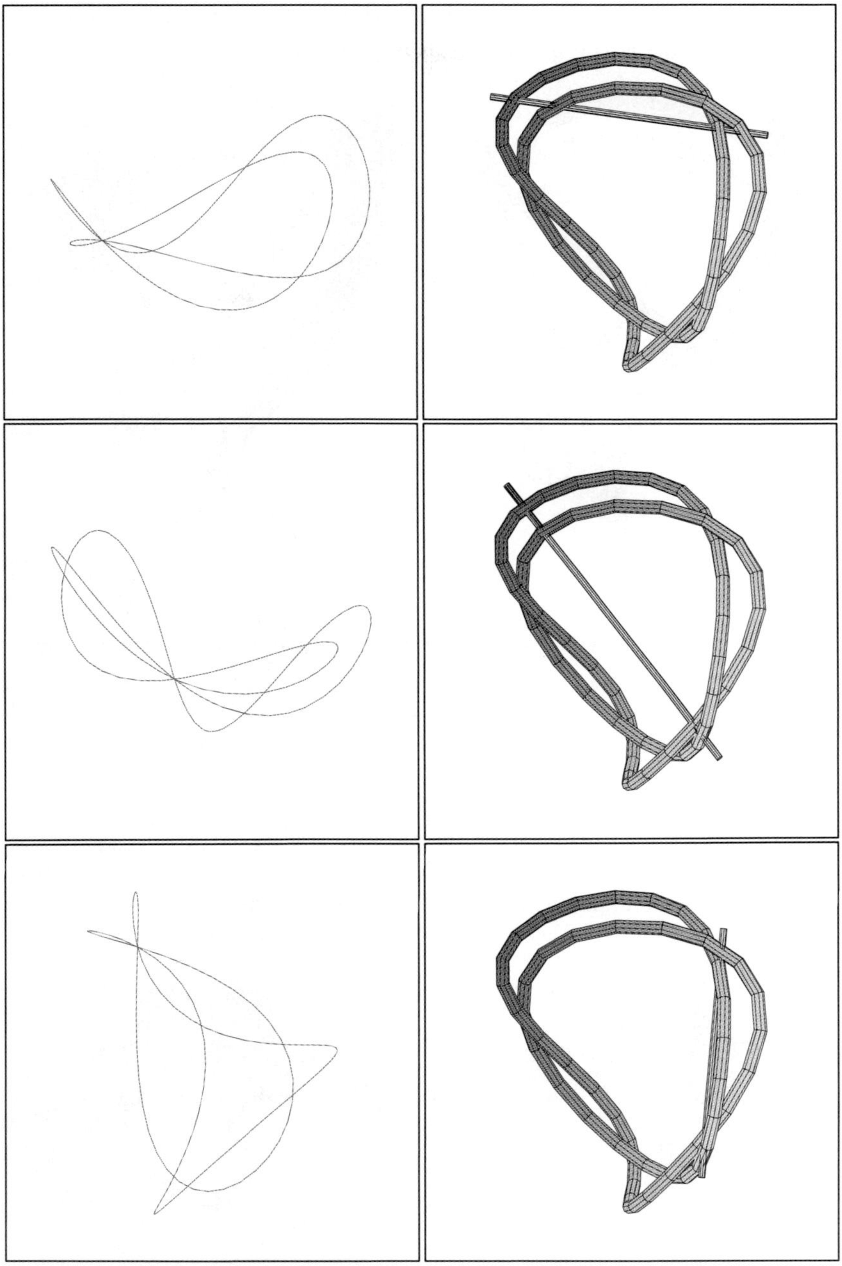

The left figure below is C together with all of its ten quadrisecants. The right figure below is the quadrisecant approximation $\widehat{C}$.

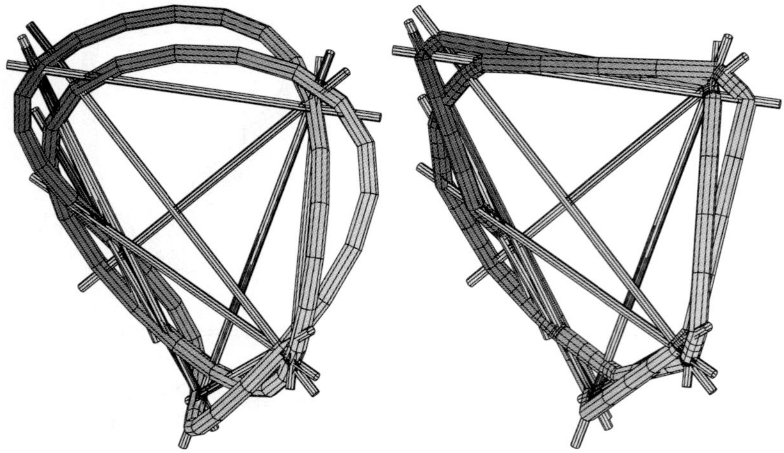

The torus knot of type (5,2) on a standard torus parameterized by

$$x(t) = (4 + \sin 5t)\cos 2t$$
$$y(t) = (4 + \sin 5t)\sin 2t$$
$$z(t) = \cos 5t$$

is shown below left with its ten quadrisecants. On the right is its quadrisecant approximation shown with the ten quadrisecants.

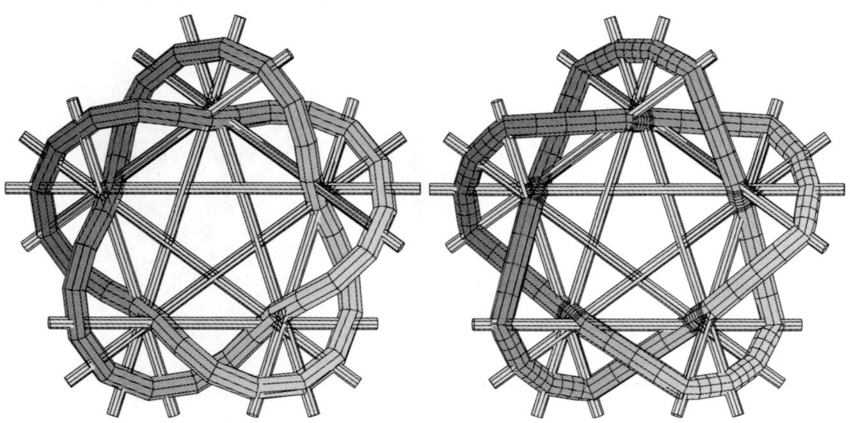

5. The knot 5_2

The knot with the following parameterization is a knot of type 5_2.

$$
\begin{aligned}
x(t) &= -33\cos t + 43\sin t + 214\sin 2t - 101\cos 3t - 47\sin 3t + 11\sin 4t \\
y(t) &= -57\cos t + 99\sin t - 54\cos 2t - 159\sin 2t - 117\cos 3t \\
&\qquad -5\sin 3t - 31\cos 4t - 45\sin 4t \\
z(t) &= 34\cos t - 21\sin t - 100\cos 2t - 93\sin 2t - 27\cos 3t \\
&\qquad -16\sin 3t + 52\cos 4t + 84\sin 4t
\end{aligned}
$$

It has ten quadrisecants. In each row of the figures below, the left figure is the projection of the knot into a plane perpendicular to the quadrisecant shown in the right figure.

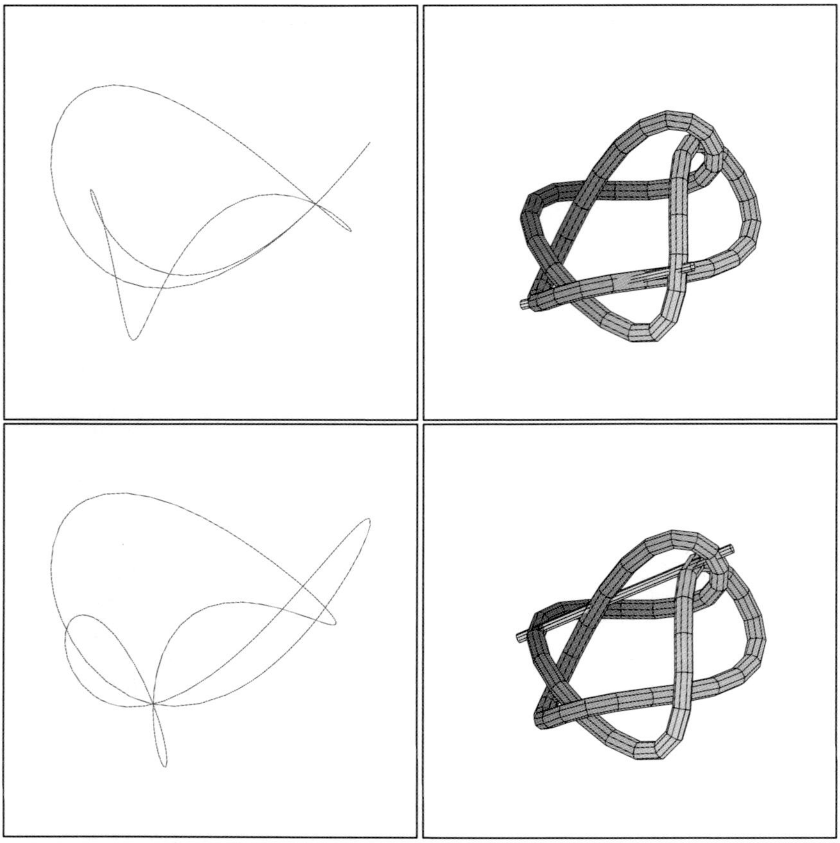

The left figure below shows the knot together with all of its ten quadrisecants. The right figure below is the quadrisecant approximation.

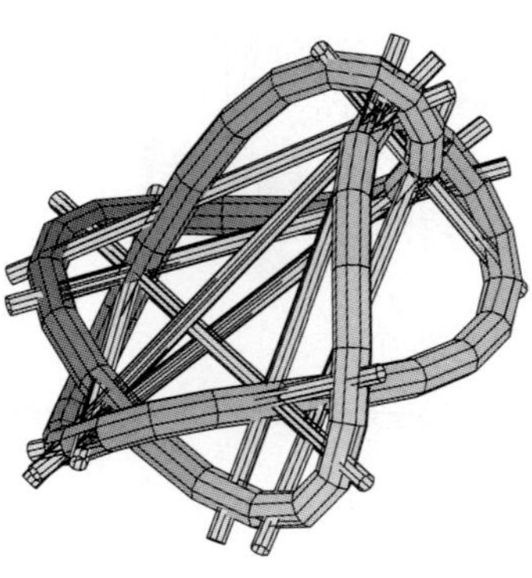

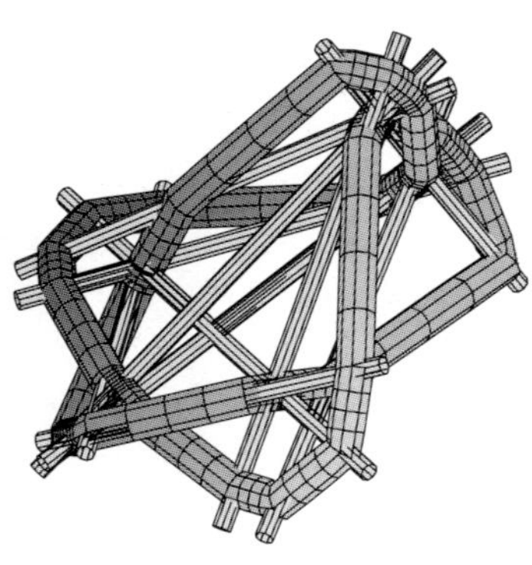

Acknowledgments

This work was partially supported by the KOSEF project 20016-101-01-2. The author would like to thank the editor and the referee for their valuable comments.

References

1. R. Budney, J. Conant, K.P. Scannell and D. Sinha, *New perspectives on self-linking*, arXiv:math.GT/0303034.
2. G. Burde and H. Zieschang, *Knots*, de Gruyter Studies in Mathematics vol. 5, Walter de Gruyter, Berlin, New York, 1985
3. G. Kuperberg,, *Quadrisecants of knots and links*, J. Knot Theory Ramifications **3**(1994) 41–50; arXiv:math.GT/9712205.
4. M. Meissen, Edge number results for piecewise-linear knots, *Knot theory* (Banach Center Publications vol. 42, Warszawa, 1998) 235–242.
5. H.R. Morton and D.M.Q. Mond, Closed curves with no quadrisecants, Topology **21**(1982) 235–243.
6. E. Pannwitz, *Eine elementargeometrisch Eigenshaft von Verschlingungen und Knoten*, Math. Ann. 108(1933) 629–672.
7. D. Rolfsen, *Knots and Links* Mathematics Lecture Series 7, Publish or Perish, 1976
8. A.K. Trautwein, *An introduction to harmonic knots*, Ideal Knots (Series on Knots and Everything vol.19, World Scientific, 1998), pp. 353–363.

CHAPTER 26

ON THE WRITHING NUMBER OF A NON-CLOSED CURVE

E.L. Starostin

Max Planck Institute for Physics of Complex Systems
Nöthnitzer Str. 38, D-01187 Dresden, Germany
E-mail: star@mpipks-dresden.mpg.de

The paper deals with the definition and computation of the writhing number of an arbitrary fragment of a space curve. The approach is based on closing the tangent indicatrix with a geodesic. A relationship connecting the writhe with the Gauß integral over the open curve is studied. Single and double helical shapes are presented as examples.

1. Introduction

The term writhing number (or simply writhe) was first proposed by Fuller[1] for a quantity $\mathcal{W}r$ that arises as a difference between the linking number $\mathcal{L}k$ and the twist (or twisting number) $\mathcal{T}w$ of a closed ribbon in the Călugăreanu-White-Fuller formula[1,2,3]

$$\mathcal{W}r = \mathcal{L}k - \mathcal{T}w.$$

The sense of this very simple-looking and very famous relationship is that the right-hand side, though defined for a ribbon, depends only on its central curve.

Since $\mathcal{W}r$ is a characteristic of geometric complexity of a spatial curve, it makes this quantity worthy of consideration when examining long physical objects.[4] In particular, values of $\mathcal{W}r$ have been computed in a number of works for various models of large-scale structure of DNA (e.g., Refs. 5, 6, 7, 8) as well as for experimental data on these molecules.[9] RNA tertiary structures and protein folding are other neighbouring areas for an application of the geometrical and topological tools developed in DNA studies.[10,11,12] However, in the strict sense, application of writhe is confined to smoothly closed shapes, though quite a lot of interesting objects have their ends not joined (or joined non-smoothly), e.g., studies based on the recently developed experimental techniques of manipulation with single DNA molecules.[13] In such problems, the model curves require an appropriately defined measure to characterize their arrangement in space. Therefore, it seems to be helpful to extend the notion of writhe to non-closed curves and their fragments. Indeed, such attempts (explicit and implicit) have been made.[9,12,14,15,16,17,18,19,20,21] It happens that the definitions of the writhe suggested by different authors are not consistent. One of the first works where it was proposed how to compute the writhing number for an open (and non-smooth) curve, is Ref. 22.

The aim of this paper is to give a consistent and natural generalization of the notion of writhe to an arbitrary fragment of a curve and to present explicit formulas for its computation as function of arc length. The basic idea consists in a construction of the closure of the fragment under consideration in such a way that it would correspond to the closure of the tangent indicatrix by an arc of a great circle, as has been proposed by Maggs.[19,20] This approach is fully consistent with the recipe, given by Le Bret[23], on how to close the tangent indicatrix of a polygonal line.

2. Notation and preliminaries

We start our consideration with a smooth non-self-intersecting curve $A = \{\mathbf{r}(s) : [0, L] \to \mathbb{R}^3\}$ of class C^2, s being the arc length. We assume that the segment has a natural orientation in the direction the arc coordinate increases. The concatenation of two segments A and B having, respectively, a common ending and starting point will be written as $A + B$.

A continuous vector function $\mathbf{u}(s) : [0, L] \to \mathbf{S}^2 \equiv \{z \in \mathbb{R}^3; \|z\| = 1\}$ may be chosen such that $\mathbf{u}(s) \cdot \mathbf{r}'(s) = 0, \forall s$; here $'$ denotes the derivative with respect to s. Let $\epsilon > 0$ be small enough so that the ribbon $R_\epsilon = \{\mathbf{r} + \mu\mathbf{u}, -\epsilon \le \mu \le \epsilon\}$ does not cross itself.

2.1. *Twist, writhe and linking number*

The *twisting number* of the ribbon (i.e., of the pair $(\mathbf{r}, \mathbf{u})$) is defined by

$$\mathcal{T}w(\mathbf{r}, \mathbf{u}) = \frac{1}{2\pi} \int_A \mathbf{r}' \times \mathbf{u} \cdot d\mathbf{u}.$$

If $\mathbf{r}$ is of class C^3 and $\mathbf{r}'' \ne \mathbf{0}$, then the twist of a ribbon may be decomposed into the twist of the Frenet frame plus the twist of the ribbon relative to the Frenet frame:[24]

$$\mathcal{T}w(\mathbf{r},\mathbf{u}) = \mathcal{T}w_F(\mathbf{r}) + \frac{1}{2\pi}\int_0^L d\phi.$$

The angle $\phi = \phi(s)$ is an angle between $\mathbf{u}$ and the principal normal. By the Frenet ribbon is meant a special one formed by the principal normal vector to the curve. The Frenet ribbon is defined by the space curve $\mathbf{r}(s)$, if $\mathbf{r}'' \neq 0$. The twist of the Frenet ribbon is

$$\mathcal{T}w_F(\mathbf{r}) = \frac{1}{2\pi}\int_0^L \tau(s)ds, \tag{1}$$

where $\tau(s)$ is the torsion of the curve $\mathbf{r}(s)$. Clearly, for planar curves, $\mathcal{T}w_F = 0$.

Now let the curve be smoothly closed: $\mathbf{r}(0) = \mathbf{r}(L), \mathbf{r}'(0) = \mathbf{r}'(L)$. For two closed curves A and B, $A \cap B = \emptyset$, the Gauß linking integral gives an integer-valued topological invariant

$$\mathcal{L}k(A,B) = \frac{1}{4\pi}\int_B\int_A \frac{(\mathbf{r}_A(s_1) - \mathbf{r}_B(s_2))\cdot(\mathbf{t}_A(s_1)\times\mathbf{t}_B(s_2))}{\|\mathbf{r}_A(s_1) - \mathbf{r}_B(s_2)\|^3}ds_1 ds_2,$$

called *linking number.*

The quantity called *writhe* may be expressed as the double integral

$$\mathcal{W}r_A = \frac{1}{4\pi}\int_0^L\int_0^L \frac{(\mathbf{r}(s_1) - \mathbf{r}(s_2))\cdot(\mathbf{t}(s_1)\times\mathbf{t}(s_2))}{\|\mathbf{r}(s_1) - \mathbf{r}(s_2)\|^3}ds_1 ds_2, \tag{2}$$

where $\mathbf{t} = \mathbf{r}'(s)$ is the tangent vector and s_1, s_2 are arc lengths. The right-hand side of Eq. (2) is the Gauß linking integral in the singular case of being over all distinct pairs of points on one curve. The writhe depends only on the shape of the curve.

2.2. *Basic relations*

The following theorem is due to Fuller[25,26] (we shall call it Fuller's first theorem).

Theorem *Let $B = \{\mathbf{r}(s)\}$ be a closed oriented space curve of class C^3 with its tangent $\mathbf{r}'(s)$, s the arc length. The tangent traces out a* closed *curve $\tilde{B}(s)$ on the unit sphere which is piecewise of class C^2. The curve $\tilde{B}(s)$ is divided into a finite family of non-self-intersecting closed piecewise C^2*

space curves. Each curve of this family then encloses a domain Ω_i defined so that the geodesic normal points into its interior. Let S_B be the sum of the areas of these domains (the components are counted with multiplicity determined by how many times the corresponding domains are encircled by the curve). Then

$$\mathcal{W}r(\mathbf{r}) = \frac{S_B}{2\pi} - 1 \qquad \mathrm{mod}\ 2.$$

Let the ribbon $(\mathbf{r}, \mathbf{u})$ also be closed: $\mathbf{u}(0) = \mathbf{u}(L)$. Denote by $\mathcal{L}k(\mathbf{r} - \epsilon\mathbf{u}, \mathbf{r} + \epsilon\mathbf{u}) \equiv \mathcal{L}k(\mathbf{r}, \mathbf{u})$ the linking number of the two boundary curves $\mathbf{r} - \epsilon\mathbf{u}$ and $\mathbf{r} + \epsilon\mathbf{u}$. For ϵ small enough, $\mathcal{L}k(\mathbf{r}, \mathbf{u})$ does not depend on ϵ. This justifies omitting ϵ in the following. In other words, we shall be dealing with arbitrarily narrow ribbons.

The famous Călugăreanu-White-Fuller theorem[1,2,3] claims that the difference of the linking and twisting numbers is the writhe:

$$\mathcal{L}k(\mathbf{r}, \mathbf{u}) - \mathcal{T}w(\mathbf{r}, \mathbf{u}) = \mathcal{W}r(\mathbf{r}).$$

3. Writhe of an arbitrary open segment

Consider a spatial curve segment $A = \{\mathbf{r}_A(s), s \in [0, L_A]\}$, with the non-vanishing tangent vector. The tangent indicatrix $\tilde{A}$ need not be closed. Let $\mathbf{t}_{A0}$ and $\mathbf{t}_{A1}$ be the tangent vectors at the beginning and end points of A, respectively.

Following Maggs,[19,20] we choose to close the tangent indicatrix $\tilde{A}$ with a geodesic $\tilde{G}$ in order to get a measure for the writhe of A (Fig. 1). This choice is natural and is supported by treatment of analogous problems in optics and quantum mechanics.[27] (In the generic case $\mathbf{t}_{A0} \neq \pm\mathbf{t}_{A1}$, there are two possible geodesics, we take either of them; the cases $\mathbf{t}_{A0} = \pm\mathbf{t}_{A1}$ will be discussed later.) Let the tangent vectors at the ends of the geodesic be denoted by $\mathbf{n}_{G0}$ and $\mathbf{n}_{G1}$.

It is only possible to define and compute the fractional part of the writhe because the choice of closing geodesic is arbitrary. Then, the writhe of an open segment A may be determined by the following relation:

$$\mathcal{W}r_A = \frac{S_{A+G}}{2\pi} \qquad \mathrm{mod}\ 1, \tag{3}$$

where S_{A+G} is the spherical area enclosed by $\tilde{A}$ and $\tilde{G}$ (in the same sense as in Fuller's first theorem). We are able to specify the area only modulo 2π.

It may be shown that the curve A can be closed with a curve having continuous tangent and such that its tangent indicatrix is $\tilde{G}$ plus possibly

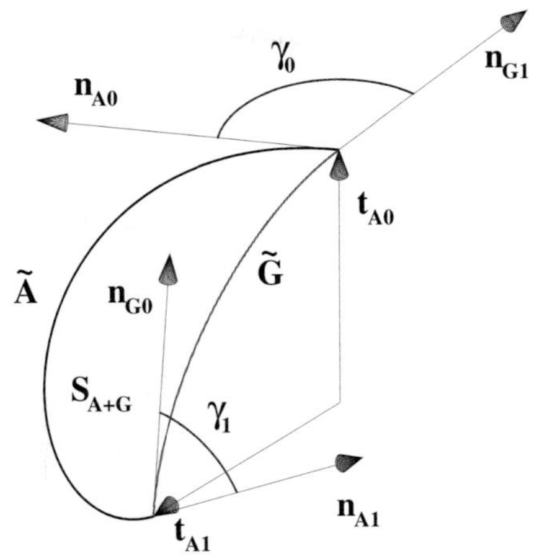

Fig. 1. The tangent indicatrix $\tilde{A}$ of the segment A closed with a geodesic $\tilde{G}$.

some pieces of great circles that do not change the enclosed spherical area modulo 2π.

Now construct a ribbon R_A for the segment A in the same manner as is done in the proof of Fuller's first theorem.[26] Generally, the ribbon R_A is determined by the unit principal normal vector $\mathbf{n}_A(s)$. Therefore, for pieces between possible inflection points of A with discontinuous normals, the twist is computed as that of the Frenet ribbon by Eq. (1). In the vicinities of the inflection points the ribbon R_A is to be arbitrary modified to make it continuous with a new modified generating vector $\mathbf{u}_A$. For the sake of simplicity, we further assume that $\mathbf{u}_{A0} = \mathbf{n}_A(0)$ and $\mathbf{u}_{A1} = \mathbf{n}_A(L_A)$. The twist $\mathcal{T}w_A$ is well-defined for the non-closed ribbon R_A, which can be closed with a ribbon R_G based on the curve G. The ribbon R_G may be constructed such that its twist equals $\mathcal{T}w_G = \frac{1}{2\pi}(\gamma_1 + \gamma_0)$, where γ_0 is an angle from the normal $\mathbf{n}_{G1}$ to $\mathbf{u}_{A0}(= \mathbf{n}_{A0})$ and γ_1 from the normal $\mathbf{u}_{A1}(= \mathbf{n}_{A1})$ to $\mathbf{n}_{G0}$.

Applying the Gauß-Bonnet theorem[28], we have

$$S_{A+G} + 2\pi\mathcal{T}w_A + \gamma_1 + \gamma_0 = 0. \tag{4}$$

Elimination of the area term from Eq. (3) and Eq. (4) results in an expression for the writhe

$$\mathcal{W}r_A = -\mathcal{T}w_A - \frac{\gamma_1 + \gamma_0}{2\pi} \mod 1. \tag{5}$$

The angles γ_1 and γ_0 are determined by the equations

$$\cos\gamma_1 = \mathbf{n}_{A1}\cdot\mathbf{n}_{G0}, \quad \sin\gamma_1 = (\mathbf{n}_{A1}\times\mathbf{n}_{G0})\cdot\mathbf{t}_{A1},$$

$$\cos\gamma_0 = \mathbf{n}_{A0}\cdot\mathbf{n}_{G1}, \quad \sin\gamma_0 = (\mathbf{n}_{G1}\times\mathbf{n}_{A0})\cdot\mathbf{t}_{A0}.$$

The vectors $\mathbf{n}_{G0}$ and $\mathbf{n}_{G1}$ may be easily expressed as functions of the initial and terminal tangents of A. Indeed, the vector $\mathbf{n}_{G0}$ lies in the plane spanned by the vectors $\mathbf{t}_{A0}$ and $\mathbf{t}_{A1}$. Besides, $\mathbf{n}_{G0}\cdot\mathbf{t}_{A1} = 0$ and we may write

$$\mathbf{n}_{G0} = \pm\frac{\mathbf{t}_{A0} - (\mathbf{t}_{A0}\cdot\mathbf{t}_{A1})\mathbf{t}_{A1}}{\sqrt{1-(\mathbf{t}_{A0}\cdot\mathbf{t}_{A1})^2}}.$$

Similarly, the vector $\mathbf{n}_{G1}$ lies in the same plane and $\mathbf{n}_{G1}\cdot\mathbf{t}_{A0} = 0$. Therefore,

$$\mathbf{n}_{G1} = \mp\frac{\mathbf{t}_{A1} - (\mathbf{t}_{A0}\cdot\mathbf{t}_{A1})\mathbf{t}_{A0}}{\sqrt{1-(\mathbf{t}_{A0}\cdot\mathbf{t}_{A1})^2}}$$

(the signs depend on the choice of the geodesic).

The angles γ_1 and γ_0 can be found from their trigonometrical functions

$$\cos\gamma_1 = \pm\frac{\mathbf{n}_{A1}\cdot\mathbf{t}_{A0}}{\sqrt{1-(\mathbf{t}_{A0}\cdot\mathbf{t}_{A1})^2}}, \quad \sin\gamma_1 = \pm\frac{\mathbf{b}_{A1}\cdot\mathbf{t}_{A0}}{\sqrt{1-(\mathbf{t}_{A0}\cdot\mathbf{t}_{A1})^2}}, \tag{6}$$

$$\cos\gamma_0 = \mp\frac{\mathbf{n}_{A0}\cdot\mathbf{t}_{A1}}{\sqrt{1-(\mathbf{t}_{A0}\cdot\mathbf{t}_{A1})^2}}, \quad \sin\gamma_0 = \pm\frac{\mathbf{b}_{A0}\cdot\mathbf{t}_{A1}}{\sqrt{1-(\mathbf{t}_{A0}\cdot\mathbf{t}_{A1})^2}}, \tag{7}$$

where $\mathbf{b}_A(s) = \mathbf{t}_A(s)\times\mathbf{n}_A(s)$ is the binormal vector and $\mathbf{b}_{A0}\equiv\mathbf{b}_A(0)$, $\mathbf{b}_{A1}\equiv\mathbf{b}_A(L_A)$.

We can conclude that Eq. (5), together with Eqs. (6), (7), provides a means to compute the fractional part of the writhe for an (almost) arbitrary curve with open ends.

It may happen that the tangent vector at the beginning is the same as one at the ending point: $\mathbf{t}_{A0} = \mathbf{t}_{A1}$. Then the tangent indicatrix is closed and it sweeps the spherical area S_A (Fig. 2). Equation (3) transforms to

$$\mathcal{W}r_A = \frac{S_A}{2\pi} \mod 1$$

and Eq. (5) to

$$\mathcal{W}r_A = -\mathcal{T}w_A - \frac{\gamma}{2\pi} \mod 1, \tag{8}$$

where γ is the angle from $\mathbf{u}_{A1}$ to $\mathbf{u}_{A0}$ (actually, from $\mathbf{n}_{A1}$ to $\mathbf{n}_{A0}$).

An analogue to Eq. (8) was used in the analysis of the elongation of a supercoiled DNA molecule carried out by Bouchiat and Mézard[17,18] (though their angle χ is measured in the opposite direction to γ).

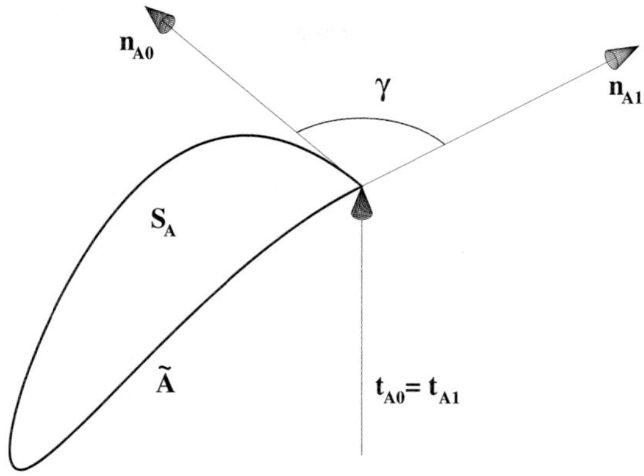

Fig. 2. The closed tangent indicatrix $\tilde{A}$ of the segment A.

Remark 1. It follows from the above that the writhe of a curve segment (closed or open) whose tangent indicatrix is geodesic and such that $\mathbf{t}_{A0} + \mathbf{t}_{A1} \neq 0$, is an integer. In other words, the writhe of any planar curve is always an integer.

Consider now the case $\mathbf{t}_{A0} = -\mathbf{t}_{A1}$. The plane of the closing geodesic G is then not determined. If we examine the behaviour of the tangent indicatrix in the vicinities $s = 0$ and $s = L_A$ (also paying attention to the neighbouring curves on $\mathbf{S}^2$ close to the critical one), then we see that the plane of the closing geodesics may rotate through $\sim \pi$ as the length of the segment changes so that the critical point $\mathbf{t}_{A0} = -\mathbf{t}_{A1}$ is passed. In mechanics terms, this phenomenon may be called flipping of the closing segment. It is the critical point where the choice of one geodesic based on a continuity argument is no longer valid.

Speaking more strictly, the value of the writhe for a segment with oppositely directed ends is not determined. It may be ascribed a value which is the average of the two limits taken as the length of the segment is pre-critical and post-critical. This means that the great circle plane for the closing geodesics in the critical point has to be taken orthogonal to the limiting positions of both of the great circle planes chosen for the growing segment $[\epsilon, L_A - \epsilon]$ and the decreasing one $[-\epsilon, L_A + \epsilon]$ as $\epsilon \to 0$. (We assume that the definition segment for the curve A may be infinitesimally extended in both directions.) Formally, this choice is a plane spanned by $\mathbf{t}_{A0}, \mathbf{n}_{A0} + \mathbf{n}_{A1}$. For such a closure, $\gamma_1 + \gamma_0 = \pi$ and Eq. (5) becomes

$$\mathcal{W}r_A = -\mathcal{T}w_A - \frac{1}{2} \mod 1,$$

while Eq. (3) does not change (as usual, S_{A+G} means the area swept out by the closed curve $A+G$ defined as above). If, in addition, $\mathbf{n}_{A0}+\mathbf{n}_{A1} = 0$, then a plane spanned by $\mathbf{t}_{A0}$ and $\mathbf{b}_{A0}$ has to be chosen.

Remark 2. In the above consideration we have used the ribbon based on the principal normal (cf. Ref. 26), though any other continuous ribbon may be taken to obtain essentially the same formula for writhe (of course, the specific expressions for the angles γ_1 and γ_0 should be appropriately modified).

In particular, the fractional part of writhe equals the twist of a special ribbon such that its generating normal vectors at the ends have the property that they could be transformed to each other by a parallel transport along the closing geodesics (i.e., $\gamma_1 + \gamma_0 = 0 \mod 2\pi$).

3.1. *Broken curve*

The above approach may be naturally extended to a sequence of disjoint segments. Let $A = \sum_{i=1}^{n} A^{(i)}$ be a set of n continuous fragments. Each $A^{(i)}$ is oriented so that $A_0^{(i)}$ and $A_1^{(i)}$ are its initial and terminal points, respectively. Based on $A^{(i)}$, a ribbon $R_A^{(i)}$ may be built as was done for a single piece of curve. We also construct n additional pieces that tie the end point of the i-th fragment to the initial point of the subsequent one. We identify formally the point $A_0^{(n+1)}$ with $A_0^{(1)}$ to make the entire curve closed. The connecting parts are built in exactly the same way as the closure of a single segment in the previous subsection. Thus, we can repeat our arguments to obtain

$$\mathcal{W}r_A = -\sum_{i=1}^{n} \mathcal{T}w_i - \frac{1}{2\pi}\sum_{i=1}^{n}(\gamma_1^{(i)} + \gamma_0^{(i)}) \mod 1, \tag{9}$$

where $\mathcal{T}w_i$ is the twist of the ribbon $R_A^{(i)}$, the angles $\gamma_1^{(i)}$ and $\gamma_0^{(i)}$ are determined by their trigonometric functions

$$\cos\gamma_1^{(i)} = \pm\frac{\mathbf{n}_{A1}^{(i)} \cdot \mathbf{t}_{A0}^{(j)}}{\sqrt{1 - \left(\mathbf{t}_{A0}^{(j)} \cdot \mathbf{t}_{A1}^{(i)}\right)^2}}, \quad \sin\gamma_1^{(i)} = \pm\frac{\mathbf{b}_{A1}^{(i)} \cdot \mathbf{t}_{A0}^{(j)}}{\sqrt{1 - \left(\mathbf{t}_{A0}^{(j)} \cdot \mathbf{t}_{A1}^{(i)}\right)^2}},$$

$$\cos\gamma_0^{(i)} = \mp\frac{\mathbf{n}_{A0}^{(j)}\cdot\mathbf{t}_{A1}^{(i)}}{\sqrt{1-\left(\mathbf{t}_{A0}^{(j)}\cdot\mathbf{t}_{A1}^{(i)}\right)^2}},\quad \sin\gamma_0^{(i)} = \pm\frac{\mathbf{b}_{A0}^{(j)}\cdot\mathbf{t}_{A1}^{(i)}}{\sqrt{1-\left(\mathbf{t}_{A0}^{(j)}\cdot\mathbf{t}_{A1}^{(i)}\right)^2}},$$

and $\mathbf{t}_{Av}^{(i)}, \mathbf{n}_{Av}^{(i)}, \mathbf{b}_{Av}^{(i)}$ are the Frenet frames at the beginning ($v = 0$) and at the end ($v = 1$) of the i-th segment ($i = 1, 2, \ldots, n,\quad j = 1 + (i \mod n)$).

Note that the value of writhe generally depends on both the order of fragments and the orientations along them.

3.2. *Non-smoothly closed loop*

A special case arises when the segment A forms a non-smoothly closed shape. This means a discontinuity of the tangent vector at the initial point and the tangent indicatrix is not closed. The whole procedure described above may be applied to such a loop although one complication appears: the resulting closed curve to which the basic Călugăreanu-White-Fuller formula is to be applied has a self-intersection point at the beginning of the loop considered. Generally, the writhe is not defined for such shapes. However, on the one hand, it can be shown that the Gauß integral exists unless the tangent at the loop starting point is exactly opposite to the end tangent. On the other hand, under the same limitation, we can restrict ourselves to consideration of two limiting curves approaching the self-intersection shape from two different sides. As is well known, the writhe jumps by 2 as a curve crosses itself.[25] Thus, the fractional part of the writhe is not affected by self-intersection and may be computed by Eq. (3) or Eq. (5) in the same way as for the open segment.

4. Writhe and the Gauß integral

The writhe of the smooth closed curve may be expressed as the double integral Eq. (2). It is evident that the writhe for an open segment of length L as defined above can no longer be computed as the Gauß integral over this segment though, in most cases, the double integral itself is also well-defined for smooth non-closed curves.

Our aim here is to obtain a formula connecting both values: on the one hand, the writhe that relates the difference between the linking number and the twisting for the ribbon built with the geodesic closure and, on the other hand, simply the double integral taken over the open segment.

4.1. *Open curve and its closure*

Consider an open smooth non-self-intersecting curve $A = \{\mathbf{r}(s) : [0, L] \to \mathbb{R}^3\}$ (Fig. 3). We assume here that the tangent vectors $\mathbf{t}(s) = \mathbf{r}'(s)$ are neither parallel nor antiparallel at the ends: $\mathbf{t}(0) \neq \pm\mathbf{t}(L)$ (we will examine these cases later). We extend the curve A with two straight line segments: $B = \{\mathbf{r}_B(s_1) = \mathbf{r}(L) + s_1\mathbf{t}(L), s_1 \in [0, l]\}$ and $C = \{\mathbf{r}_C(s_2) = \mathbf{r}(0) + s_2\mathbf{t}(0), s_2 \in [-l, 0]\}$. Note that both segments have the same length l. Now connect the end points of B and C with the straight line segment $D_l = \{\mathbf{r}_D(\xi) = (1-\xi)\mathbf{r}_B(l) + \xi\mathbf{r}_C(-l) = (1-\xi)(\mathbf{r}(L) + l\mathbf{t}(L)) + \xi(\mathbf{r}(0) - l\mathbf{t}(0))\}$. The direction of D_l is determined by its tangent

$$\mathbf{t}_D(l) = \frac{\frac{d\mathbf{r}_D(\xi)}{d\xi}}{\left\|\frac{d\mathbf{r}_D(\xi)}{d\xi}\right\|} = \frac{\mathbf{r}_C(-l) - \mathbf{r}_B(l)}{\|\mathbf{r}_C(-l) - \mathbf{r}_B(l)\|}.$$

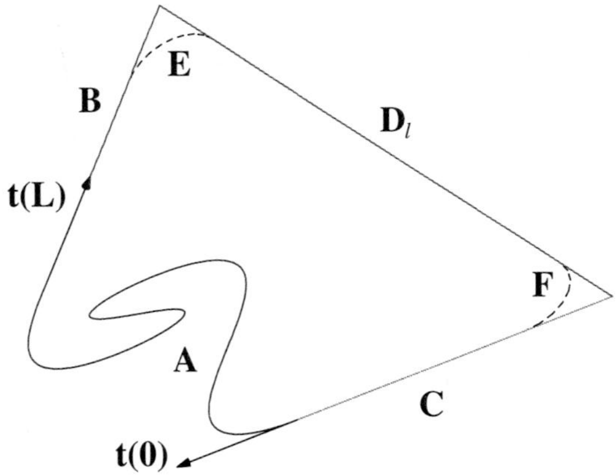

Fig. 3. The curve A is extended with the straight line segments B and C and closed with D_l.

Now let the lengths of B and C increase to infinity and compute the limiting orientation of the tangent $\mathbf{t}_D$:

$$\mathbf{t}_{D_\infty} = \lim_{l\to\infty} \mathbf{t}_D(l) = -\frac{\mathbf{t}(L) + \mathbf{t}(0)}{\|\mathbf{t}(L) + \mathbf{t}(0)\|}. \tag{10}$$

Thus, we see that in the limit $l \to \infty$, D_∞ lies in the plane defined by the initial and end tangents of the segment A.

In the case when $\mathbf{t}(0) = \pm\mathbf{t}(L)$, we can also attach two straight line segments. If $\mathbf{t}(0) = -\mathbf{t}(L)$, then all straight lines connecting these segments belong to the same plane defined by $\mathbf{t}(0)$ and $\mathbf{r}(L) - \mathbf{r}(0)$. The case $\mathbf{t}(0) = \mathbf{t}(L)$ requires a special consideration.

What we have now is a closed circuit $A + B + D_l + C$. It is smooth except for two points at the beginning and at the end of D_l. We modify B and D_l in the small vicinity of where they join themselves by introducing a planar curvilinear segment E with the tangent varying from $\mathbf{t}(L)$ to $\mathbf{t}_D$. All three segments involved belong to the same plane spanned by $\mathbf{t}(L)$ and $\mathbf{t}_D$. We can assume that the length of E does not depend on l. The length of the shortened segment B_* is decreased to be l_*.

A similar procedure may be carried out to smooth the join of the segments D_l and C. The new planar curvilinear segment F belongs to the plane spanned by $\mathbf{t}(0)$ and $\mathbf{t}_D$. The length of F is the same for every l; without loss of generality, we assume that the length of the shortened segment C_* equals l_*, as well.

We have come to the smooth closed curve $A+B_*+E+D_*+F+C_*$. We are interested in the limiting case when $l_* \to \infty$. The tangent indicatrix of the initial curve A is then closed by a geodesic corresponding to the limiting curve $E_\infty + D_\infty + F_\infty$. This follows from the construction of the curves and from Eq. (10). The limiting curve $B_\infty + E_\infty + D_\infty + F_\infty$ may be considered as an implementation of the first part G of the closure suggested in Sec. 3. Thus, the writhe of the open segment A may be computed as the writhe of the limiting closed curve $A + B_\infty + E_\infty + D_\infty + F_\infty + C_\infty$ and its fractional part satisfies Eq. (3) and Eq. (5).

However, for the smooth closed curve $A + B_\infty + E_\infty + D_\infty + F_\infty + C_\infty$, the writhe may be obtained independently by the double integral formula. Since the circuit consists of 6 parts, we have to consider all pairs involved in the double integration. For brevity, we will denote an integral over a pair of curves P and Q by (P, Q). Clearly, (P, Q) equals (Q, P).

Before proceeding, we obtain some simple estimate for the value of the double integral

$$I_2 = \int_0^{\mathcal{L}_2} \int_0^{\mathcal{L}_1} I_{\mathcal{W}r}(\sigma_1, \sigma_2) \, d\sigma_1 \, d\sigma_2,$$

where

$$I_{\mathcal{W}r}(\sigma_1, \sigma_2) = \frac{(\mathbf{r}_1(\sigma_1) - \mathbf{r}_2(\sigma_2)) \cdot (\mathbf{t}_1(\sigma_1) \times \mathbf{t}_2(\sigma_2))}{\|\mathbf{r}_1(\sigma_1) - \mathbf{r}_2(\sigma_2)\|^3}.$$

The integral I_2 is taken over two smooth curves $\mathbf{r}_1(\sigma_1), \sigma_1 \in [0, \mathcal{L}_1]$ and $\mathbf{r}_2(\sigma_2), \sigma_2 \in [0, \mathcal{L}_2]$. Let $\Delta \equiv \min_{\sigma_1,\sigma_2} \|\mathbf{r}_1(\sigma_1) - \mathbf{r}_2(\sigma_2)\| > 0$. Then

$$|I_2| \leq \int_0^{\mathcal{L}_2}\int_0^{\mathcal{L}_1} |I_{\mathcal{W}r}(\sigma_1,\sigma_2)|\, d\sigma_1\, d\sigma_2 \leq \int_0^{\mathcal{L}_2}\int_0^{\mathcal{L}_1} \frac{d\sigma_1\, d\sigma_2}{\|\mathbf{r}_1(\sigma_1) - \mathbf{r}_2(\sigma_2)\|^2} \leq \frac{\mathcal{L}_1\mathcal{L}_2}{\Delta^2}. \tag{11}$$

Eq. (11) implies that $\lim_{\Delta\to\infty} I_2 = 0$ for any two curves of finite length. If one of the curves has its length of order Δ or less, i.e., $\mathcal{L}_i = \mathcal{O}(\Delta), i = 1, 2$, and the other has finite length $\mathcal{L}_{3-i}$, then the integral I_2 vanishes as $\Delta \to \infty$.

We now return to the integral over pairs of curves. The integrals (B_*, B_*), (B_*, E), (B_*, D_*), (E, E), (E, D_*), (D_*, D_*), (D_*, F), (D_*, C_*), (F, F), (F, C_*), (C_*, C_*) equal zero because the integrand vanishes for coplanar curves. If $\mathbf{t}(0) = -\mathbf{t}(L)$, then the whole closure is planar and the integrals (B_*, F), (E, F), (E, C_*) also vanish for every l_*. If $\mathbf{t}(0) \neq \pm\mathbf{t}(L)$, then the length of D_* is of order l_* for large l_*. Therefore, by applying Eq. (11), we conclude that the integrals (B_*, F), (E, F), (E, C_*) as well as (A, E), (A, D_*), (A, F) all approach zero as $l_* \to \infty$. We denote the remaining possibly non-zero integrals as follows:

$$\mathcal{W}y = \frac{1}{4\pi}\int_A\int_A I_{\mathcal{W}r}(s,\tilde{s})\, ds\, d\tilde{s},$$

$$\mathcal{S}w_1 = \frac{1}{2\pi}\int_A\int_{B_\infty} I_{\mathcal{W}r}(s,s_1)\, ds_1\, ds, \quad \mathcal{S}w_2 = \frac{1}{2\pi}\int_A\int_{C_\infty} I_{\mathcal{W}r}(s,s_2)\, ds_2\, ds,$$

$$\mathcal{S}q = \frac{1}{2\pi}\int_{C_\infty}\int_{B_\infty} I_{\mathcal{W}r}(s_1,s_2)\, ds_1\, ds_2.$$

We call them the wry, the swirl and the squint, respectively.

Thus, the writhe of A may be represented as a sum

$$\mathcal{W}r = \mathcal{W}y + \mathcal{S}w_1 + \mathcal{S}w_2 + \mathcal{S}q. \tag{12}$$

The first summand $\mathcal{W}y$ is simply the double integral over the open segment under consideration. Therefore, Eq. (12) provides a connection between this integral and the writhe.

Let us now examine the case of parallel tangents $\mathbf{t}(0) = \mathbf{t}(L)$. The two attached segments B and C then have opposite directions and, instead of the straight line D, we connect them by a circular arc $\breve{D}$ joining B and C at the same distance l from the ends of A. The arc $\breve{D}$ lies in the plane spanned by the vectors $\mathbf{r}(L) - \mathbf{r}(0)$ and $\mathbf{t}(0)$ and its length is of order l for large l. Again, the smoothing curves E and F can be constructed in a similar way as it is done in the regular case. Thus, we obtain the smooth planar closure of the curve A.

After letting the lengths of B_* and C_* go to infinity and analysing the double integral components in the expression for the writhe, we come to the same Eq. (12) with the right-hand terms defined as above.

It may occur that the ray B or C intersects the curve A. Then, generally, the writhe of the whole closed curve $A+B+E+D+F+C$ is not defined. The situation is the same as for a non-smoothly closed loop (Sec. 3.2). In the generic case, when the tangents in the point of the intersection are neither coincident nor of opposite direction, the fractional part of the writhe still can be found by examination of the two limiting positions of the curves in the vicinity of the intersection point. Since the writhe jumps by 2 as the curve goes through itself, the half-sum of the writhes for those curves may be taken as the value of the writhe. The same approach may be applied to another singular case when the rays B and C cross each other. Moreover, the constraint of non-self-intersection of the initial open fragment A may also be weakened in the similar fashion.

Next we clarify the structure of the integrals $\mathcal{S}w$ and $\mathcal{S}q$.

4.2. *Swirl*

Consider

$$\mathcal{S}w_1 = \frac{1}{2\pi}\int_0^L\int_0^\infty \frac{(\mathbf{r}(s) - \mathbf{r}(L) - s_1\mathbf{t}(L))\cdot(\mathbf{t}(s)\times\mathbf{t}(L))}{\|\mathbf{r}(s) - \mathbf{r}(L) - s_1\mathbf{t}(L)\|^3}\,ds_1\,ds$$

$$= \frac{1}{2\pi}\int_0^L (\mathbf{R}(s)\cdot(\mathbf{t}(s)\times\mathbf{t}_1)\hat{I}_{\mathcal{S}w}(s)\,ds. \quad (13)$$

Here we denote

$$\hat{I}_{\mathcal{S}w}(s) = \int_0^\infty \frac{ds_1}{[(\mathbf{R} - (\mathbf{R}\cdot\mathbf{t}_1)\mathbf{t}_1)^2 + (\mathbf{R}\cdot\mathbf{t}_1 - s_1)^2]^{\frac{3}{2}}}$$

and $\mathbf{R} \equiv \mathbf{R}(s) \equiv \mathbf{r}(s) - \mathbf{r}(L)$, $\mathbf{t}_1 \equiv \mathbf{t}(L)$. We can represent $\hat{I}_{\mathcal{S}w}(s)$ as

$$\hat{I}_{\mathcal{S}w}(s) = \int_0^\infty \frac{ds_1}{[a^2 + (b - s_1)^2]^{\frac{3}{2}}}, \quad b \equiv b(s) \equiv \mathbf{R}\cdot\mathbf{t}_1, \quad a^2 \equiv a^2(s) \equiv (\mathbf{R}-b\mathbf{t}_1)^2,$$

and carry out the integration to get

$$\hat{I}_{\mathcal{S}w}(s) = \left.\frac{s_1 - b}{a^2\sqrt{a^2 + (b - s_1)^2}}\right|_0^\infty = \frac{1}{\sqrt{a^2 + b^2}(\sqrt{a^2 + b^2} - b)}. \quad (14)$$

Substituting $a(s), b(s)$ into Eq. (14) and the result into Eq. (13) yields

$$\mathcal{S}w_1 = \frac{1}{2\pi}\int_0^L \frac{\mathbf{R}(s)\cdot(\mathbf{t}(s)\times\mathbf{t}_1)}{\|\mathbf{R}(s)\|(\|\mathbf{R}(s)\| - \mathbf{R}(s)\cdot\mathbf{t}_1)}\, ds. \quad (15)$$

Let us introduce the spherical coordinate system with the origin at the point $\mathbf{r}(L)$ and let the z-axis be directed along the ray B (Fig. 4). Then $\mathbf{R}(s) = (\rho\cos\psi\cos\phi, \rho\cos\psi\sin\phi, \rho\sin\psi)$, $\mathbf{t}_1 = (0, 0, 1)$, and $\rho \equiv \rho(s)$, $\phi \equiv \phi(s)$, $\psi \equiv \psi(s)$ are the functions describing the curve A.

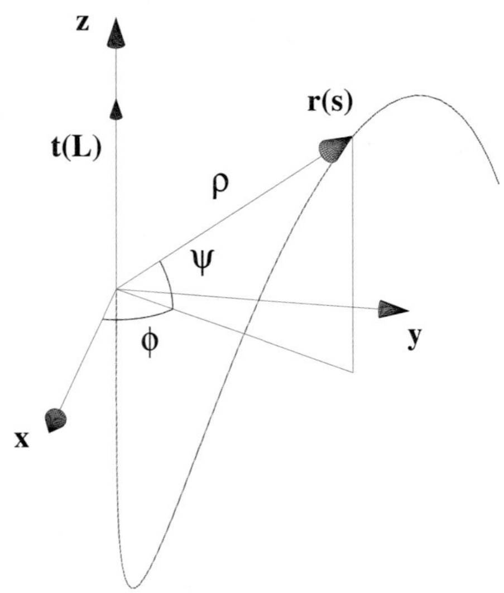

Fig. 4. The spherical coordinates ρ, ϕ, ψ. The z-axis is directed along the tangent $\mathbf{t}(L)$ at the end point of the curve $\mathbf{r}(s)$.

In these coordinates, Eq. (15) takes the form

$$\mathcal{S}w_1 = \frac{1}{2\pi}\int_0^L \phi'(1 + \sin\psi)\, ds.$$

Note that the swirl is zero if $\phi(s) = const$, i.e., if the whole curve A is planar. The swirl also vanishes when $\psi(s) = -\frac{\pi}{2}$ which means that the curve A is a straight line continuation of the ray B. It is natural that the swirl is scale-invariant: it does not depend explicitly on how far the points of the curve are from the ray. The second integral $\mathcal{S}w_2$ over the ray C_∞ has the same structure.

4.3. *Squint*

Now consider the integral $\mathcal{S}q$ over the two rays B_∞ and C_∞. It is convenient to introduce special Cartesian coordinates with origin at the point $\mathbf{r}(0)$ and the x-axis directed along $\mathbf{r}(L)-\mathbf{r}(0)$ (Fig. 5). Let the y-axis lie in the plane of the ray B_∞ and the z-axis be chosen such that the whole coordinate system is right-handed. Denote by $\phi \in [0,\pi]$ the angle from the x-axis to the direction of $\mathbf{t}(L)$. The orientation of the ray C_∞ is defined by two angles: $\psi \in [0,\pi]$ between the x-axis and $\mathbf{t}(0)$ and $\theta \in [0,2\pi]$ between the xy-plane and the plane spanned by the x-axis and $\mathbf{t}(0)$. In this coordinate system we may represent both rays as follows:

$$
\begin{aligned}
\mathbf{r}_B(s_1) = (g + s_1\cos\phi, s_1\sin\phi, 0), \quad g \equiv \|\mathbf{r}(L) - \mathbf{r}(0)\|,\\
\mathbf{t}_B = (\cos\phi, \sin\phi, 0), \quad s_1 \in [0,\infty],\\
\mathbf{r}_C(s_2) = (s_2\cos\psi, s_2\sin\psi\cos\theta, s_2\sin\psi\sin\theta),\\
\mathbf{t}_C = (\cos\psi, \sin\psi\cos\theta, \sin\psi\sin\theta), \quad s_2 \in [-\infty, 0].
\end{aligned}
$$

We wish to compute the integral

$$
\begin{aligned}
\mathcal{S}q &= \frac{1}{2\pi}\int_{-\infty}^{0}\int_{0}^{\infty} \frac{(\mathbf{r}_B(s_1) - \mathbf{r}_C(s_2))\cdot(\mathbf{t}_B(s_1)\times\mathbf{t}_C(s_2))}{\|\mathbf{r}_B(s_1) - \mathbf{r}_C(s_2)\|^3}\, ds_1\, ds_2\\
&= \frac{1}{2\pi}(\mathbf{t}_B \times \mathbf{t}_C)\int_{-\infty}^{0}\int_{0}^{\infty} \frac{\mathbf{r}_B(s_1) - \mathbf{r}_C(s_2)}{\|\mathbf{r}_B(s_1) - \mathbf{r}_C(s_2)\|^3}\, ds_1\, ds_2\\
&= \frac{g}{2\pi}\sin\psi\sin\phi\sin\theta\int_{-\infty}^{0}\int_{0}^{\infty} \frac{1}{(s_1^2 + 2ps_1 + q^2)^{\frac{3}{2}}}\, ds_1\, ds_2,
\end{aligned}
$$

where $p \equiv p(s_2) \equiv g\cos\phi - s_2(\cos\psi\cos\phi + \sin\psi\sin\phi\cos\theta)$, $q^2 \equiv q^2(s_2) \equiv s_2^2 - 2gs_2\cos\psi + g^2$.

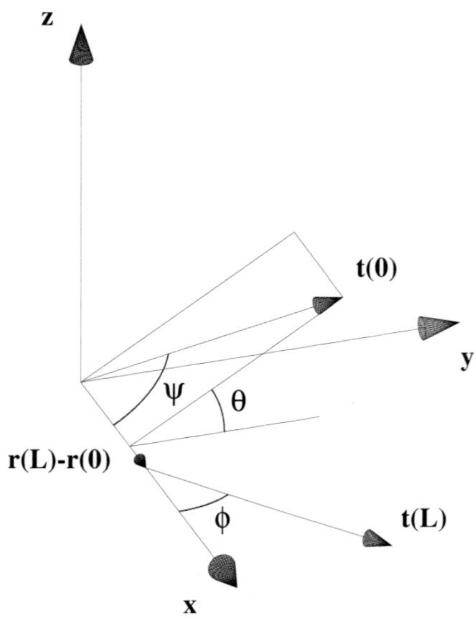

Fig. 5. The angles ϕ, θ, ψ determine the orientation of the tangents $\mathbf{t}(0)$ and $\mathbf{t}(L)$ at the ends of the curve segment.

It is easy to perform the first integration:

$$\mathcal{S}q = \frac{g}{2\pi}\sin\psi\sin\phi\sin\theta\int_{-\infty}^{0}\left(\left.\frac{s_1+p}{(q^2-p^2)\sqrt{s_1^2+2ps_1+q^2}}\right|_0^{\infty}\right)ds_2$$

$$= \frac{g}{2\pi}\sin\psi\sin\phi\sin\theta\int_{-\infty}^{0}\frac{ds_2}{q(p+q)}$$

$$= \frac{g}{2\pi}\sin\psi\sin\phi\sin\theta\int_{-\infty}^{0}\frac{1}{\sqrt{s_2^2-2gs_2\cos\psi+g^2}}$$

$$\times\frac{ds_2}{g\cos\phi - s_2(\cos\psi\cos\phi+\sin\psi\sin\phi\cos\theta)+\sqrt{s_2^2-2gs_2\cos\psi+g^2}}.$$

The last integral can also be done and the result may be presented as an algebraic formula which does not depend on g. However, the derivation of the final expression involves complicated algebra, and we instead prefer to obtain $\mathcal{S}q$ in a different way.

We can consider both rays and the straight line connecting them as an (infinite) polygonal line with three links. The writhe of this line, as defined in the previous section, is exactly equal to $\mathcal{S}q$. Then the squint is essentially proportional to the signed area of the spherical triangle formed by the geodesics that join the vertices corresponding to the vectors $\mathbf{t}(0)$, $\mathbf{r}(L) - \mathbf{r}(0)$, and $\mathbf{t}(L)$. The triangle has its two sides equal to ψ and ϕ and the angle between them θ. By the cosine rule for sides of spherical triangles, we find the third side χ from

$$\cos\chi = \cos\psi\cos\phi + \sin\psi\sin\phi\cos\theta,$$

and the signed area of the triangle can be calculated by l'Huilier's theorem[29] as

$$S = 4\nu\arctan\sqrt{\tan\frac{\Sigma}{2}\tan\frac{\Sigma-\phi}{2}\tan\frac{\Sigma-\psi}{2}\tan\frac{\Sigma-\chi}{2}},$$

where $\nu = \operatorname{sign}((\mathbf{r}(0) - \mathbf{r}(L))\cdot(\mathbf{t}(0)\times\mathbf{t}(L))) = \operatorname{sign}(\sin\psi\sin\phi\sin\theta)$ and $\Sigma = \frac{1}{2}(\phi+\psi+\chi)$. Then the squint is $\mathcal{S}q = \frac{S}{2\pi}$.

5. Example: helical shapes of arbitrary length

Helical structures are common in DNA modelling. They are also often met in various physical and, in particular, biomechanical models. Besides, the computation of the writhe (and twist) of a helix (a helical ribbon) is a favourite example of an application of Fuller's first theorem (e.g., see Refs. 24, 25, 30) though the author is not aware of works where the writhe is computed for a piece of helix with a *non-integer* number of turns.

In this section we shall be dealing with a circular right-handed helix: $\mathbf{r}(s) = (\cos as, \sin as, \sqrt{1-a^2}s)$, $s \in [0, L]$, s arc length, L the length of the segment and $a \in [0, 1]$. The limiting values of the parameter a correspond to a straight line ($a = 0$) and a circle of the unit radius ($a = 1$). The curve is periodic with period $T = \frac{2\pi}{a}$. The curvature and the torsion are $\kappa = a^2$ and $\tau = a\sqrt{1-a^2}$, respectively. The tangent indicatrix of the helix is a circular arc on $\mathbf{S}^2$ of radius a.

5.1. *Writhe of a single helix*

Applying Eq. (5) to the single helix yields

$$\mathcal{W}r(L) = \frac{1}{\pi}\left[\arctan\left(\sqrt{1-a^2}\sin\frac{aL}{2}, \cos\frac{aL}{2}\right) - \frac{L}{2}a\sqrt{1-a^2}\right] + 2\,\text{round}\left(\frac{aL}{4\pi}\right), \quad (16)$$

where we denote by $z = \arctan(x, y)$ a function such that $\sin z = \frac{x}{\sqrt{x^2+y^2}}$, $\cos z = \frac{y}{\sqrt{x^2+y^2}}$, $-\pi < z \leq \pi$ and round(x) is the function that gives an integer nearest to x; for half-integers round$(n + \frac{1}{2}) = n$ for negative $n \in \mathbb{Z}$ and round$(n+\frac{1}{2}) = n+1$ for non-negative $n \in \mathbb{Z}$. The last term in the right-hand side of Eq. (16) is added to make the writhe a continuous function of arc length L. Fig. 6 shows the writhe vs. number of periods L/T for three different values of the parameter a.

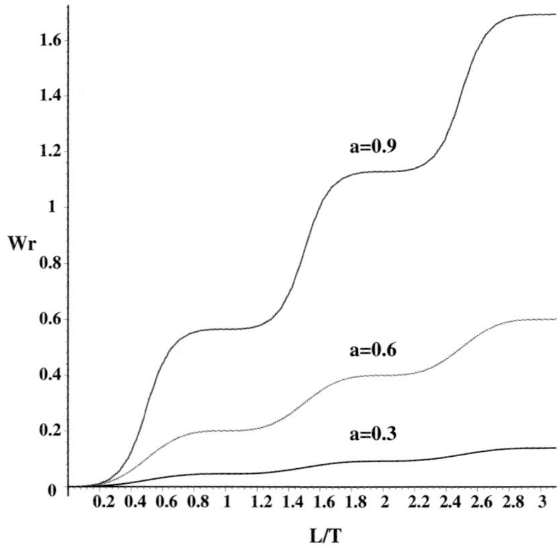

Fig. 6. The writhe $\mathcal{W}r$ of the circular helix as a function of the arc length L normalized on the period $T = 2\pi/a$, for three different values of the parameter a.

We may use Eq. (16) to compute the writhe for the particular lengths $L = \frac{\pi m}{a}$ (m is the number of half-periods):

$$\mathcal{W}r\left(\frac{mT}{2}\right) = \frac{m}{2}(1 - \sqrt{1-a^2}). \tag{17}$$

For an integer number of periods, i.e., for even m, Eq. (17) coincides with the expressions presented in Refs. 25, 30.

In the limiting case $a \to 0$, Eq. (16) produces $\mathcal{W}r \to 0$ as it should be for a straight line. Another limiting value of the parameter is $a \to 1$, the helix approaches a circle $\mathbf{r}(s) = (\cos s, \sin s, 0)$, covered $L/2\pi$ times. Then Eq. (16) gives $\mathcal{W}r \to \text{round}\left(\frac{L}{2\pi}\right)$ and, in particular, $\mathcal{W}r \to \text{round}\left(\frac{m}{2}\right)$ for m half-periods. The latter should not be a surprise in case of even m, because the closure chosen does not shrink to a point as the ends of the helical curve approach each other in the limit after an integer number of periods.

5.2. *Writhe of a double helix*

The procedure for calculation of the writhe of a double helix is very similar to that for a single helical shape. The helix is assumed to be closed at both ends in the same manner as is described above for an arbitrary open curve. The difference is that each of the two closing curves joins two different helices. Each of the two helical curves will be called a strand. The length of each strand is the same and is denoted by L. Both strands have the same axis, otherwise they may be located arbitrarily with respect to each other (as happens in B-form DNA, for example).

We apply Eq. (9) to two disjoint strands of the double helix and obtain

$$\mathcal{W}r = -\frac{L}{\pi}a\sqrt{1-a^2} \mod 1.$$

If we think about the double helix as a continuously growing structure, then it is evident that the last equation gives not only the fractional part of $\mathcal{W}r$, but its exact value as function of one strand length. Note also the negative sign of the writhe for the right-handed ($a > 0$) double helix.

Let $h = L\sqrt{1-a^2}$ be the length of the axis of the double helix. Then

$$\mathcal{W}r = -\frac{ha}{\pi}.$$

We see that the growing double helix delivers an example of a family of curves $A(h)$, parametrized with a continuous parameter h, such that the writhe is a linear function of length. Clearly, the writhe per unit length of the (double) helix axis is constant.

Note also that the writhe, rather unexpectedly, does not depend on the offset angle φ which controls the mutual location of the strands (in terms of DNA we may reformulate the last observation as an invariance property of writhe with respect to the widths of the minor or major grooves). The symmetric case of $2\varphi = \pi$ was considered in Ref. 30, where a formula was obtained for the limiting value of the writhe per unit length when the integer number of turns tends to infinity.

We remark that the writhe for n-strand helical shapes ($n \geq 3$) may be easily computed on the basis of the results derived for the single and double helices.

6. Conclusion

We have considered the generalization of the notion of writhing number for an arbitrary space curve and we have obtained effective formulas for its computation. The writhe of an open curve is defined here as a difference between the linking number and the twist of a ribbon based on a curve closure such that the tangent indicatrix is closed with a geodesic. The approach allows us to represent the writhe as a function of the arc coordinate measured along a curve. In some sense, we can then think about the writhe as being locally defined (cf. Ref. 31). The technique was extended to the sequences of the disjoint fragments of curves in space.

A relation was established between the writhe and the Gauß integral taken over the open fragment. It is shown, that the difference between these two quantities may be represented as three single integrals. We clarified the structure and meaning of these integrals.

Application of the formulas presented was illustrated on single and double helices of arbitrary length. In particular, the writhe as a continuous function of arc length is defined for a helix. A double helix of finite length, with "geodesic" closures at the ends, provides an example of a one-parameter family of curves that realize linear dependence of writhe on length, and the writhe is invariant with respect to the value of the offset between the strands.

Acknowledgements

Most of this work was carried out during the author's stay at the Institute of Technical Mechanics of the University of Karlsruhe. Support from the Alexander von Humboldt Foundation is gratefully acknowledged. The author would like to express his thanks to Jens Wittenburg for hospitality and attention to the work. The author is also pleased to thank John Maddocks for helpful discussions and support.

References

1. F. B. Fuller, *Proc. Natl. Acad. Sci. USA* **68**, 815 (1971).
2. G. Călugăreanu, *Czechoslovak Mathematical Journal* **11**, 588 (1961).
3. J. H. White, *Am. J. Math.* **91**, 693 (1969).
4. C. F. Barenghi, R. L. Ricca and D. C. Samuels, *Physica D* **157**, 197 (2001).
5. E. L. Starostin, *Meccanica* **31**, 235 (1996).
6. J. H. White and W. R. Bauer, *J. Mol. Biol.* **189**, 329 (1986).
7. D. A. Beard and T. Schlick, *J. Chem. Phys.* **112**, 7323 (2000).
8. M. Barbi, J. Mozziconacci and J.-M.Victor, *arXiv: q-bio.SC/0404037* (2004) 4 pages.
9. B. Fain and J. Rudnick, *Phys. Rev. E* **55**, 7364 (1997).
10. V. R. Chechetkin and V. V. Lobzin, *Phys. Lett. A* **250**, 443 (1998).
11. P. Røgen and H. Bohr, *Mathematical Biosciences* **182**, 167 (2003).
12. P. K. Agarwal, H. Edelsbrunner and Y. Wang, *Discrete & Computational Geometry* **32**, 37 (2004).
13. C. Bustamante, S. B. Smith, J. Liphardt and D. Smith, *Current Opinion in Structural Biology* **10**, 279 (2000).
14. E. Orlandini, M. C. Tesi, S. G. Whittington, D. W. Sumners and E. J. Janse van Rensburg, *J. Phys. A: Math. Gen.* **27**, L333 (1994).
15. J. F. Marko, *Phys. Rev. E* **55**, 1758 (1997).
16. A. V. Vologodskii and J. F. Marko, *Biophysical Journal* **73**, 123 (1997).
17. C. Bouchiat and M. Mézard, *Phys. Rev. Lett.* **80**, 1556 (1998).
18. C. Bouchiat and M. Mézard, *Eur. Phys. J. E* **2**, 377 (2000).
19. A. C. Maggs, *Phys. Rev. Lett.* **85**, 5472 (2000).
20. A. C. Maggs, *J. Phys. Chem.* **114**, 5888 (2001).
21. S. Kutter and E. M. Terentjev, *Eur. Phys. J. B* **21**, 455 (2001).
22. C. J. Benham, *J. Mol. Biol.* **123**, 361 (1978).
23. M. Le Bret, *Biopolymers* **18**, 1709 (1979).
24. J. J. Tyson and S. H. Strogatz, *Int. J. Bifurcation and Chaos* **1**, 723 (1991).
25. F. B. Fuller, *Proc. Natl. Acad. Sci. USA* **75**, 3557 (1978).
26. J. Aldinger, I. Klapper and M. Tabor, *Journal of Knot Theory and Its Ramifications* **4**, 343 (1995).
27. J. Samuel and R. Bhandari, *Phys. Rev. Lett.* **60**, 2339 (1988).
28. M. Spivak, *A Comprehensive Introduction to Differential Geometry* (Publish or Perish, Berkeley, 1979), vol. 3, chapter 6.
29. E. W. Weisstein, *The CRC Concise Encyclopedia of Mathematics* (CRC Press, 1999), p. 1012.
30. R. L. Ricca, *J. Phys. A: Math. Gen.* **28**, 2335 (1995).
31. R. D. Kamien, *Eur. Phys. J. B* **1**, 1 (1998).

CHAPTER 27

ON A MATHEMATICAL MODEL FOR THICK SURFACES

Paweł Strzelecki

Institute of Mathematics, Warsaw University
ul. Banacha 2, PL–02–097 Warsaw, Poland;
E-mail: pawelst@mimuw.edu.pl

Heiko von der Mosel

Institut für Mathematik, RWTH-Aachen,
Templergraben 55, 52062 Aachen, Germany
E-mail: heiko@instmath.rwth-aachen.de

Motivated by previous work on elastic rods with self-contact, involving the concept of the global radius of curvature for curves (as defined by Gonzalez and Maddocks), we define the *global radius of curvature* $\triangle[X]$ for a wide class of parametric surfaces. It turns out that a positive lower bound $\triangle[X] \geq \theta > 0$ provides, naively speaking, the surface with a thickness of magnitude θ; it serves as an excluded volume constraint for X, prevents self-intersections, and implies that the image of X is an embedded C^1-manifold with a Lipschitz continuous normal. Taking into account possible applications to variational problems for embedded objects, we also obtain a convergence and a compactness result for such thick surfaces.

The main object of this note is to introduce and explain the crucial notions and results. Thus, we defer almost all proofs to our forthcoming paper.[22]

1. Introduction

Physical surfaces such as sheets of paper, thin elastic plates, pieces of cloth, or aluminium foil often undergo large deformations in space so that different parts of the same object touch each other. These self-contact phenomena can also be observed on various smaller length scales, especially in biological systems, e.g., pinched skin tissue, buckled membranes, or conformations of lipid vesicles under thermal influence. The underlying common feature of all these examples is that of a surface with a small but positive thickness reflecting the fact that interpenetration of matter is impossible. The mathematical modelling of the intuitively obvious mechanism of self-avoidance is a challenging task: one needs an analytically tractable notion of thickness for surfaces, which in particular should be accessible to variational methods in order to deal with energy minimization problems in the framework of nonlinear elasticity. Moreover, surfaces with positive thickness are embedded; hence a suitable notion of self-avoidance should also lead to a novel treatment of classical geometric boundary value problems such as the Plateau problem or free and semi-free problems in the class of *embeddings*. This would produce physically relevant solutions of fixed topological type without self-intersections — in contrast to the classical solutions, where one frequently encounters non-embedded solutions due to the geometry of the boundary configurations. (See the discussion on minimal surfaces in Ch. 4.10 of Dierkes et al.[4])

Our aim is to introduce a purely geometric notion of thickness for a large class of (nonsmooth) parametric surfaces suitable for the calculus of variations. Motivated by the second author's previous cooperations on elastic rods with self-contact[10,18,19,20], which involved the concept of the global radius of curvature for curves as suggested by Gonzalez and Maddocks,[9] we define the *global radius of curvature for surfaces.*

The idea can be sketched as follows. Take a continuous parametric surface $X : \mathbb{R}^2 \supset \mathbb{B}^2 \to \mathbb{R}^3$ of disk-type which possesses a tangent plane on a dense subset $G \subset \mathbb{B}^2$. (Note carefully that this is a very wide class of surfaces: the area of X can be infinite, and the "good" set G where the tangent plane exists might just be countable.) Consider the radii of all spheres touching the image $X(\overline{\mathbb{B}}^2)$ of X in one of the "good" points $X(w), w \in G$ and containing at least one other point of the surface. We define the infimum of these radii as the global radius of curvature $\triangle[X]$ of the surface X. It turns out that a positive lower bound on $\triangle[X]$ serves as an *excluded volume constraint* for the surface X.

In fact, one of our main results is that $\triangle[X] \geq \theta > 0$ implies that $X(\overline{\mathbb{B}}^2)$ is a $C^{1,1}$-manifold with boundary, where the domain size and the $C^{1,1}$-norms of the local graph representation of $X(\overline{\mathbb{B}}^2)$ are uniform and depend solely on the constant θ (Theorems 9 and 10). This result requires careful analysis of the normal in the interior and near the boundary, since the set $\mathbb{B}^2 \setminus G$ of bad points without a tangent plane is allowed to have full measure. In view of applications in the calculus of variations we prove that the excluded volume constraint in terms of the global radius of curvature is stable under pointwise convergence of parametrizations (see Theorem 13). Moreover, assuming a uniform upper bound on area and a uniform positive lower bound on the global radius of curvature for a family of surfaces we can prove the existence of a C^1-convergent subsequence to a limit manifold of class $C^{1,1}$ again with uniform control on the local graph representations (Theorem 14). This compactness result may be a crucial step towards the study of variational problems for embedded surfaces in geometry and non-linear elasticity. Let us also mention that our results carry over to arbitrary co-dimension and are not restricted to disk-type surfaces.

An alternative approach to prevent a surface from self-intersecting is to introduce explicit repulsive forces between pairs of points on the surface. Based on this idea Kusner and Sullivan[12] suggested a Möbius invariant *knot energy* for k-dimensional submanifolds in $\mathbb{R}^d$ without boundary. These highly singular potential energies, however, require some regularization to account for adjacent points on the surface and, apart from the one-dimensional case of knot energies for curves[17,7,11], there are no analytical results regarding existence of minimizers or their regularity. Banavar, Gonzalez, Maddocks and Maritan[1] proposed so-called *many-body potentials*, replacing the Euclidean distance between two points by geometric multipoint functions on curves, or *tangent-point distances* for surfaces as Lagrangians for multiple integrals, in order to avoid the technical difficulties arising from the singularities in the potential, and to introduce an intrinsic length-scale for thickness. Although not stated explicitly in their paper[1] Banavar et. al clearly had the concept of global curvature for smooth surfaces based on tangent-point distances in mind from which their many-body potentials arise. Apart from numerical investigations for tube-like chains in the protein science[2] based on this idea, however, there presently are, to the best of our knowledge, no analytical results in this direction, with one exception: For a particular example of a three-body potential, the so-called *total Menger curvature* on one-dimensional sets, there is a remarkable regularity result of Léger[13] motivated by removability problems for bounded analytical functions in the complex plane (i.e., Vitushkin's conjecture and its solution by David). Léger proved that a Borel set E with bounded total Menger curvature and with positive and finite one-dimensional Hausdorff measure is actually the union of Lipschitz graphs apart from a measure zero set. He also claimed an analogous result for higher dimensional objects without giving the proof. Another contribution to thickness of surfaces in terms of the classical injectivity radius and the geometric focal distance for $C^{1,1}$-smooth submanifolds without boundary is given by the work of Durumeric[5]. He proves among other things a compactness result based on Gromov's compactness theorem, and he provides upper bounds on the diffeomorphism and isotopy types for $C^{1,1}$-submanifolds with a uniform lower bound on the injectivity radius.

There are other papers that investigate the structure of surfaces under various weak assumptions imposed on geometric quantities. Semmes[21] considered hypersurfaces M^d in $\mathbb{R}^{d+1}$ whose normals have small norm in the space BMO of functions of bounded mean oscillation (such surfaces can twist and spiral, and be far from being graphs). He proved that each such M is a chord-arc surface with small constant, i.e. for each $x \in M$ and each $R > 0$, the intersection of the ball $B_R(x) := \{y \in \mathbb{R}^{d+1} : |y - x| < R\}$ with M stays close to the hyperplane that passes through x and is perpendicular to the mean value of the normal, $n_{x,r} = ⨍_{B_R(x)} n(y)dy$, taken w.r.t. the surface measure on M. Toro[23] proved that surfaces with generalized fundamental form in L^2 are Lipschitz manifolds (as a consequence, the graph of every function $u \in W^{2,2}(\Omega)$, where $\Omega \subset \mathbb{R}^2$, can be parametrized by a bi-Lipschitz map). Her work was later generalized by Müller and Šverák[16] who gave a sharp condition on the L^2-norm of the second fundamental form, guaranteeing that a complete, connected, noncompact surface immersed in $\mathbb{R}^d$ is embedded.

For surfaces S homeomorphic to $\mathbb{R}^2$ these results where sharpened by Bonk and Lang[3], who, to answer a conjecture of Fu[8], considered a very rich class of Alexandrov surfaces, with a notion of integral curvature defined as a signed measure μ on S (if S is smooth, then for each $A \subset S$ the value $\mu(A)$ is equal to the integral of Gaussian curvature over A w.r.t. the surface measure). They proved that if $\mu^+(S) < 2\pi$ and $\mu^-(S)$ is finite, then S is bi-Lipschitz equivalent to the plane. The bound 2π is sharp.

Our work is also related to Federer's notion of *sets of positive reach* introduced in his seminal paper[6] on curvature measures. In fact, Section 4 of that paper provides valuable tools for the proofs of our convergence and compactness results, see Theorems 13 and 14.

The presentation is structured as follows: In Section 2 we give the precise definitions of the class of admissible surfaces, of the global radius of curvature for surfaces and provide simple analytical and topological consequences. In Section 3 we first discuss a priori estimates for the normal line depending only on a positive lower bound for the global radius of curvature, and next use these estimates to describe the structure of the image of a thick surface X; this image turns out to be a $C^{1,1}$-manifold. Finally,

Section 4 contains the convergence and the compactness results.

2. Basic definitions

2.1. *Admissible mappings*

We work with parametric surfaces $X\colon \overline{\mathbb{B}}^2 \to \mathbb{R}^3$, continuous up to the boundary of the unit disk. We also require X to be differentiable in the classical sense at all points $w \in G$, where G is a dense subset of $\mathbb{B}^2$ (which of course may depend on X), and we impose the condition

$$\operatorname{rank} DX(w) = 2 \qquad \text{for all } w \in G, \tag{2.1}$$

so that the tangent plane $T_wX := X(w)+DX(w)(\mathbb{R}^2)$ is a well defined two-dimensional affine plane. Each such mapping X will be called *admissible*. The class of all admissible mappings is denoted by $\mathcal{A}(\mathbb{B}^2,\mathbb{R}^3)$. It is clear that $X \in \mathcal{A}(\mathbb{B}^2,\mathbb{R}^3)$ could *a priori* have infinite area. On the other hand, if $X \in C^1(\mathbb{B}^2,\mathbb{R}^3) \cap C^0(\overline{\mathbb{B}}^2,\mathbb{R}^3)$ is an immersion, then X is admissible.

If $X \in \mathcal{A}(\mathbb{B}^2,\mathbb{R}^3)$ is differentiable in the classical sense at w and $\operatorname{rank} DX(w) = 2$, then w is called a *good parameter*; hence $G \subset \mathbb{B}^2$ is the set of good parameters. Note that if Σ is an arbitrary two-dimensional Riemannian manifold with or without boundary then the class $\mathcal{A}(\Sigma,\mathbb{R}^3)$, and in fact also $\mathcal{A}(\Sigma,\mathbb{R}^d)$, where $d \geq 3$, can be defined in a similar way.

Remark. To give an example of a well investigated class of (nonsmooth) mappings where most of the above assumptions are automatically satisfied, we recall here the definition and a handful of properties of n-absolutely continuous functions (the class of these functions is denoted by AC^n). In our setting, $n = 2$. Let $\Omega \subset \mathbb{R}^n$. One says (cf. Malý[14]) that $f \in AC^n(\Omega,\mathbb{R}^d)$ whenever for every $\varepsilon > 0$ there exists a $\delta > 0$ such that for every $k \in \mathbb{N}$ and every finite family of pairwise disjoint balls $B_1,\dots,B_k$ in Ω the following is satisfied:

$$\sum_{i=1}^{k} \mathcal{L}^n(B_i) < \delta \quad \Rightarrow \quad \sum_{i=1}^{k} \Big(\operatorname*{osc}_{B_i} f\Big)^n < \varepsilon,$$

where osc f stands for the sum of oscillations of all coordinates of f, and $\mathcal{L}^n$ denotes the Lebesgue measure.

Obviously, such mappings are continuous. Malý proves that such f's are also almost everywhere differentiable in the classical sense and have weak gradients in L^n (so that for $n = 2$ the area of f is finite!). Moreover, for $d \geq n$ the Lusin condition is satisfied, i.e. $\mathcal{H}^n(f(E)) = 0$ whenever $\mathcal{L}^n(E) = 0$, and the area formula holds.

Thus, for $\Omega \subset \mathbb{R}^2$, $AC^2(\Omega, \mathbb{R}^3)$ is a proper subspace of $W^{1,2}(\Omega, \mathbb{R}^3)$. The latter space contains discontinuous mappings; it also contains mappings which are continuous but nowhere differentiable in the classical sense. On the other hand, for bounded Ω we have

$$\bigcup_{p>2} W^{1,p}(\Omega, \mathbb{R}^3) \subset AC^2(\Omega, \mathbb{R}^3), \tag{2.2}$$

so that the class AC^2 is indeed larger than any of the Sobolev spaces $W^{1,p}$, $p > 2$. In fact, this inclusion is proper: consider the map $f\colon B_1(0) \to \mathbb{R}^2 \subset \mathbb{R}^3$ defined by $w \mapsto f(w) = \big(|w| \log(1 + |w|^{-1})\big)^{-1} w$. One can check that $f \in AC^2 \setminus W^{1,p}$ for every $p > 2$.

2.2. *Global radius of curvature for surfaces*

Let X be an admissible mapping, i.e., $X \in \mathcal{A}(\mathbb{B}^2, \mathbb{R}^3)$. For all *good* $w' \in \mathbb{B}^2$ we set

$$n(w') := \frac{X_u(w') \wedge X_v(w')}{|X_u(w') \wedge X_v(w')|}.$$

Then for $w \in \mathbb{B}^2$ and $w' \in G \subset \mathbb{B}^2$ we define

$$\begin{aligned} &r(X(w); X(w'), DX(w')) \\ &: = \begin{cases} 0 & \text{if } X(w) = X(w'), \\ \infty & \text{if } X(w) \in T_{w'}X \text{ and } X(w) \neq X(w'), \\ \dfrac{|X(w) - X(w')|}{2\left| n(w') \cdot \frac{X(w)-X(w')}{|X(w)-X(w')|}\right|} & \text{in the remaining cases.} \end{cases} \end{aligned} \tag{2.3}$$

In plain words, $r(x, y, p)$ is the radius of the unique sphere through the points $x, y \in \mathbb{R}^3$ tangent to the affine plane $y + p(\mathbb{R}^2)$, where p is a linear map from $\mathbb{R}^2$ to $\mathbb{R}^3$ with $\operatorname{rank} p = 2$. This radius becomes infinite when the vector $x - y \neq 0$ lies in the plane $p(\mathbb{R}^2)$, and is set to be zero if $x = y$.

Definition 1: For arbitrary $w \in \mathbb{B}^2$ we call

$$\rho[X](w) \colon = \inf_{\text{good } w' \in \mathbb{B}^2} r(X(w); X(w'), DX(w'))$$

the *global radius of curvature of* X *at* w, and

$$\triangle[X]\colon = \inf_{w\in\mathbb{B}^2} \rho[X](w)$$

the *global radius of curvature of* X.

The intuitive idea behind this concept is that a positive lower bound $\theta > 0$ on $\triangle[X]$ will allow us to place a pair of open balls of radius at least θ at each point of the surface "touching" the surface from both sides without intersecting it. From this we can infer that any surface X with $\triangle[X] \geq \theta$ satisfies the *excluded volume constraint* as described in the introduction. In particular, $X(\mathbb{B}^2)$ is an embedded surface in $\mathbb{R}^3$. Of course, all this requires proof, especially since *a priori* only good points $X(w)$, $w \in G$, — hence a possibly only countable set of surface points! — can be used in this construction. In the next section we describe just the backbone of the reasoning and refer to our paper[22] for more details.

As a first consequence of Definition 1 let us note that for any $w \in \mathbb{B}^2$ with $\rho[X](w) > 0$ one has $X(w) \neq X(w')$ for all $w' \in G$. Consequently, if $\triangle[X] > 0$, then $X(w) = X(\tilde{w})$ implies that either $w = \tilde{w}$ or that both w and $\tilde{w}$ are "bad" parameters (i.e., $w, \tilde{w} \in \mathbb{B}^2 \setminus G$).

Moreover, if $w' \in G$ and $\triangle[X] \geq \theta > 0$ then the two open balls B_1, B_2 of radius θ centered at $X(w') \pm \theta n(w')$ do not intersect the surface $X(\mathbb{B}^2)$, since otherwise we could find a point $X(w)$ such that $r(X(w); X(w'), DX(w')) < \theta$ contradicting our assumption on $\triangle[X]$. We shall sometimes call B_1, B_2 *excluded* or *forbidden* balls.

Remark. Let us note that for admissible mappings $X \in \mathcal{A}(\mathbb{B}^2, \mathbb{R}^d)$, $d > 3$, the global radius of curvature $\triangle[X]$ can be defined analogously. The definition of $r(x, y, p)$ remains unchanged. There is, however, one notable difference. For every good parameter w' in the domain we have now — instead of two excluded touching balls centered at $X(w') \pm \theta n(w')$ — a forbidden region

$$U_{w'} = \bigcup_{q\in S_{\theta,w'}} B_\theta(q),$$

where the set of centers

$$S_{\theta,w'} := \mathbb{S}_\theta^{d-1}(X(w')) \cap N_{w'}X$$

is given by the intersection of the round $(d-1)$-dimensional sphere $\mathbb{S}_\theta^{d-1}(X(w')) = \{s \in \mathbb{R}^d : |s - X(w')| = \theta\}$ with the affine normal space

$N_{w'}X = X(w') + \big(DX(w')(\mathbb{R}^2)\big)^{\perp}$. Thus, $U_{w'}$ looks, roughly speaking, like a thick degenerate doughnut. We have $\dim N_{w'}X = d-2$ for good w', therefore $S_{\theta,w'}$ is in fact a $(d-3)$-dimensional sphere in $N_{w'}X$. (Note that for $d=3$ the centers of B_1, B_2 do form a zero dimensional sphere contained in the normal line.) Analogously to the codimension 1 case, $X(\mathbb{B}^2)\cap U_{w'}$ is empty.

3. Structure of thick surfaces

3.1. *Interior continuity of the normal*

Let $X\colon \mathbb{B}^2 \to \mathbb{R}^3$ be an admissible surface with $\triangle[X] \geq \theta > 0$, and let $\varrho \in (0,\theta)$. Assume that $w \in \mathbb{B}^2$ is a good parameter, i.e., $w \in G$. Let

$$\ell(w) \colon = \big\{X(w) + tn(w) \mid t \in \mathbb{R}\big\}$$

be the (affine) normal line to X at w. We set

$$\begin{aligned} d(w) \ &: = \operatorname{dist}\big(X(w), X(\partial\mathbb{B}^2)\big),\\ C_\varrho(w) \ &: = \{p \in \mathbb{R}^3 \mid \operatorname{dist}(p, \ell(w)) = \varrho\}, \end{aligned}$$

and write π_w to denote the orthogonal projection onto the affine tangent plane T_wX.

To prove that the normal direction to X is uniformly continuous on compact subsets of the disk, we show first that at every good point w the image of X stretches away from $X(w)$ in all directions parallel to T_wX, as long as the distance from $X(w)$ is comparable to θ. Intuitively, the surface cannot fold abruptly at length scales much smaller than θ: close to *every* straight line through $X(w)$ in T_wX intersected with $B_r(X(w))$ we see points of the surface, as long as $r < \theta \leq \triangle[X]$ and the boundary $X(\partial\mathbb{B}^2)$ is far away.

Definition 2: We say that X has the *ϱ-stretching property* at $w \in G \subset \mathbb{B}^2$ iff

$$\pi_w\big(C_\varrho(w) \cap X(\mathbb{B}^2) \cap B_{2\varrho}(X(w))\big)$$

is a circle of radius ϱ in the tangent plane T_wX.

Lemma 3: *Assume that $\triangle[X] \geq \theta > 0$. If w is a good parameter and $\varrho \in (0, \varrho_0]$, where $\varrho_0 \colon = \min\big(\theta, d(w)/2\big)$, then X has the ϱ-stretching property at w.*

Since the proof of this lemma is short and contains a topological argument which (a) shows how the definition of good parameters is used and (b) is applied later, in a modified version, to control the behavior of the normal at the boundary, we give the details.

Proof. Fix $w \in G$ and $\varrho \in (0, \varrho_0]$. Without loss of generality we assume that $X(w) = O = (0,0,0) \in \mathbb{R}^3$ and the normal $n(w) = (0,0,1)$. Let $B_1 = B_\theta(0,0,\theta)$ and $B_2 = B_\theta(0,0,-\theta)$; then $X(\mathbb{B}^2) \cap (B_1 \cup B_2) = \emptyset$. Since rank$DX(w) = 2$, the curve $X(\partial B_\delta(w))$ is, for some sufficiently small $\delta \in (0, \varrho)$, linked with the normal line $\ell(w)$. Now, let $I \subset C_\varrho(w) \backslash (B_1 \cup B_2)$ be a fixed (but otherwise arbitrary) vertical line segment contained in $B_{2\varrho}(O)$ and having its endpoints on ∂B_1 and ∂B_2.

To show that $X(\mathbb{B}^2) \cap I$ is nonempty, consider a homotopy $(\gamma_s)_{s\in[0,1]}$ from $\gamma_0 = X(\partial B_\delta(w))$ to $\gamma_1 = X(\partial \mathbb{B}^2)$, defined as a composition of X with a homotopy from $\partial B_\delta(w)$ to $\partial\mathbb{B}^2$ in $\overline{\mathbb{B}}^2 \setminus B_\delta(w)$. Let σ be the closed curve consisting of I and two straight segments that join the endpoints of I to $O = X(w)$. The curves γ_0 and σ are linked, whereas γ_1 and σ are not linked, for otherwise we would have dist $(X(w), \gamma_1) < \varrho$, contrary to the definition of ϱ_0. (See the figure below.)

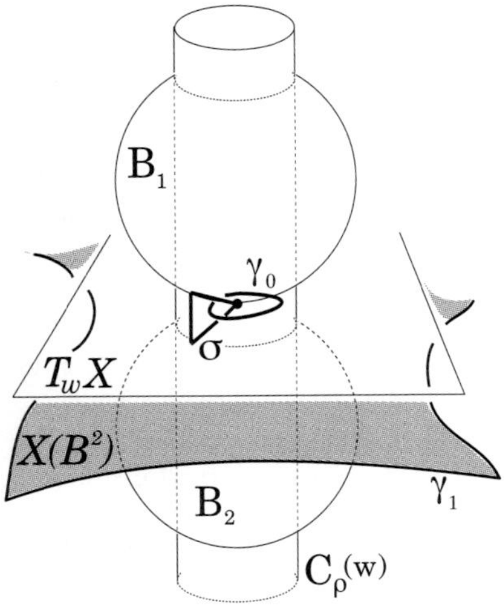

Figure 1. Touching tangent balls B_1 and B_2, and the curves $\sigma, \gamma_0, \gamma_1$.

It follows that γ_s must, for some $s \in (0,1)$, contain some point $p \in \sigma$. Certainly $p \notin B_1 \cup B_2$. Moreover, $p \neq 0 = X(w)$ since w is a good parameter. Thus, $p \in I$. This completes the proof of the lemma. □

Next, we obtain the following.

Lemma 4: *Let $\triangle[X] \geq \theta > 0$. If $w, w' \in \mathbb{B}^2$ are good parameters such that*

$$|X(w) - X(w')| < \min\bigl(\theta, d(w)/2\bigr)$$

and $\alpha(w, w') \in [0, \frac{\pi}{2}]$ is the angle between the normal directions at w and w', then

$$\alpha(w, w') \leq \frac{5\pi}{\theta}|X(w) - X(w')|. \tag{3.1}$$

The proof is lengthy but elementary. It follows from the stretching property that $\alpha(w, w')$ cannot be too large, for otherwise the forbidden balls associated to w' would contain some point in $X(\mathbb{B}^2)$ close to $T_w X$.

Since the estimate of Lemma 4 is uniform, and the set of good parameters is dense, we immediately obtain the following.

Corollary 5: *The normal direction has a continuous extension to all $w \in \mathbb{B}^2$ and the estimate* (3.1) *holds for all w, w' such that $|X(w) - X(w')| \leq \min\bigl(\theta, d(w)/2\bigr)$.*

Since now we can speak of an affine normal line $\ell(w)$ at every point $X(w)$, $w \in \mathbb{B}^2$, we can associate to *each* point on the surface a pair of open balls of radius θ touching the surface without intersecting it:

Corollary 6: *Let $w \in \mathbb{B}^2$. Then*

$$X(\mathbb{B}^2) \cap B_1 = X(\mathbb{B}^2) \cap B_2 = \emptyset$$

for the two open balls B_1, B_2 centered on the normal line $\ell(w)$ with radius θ, and touching each other at $X(w)$.

The proof is a simple *reductio ad absurdum* based on Hausdorff convergence of excluded balls: if there were some points of the surface in B_1 or B_2, then one of the excluded balls B_1^j, B_2^j associated to $w_j \in G$, where $w_j \to w$ as $j \to \infty$, would contain these points for j sufficiently large, a contradiction. As in Corollary 5 we need here that the set G of good parameters be dense in $\mathbb{B}^2$.

3.2. *Continuity of the normal at the boundary*

From now on we assume that X is an admissible mapping with a rectifiable boundary contour $X(\partial\mathbb{B}^2)$ and with $\triangle[X] \geq \theta > 0$. Moreover we suppose that the global radius of curvature $\triangle[X|_{\partial\mathbb{B}^2}]$ of the curve $X(\partial\mathbb{B}^2)$ (as defined in Gonzalez et al.[10], p. 35) is bounded below by θ. Note carefully that from now on $\triangle[\cdot]$ is used to denote two closely related but formally different notions. We always distinguish the argument in brackets, to avoid misunderstanding.

Theorem 7: Let $w \in \partial\mathbb{B}^2$. If $(w_j)_{j=1,2,\ldots} \subset G \subset \mathbb{B}^2$ is a sequence of good parameters such that $w_j \to w$ as $j \to \infty$ and the normal vectors

$$n(w_j) \colon = \frac{X_u \wedge X_v(w_j)}{|X_u \wedge X_v(w_j)|} \overset{j\to\infty}{\longrightarrow} \nu \in S^2,$$

then for every good parameter $w' \in \mathbb{B}^2$ such that $|X(w') - X(w)| \leq \theta/100$ we have

$$\alpha(w, w') \leq \frac{100}{\theta}|X(w) - X(w')|, \tag{3.1}$$

where $\alpha(w, w') \in [0, \frac{\pi}{2}]$ is the angle between the affine normal line $\ell(w')$ and the line $\ell(w) = \{X(w) + t\nu \mid t \in \mathbb{R}\}$. In particular, $\ell(w)$ does not depend on the choice of the sequence $(w_j) \subset G$.

As before, by density, this theorem implies the following.

Corollary 8: *The normal direction (and therefore $T_w X$) has a continuous extension to all $w \in \overline{\mathbb{B}}^2$ and the estimate*

$$\alpha(w, w') \leq \frac{500}{\theta}|X(w) - X(w')| \tag{3.2}$$

holds for all $w, w' \in \overline{\mathbb{B}}^2$ such that $|X(w) - X(w')| \leq \theta/400$.

Moreover, for all $w \in \overline{\mathbb{B}}^2$ we have

$$X(\mathbb{B}^2) \cap B_1 = X(\mathbb{B}^2) \cap B_2 = \emptyset$$

for the two open balls B_1, B_2 centered on the normal line $\ell(w)$ with radius θ, and touching each other at $X(w)$.

We omit the proofs. The main idea behind the proof of Theorem 7 is that if the surface contains a point $X(w')$, $w' \in \mathbb{B}^2$, such that $\delta := |X(w) - X(w')|$ does not exceed, say, $\theta/100$, then it contains also lots of other points $X(w'')$ lying very close to some half-circle of radius δ, centered at $X(w)$ and perpendicular to ν. This is the boundary counterpart of the stretching property from the previous section. Next, one shows that the normal direction at $X(w')$ must be close to ν, for otherwise the excluded balls associated to w' would contain one of the points $X(w'')$ constructed in the first step of the proof. This part of the argument is tedious but reduces to elementary geometry.

3.3. *Structure of the image*

For $p = X(w) \in X(\overline{\mathbb{B}}^2)$, $\pi_w \colon \mathbb{R}^3 \to T_w X$ the orthogonal projection, and $\ell(w)$ the affine normal line passing through p let

$$V_\varrho(p) \colon = \{q \in \mathbb{R}^3 \mid \operatorname{dist}(q, \ell(w)) < \varrho,\ |q - \pi_w(q)| < \varrho\}$$

denote a solid open cylinder with axis parallel to $\ell(w)$, centered at $p = X(w)$, with radius $\varrho > 0$ and height 2ϱ.

Theorem 9: Let $\triangle[X] \geq \theta > 0$ and let $w \in G \subset \mathbb{B}^2$ be such that

$$\operatorname{dist}(X(w), X(\partial\mathbb{B}^2)) > 2\sigma\theta\,,$$

where $\sigma \in (0, 1/100]$ can be chosen at will. Then $X(\mathbb{B}^2) \cap V_{\sigma\theta}(X(w))$ is a graph of a function $g \in C^{1,1}(B^2_{\sigma\theta}(0), \mathbb{R})$ with $\|g\|_{C^{1,1}} \leq C/\theta$, where C is some absolute constant.

Thus, loosely speaking, a portion of the surface contained in a cylinder of size comparable to θ is a graph of a $C^{1,1}$ function. The norm of this function is estimated inversely proportional to θ.

Remark. In fact, the assumption $\triangle[X] \geq \theta$ is not applied directly in the proof. What matters is the existence of excluded touching balls for every point in $X(\mathbb{B}^2)$, as given in Corollary 6, and Lipschitz continuity (w.r.t. to distances measured in the image) of the normal direction $\ell(w)$. These two facts imply that the intersection of $X(\mathbb{B}^2)$ with a neighbourhood of $X(w)$ is a graph of some function g, and that g is everywhere differentiable and Lipschitz. Lipschitz continuity of Dg follows then from a simple trick, using again Lemma 4. Thus, Theorem 9 applies to any continuous surface for which the excluded balls exist at every point in the image such that the line joining their centers varies in a Lipschitz continuous way. The original parametrization is not really important here.

Theorem 10: Let $p \in X(\partial\mathbb{B}^2)$, where $X : \overline{\mathbb{B}}^2 \to \mathbb{R}^3$ is an admissible surface with $\triangle[X] \geq \theta$ such that $\triangle[X|_{\partial\mathbb{B}^2}] \geq \theta$ for some $\theta > 0$. Assume that $X|_{\partial\mathbb{B}^2}$ is weakly monotone. Then, there exists a function $g \in C^{1,1}(B^2_{\theta/300}(0))$ such that $\|g\|_{C^{1,1}} \leq C/\theta$,

$$X(\overline{\mathbb{B}}^2) \cap V_{\theta/300}(p) = \mathrm{Graph}\,(g|_{\Omega^+}),$$

where $B^2_{\theta/300}(0)$ is a disk in T_wX, and $\Omega^+ = \{(x,y) \in B^2_{\theta/300}(0) \mid y > \psi(x)\}$ for some function ψ of class $C^{1,1}(\mathbb{R})$ with $\psi(0) = \psi'(0) = 0$ and $\|\psi\|_{C^{1,1}} \leq C/\theta$.

(A familiar) example. An open rotational cylinder with two hemispheres of the same radius attached at both ends shows that $C^{1,1}$ regularity is optimal for thick surfaces (with pairs of excluded touching balls existing at every point in the image). This particular surface fails to be C^2 at all points where the hemispheres meet the cylinder.

The theorems of this section show how strong in fact the assumption $\triangle[X] \geq \theta > 0$ is. Even if the global curvature radius of the boundary curve $\gamma := X(\partial\mathbb{B}^2)$ is positive, γ can be badly knotted. However, if we know *in addition* that γ bounds a surface $X \in \mathcal{A}(\mathbb{B}^2, \mathbb{R}^3)$ with $\triangle[X] \geq \theta > 0$, then it follows from Theorems 9 and 10 that γ cannot be knotted! This is vaguely reminiscent of the famous Fáry–Milnor theorem[15] relating curvature to topology.

4. Convergence and compactness

Following Federer[6], we define the *reach* of a set $A \subset \mathbb{R}^d$ as the supremum of those $r \in \mathbb{R}$ for which every $x \in B_r(A)$ has a unique next point $a \equiv \Pi_A(x) \in A$, such that $\mathrm{dist}\,(x, A) = |x - \Pi_A(x)|$. If the set $A \subset \mathbb{R}^d$ is closed, then the map $\Pi_A(.) : B_{\mathrm{reach}\,(A)}(A) \to A$ is continuous (cf. Thm. 4.8(4) in Federer[6]). For $A \subset \mathbb{R}^d$ and $a \in A$ one defines the *tangent cone* T_aA as

$$\begin{aligned} T_aA := \{v \in \mathbb{R}^d \,|\, v = 0 \\ \text{or } \forall \epsilon > 0\ \exists b \in A \cap B_\epsilon(a) \setminus \{a\} \text{ such that } \left|\tfrac{b-a}{|b-a|} - \tfrac{v}{|v|}\right| < \epsilon\}, \end{aligned}$$

which reduces to the classical (linear, not affine!) tangent plane $T_a\Sigma$, if $A = \Sigma \subset \mathbb{R}^d$ is a C^1-submanifold in $\mathbb{R}^d$. Thm. 4.18 in Federer[6] characterizes closed sets of positive reach by an inequality reflecting a uniform second order contact between the set and its tangent cone:

Theorem 11: For a closed set $A \subset \mathbb{R}^d$ and $t > 0$ one has $\mathrm{reach}(A) \geq t$ if and only if $2\,\mathrm{dist}\,(b-a, T_aA) \leq |b-a|^2/t$ for all $a, b \in A$.

Returning to our setting of admissible surfaces $X \in \mathcal{A}(\mathbb{B}^2, \mathbb{R}^3)$, and invoking Lemma 3 (iii) of Gonzalez et al.[10], we easily prove the following characterization.

Lemma 12: *Let $X \in \mathcal{A}(\mathbb{B}^2, \mathbb{R}^3)$ with $\triangle[X] > 0$ and $\triangle[X|_{\partial\mathbb{B}^2}] > 0$ and $\theta > 0$. Then the following two statements are equivalent:*

(i) $\triangle[X] \geq \theta$ *and* $\triangle[X|_{\partial\mathbb{B}^2}] \geq \theta$,
(ii) $\mathrm{reach}(X(\overline{\mathbb{B}}^2)) \geq \theta$ *and* $\mathrm{reach}(X(\partial\mathbb{B}^2)) \geq \theta$.

Federer proves the following: *if $K \subset \mathbb{R}^d$ is compact, then for each $t > 0$ the set*

$$\{A \subset K, A \neq \emptyset, \mathrm{reach}(A) \geq t\}$$

is compact with respect to the Hausdorff distance. (See Remark 4.14 in his paper[6].) Applying this compactness theorem, we prove the following two results.

Theorem 13: Let a sequence $(X_j)_j \subset \mathcal{A}(\mathbb{B}^2, \mathbb{R}^3)$ satisfy $\triangle[X_j] \geq \theta > 0$ and $\triangle[X_j|_{\partial\mathbb{B}^2}] \geq \theta$, and let $X_j|_{\partial\mathbb{B}^2}$ be weakly monotone parametrizations of the boundary curves $X_j(\partial\mathbb{B}^2)$. Assume also that $X \in C^0(\overline{\mathbb{B}}^2, \mathbb{R}^3)$ and that $X_j(\partial\mathbb{B}^2)$ converges to $X(\partial\mathbb{B}^2)$ in Hausdorff distance.

If $X_j(w) \to X(w)$ as $j \to \infty$ for all w belonging to some dense subset of $\mathbb{B}^2$, then

(1) the X_j are uniformly bounded;
(2) the sets $A_j := X_j(\overline{\mathbb{B}}^2)$ converge in Hausdorff distance to a $C^{1,1}$-manifold A with $\mathrm{reach}\,A \geq \theta$, and we have $A = X(\overline{\mathbb{B}}^2)$.

Theorem 14: Let $X_j : \mathbb{B}^2 \to \mathbb{R}^3$ be a sequence of admissible surfaces with $\triangle[X_j] \geq \theta > 0$. Assume moreover that:

(1) $\sup_j \mathcal{H}^2(X_j(\overline{\mathbb{B}}^2)) \leq M < +\infty$;

(2) $X_j\big|_{\partial\mathbb{B}^2}$ are weakly monotone parametrizations of rectifiable Jordan curves with global radius of curvature uniformly bounded below by θ, and there exists some $R > 0$ such that each of the curves $X_j(\partial\mathbb{B}^2)$ contains a point $p_j \in B_R(0)$.

Then one can select a subsequence j' such that $A_{j'} := X_{j'}(\overline{\mathbb{B}}^2)$ converge in Hausdorff distance to A, A is a $C^{1,1}$-manifold with boundary, and the nearest point projection $\pi_A \colon B_\theta(A) \to A$ is well defined.

Remarks. 1. It follows from our proofs that in both theorems above the limit manifold A is also equipped with local graph representations whose norms and sizes are uniformly controlled by θ, as described in Section 3.3. Moreover, Corollary 6 holds for A, i.e., we have two touching balls B_1 and B_2 at every point of A, and $A \subset \mathbb{R}^3 \setminus (B_1 \cup B_2)$.
2. The uniform area bound in the assumptions of Theorem 14 is satisfied when $\sup_j \|X_j\|_{W^{1,2}} \le M < +\infty$. The second part of assumption (2), i.e. the existence of a point $p_j \in X_j(\partial\mathbb{B}^2) \cap B_R(0)$, is obviously satisfied when the boundary contours converge to a fixed curve or are themselves fixed (as often encountered in the calculus of variations).
3. Using variants of Theorem 14 for closed surfaces of arbitrary fixed genus, one can prove that in the class of all $C^{1,1}$-surfaces Σ_g that have thickness (i.e., reach in the sense of Federer) greater than or equal to $\theta > 0$ there exists an *ideal* representative, i.e. a surface with minimal area.

Sketch of proof of Theorem 14.
Step 1. All $A_j = X_j(\overline{\mathbb{B}}^2)$ are contained in a fixed ball $B_R(0) \subset \mathbb{R}^3$. To see this, fix j and consider the covering $\{B_{\sigma_0\theta}(p) \mid p \in A_j\}$ of A_j. (Since each A_j is a $C^{1,1}$-manifold with uniform control on the local graph representation, cf. Theorems 9 and 10, we can find two absolute constants K and σ_0 such that, for each $j = 1, 2, \ldots$,

$$K^{-1}\sigma^2\theta^2 \le \mathcal{H}^2\big(B_{\sigma\theta}(p) \cap A_j\big) \le K\sigma^2\theta^2 \tag{4.1}$$

whenever $\sigma \le \sigma_0$. In plain words, pieces of the surface in a ball of radius $\delta \lesssim \theta$ have their area comparable to the area of a flat disk of radius δ.) Apply Vitali's lemma to this covering, to obtain a (possibly finite) sequence of pairwise disjoint balls $B_{\sigma_0\theta}(p_k)$, where $p_k \in A_j$, such that $\{B_{5\sigma_0\theta}(p_k) \mid k = 1, 2, \ldots\}$ is a covering of A_j. Take N of these balls. Invoking (4.1), we obtain

$$NK^{-1}\sigma_0^2\theta^2 \le \sum_{k=1}^{N} \mathcal{H}^2(B_{\sigma_0\theta}(p_k) \cap A_j) \le \mathcal{H}^2(X_j(\overline{\mathbb{B}}^2)) \le M.$$

This yields $N \le KM\sigma_0^{-2}\theta^{-2}$, and a uniform bound for diam $(X_j(\overline{\mathbb{B}}^2))$ follows. Together with assumption (2) this implies that there is an $\tilde{R} > 0$ such that $\bigcup_j X_j(\overline{\mathbb{B}}^2) \subset B_{\tilde{R}}(0)$.

Step 2. Applying Federer's compactness theorem for A_j with reach$(A_j) \ge \theta$ by Lemma 12, we select a subsequence such that $A_j \to A$ in Hausdorff distance and the reach of A is greater or equal to θ.

Step 3. Take a small number $\varepsilon > 0$ and consider the covering of $B_{\varepsilon\theta}(A)$ given by $\{B_{\varepsilon\theta}(p) \mid p \in A\}$. Apply Vitali's lemma and select a (possibly finite) sequence $B_{\varepsilon\theta}(p_k)$, $k \ge 1$, such that

$$A \subset \bigcup_{k\ge 1} B_{5\varepsilon\theta}(p_k)\,.$$

We can assume that dist $(A_j, A) \le \varepsilon\theta/2$ for all j. Moreover, we can find points $p_{jk} \in A_j$, $j = 1, 2, \ldots$, such that $p_{jk} \to p_k$ as $j \to \infty$. For each j and each k, we have $B_{\varepsilon\theta/2}(p_{jk}) \subset B_{\varepsilon\theta}(p_k)$; thus, for each fixed j the balls $B_{\varepsilon\theta/2}(p_{jk})$, where $k \ge 1$, are pairwise disjoint. An argument analogous to the one carried out in Step 1 above shows that for a fixed small ε the index k can take only finitely many values, say $k = 1, \ldots, N$ with some $N = N(\varepsilon, M, \theta)$.

For each fixed j the balls $B_{6\varepsilon\theta}(p_{jk})$, $k = 1, \ldots, N$, form a covering of $B_{\varepsilon\theta}(A)$ and of $B_{\varepsilon\theta/2}(A_j)$. Let $n_{jk} \in \mathbb{S}^2$ be normal to $A_j = X_j(\mathbb{B}^2)$ at $p_{jk} =: X_j(w_{jk})$. Selecting finitely many subsequences, we may assume that

$$n_{jk} \to \nu_k \in \mathbb{S}^2 \qquad \text{as } j \to \infty,\ k = 1, \ldots, N,$$

and moreover that

$$|n_{jk} - \nu_k| < \delta \qquad \text{for all } j = 1, 2 \ldots \text{ and } k = 1, \ldots, N.$$

Now, fixing δ and ε sufficiently small, we can apply the results of the previous section to conclude that $A_j \cap B_{10\varepsilon\theta}(p_k)$ is contained in a graph of a function $g_{jk}\colon B^2_{20\varepsilon\theta} \to \mathbb{R}$ such that $g_{jk} \in C^{1,1}$, $\|g_{jk}\|_{C^{1,1}} \le \text{const}/\theta$, where $B^2_{20\varepsilon\theta}$ is a fixed disk in the plane passing through p_k and perpendicular to ν_k (i.e. in the plane which is the limit of $T_{w_{jk}}X_j$ as $j \to \infty$), $k = 1, \ldots, N$.

Since the Lipschitz norms of all Dg_{jk} are uniformly bounded, by the Arzela–Ascoli theorem we may assume that $g_{jk} \to g_k$ as $j \to \infty$ in the

C^1-topology, for each fixed $k = 1, \dots, N$. Moreover, each g_k is of class $C^{1,1}$ since the Lipschitz condition for Dg_{jk} is preserved in the limit $j \to \infty$.

Now, an arbitrary point $p \in A$ belongs to $B_{5\varepsilon\theta}(p_k)$ for some k. It follows from previous considerations that $A \cap B_{5\varepsilon\theta}(p)$ is a $C^{1,1}$-graph. □

Acknowledgements. This work has been started while the first author was enjoying the freedom of being a Humboldt Fellow in Bonn; he gratefully acknowledges the generosity of the Alexander von Humboldt Foundation. The first author is also partially supported by KBN grant no. 2 PO 3A 028 22. The second author enjoyed the hospitality of the Centre Bernoulli at the EPFL Lausanne during March and April 2003. We would also like to thank the Sonderforschungsbereich 611 at the University of Bonn for supporting our research.

References

1. Banavar, J.R., Gonzalez, O, Maddocks, J.H., Maritan, A. Self-interactions of strands and sheets. J. Statist. Phys. **110** (2003), 35–50.
2. Banavar, J.R., Maritan, A., Micheletti, C., Trovato, A. Geometry and physics of proteins. Proteins **47** (2002) 315–322.
3. Bonk, M., Lang, U. Bi-Lipschitz parameterization of surfaces. Math. Annalen **327** (2003), 135–169.
4. Dierkes, U., Hildebrandt, S., Küster, A., Wohlrab, O. Minimal Surfaces I, II. Grundlehren math. Wiss. **295** & **296**. Springer, Berlin 1992.
5. Durumeric, O.C. Thickness formula and C^1 compactness for $C^{1,1}$ Riemannian submanifolds. ArXiv: math.DG/0204050 (2002).
6. Federer, H. Curvature measures. Trans. Amer. Math. Soc. **93** (1959), 418–491.
7. Freedman, M.H., He, Zheng-Xu, Wang, Zhenghan. Möbius energy of knots and unknots. Ann. of Math. (2) **139** (1994), 1–50.
8. Fu, J.H.G. Bi-Lipschitz rough normal coordinates for surfaces with an L^1 curvature bound. Indiana Univ. Math. J. **47** (1998), 439–453.
9. Gonzalez, O., Maddocks, J.H. Global Curvature, Thickness, and the Ideal Shape of Knots. The Proceedings of the National Academy of Sciences, USA **96** no. 9, (1999) 4769–4773.
10. Gonzalez, O., Maddocks, J.H., Schuricht, F., von der Mosel, H. Global curvature and self-contact of nonlinearly elastic curves and rods. Calc. Var. Partial Differential Equations **14** (2002), 29–68.
11. He, Zheng-Xu. The Euler-Lagrange equation and heat flow for the Möbius energy. Comm. Pure Appl. Math. **53** (2000), 399–431.
12. Kusner, R.B., Sullivan, J.M. Möbius-invariant knot energies. *Ideal knots*, 315–352, Ser. Knots Everything, 19, World Sci. Publishing, River Edge, NJ, 1998.
13. Léger, J.C. Menger curvature and rectifiability. Ann. of Math. (2) **149** (1999), 831–869.

14. Malý, J. Absolutely continuous functions of several variables. J. Math. Anal. Appl. **231** (1999), no. 2, 492–508.
15. Milnor, J. W. On the total curvature of knots. Ann. of Math. **52** (1950), 248–257.
16. Müller, S., Šverák, V. On surfaces of finite total curvature. J. Differential Geom. **42** (1995), 229–258.
17. O'Hara, J. Energy of a knot. Topology **30** (1991), 241–247.
18. Schuricht, F., von der Mosel, H. Global curvature for rectifiable loops. Math. Z. **243** (2003), 37–77.
19. Schuricht, F., von der Mosel, H. Euler-Lagrange equations for nonlinearly elastic rods with self-contact. Arch. Rat. Mech. Anal. **168** (2003), 35–82.
20. Schuricht, F., von der Mosel, H. Characterization of ideal knots. Calc. Var. Partial Differential Equations **19** (2004), 281–305.
21. Semmes, S. Hypersurfaces in R^n whose normal has small BMO norm. Proc. Amer. Math. Soc. **112** (1991), 403–412.
22. Strzelecki, P., von der Mosel, H. Global curvature for surfaces and area minimization under a thickness constraint. Preprint 189 SFB 611 Univ. Bonn (2004), to appear in Calc. Var. Partial Differential Equations.
23. Toro, T. Surfaces with generalized second fundamental form in L^2 are Lipschitz manifolds. J. Differential Geom. **39** (1994), 65–101.

CHAPTER 28

SOME ROPELENGTH-CRITICAL CLASPS

John M. Sullivan and Nancy C. Wrinkle

Institut für Mathematik, Technische Universität Berlin, MA 3-2, Str. des 17. Juni 136, 10623 Berlin, Germany
and
Department of Mathematics, University of Illinois, 1409 W. Green St., Urbana IL 61801, USA
E-mail: jms@isama.org

Department of Mathematics, Northeastern Illinois University, Science Building 214C, 5500 N. St. Louis Av., Chicago IL 60625, USA
E-mail: n-wrinkle@neiu.edu

We describe several configurations of clasped ropes which are balanced and thus critical for the Gehring ropelength problem.

1. Introduction

The ropelength of a link is given by the ratio of its length to its thickness. There are many ways to measure the thickness of a space curve, but for links one particularly simple notion, the *Gehring thickness*, is simply the minimum distance between different components. With Cantarella, Fu and Kusner we introduced a theory of criticality[2] for the Gehring ropelength problem. Our necessary and sufficient conditions for ropelength criticality take the form of a balance criterion which says that the tension force trying to reduce the length of the curves must be balanced by contact forces acting at points achieving the minimum distance. One simple example, with surprising intricacy for its solution,[2] is the *clasp*. This is a generalized link whose components are not closed curves but instead have constrained endpoints. In the clasp, one rope whose ends are attached to the ceiling is looped around another whose ends are attached to the floor. The ropelength-critical clasp we described is presumably the minimizer for ropelength, and is surprising in several ways: the tips of the two components are 6% further apart than they need to be—leaving a small gap between the ropes—and the curvature of the core curves blows up at the tips.

Here, we describe critical configurations of several generalized clasps with two or more components. Our new examples include clasps with one or both curves doubled, and connect sums of clasps with Hopf links. All of our examples are proven critical by the balance criterion. Many of them we expect are global minimizers (though we know no way to show this), but others are clearly not minimizing and are presumably unstable equilibria.

2. Background

Here we recall the necessary definitions and theorems from our work[2] with Cantarella, Fu and Kusner. These will show the balance criterion for Gehring ropelength in the form in which we apply it to our new examples.

Definition 1: A *generalized link* L is a curve L (with disjoint components) together with obstacles and endpoint constraints. In particular, each endpoint $x \in \partial L$ is constrained to stay on some affine subspace $M_x \subset \mathbb{R}^3$. Furthermore, there is a finite collection of *obstacles* for the link, each obstacle

$$\{p \in \mathbb{R}^3 : g_j(p) < 0\}$$

being given in terms of a C^1 function g_j having 0 as a regular value. By calling these sets obstacles, we mean that L is constrained to stay in the region where $\min g_j \geq 0$.

Definition 2: The *Gehring thickness* $\mathrm{GThi}(L)$ of a curve L is the minimal distance between points on different components of L. This is the supremal τ for which the $(\tau/2)$-neighborhoods of the components of L are disjoint.

We formulate the ropelength problem as minimizing the length of a generalized link subject to the constraint that its Gehring thickness remains at least 1. The contact points of the different components of the link are of primary importance, and are called *struts*.

Definition 3: An (unordered) pair of points x and y on different components of L is a *Gehring strut* if $|y - x| = \mathrm{GThi}(L)$. The set of all Gehring struts of L is denoted $\mathrm{GStrut}(L)$.

Given a generalized link L, only variations preserving the endpoint constraints should be allowed. A continuous vectorfield ξ along L is said to be *compatible* with these constraints if it is tangent to M_x at each endpoint $x \in \partial L$. We write $\mathrm{VF}_c(L)$ for the space of all compatible vectorfields.

Given a set of obstacles $g_j < 0$ for a link L, we write

$$O(L) := \min_j \min_{x \in L} g_{j(x)}.$$

Then L avoids the obstacles g_j if and only if $O(L) \geq 0$. We define the wall struts of L by

$$\mathrm{Wall}_j(L) := L \cap \{g_j = 0\}, \quad \mathrm{Wall}(L) = \bigsqcup_j \mathrm{Wall}_j(L).$$

This incorporates those parts of L which are on the boundary of the obstacles.

Usually we will be minimizing the length $\mathrm{Len}(L)$ of a link while constraining $\mathrm{GThi}(L)$ to be at least 1. Sometimes, however, we wish to consider a slightly more general objective functional, a weighted sum $\mathrm{Len}^w(L) := \sum w_i \mathrm{Len}(L_i)$ of the lengths of the different components. Here $w_i \geq 0$ can be viewed as the elastic tension within component L_i.

It is known[2] that critical links for this (weighted) Gehring ropelength problem are curves of finite total curvature. (See Ref. 3 for an expository account of such curves.) This means that the first variation of length under a compatible ξ is given by $\delta_\xi \mathrm{Len}(L) = -\int_L \langle \xi, \mathcal{K} \rangle$, where $\mathcal{K}$ is a vector-valued Radon measure along L (what we call a *force* along L) called the curvature force. For a C^2 link, we have $\mathcal{K} = \kappa N\, ds$ in terms of the Frenet frame. For weighted length, it follows that $\delta \mathrm{Len}^w = -\mathcal{K}^w$ where $\mathcal{K}^w = w_i \mathcal{K}$ along component L_i.

A variation of L will change the length of the struts and change the values of the obstacle functions g_j. We collect these changes into the *rigidity operator* A. If L is varied with initial velocity $\xi \in \mathrm{VF}_c(L)$, then $A(\xi)$ is by definition a continuous function on $\mathrm{GStrut}(L) \sqcup \mathrm{Wall}(L)$; its value on a strut $\{x, y\}$ of length 1 is $\delta_\xi |x - y| = (\xi_x - \xi_y) \cdot (x - y)$, and its value on a wall strut x (where $g_j(x) = 0$) is $\delta_\xi g_j(x) = \xi_x \cdot \nabla g_j$.

The (one-sided) first variation $\delta_\xi^+ \mathrm{GThi}(L)$ of Gehring thickness is (by Clarke's differentiation theorem for min-functions, see Ref. 2) the minimum of $A(\xi)$ over all struts; similarly the first variation $\delta_\xi^+ O(L)$ of the obstacle function is the minimum over all wall struts.

Intuitively, we expect a ropelength-critical configuration to be one whose length cannot be reduced without also reducing thickness. For technical reasons[2] we define strong criticality to require this reduction to happen at a definite rate:

Definition 4: We say that a generalized link L is *strongly critical* for minimizing weighted length when constrained by GThi if there is an $\varepsilon > 0$ such that, for all compatible ξ with $\delta_\xi \mathrm{Len}^w = -1$, we have

$$\min\left(\delta_\xi^+ \mathrm{GThi}(L),\ \delta_\xi^+ O(L)\right) \leq -\varepsilon.$$

The adjoint A^* of the rigidity operator takes a Radon measure on GStrut $\sqcup$ Wall and gives a force along L. This adjoint is what appears in our balance criterion,[2] which in turn allows us to explicitly solve for the shapes of critical configurations of various links.

Theorem 5: A generalized link L is strongly critical for weighted Gehring ropelength if and only if there is a positive Radon measure μ on $\mathrm{GStrut}(L) \sqcup \mathrm{Wall}(L)$ such that $-\mathcal{K}^w = A^*\mu$ as linear functionals on $\mathrm{VF}_c(L)$. That is, $-\mathcal{K}^w$ and $A^*\mu$ agree as forces along L except at endpoints $x \in \partial L$, where they may differ by an atomic force in the direction normal to M_x. □

Most of the configurations we care about here consist – as did the tight simple clasp[2] – of convex curves in perpendicular planes. Let P_i $(i = 1, 2)$ be the $x_i z$-coordinate plane in $\mathbb{R}^3 = \{(x_1, x_2, z)\}$. If $\gamma_i \subset P_i$ are two components of our generalized link, then we parametrize each curve γ_i by the z-component, u_i, of its unit tangent vector. If there is a strut $\{p_1, p_2\}$ of length 1 connecting these two components γ_i then the curves are both perpendicular to this strut, so elementary trigonometry gives the following lemma:[2]

Lemma 6: *Let γ_1 and γ_2 be two components of a link L, lying in perpendicular planes. Suppose there is a strut $\{p_1, p_2\}$ of length 1 connecting these components. Then in the notation of the previous paragraph, the parameters x_i and u_i for the points p_i satisfy $0 \leq x_i \leq u_i \leq 1$, and any two of the numbers x_1, x_2, u_1, u_2 determine the other two (up to sign) according to the formulas*

$$x_i^2 = 1 - \frac{x_j^2}{u_j^2} = \frac{u_i^2(1-u_j^2)}{1-u_i^2u_j^2},$$

$$u_i^2 = \frac{1 - x_j^2/u_j^2}{1-x_j^2} = \frac{x_i^2}{1-x_j^2},$$

where $j \neq i$. □

Note that the lemma above says nothing about balancing of forces, but is merely a geometric fact about curves in perpendicular planes that stay distance 1 apart. To balance symmetric planar curves, we make use of the following lemma:[2]

Lemma 7: *Suppose $\gamma_i \in P_i$ is symmetric across the z-axis, and parametrized by u_i as above. Consider the net curvature force of a mirror-image pair of infinitesimal arcs of γ. This net force acts in the vertical direction with magnitude* $2|\mathrm{d}u_i|$. □

3. Weighted clasps

The *τ-clasp* $C(\tau)$ is a generalized link consisting of two clasped ropes arranged according to the following description. We fix four planes, in a tetrahedral pattern as shown in Figure 1, each making an angle of $\arcsin\tau$ with the vertical. The τ-clasp consists of two unknotted arcs α_1 and α_2 whose endpoints are constrained to the four planes; the complement of the tetrahedron also serves as an obstacle for the link. The isotopy class of the link is specified so that closing each arc within the planes of its endpoints would produce a Hopf link.

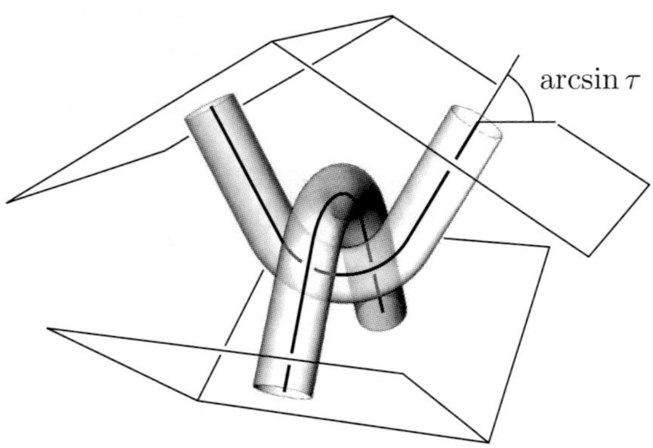

Fig. 1. In the τ-clasp problem, the endpoints of two ropes are constrained to lie in four planes whose normals make angle $\arcsin\tau$ with the horizontal. The parameter u_i ranges from $-\tau$ to τ along each arc, as shown at the end of the top right arc. If extended, the four planes shown would form the sides of a tetrahedron; both arcs are constrained to stay within this tetrahedron.

If the two components α_1 and α_2 of a clasp have different tensions, the configuration can only be balanced if the opening angles also differ. For $\tau_1, \tau_2 \in (0, 1]$, we now define the weighted clasp $C(\tau_1, \tau_2)$. It is just like

$C(\tau)$ except that its component α_i is attached to planes at angle $\arcsin \tau_i$. For this generalized link, we will describe a critical configuration for the weighted Gehring ropelength problem, where the weights are $w_i = 1/\tau_i$ on the two components. These tensions are chosen to ensure a net balance of vertical forces $2w_1\tau_1 = 2 = 2w_2\tau_2$ at the ends of the clasp.

It follows from Lemma 7 that the clasped contact between α_1 and α_2 is determined by the balance equation $u_1/\tau_1 + u_2/\tau_2 = 1$. Plugging this into Lemma 6 gives us equations for the shapes. Namely, we get an explicit formula for the x_i-coordinate of α_i:

$$x_i(u_i) = \sqrt{\frac{u_i^2(1-u_j^2)}{1-u_i^2u_j^2}} = \sqrt{\frac{u_i^2\big(1-\tau_j^2(1-u_i/\tau_i)^2\big)}{1-u_i^2\tau_j^2(1-u_i/\tau_i)^2}},$$

where $j \neq i$. As for the critical τ-clasp of Ref. 2, the z-coordinate is then determined as a hyperelliptic integral, through the relation $\mathrm{d}z/\mathrm{d}x_i = u_i/\sqrt{1-u_i^2}$ that defines u_i. This integral, and the similar one for the total arclength of the curve, can easily be computed numerically.

Note that when $\tau_1 = \tau_2$, the curves arising here are exactly the symmetric τ-clasp curves. But when $\tau_1 \neq \tau_2$, the two touching curves have shapes different from each other and from any symmetric clasps.

The calculations above, combined with Theorem 5, serve to prove:

Theorem 8: The configuration of $C(\tau_1, \tau_2)$ described above is critical for weighted ropelength. □

We expect these configurations are in fact the ropelength minimizers.

4. Clasps with parallels

We next describe a family of examples based on the connect sum of the τ-clasp with the Hopf link. We describe configurations which we show are balanced and thus critical points, but they are not minimizers and are presumably very unstable equilibria. Although we use unweighted length as our objective functional here, some of the shapes that appear are those of the weighted clasp curves we described above.

We first consider a configuration $C^1_{1,1}(\tau)$ which can be defined as the connect sum of the τ-clasp $C(\tau)$ with a Hopf link. Letting β and γ denote the two components of $C(\tau)$, we add a third component α, which is linked to γ but not to β. We guess that in the ropelength minimizer for this link, the components β and γ retain the shapes they have in the minimizing clasp, and α is a round circle around a point on a straight end of γ, far from β.

Here, however, we will be interested in describing a critical configuration with more symmetry, where α lies in the plane of β.

Definition 9: The generalized link $C^m_{k,l}(\tau)$ is defined from $C^1_{1,1}(\tau)$ by replacing α by k parallel copies, β by l parallel copies, and γ by m parallel copies. We also adjust the angles of the bounding planes: those containing the ends of the l copies of β now lie at angle $\arcsin(\tau/l)$ from the vertical, and those with γ at angle $\arcsin(\tau/m)$.

In the configurations we describe, the copies of α and β all lie nested with one another in a single vertical plane P_1, and the copies of γ are nested in the perpendicular vertical plane P_2. (Figure 2 sketches what such a configuration for $C^1_{2,1}(\tau)$ might look like.) In particular, the shapes of all

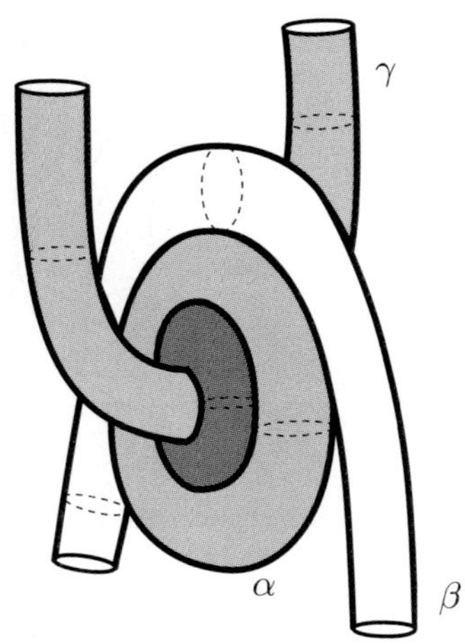

Fig. 2. The generalized link $C^1_{2,1}(\tau)$ has a critical (but not minimizing) configuration as shown here, with the two copies of α nested in a common plane with β.

components are determined by those of the innermost α and γ: the other copies of α are successive outer parallels at distance 1 and the copies of β are further outer parallels except that they peel off at the angle corresponding to $u_1 = \pm\tau/l$, to proceed straight out and down to meet their bounding planes perpendicularly. Similarly, the copies of γ are determined as successive outer parallels at distance 1 to the innermost copy; all of them have straight segments out and up to the bounding planes at $u_2 = \pm\tau/m$ and have a curved arc parametrized by $u_2 \in [-\tau/m, \tau/m]$. Here τ can be any nonnegative number not exceeding $\min(m, n)$. The angles are again determined by an overall balance of vertical forces: the l strands of β exert a vertical force of $2|u_1| = 2\tau/l$ each, while the m strands of γ exert a vertical force of $2|u_2| = 2\tau/m$ each.

Our entire configuration, like the τ-clasp, has mirror symmetry across the planes P_1 and P_2, but unless we have $k = 0$ and $l = m$, there is no longer the extra symmetry interchanging top and bottom.

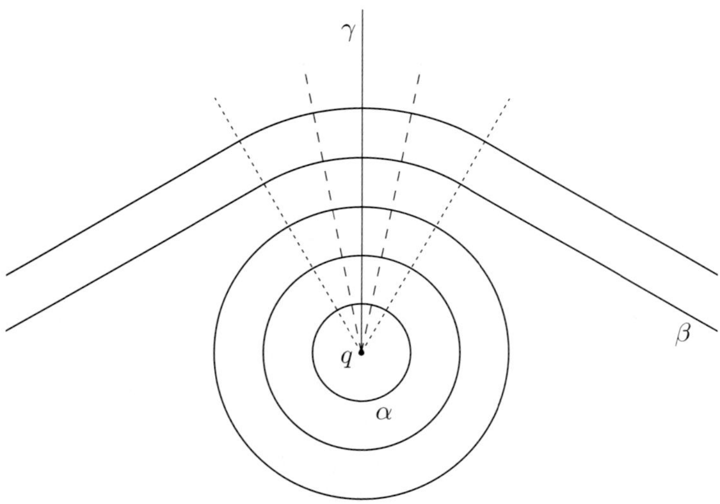

Fig. 3. In this variant $C^1_{3,2}(1)$ of the clasp problem, there are $k = 3$ nested parallel copies of a closed arc α inside $l = 2$ parallel copies of a clasp arc β, here proceeding out and down towards bounding planes below the figure. These are linked to $m = 1$ copies of a clasp arc γ which (in the case $\tau = 1$ illustrated) is attached to a horizontal ceiling. (We see γ in projection to the plane P_1 of the figure only as a vertical line; the dot at the center is its tip q.) The arcs β peel away from the arcs α at $u_1 = \tau/k$, at the dotted line shown at 30°; the arcs α and β are clasped to γ for $|u_1| \le \tau/(k+l)$ (above the dashed lines shown), and then continue as circular arcs (between the dashed and dotted lines). The figure has been drawn entirely with circles, but the true analytic arcs near the tips would be imperceptibly different. The arc γ is an analytic clasp for its entire curved region $|u_2| \le 1$, and then (since $\tau/m = 1$) proceeds vertically to the ceiling.

It remains to describe the shapes of the innermost copies α and γ exactly, and to show the resulting configuration is balanced. (See Figure 3, which shows $C^1_{3,2}(1)$.) We will find it useful to use the abbreviation $n := k + l$ for the total number of curves in the plane P_1 containing α and β and their parallels. Note that the m copies of γ intersect the plane P_1 in a series of

m points q_i spaced at distance 1 along the z-axis. The inner copy α would naively be a stadium curve looping around these points at distance 1. Our configuration is close to this, changing only part of the upper semicircle. That is, α consists of a semicircle around the bottommost q_i, joined to vertical segments of length m, and to an upper arc parametrized by $u_1 \in [-1, 1]$. For $|u_1| \geq \tau/n$ this upper arc is also part of the unit circle around the topmost q_i, but for $u_1 \in [-\tau/n, \tau/n]$ it is an analytic arc determined below. Note that the parallel copies of β include outer parallels of this analytic arc in the range $|u_1| \leq \tau/n$, but also outer parallels to the circle (that is, larger circular arcs of integer radius around the topmost q_i) in the range $|u_1| \in [\tau/n, \tau/l]$, before peeling off straight at $|u_1| = \tau/l$.

The innermost γ consists more simply of an analytic arc parametrized by $|u_2| \leq \tau/m$ joined to the straight segments out to the bounding planes. It now remains merely to describe the analytic arcs of the innermost α and γ: these form a weighted clasp $C(\tau/n, \tau/m)$ as in Theorem 8.

The important observation is that in our situation we have a convex planar curve with nested outer parallels. Each of the parallels has struts only to the next ones inward and outward, and these touch at points with equal direction u_i. Thus the innermost curve behaves like a curve with increased tension, proportional to the total number of parallel curves. If this increased force is balanced by struts to the other innermost curve, then the struts between the parallels distribute this balancing force outwards to balance the equal tension on each of the parallel strands.

Theorem 10: The configuration of $C^m_{k,l}(\tau)$ described above is critical for ropelength.

Proof: The lower semicircle of α and its parallels exert a total force $2k$ upwards on the bottommost q_i. This is transmitted upwards to the topmost q_i by atomic strut forces in the struts connecting the q_i. The remaining circular pieces around this topmost q_i, namely k circles for $|u_1| \in [\tau/l, 1]$ and a total of n for $|u_1| \in [\tau/n, \tau/l]$, exert a balancing downwards force of $2k(1 - \tau/l) + 2n(\tau/l - \tau/n) = 2k$ on this q_i. The analytic arcs of the innermost α and γ stay at constant distance 1 from each other; since they are the curves of the critical weighted clasp $C(\tau/n, \tau/m)$, they balance each other with weights n and m, or equivalently, with unit weights on each of them and their respective n and m parallels. □

We now consider in more detail the specific example of $C^1_{1,1}(1)$. This is the ordinary simple clasp $C(1)$—whose endpoints are attached to horizontal planes—with the addition of a closed component α. The lower half of α is a semicircle of radius 1 centered at the tip of γ. The upper half of α consists of three parts: two circular arcs of angle 30° and an arc clasped to γ. The curve β consists of two vertical segments from the floor up to the height of the tip of γ, connected by an arc that is an outer parallel to the upper half of α. The curve γ includes no circular arcs, but only the analytic arc determined above and straight segments up to the ceiling.

In general, note that our critical configuration of $C^m_{k,l}(\tau)$ includes circular arcs in the β components when $k > 0$, but not when $k = 0$. For $k = 0$ all the force balancing happens between the analytic arcs described above.

5. Conjectured minimizers for two cases

For the two cases $k = 0$, $l = 2$, $m = 1, 2$, we now describe a different critical configuration $L^m(\tau)$ for the generalized link $C^m_{0,2}(\tau)$. In both cases, we conjecture that this configuration (unlike the one described above) is the minimizer for Gehring ropelength. The drawings in Figure 4 indicate the relative positions of the components in these configurations, without showing the exact geometric features described below.

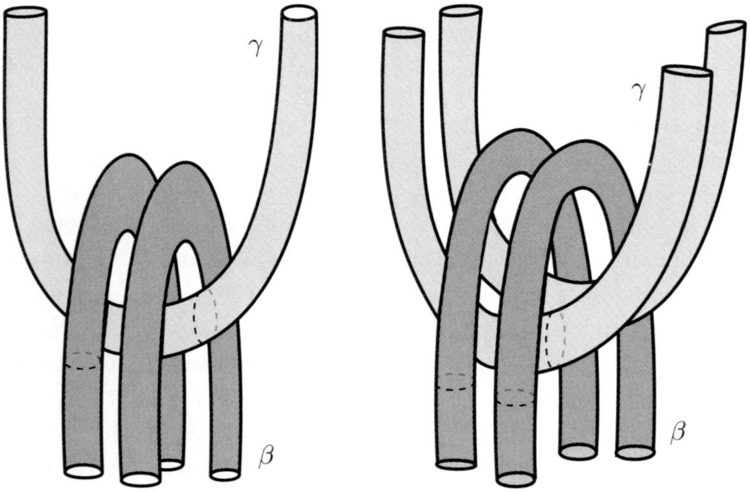

Fig. 4. The configurations $L^m(\tau)$ are presumably the minimizers for the clasp problems $C^m_{0,2}(\tau)$ for $m = 1$ (left) and $m = 2$ (right). They involve congruent curves in parallel planes. The exact geometry of the critical configurations—including unit-length horizontal segments at the tips of most components, and the fact that the lower ends on the left must proceed out within 30° of the horizontal—is not shown in these sketches.

For $C^1_{0,2}(\tau)$ the configuration we described above consists of noncongruent curves β_0 and γ_0, together with an outer parallel to β_0. In our new configuration $L^1(\tau)$, both β curves are congruent to β_0, but translated out to lie in parallel planes $x_2 = \pm\frac{1}{2}$. The γ curve is a copy of γ_0, but split apart at the tip, with a unit-length straight segment inserted from $x_2 = -\frac{1}{2}$ to $x_2 = \frac{1}{2}$.

For $C^2_{0,2}(\tau)$ the configuration we described above consists of two ordinary τ-clasp curves together with their outer parallels. In our new configuration $L^2(\tau)$, all four curves are congruent, and each looks like the ordinary τ-clasp but split at the tip, with a unit-length segment inserted. The two copies of β lie in the planes $x_2 = \pm\frac{1}{2}$, and are translates of each other perpendicular to these planes. Similarly the two copies of γ lie in the planes $x_1 = \pm\frac{1}{2}$. Note that for $\tau = 1$, the curvature of each curve is unbounded near the tip; $L^2(\tau)$ gives an example of a Gehring-critical configuration in which curvature approaches infinity but then immediately jumps to zero.

Theorem 11: The configurations $L^1(\tau)$ and $L^2(\tau)$ described above are critical for Gehring ropelength.

Proof: Consider the case $m = 2$. In the usual τ-clasp, there is 2–2 strut contact, with the four struts in any given set lying over the four quadrants of the x_1x_2-plane. Here, those struts are pulled apart from each other. They carry exactly the same balancing forces as before, but the horizontal components of those forces need to be balanced by further horizontal struts. These connect corresponding points on the two copies of β, and similarly for γ. Each quadrilateral of struts has now been replaced by an octagon (with four translated copies of the original diagonal edges plus four new horizontal edges). But the forces still balance within each such octagon.

The case $m = 1$ is similar but even simpler. In this case the quadrilateral of struts is split apart only in one direction. Again, new horizontal struts connecting the copies of β allow the horizontal components of force to be properly balanced. The struts here close into hexagons. □

6. A chained clasp

Our next example is that of two clasps joined end to end, as in Figure 5. That is, we have three components: the open arcs β and γ are attached to the floor and ceiling, respectively, and are unlinked with each other; each, however, is linked to the closed component α. The configurations we consider have reflection symmetry across a horizontal plane, interchanging

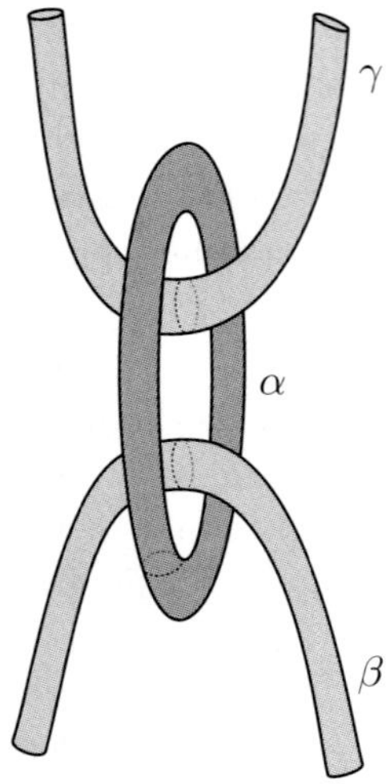

Fig. 5. Two clasps, when joined end to end, form a generalized link with three components. For the case $\tau = 1$ of vertical ends, the middle closed component α has varying length in our family of critical configurations.

β and γ while preserving α.

If we make such a configuration with $\tau < 1$, the junctions (where α clasps to β and to γ) will move towards each other, until the tips of β and γ touch at distance 1 producing an isolated strut. Here β and γ are congruent τ-clasp curves. The curve α is like a stadium curve, but its tips (for $|u| \leq \tau$) are τ-clasps. These are followed by circular arcs around the tips of β and γ, which are finally joined by unit-length vertical segments. Assuming the top/bottom symmetry, this critical configuration is uniquely determined.

On the other hand, when $\tau = 1$, the critical configuration has one simple clasp at the junction of α with β, and another symmetric one between α and γ. These clasps can be close to each other (with the tips of β and γ as close as distance 1) or can move farther apart. There is a one-parameter family of equal-length symmetric critical configurations. In this family, the length of the closed component α varies.

Theorem 12: The configurations of the chained clasps described above are critical for Gehring ropelength.

Proof: For $\tau = 1$ the balancing is just that for the α, β and α, γ clasps separately.

For $\tau < 1$ we need to combine the balancing for the clasp with that for the simple closed chain described in Ref. 1, 2. The curves β and γ are ordinary τ-clasp curves, positioned so their tips are unit distance apart. These are balanced by clasp arcs of α. The circular arcs of α focus net force $2 - 2\tau$ downwards on the tip of γ and upwards on the tip of β. These forces are balanced by a force on the isolated strut connecting these two tips. □

Note that we could build similar configurations with several chained components α_i in between β and γ. For $\tau < 1$ the α_i are all congruent to the curve α described above. For $\tau = 1$, each would be a stadium curve, but they could have differing lengths. Again, we expect that all of these are ropelength minimizers.

7. The granny clasp

Our final example generalizs the simple clasp not by introducing extra components, but by clasping the two components in a more intricate way. The connect sum of two trefoil knots of the same handedness is called the granny knot. We define the granny clasp G_1 to be the generalized link shown in Figure 6 (left), a clasp of two ropes based on this granny knot.

Fig. 6. This generalized link is called the granny clasp. In the first configuration shown, if the two free ends at the left were joined and likewise the two on the right, we would get a granny knot, the connect sum of two trefoils of the same handedness. In the middle, we see an isotopic configuration, presumably close to a minimizer for the ordinary ropelength problem[1] where each component avoids itself as well as the others. Note that, compared to the first picture, the upper two strands have crossed, as have the lower two: the endpoints of the generalized link are free to move within the floor and ceiling bounding planes. For the Gehring ropelength we consider here, however, the appropriate setting is Milnor's link-homotopy. Our critical configuration (and presumed minimizer) has components which are planar but fail to be embedded curves, as on the right.

It seems clear that if this configuration of the granny clasp were tied tight in rope, each component would contact itself as well as the other, as suggested in Figure 6 (center). But our constraint on the Gehring thickness GThi ≥ 1 does not see self-contact of a single component. It is thus important to remember that the natural setting for Gehring ropelength problems[2] is Milnor's link homotopy. Two configurations are link-homotopic if there is a homotopy between them where the components stay disjoint but self-intersections of any given component are allowed.

In the critical configuration we describe for the granny clasp G_1, each component does have a point of self-intersection, like those shown in Figure 6 (right). More precisely, if β denotes the component attached to the floor, it—like the curves in all our previous examples—lies in a vertical plane and has mirror symmetry across a vertical line ℓ in that plane. In Figure 7 we see (as the solid line) one symmetric half of β, consisting of four analytic pieces joined in a C^1 fashion: first a vertical segment up from the floor, then a $\tau = 1$ clasp arc leading to the point p of self-intersection, then continuing on the other side of ℓ with two more clasp arcs leading to the tip t of β. We expect that this configuration is the minimizer, but as usual will prove only that it is critical.

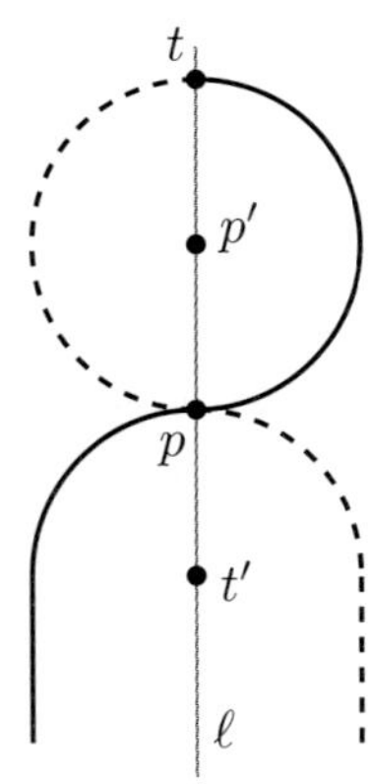

Fig. 7. In the critical configuration of the granny clasp G_1, each component is planar and built from two symmetric halves, here shown as solid and dashed lines. Here p is the point of self-intersection and t is the tip. The other component is congruent to this one, and lies in a perpendicular plane; the two places where it intersects the plane of the figure are the points p' and t' corresponding to p and t. Each half of the component shown consists of a straight arc from the floor up to the level of t', followed by three $\tau = 1$ clasp arcs, the first up to p, the next to the level of p' and the last up to t. (This figure has been drawn with six quarter circles in place of the $\tau = 1$ clasp arcs.)

Theorem 13: The (nonembedded) configuration of the generalized link-homotopy class G_1 built from clasp arcs as described above is critical for the Gehring ropelength problem.

Proof: The configuration can be balanced as follows: Cut space with horizontal planes at the heights of t', p, p' and t. Below t' and above t we have only vertical straight segments. In each of the three intermediate slabs we see exactly a $\tau = 1$ clasp. (Even though the symmetric halves of β do not connect to each other through p as in a clasp, they still balance the other component in the same way.) □

We generalize this example as follows: let G_n be the generalized link of two components obtained from the connect sum of two $(2, 2n+1)$–torus knots of the same handedness. That is, each chain of three half-twists in our first picture of G_1 is replaced by $2n+1$. The link G_2 is shown in Figure 8 (left); note that G_0 is the ordinary clasp $C(1)$.

Fig. 8. This higher-order granny clasp G_2 is shown in three views. It is derived (left) from the connect sum of two $(2, 5)$–torus knots. Our conjectured minimizer for the Gehring ropelength problem is a nonembedded representative (right) of this generalized link-homotopy class, built from several clasp arcs. The link homotopy from the initial configuration to this critical configuration involves twisting the pairs of endpoints one full turn in opposite directions. In the center we see the configuration after each has been twisted one half turn.

Again we can describe a critical configuration for G_n—as shown in Figure 8 (right)—which we expect is the mininimizer. We obtain this configuration from the initial one by twisting the pair of endpoints in the ceiling around each other n full turns (relative to the pair in the floor) and then letting the n points of self-contact of each component become self-intersections. These n self-intersections of each component occur where it crosses its plane of symmetry. Each half of each component is built from $2n+1$ clasp arcs plus a straight vertical segment.

A final generalization $G_n(\tau)$ would allow the ends of the clasp to be attached to slanted planes. Here the critical configuration would be built from τ-clasp arcs (for $|u| \leq \tau$) near each self-intersection point, and arcs of circles (for $|u| \geq \tau$) centered at the self-intersections of the other component.

Acknowledgements

We wish to thank Rob Kusner for helpful suggestions, Jason Cantarella for the use of his τ-clasp figure, and TU Berlin and the DFG Research Center MATHEON for hosting Wrinkle in January 2004 when this work was done.

References

1. J. Cantarella, R. Kusner, J. Sullivan. *On the minimum ropelength of knots and links.* Inventiones Math. **150**:2, 2002, pp 257–286. `ArXiv:math.GT/0103224`.
2. J. Cantarella, J.H.G. Fu, R. Kusner, J. Sullivan, N. Wrinkle. *Criticality for the Gehring link problem.* Preprint, 2004. `ArXiv:math.DG/0402212`.
3. J. Sullivan. *Curves of finite total curvature.* Lecture notes from the May 2004 Oberwolfach Seminar "Discrete Differential Geometry". Birkhäuser, 2005, to appear.

CHAPTER 29

REMARKS ON SOME HYPERBOLIC INVARIANTS OF 2-BRIDGE KNOTS

Jim Hoste

Department of Mathematics, Pitzer College
Claremont, CA 91711, USA
jhoste@pitzer.edu

Patrick D. Shanahan

Department of Mathematics, Loyola Marymount University
Los Angeles, CA 90045, USA
pshanahan@lmu.edu

In this paper we describe a recursive technique that can be used to investigate a number of hyperbolic invariants of certain classes of 2-bridge knots. We explicitly compute the representation and character varieties, the trace and cusp fields, and A-polynomials for these knots. Using these computations we obtain information about these knots' commensurability classes. Experimental results lead naturally to two conjectures regarding trace fields and commensurability classes of 2-bridge knots. We conclude with a discussion of a result of Hodgson which generalizes other results of this paper.

1. Introduction

The study of hyperbolic 3-manifolds has been a major focus of research in 3-manifold topology motivated in large part by the Geometrization Conjecture of W. P. Thurston[12]. Even with a possible proof of the conjecture provided by Perelman, hyperbolic 3-manifolds will likely remain an active area of research since hyperbolic geometry is the least understood of the eight possible geometries for 3-manifolds and by far the most prevalent. (See Anderson[1] for an overview of the Geometrization Conjecture and the recent work of Perelman.) A particularly tractable class of hyperbolic 3-manifolds are hyperbolic knot complements in $\mathbb{S}^3$. If K is a knot in $\mathbb{S}^3$ then we say K is *hyperbolic* if the complement $\mathbb{S}^3 - K$ admits a complete hyperbolic metric of finite volume. This is equivalent to having the complement be homeomorphic to the quotient space $\mathbb{H}^3/\Gamma$ where Γ is a discrete and torsion free subgroup of $\text{Isom}_+(\mathbb{H}^3)$, the group of orientation preserving isometries of $\mathbb{H}^3$. Since $\text{Isom}_+(\mathbb{H}^3)$ is isomorphic to the matrix group $\text{PSL}_2\mathbb{C}$, a hyperbolic knot K must have a discrete and faithful representation $\rho_0 : \pi_1(\mathbb{S}^3 - K) \to \text{PSL}_2\mathbb{C}$. Recall that a *representation* $\rho : \pi_1(\mathbb{S}^3 - K) \to \text{PSL}_2\mathbb{C}$ is simply a group homomorphism. By the Mostow-Prasad Rigidity Theorem the representation ρ_0 is unique up to conjugation. Moreover, it is not hard to show (using the Wirtinger presentation for example) that any representation $\rho' : \pi_1(\mathbb{S}^3 - K) \to \text{PSL}_2\mathbb{C}$ lifts to a representation $\rho : \pi_1(\mathbb{S}^3 - K) \to \text{SL}_2\mathbb{C}$. In this article, we will investigate some invariants of 2-bridge knots associated to such representations. The first part of the paper summarizes work that appears in Hoste and Shanahan.[5,6,7]

We begin by focusing on a special subclass of 2-bridge knots. Let $J(m, n)$ be the knot pictured in Figure 1, where m and n are the number of right-handed half-twists contained in each box. (A right handed half-twist coincides with a right handed crossing if the strands that are twisted have parallel orientations.) Note that $J(m, n)$ is a knot precisely when at least one of m or n is even, and a two-component link if both are odd. For example, $J(-1, 2)$ is the right-handed trefoil, $J(2, -2)$ is the figure eight knot, and $J(-1, 1)$ is the Hopf link. All of these knots are 2-bridge knots and they include the collection of $(2, 2k-1)$-torus knots (for $m = 1$) and twist knots (for $m = 2$).

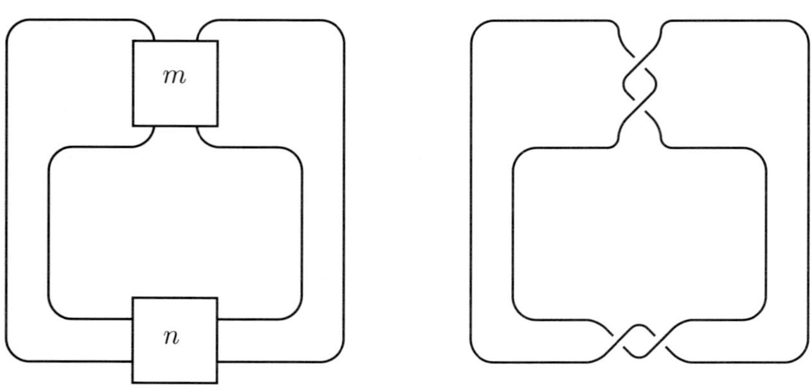

Fig. 1. $J(m, n)$ and the figure eight knot $J(2, -2)$.

Clearly, $J(m,n)$ is symmetric in m and n. If $J(m,n)$ is a knot, then its orientation is immaterial, as there exists an obvious rotation of $\mathbb{S}^3$ carrying the knot to its reverse. Furthermore, $J(-m,-n)$ is the mirror image of $J(m,n)$. Since we will make little distinction between a knot and its reflection we consider only those m and n for which $m > 0$ and n is even. Our analysis begins with the following proposition describing a presentation of the fundamental groups of these knots, a proof of which can be found in Hoste and Shanahan.[6]

Proposition 1: *The fundamental group of* $\mathbb{S}^3 - J(m,2n)$ *has a presentation of the form* $< a, b \mid aW = Wb >$, *where*

$$W = \begin{cases} \left[(ab^{-1})^k(a^{-1}b)^k\right]^n, & \text{if } m = 2k; \\ \left[(ab^{-1})^k ab(a^{-1}b)^k\right]^n, & \text{if } m = 2k+1. \end{cases}$$

Furthermore, a meridian μ *and the preferred longitude* λ *are given by*

$$\mu = a \quad \text{and} \quad \lambda = WW^* a^{-2e(W)}.$$

Here W^* *is the word* W *read backwards and* $e(W)$ *is the sum of the exponents in* W. *Therefore,* $e(W) = 0$ *if* m *is even and* $e(W) = 2n$ *if* m *is odd.*

Recall that the discrete faithful representation ρ_0 of a hyperbolic knot group must be *parabolic*, that is each meridian of the knot has parabolic image (an element with trace ± 2) in $\mathrm{SL}_2\mathbb{C}$. Typically this representation is just one of many parabolic representations. If $\Gamma \subset \mathrm{SL}_2\mathbb{C}$ is the image of ρ_0, then the *trace field* is defined to be the extension of $\mathbb{Q}$ obtained by adjoining the traces of all the elements in Γ. This field is well-defined and known to be an algebraic number field.

Theorem 2: The parabolic representations of the knot $J(m,2n)$ are determined by the roots of the polynomial $\Phi_{m,n}(z)$ defined recursively by

$$\Phi_{m,n+1}(z) - \tau_m(z)\Phi_{m,n}(z) + \Phi_{m,n-1}(z) = 0.$$

For a given value of m, $\tau_m(z)$ and the initial values $\Phi_{m,0}(z)$ and $\Phi_{m,1}(z)$ are polynomials derived from a certain matrix w_m which is described in Section 2.

Corollary 3: *If* $J(m,2n)$ *is hyperbolic and* z_0 *is the root of* $\Phi_{m,n}(z)$ *corresponding to* ρ_0, *then the trace field of* $J(m,2n)$ *is* $\mathbb{Q}(z_0)$. *The degree of the* *trace field of* $J(m, 2n)$ *is less than or equal to*

$$deg(\Phi_{m,n}(z)) = \begin{cases} mn - 1, \text{ if } & mn > 0; \\ -mn, \quad \text{if} & mn \leq 0. \end{cases}$$

Because of this corollary it is clearly of interest whether or not the polynomials $\Phi_{m,n}(z)$ are irreducible. In the case of twist knots we can settle this question.

Theorem 4: For $m = 2$, $\Phi_{2,n}$ is $\mathbb{Z}$-irreducible for all n.

Two hyperbolic knots are *commensurable* if their complements share a common finite-sheeted cover. For knots in $\mathbb{S}^3$ it is known that the trace field is a commensurability invariant. This gives the following corollary.

Corollary 5: *No two hyperbolic twist knots are commensurable.*

If μ and λ are the meridian and preferred longitude for a hyperbolic knot K, then one can conjugate ρ_0, if necessary, so that

$$\rho_0(\mu) = \begin{pmatrix} 1 & 1 \\ 0 & 1 \end{pmatrix} \quad \text{and} \quad \rho(\lambda) = \begin{pmatrix} 1 & \alpha \\ 0 & 1 \end{pmatrix}.$$

The *cusp field* of K is defined to be $\mathbb{Q}(\alpha)$ and is known to be a subfield of the trace field of K.

Theorem 6: For $m = 2$ or 3, the cusp and trace fields of $J(m, 2n)$ are identical.

Given a representation ρ into $\mathrm{PSL}_2\mathbb{C}$, its associated *character* is the map

$$\chi_\rho : \pi_1(\mathbb{S}^3 - K) \to \mathbb{C}$$

given by $\chi_\rho(g) = \mathrm{tr}(\rho(g))$. The set of all characters form an affine algebraic set called the *$PSL_2\mathbb{C}$-character variety* which encodes a wealth of topological and geometric information about the knot complement. The irreducible component of the character variety which contains χ_{ρ_0} is called the *geometric component*.

Theorem 7: The $\mathrm{PSL}_2\mathbb{C}$-character variety of $J(m, 2n)$ is defined by the polynomial $(x - y)r_{m,n}(x, y)$ where $r_{m,n}(x, y)$ is a polynomial defined recursively by

$$r_{m,n+1}(x, y) - t_m(x, y)r_{m,n}(x, y) + r_{m,n-1}(x, y) = 0. \tag{1}$$

As in Theorem 2, the polynomials t_m and the initial values of r are derived from the matrix w_m described in Section 2.

Calegari and Dunfield[2] proved that the presence of certain types of characters gives information on the commensurability class of the knot. Their result provides the motivation for the following theorem and corollary.

Theorem 8: For $m = 2$, let X_n be the $\mathbb{Q}$-irreducible component of the $\mathrm{PSL}_2\mathbb{C}$-character variety of the hyperbolic twist knot $J(2, 2n)$ which contains the geometric component. Then for every $n \neq 0, \pm 1$ there exists a character of a non-integral reducible representation in X_n.

Corollary 9: *The complement of a twist knot $J(2, 2n)$ is commensurable to a fibered knot in a $\mathbb{Z}/2\mathbb{Z}$-homology sphere if and only if it is fibered itself, that is, if it is the unknot, the trefoil, or the figure eight knot.*

An important invariant of knots is the A-polynomial introduced by Cooper, Culler, Gillet, Long, and Shalen.[3] In Section 3 we define a generalization of this invariant for links. In general, it is difficult to compute the A-polynomial for all but the simplest knots and links. However, our techniques lead to the following theorem which allows one to compute the A-polynomial for $J(m, 2n)$ with $m = 1, 2$, or 3. These are the only infinite families of knots for which formulae for the A-polynomials have been found.

Theorem 10: The A-polynomial of $J(m, 2n)$ is the z-resultant of the two polynomials $R_{m,n}(M, z)$ and $S_{m,n}(L, M, z)$ where $R_{m,n}$ is defined recursively by

$$R_{m,n+1}(M, z) - T_m(M, z)R_{m,n}(M, z) + R_{m,n-1}(M, z) = 0. \qquad (2)$$

The polynomials T_m, S, and the initial values of R are all derived from the matrix w_m as described in Section 2.

Corollary 11: *For $m = 1$, 2, or 3 the equation $S_{m,n}(L, M, z) = 0$ can be solved for z as a rational function of L and M. Substituting this expression for z in (2) gives a recursive formula for the A-polynomial.*

2. The Recursive Technique

Every 2-bridge knot, $K_{p/q}$, in $\mathbb{S}^3$ corresponds to a pair of co-prime, odd integers, (p, q), with $-q < p < q$. The fundamental group of the complement of $K_{p/q}$ always has a presentation of the form

$$\pi_1(\mathbb{S}^3 - K_{p/q}) = < a, b \mid aW = Wb > \qquad (3)$$

where W has the form

$$W = b^{\epsilon_1} a^{\epsilon_2} \dots b^{\epsilon_{q-2}} a^{\epsilon_{q-1}},$$

with each $\epsilon_i = (-1)^{[\![i|p|/q]\!]}$. Here $[\![x]\!]$ is the greatest integer less than or equal to x. It follows that $\epsilon_i = \epsilon_{q-i}$ for all i. Furthermore, the meridian μ and the preferred longitude λ of $K_{p/q}$ are represented by the words

$$\mu = a \quad \text{and} \quad \lambda = WW^*a^{-2e(W)},$$

where $W^* = a^{\epsilon_{q-1}}b^{\epsilon_{q-2}}\dots a^{\epsilon_2}b^{\epsilon_1}$ is W written backwards and $e(W) = \sum \epsilon_i$.

The presentation (3) is not unique, but the existence of at least one such presentation follows directly from Wirtinger's algorithm applied to Schubert's canonical 2-bridge diagram of $K_{p/q}$. The generators a and b come from the two bridges (overpasses) and represent (up to conjugation) the meridian μ.

Suppose now that $\rho : \pi_1(\mathbb{S}^3 - K_{p/q}) \to \mathrm{SL}_2\mathbb{C}$ is an irreducible representation. After conjugating, if necessary, we may assume that

$$\rho(a) = \begin{pmatrix} M & 1 \\ 0 & M^{-1} \end{pmatrix} \quad \text{and} \quad \rho(b) = \begin{pmatrix} M & 0 \\ -z & M^{-1} \end{pmatrix}$$

where $M \neq 0$ and $z \neq 0$. (The choice of $-z$ as opposed to z is entirely arbitrary.)

In order to get a representation we require that the relation $aW = Wb$ continue to hold in $\mathrm{SL}_2\mathbb{C}$. It is shown in Riley[11] that the four equations determined by this relation reduce to the single equation

$$R = (M - M^{-1})W_{12} + W_{22} = 0. \tag{4}$$

Here W_{ij} denotes the (i, j)-entry of W.

We now wish to study the bi-infinite sequence of 2-bridge knots $\{J(m, 2n)\}_{n=-\infty}^{\infty}$. For these knots, the word W has the form $W = w_m^n$, where w_m is the word described in Proposition 1. It follows from the Cayley-Hamilton identity and the fact that $\det w_m = 1$, that

$$w_m^{n+1} - \mathrm{tr}(w_m)w_m^n + w_m^{n-1} = 0. \tag{5}$$

Moreover

$$(w_m^{n+1})_{ij} - \mathrm{tr}(w_m)(w_m^n)_{ij} + (w_m^{n-1})_{ij} = 0.$$

Relation (4) now becomes

$$R_{m,n} = (M - M^{-1})(w_m^n)_{12} + (w_m^n)_{22} = 0. \tag{6}$$

For a fixed value of m, we now have the recursive formula

$$R_{m,n+1} - \mathrm{tr}(w_m)R_{m,n} + R_{m,n-1} = 0. \tag{7}$$

We are now prepared to prove Theorem 2. The representation ρ is parabolic if and only if $M = 1$. Letting $\Phi_{m,n}(z) = R_{m,n}|_{M=1}$ immediately gives the recursive formula of Theorem 2. If $J(m, 2n)$ is hyperbolic and z_0 is the root of $\Phi_{m,n}(z)$ corresponding to the discrete faithful representation ρ_0, then it is easy to see that the trace field is $\mathbb{Q}(z_0)$, since $\mathrm{tr}(\rho_0(ab)) = 2 - z_0$. Finally, a simple inductive argument gives the degree of $\Phi_{m,n}(z)$. (See Hoste and Shanahan[5] for more details.)

Continuing to focus on the recursion (7) leads to a proof of Theorem 7. Hilden, Lozano and Montesinos[4] show that the $\mathrm{PSL}_2\mathbb{C}$-character variety is parameterized by the variables $x = \mathrm{tr}(\rho(a^2)) = M^2 + M^{-2}$ and $y = \mathrm{tr}(\rho(ab)) = M^2 + M^{-2} - z$. In order to prove Theorem 7 we must first show that it is possible to change variables from $\{M, z\}$ to $\{x, y\}$ in $R_{m,n}(M, z)$ to produce the polynomial $r_{m,n}(x, y)$. Because $z = y - x$ it remains to consider how to eliminate M in favor of x and y. Here we make use of the following lemma.

Lemma 12: *If $W = b^{\epsilon_1} a^{\epsilon_2} \dots a^{\epsilon_s}$ is any word in a and b such that the sequence $\{\epsilon_1, \epsilon_2, \dots, \epsilon_s\}$ of exponents is palindromic and if*

$$R(W) = (M - M^{-1})W_{12} + W_{22},$$

then both tr(W) and $R(W)$ are symmetric in M^2.

Proof: It is easy to show that $W_{21} = -zW_{12}$ and as a consequence both $\mathrm{tr}(W)$ and $R(W)$ are in fact polynomials in $M^{\pm 2}$. Furthermore, it is shown in Hoste and Shanahan[6] that $\mathrm{tr}(W)$ is symmetric in M. Now if $W = a^{\epsilon}\overline{W}b^{\epsilon}$ with $\epsilon = \pm 1$, then one can check that $R(W) = R(\overline{W})$. If instead, $W = b^{\epsilon}\overline{W}a^{\epsilon}$, then

$$R(W) = \begin{cases} (1+z)(R(\overline{W}) - \mathrm{tr}(\overline{W})) + \mathrm{tr}(a^2\overline{W})) & \text{if } \epsilon = 1, \\ (1+z)(R(\overline{W}) - \mathrm{tr}(\overline{W})) + \mathrm{tr}(b^{-2}\overline{W})) & \text{if } \epsilon = -1. \end{cases}$$

Since $\overline{W}$ is shorter than W, the result follows by induction. □

Since $R_{m,n}(M, z)$ is symmetric in M^2 it can be rewritten as

$$R_{m,n}(M, z) = \sum_{i,j} c_{ij}(M^2 + M^{-2})^i z^j.$$

Thus, irreducible $\mathrm{PSL}_2\mathbb{C}$-characters correspond to roots of

$$r_{m,n}(x, y) = \sum_{i,j} c_{ij} x^i (x - y)^j.$$

Using the lemma a second time, we may convert the polynomial $\mathrm{tr}(w_m)$ to $t_m(x,y)$, therefore giving the recursive formula (1) for $r_{m,n}(x,y)$. Finally, the factor of $x-z$ in Theorem 7 accounts for the abelian representations. Additional details may be found in Hoste and Shanahan.[7]

Another consequence of the recursive formula (7) is Theorem 10. The polynomial $T_m(M,z)$ is simply $\mathrm{tr}(w_m)$. To obtain the relation S_m we must consider the image of the preferred longitude λ of $J(m,n)$ under ρ. As already mentioned, $\lambda = WW^*a^{-2e(W)}$. Since ρ is a representation and since λ and μ commute, we must have

$$\rho(\lambda) = \begin{pmatrix} L & * \\ 0 & L^{-1} \end{pmatrix}.$$

It is shown in Hoste and Shanahan[6] that computing $\rho(\lambda)$ in terms of W gives the relation

$$S_{m,n} = (w_m)_{12}L + (\tilde{w}_m)_{12}M^{-2ne(w_m)} = 0, \tag{8}$$

where $(\tilde{w}_m)_{12}$ is obtained from $(w_m)_{12}$ by replacing M with M^{-1}. Furthermore, note that $e(w_m)$ equals zero if m is even and 2 otherwise. Thus $S_{m,n}$ has no dependance on n when m is even, and only a relatively minor dependance on n otherwise. In fact, using a non-preferred longitude, even this dependance may be eliminated. The A-polynomial of $J(m,n)$ is now obtained by computing the z-resultant of the relations $R_{m,n}(M,z)$ and $S_{m,n}(L,M,z)$.

So far each of Theorems 2, 7, and 10 follow from the basic recursion formula (7). Further investigation into the class of knots $J(m,n)$ will lead to the other results stated in the Introduction, which we will presently discuss. In Section 3 we will explore the extent to which these results might apply to all 2-bridge knots.

For any 2-bridge knot, equation (4) reduces to $W_{22} = 0$ if $M = 1$, which is the case for parabolic representations. It is straightforward to compute W_{22} for any 2-bridge knot $K_{p/q}$ directly from p and q (see for example page 140 of Maclachlan and Reid[8]). Using a computer algebra system such as *Mathematica*, it is easy to determine if W_{22} is $\mathbb{Z}$-irreducible for small values of p and q. In Section 3 we will describe the outcome of such computer experiments. Proving that an infinite family of such polynomials are all irreducible is, however, a formidable task. Remarkably, this can be accomplished for the twist knots $J(2,2n)$ as stated in Theorem 4 and proven in Hoste and Shanahan.[5]

Proving Theorem 4 relies on the change of variable $z = x - x^{-1}$ that amazingly gives a simple "closed" formula for the polynomial

$$\Phi_{2,n}(x - x^{-1}) = \frac{x^{4n+2} + x^{4n+1} - x + 1}{x^{2n-1}(x^2 + 1)}$$

if $n > 0$. (A similar formula exists for $n < 0$.) One is led somewhat naturally to this substitution by the geometric point of view taken in Neumann and Reid.[10] Using this formula it is possible to carefully describe where all the roots lie in the complex plane, and this in turn leads to the irreducibility of $\Phi_{2,n}$ by consideration of its coefficients and the possible sums of subsets of the roots.

Combining Theorems 2 and 4 determines that the degree of the trace field of any twist knot $J(2, 2n)$ is either $2n - 1$ if $n > 0$ or $-2n$ if $n < 0$. This can be restated as 2 less than the crossing number of the twist knot. Because the trace field is a commensurability invariant we see that no two hyperbolic twist knots are commensurable, as stated in Corollary 5.

In general, the cusp field of a hyperbolic knot is always a subfield of the trace field. Therefore for the knots $J(m, 2n)$ the degree of the cusp field is also bounded by $mn - 1$ if $mn > 0$ and $-mn$ otherwise. If $m = 2$ or 3, we can actually prove that the cusp and trace fields coincide. This is done by finding a formula for $z_{m,n}$, the root of $\Phi_{m,n}(z)$ giving the hyperbolic structure, as a rational function of $\alpha_{m,n}$, the cusp coefficient. For $m = 2$ this formula was first observed by Neumann and Reid.[10] It seems likely that the cusp and trace fields of $J(m, 2n)$ coincide for all m and n. (See Hoste and Shanahan[7] for more details.)

In a similar manner, small values of m, in particular $m = 1, 2$ or 3, also allow us to prove Corollary 11. In this case the relation $S_{m,n}(L, M, z) = 0$ is simple enough to solve for z in terms of L and M. This value of z can then be substituted into the relation $R_{m,n}(M, z)$ to obtain the A-polynomial, without using resultants to eliminate z.

Theorem 4 plays an important role in the proof of Theorem 8. Using the recursion (1) it is easy to identify the character of a non-integral reducible representation on the character variety of the twist knot $J(2, 2n)$ for $n \neq 0, \pm 1$. The irreducibility of $\Phi_{2,n}(z)$ can be used to show that $r_{2,n}(x, y)$ is also irreducible and thus this character must lie on the component X_n described in Theorem 8. The presence of such a character, together with Theorem 6 is sufficient to prove Corollary 9 using the criteria developed by Calegari and Dunfield.[2]

3. Extensions and Conjectures

From Theorem 2 and Theorem 4 we see that the degree of the trace field of a hyperbolic twist knot increases monotonically as a function of the number of twists. This phenomenon extends more generally to all 2-bridge knots. Recall that for any 2-bridge knot $K_{p/q}$ the polynomial R of (4) parametrizes the set of irreducible representations. Thus, the roots of the polynomial

$$\Phi_{p/q}(z) = R|_{M=1} = W_{22}|_{M=1}$$

correspond to all parabolic representations of $\pi_1(\mathbb{S}^3 - K_{p/q})$. Experimentation with *Mathematica* reveals the following result.

Proposition 13: *For all* $1 < p < q < 501$ *with* q *an odd prime or the square of an odd prime, the polynomial* $\Phi_{p/q}(z)$ *is irreducible and of degree* $(q-1)/2$. *Thus, the degree of the trace field of* $K_{p/q}$ *is* $(q-1)/2$.

We note that there are 5570 distinct such 2-bridge knots. The condition that q be prime or the square of a prime is certainly not necessary for $\Phi_{p/q}$ to be irreducible. For example, the twist knots are given by the fractions $\frac{2n+1}{4n+1}$ and $\frac{2n-1}{4n-1}$ for $n = 1, 2, \ldots$. It seems reasonable to make the following conjecture.

Conjecture 1: *For all hyperbolic 2-bridge knots* $K_{p/q}$ *with* q *prime or the square of a prime, the degree of the trace field is* $(q-1)/2$.

We have been unable to extend the techniques which were used to prove Theorem 4 to other classes of 2-bridge knots. In general, it is hard to prove that an infinite family of polynomials are all irreducible. It is interesting to note that Eisenstein's criterion can never be directly used to show that $\Phi_{p/q}(z)$ is irreducible. To see that this is true first write the word W of (3) as $W = V\overline{V}^*$ where $\overline{V}^*$ is the word V written backwards and with a and b interchanged. Setting $M = 1$, it is not difficult to determine the entries of $\overline{V}^*$ in terms of those of V and thus obtain that

$$\Phi_{p/q}(z) = (V_{22})^2 - \frac{(V_{21})^2}{z}.$$

From this it follows that $\Phi_{p/q}(z^2)$ factors nontrivially as

$$\Phi_{p/q}(z^2) = \left(V_{22} - \frac{V_{21}}{z}\right)\left(V_{22} + \frac{V_{21}}{z}\right).$$

Therefore, since the polynomials $\Phi_{p/q}(z)$ and $\Phi_{p/q}(z^2)$ have the same coefficients, Eisenstein's criterion cannot prove directly that $\Phi_{p/q}(z)$ is irreducible.

Further experimentation reveals the following.

Proposition 14: *For all $3 < q < 101$ with q a prime or the square of an odd prime, the discriminants of the trace fields of distinct hyperbolic knots $K_{p/q}$ and $K_{p'/q}$ are different. Thus, $\mathbb{S}^3 - K_{p/q}$ and $\mathbb{S}^3 - K_{p'/q}$ are commensurable if and only if the knots are equivalent.*

This leads to the following conjecture.

Conjecture 2: *If q and q' are odd primes or the squares of odd primes, then the hyperbolic manifolds $\mathbb{S}^3 - K_{p/q}$ and $\mathbb{S}^3 - K_{p'/q'}$ are commensurable if and only if the knots are equivalent.*

Returning to the special case of twist knots $J(2, 2n)$, there is a more geometric reason to expect that the degree of the trace field approaches infinity as n approaches infinity. Since the twist knot complement $\mathbb{S}^3 - J(2, 2n)$ may be obtained by $1/n$ Dehn filling on one component of the Whitehead link, these manifolds provide a nice example of the following unpublished result of Craig Hodgson.

Theorem 15: If $\mathcal{L} = \{K_1, K_2, \ldots, K_k\}$ is a k-component hyperbolic link in $\mathbb{S}^3$ and if $M_{1/q}$ is the manifold obtained by $1/q$ Dehn filling on K_1, then the degree of the trace field of $M_{1/q}$ approaches infinity as $|q|$ approaches infinity.

We finish this paper with a sketch of a proof of this result. We believe that this proof is essentially the same as the one Hodgson proposed. An interesting feature of the proof is that it relies on the notion of a *relative* A-polynomial of a link $\mathcal{L}$ with respect to its j-th component. Suppose that $\mathcal{L} = \{K_1, K_2, \ldots, K_n\}$ is a link in $\mathbb{S}^3$ and fix a preferred meridian and longitude, $\{\mu_i, \lambda_i\}$, for each component K_i. Let R be the set of all representations of $\pi_1(\mathbb{S}^3 - \mathcal{L})$ into $\mathrm{PSL}_2\mathbb{C}$. Then R is an affine algebraic set, and by adjoining additional polynomials to the ideal of R we obtain an algebraic subset R_j of R with the property that $[\mathrm{tr}(\rho(\mu_i))]^2 = 4$ for all $\rho \in R_j$ and $i \neq j$. Notice that any representation which corresponds to a hyperbolic structure complete at the ends associated with K_i, $i \neq j$, is in R_j. As with the A-polynomial of a knot, the basis $\{\mu_j, \lambda_j\}$ associated to the end K_j provides a way of projecting R_j onto $\mathbb{C}^2$. This projection is defined as follows. Consider the algebraic subset U_j of R_j consisting of all representations $\rho \in R_j$ for which $\rho(\mu_j)$ and $\rho(\lambda_j)$ are upper triangular. Notice that since μ_j and λ_j commute, every representation in R_j is conjugate to one in U_j. Thus if

$\rho \in U_j$, then we have

$$\rho(\mu_j) = \begin{pmatrix} M & * \\ 0 & 1/M \end{pmatrix} \text{ and } \rho(\lambda_j) = \begin{pmatrix} L & * \\ 0 & 1/L \end{pmatrix}.$$

Define $\xi : U_j \to \mathbb{C}^2$ by $\xi(\rho) = (L, M)$. Since ξ is a regular map, the Zariski closure of the image $\xi(U_j)$ is an algebraic subset C of $\mathbb{C}^2$. Let $C_1, C_2, \ldots, C_p$ be the distinct irreducible complex-dimension 1 components of C and, for $1 \le i \le p$, let $F_i(L, M)$ be the irreducible polynomial defining C_i. We define the *A-polynomial of $\mathcal{L}$ relative to K_j* to be the polynomial

$$A^j_{\mathcal{L}}(L, M) = \prod_{i=1}^{p} F_i(L, M).$$

Clearly the relative A-polynomial is only well-defined up to multiplication by a nonzero complex number. There are a number of quite beautiful and remarkable properties of the A-polynomial which remain true for the relative version. One essential observation is that $A^j_{\mathcal{L}} \in \mathbb{Z}[L, M]$. This result follows directly from the fact that the relative A-polynomial can be computed (at least in theory) by repeatedly taking resultants of integer-coefficient polynomials, eliminating all variables except L and M.

In order to proceed it is necessary that we review some properties of ends of hyperbolic 3-manifolds. Recall that if M^3 is a finite volume hyperbolic 3-manifold whose ends are homeomorphic to $T^2 \times [0, \infty)$, then an end E can be of two types. If E is complete, then E is the quotient of $\mathbb{C} \times [c, \infty)$ by a group $\Delta \cong \mathbb{Z} \times \mathbb{Z}$ generated by a pair of parabolic elements with common fixed point at infinity. (Here we are using the upper half-space model of $\mathbb{H}^3$.) These parabolic isometries act as Euclidean translations in linearly independent directions on the cross sections $\mathbb{C} \times \{t\}$ of $\mathbb{C} \times [c, \infty)$ and the quotients of these cross sections are torus cross sections in E. If the end E is not complete, then E is the quotient of $\{(z, t) : 0 < |z| \le Ct, t > 0\}$ for some positive constant C by a group $\Delta \cong \mathbb{Z} \times \mathbb{Z}$ generated by a pair of loxodromic isometries with common axis $z = 0$. The quotient of each cone $|z| = ct$, with $c \le C$, gives a torus cross section in E.

Suppose E is incomplete but its metric completion corresponds to p/q Dehn surgery with respect to generators $X = \begin{pmatrix} x & 0 \\ 0 & 1/x \end{pmatrix}$ and $Y = \begin{pmatrix} y & 0 \\ 0 & 1/y \end{pmatrix}$ of Δ. Then $X^p Y^q = I$ and the action of $X^p Y^q$ on $\mathbb{H}^3$ is a rotation of $\pm 2\pi$ around the axis $z = 0$. The linear fractional transformations of $\mathbb{C}$ corresponding to X and Y are $z \mapsto x^2 z$ and $z \mapsto y^2 z$ respectively. Thus we have $2p \log x + 2q \log y = \pm 2\pi i$. Equating the real and imaginary parts of this equation gives

$$p \log|x| + q \log|y| = 0$$
$$p \arg x + q \arg y = \pm\pi,$$

where $\arg z$ is the principal argument of z. Solving for p and q in terms of x and y gives

$$p = \frac{\mp\pi \log|y|}{\log|x| \arg y - \arg x \log|y|}$$
$$q = \frac{\pm\pi \log|x|}{\log|x| \arg y - \arg x \log|y|}. \tag{9}$$

Notice that if there are only finitely many possibilities for x and y, then there are only finitely many possibilities for p and q. We will use this fact in the proof of Theorem 15.

Another important feature of the proof is the notion of Mahler measure of a polynomial. The *Mahler measure* $\mathcal{M}(f)$ of a polynomial with integer coefficients

$$f(z) = a_n z^n + a_{n-1}x^{n-1} + \cdots + a_0 = a_n(z - r_1)(z - r_2)\ldots(z - r_n)$$

where $a_n \neq 0$ is defined as

$$\mathcal{M}(f) = |a_n| \prod_{i=1}^{n} \max(|a_i|, 1).$$

A related measure of $f(z)$ is its *length* $\ell(f)$ given by

$$\ell(f) = |a_n| + |a_{n-1}| + \cdots + |a_0|.$$

An important relation between these two measures is given by the following lemma. (See Mahler[9] for a proof.)

Lemma 16: *Let f be a polynomial with integer coefficients. Then*

(i) $1 \leq \mathcal{M}(f) \leq \ell(f)$, *and*
(ii) $\mathcal{M}(f_1 f_2) = \mathcal{M}(f_1)\mathcal{M}(f_2)$.

We will also need the following lemma.

Lemma 17: *There are only finitely many polynomials with integer coefficients, bounded degree, and bounded Mahler measure. In particular, the* *number of such polynomials with degree not more than D and Mahler measure not more than R is no greater than*

$$\sum_{d=1}^{D}\prod_{i=0}^{d}\left(2\binom{d}{i}[\![R]\!]+1\right).$$

Here $[\![R]\!]$ is the greatest integer less than or equal to R.

Proof. Suppose $f(z)=\sum_{i=0}^{n}a_iz^i=a_n\prod_{i=1}^{n}(z-r_i)$. If $I\subset\{1,2,\dots,n\}$ is any subset then

$$\begin{aligned}|a_n|\prod_{i\in I}|r_i| &\le |a_n|\prod_{i\in I}\max(|r_i|,1),\\ &\le |a_n|\prod_{i=1}^{n}\max(|r_i|,1),\\ &\le \mathcal{M}(f).\end{aligned}$$

Now each coefficient a_i of f is the product of a_n with a symmetric polynomial in the roots $\{r_i\}$. From this is follows that

$$|a_i|\le\binom{n}{i}\mathcal{M}(f)\le\binom{n}{i}R\quad\text{for all i.}$$

Since each coefficient a_i is an integer which satisfies $|a_i|\le\binom{n}{i}R$ we see that there are only $2\binom{n}{i}[\![R]\!]+1$ possibilities for a_i. Multiplying these together and adding over all possible degrees gives the desired bound. □

We are now prepared to prove Theorem 15. Let D be any positive integer. Suppose $\mathcal{L}=\{K_1,K_2,\dots,K_k\}$ and $M_{1/q}$ are defined as in the theorem. Let $\rho_{1/q}:\pi_1(\mathbb{S}^3-\mathcal{L})\to\mathrm{PSL}_2\mathbb{C}$ be the discrete representation with image $\Gamma_{1/q}$ where the metric completion of $\mathbb{H}^3/\Gamma_{1/q}$ is the manifold $M_{1/q}$ obtained by $1/q$ Dehn surgery on K_1. Let μ_1 and λ_1 be the meridian and longitude respectively of K_1. Now $\pi_1(M_{1/q})$ is the quotient of $\pi_1(\mathbb{S}^3-\mathcal{L})$ obtained by adjoining the relation $\mu_1\lambda_1^q=1$. Since $\rho_{1/q}(\mu_1\lambda_1^q)=I$ it follows that $\rho_{1/q}$ descends to a discrete faithful representation $\overline{\rho}_{1/q}$ of $\pi_1(M_{1/q})$. Furthermore, both representations generate the same trace field since their images share the same generators. Therefore $\mathbb{Q}(\mathrm{tr}\,\Gamma_{1/q})$ is the trace field of $M_{1/q}$. Finally, we assume that this trace field has degree less than D.

After conjugating, if necessary, we may assume that

$$X_{1/q}=\rho_{1/q}(\mu_1)=\begin{pmatrix}x_{1/q} & 0\\ 0 & x_{1/q}^{-1}\end{pmatrix}\ \text{and}\ Y_{1/q}=\rho_{1/q}(\lambda_1)=\begin{pmatrix}y_{1/q} & 0\\ 0 & y_{1/q}^{-1}\end{pmatrix}$$

where $x_{1/q} \neq \pm 1$ and $y_{1/q} \neq \pm 1$. Furthermore, $\mu_1 \lambda_1^q = 1$ so $X_{1/q} = Y_{1/q}^{-q}$ and $x_{1/q} = y_{1/q}^{-q}$

Let $A^1_{\mathcal{L}}(L, M)$ be the A-polynomial of $\mathcal{L}$ relative to K_1. By definition $\rho_{1/q}$ gives a complete structure on the end of $\mathcal{L}$ associated with K_i for all $i > 1$. Hence, $[\mathrm{tr}(\rho_{1/q}(\mu_i))]^2 = 4$ for all $i > 1$ and therefore $A^1_{\mathcal{L}}(y_{1/q}, x_{1/q}) = 0$ for all q. For each q, define $f_{1/q}(t)$ as $f_{1/q}(t) \doteq A^1_{\mathcal{L}}(t, t^{-q})$, where $f_{1/q}(t) \in \mathbb{Z}[t]$ and $\doteq$ denotes equality up to multiplication by powers of t. The coefficients of $f_{1/q}$ are sums of coefficients of $A^1_{\mathcal{L}}(L, M)$ and hence

$$\mathcal{M}(f_{1/q}) \leq \ell(f_{1/q}) \leq \ell(A^1_{\mathcal{L}}) \text{ for all } q.$$

Thus, each $f_{1/q}$ has bounded Mahler measure.

Let $t_{1/q} = y_{1/q}$ and $x_{1/q} = t_{1/q}^{-q}$ and hence $f_{1/q}(t_{1/q}) = 0$. Let $g_{1/q}$ be the irreducible factor of $f_{1/q}$ having $t_{1/q}$ as a root. Now $t_{1/q}$ has degree at most 2 over $\mathbb{Q}(\mathrm{tr}\,\Gamma_{1/q})$ since $t_{1/q}$ is a root of $Z^2 - (t_{1/q} + 1/t_{1/q})Z + 1 = Z^2 - (\mathrm{tr}\ Y_{1/q})Z + 1$. Since $[\mathbb{Q}(\mathrm{tr}\,\Gamma_{1/q}) : \mathbb{C}] < D$ we have $\deg g_{1/q} \leq 2D$. By Lemma 16, we have that $\mathcal{M}(g_{1/q}) \leq \mathcal{M}(f_{1/q})$ and so there are only finitely many possibilities for $g_{1/q}$ by Lemma 17 . This implies there are only finitely many possibilities for $t_{1/q}$, hence, there are only finitely many possibilities for $x_{1/q}$ and $y_{1/q}$. Therefore, from Equations (9) there are only finitely many possibilites for q. Finally, we note that the entire argument can easily be generalized to p/q surgery if we assume that $|p| < E$. The degree of $g_{p/q}(t)$ is then bounded by $2DE$ for all q. Thus, for p bounded, the degree of the trace field of the manifold obtained by p/q surgery on one component of $\mathcal{L}$ goes to infinity as $|q|$ goes to infinity.

References

1. M. T. Anderson, *Geometrization of 3-manifolds via the Ricci Flow*, Notices A. M. S. **51** no. 2 (2004) 184–193.
2. D. Calegari and N. Dunfield, *Commensurability of 1-cusped hyperbolic 3-manifolds*, Trans. Amer. Math. Soc. **354** no. 7 (2002) 2955–2969.
3. D. Cooper, M. Culler, H. Gillet, D. D. Long, and P. B. Shalen, *Plane curves associated to character varieties of 3-manifolds*, Invent. Math. **118** (1994) 47–84.
4. H. M. Hilden, M. T. Lozano, and J. M. Montesinos, *On the arithmetic 2-bridge knots and link orbifolds and a new knot invariant*, J. Knot Theory and its Ramifications **4** no. 1 (1995) 81–114.
5. J. Hoste and P. D. Shanahan, *Trace fields of twist knots*, J. Knot Theory and its Ramifications **10** no. 4 (2001) 625–639.
6. J. Hoste and P. D. Shanahan, *A formula for the A-polynomial of twist knots*, J. Knot Theory and its Ramifications **13** no. 2 (2004) 193–209.

7. J. Hoste and P. D. Shanahan, *Commensurability classes of twist knots*, arXiv:math.GT/031105, to appear in J. Knot Theory and its Ramifications.
8. C. Maclachlan and A. W. Reid, The arithmetic of hyperbolic 3-manifolds, (Springer-Verlag, New York, 2003).
9. K. Mahler, Lectures on transcendental numbers, LNM **546**, (Springer-Verlag, New York, 1976).
10. W. D. Neumann and A. W. Reid, *Arithmetic of hyperbolic manifolds*, Topology '90, de Gruyter, Berlin (1992) 17–26.
11. R. Riley, *Nonabelian representations of 2-bridge knot groups*, Quart. J. Math. Oxford (2) **35** (1984) 191–208.
12. W. P. Thurston, *Three dimensional manifolds, Kleinian groups, and hyperbolic geometry*, Bull. Amer. Math. Soc. **6** no. 3 (1982) 357–381.

CHAPTER 30

CONJECTURES ON THE ENUMERATION OF ALTERNATING LINKS

Paul Zinn-Justin

LPTMS Université Paris-Sud
F-91405 ORSAY Cedex, FRANCE
E-mail: pzinn@lptms.u-psud.fr

We show how some issues of enumeration in knot theory are related to statistical models on random lattices. The insight it gives us leads to conjectures on asymptotic counting as the number of crossings becomes large. We describe some numerical work to test these conjectures. We briefly outline a generalization to virtual knot theory.

1. Introduction

The mathematical description and classification of knots and related objects of knot theory is a subject with a history of several centuries. We refer to Ref. 1 for an introduction to the existing literature. We are concerned here with one of the most basic questions one can ask about these: their *enumeration*. For technical reasons which will be explained below, we concentrate here on a specific type of objects known as *alternating* links or tangles. The alternating property means that such objects possess a representation, once projected onto a plane, as a diagram in which over- and underpasses alternate.

In general, we wish to enumerate classes of alternating links or tangles specified by some parameters such as the (minimal) number of crossings and the number of connected components. There is no known formula to compute these numbers. The approach we use here is to consider the generating function of such numbers as the partition function of a stastical model living on random dynamical lattices. This is described in Sec. 2. We then show how the insight gained in this reformulation allows us to predict the form of the asymptotic behavior of this series. The most important quantity from a physical point of view is the power-law exponent ("critical exponent"); we give in Sec. 3 some arguments in favor of a conjecture for its expression and specialize it to the case of knots. In Sec. 4 we present two attempts at checking numerically this conjecture: a transfer matrix-based approach to compute the first few terms of the generating series; and a random sampling approach. Finally, we discuss in Sec. 5 how these results generalize to virtual alternating links or tangles.

2. A model for the enumeration of alternating links

Our whole approach in based on the representation of knots, links and tangles as diagrams. We do not give rigorous mathematical definitions here. Let us simply recall that knots are circles embedded in 3-dimensional space, links are collections of intertwined knots and tangles are links with some open strings with fixed ends. In order to avoid unwanted symmetry factors in the counting, tangles are usually preferred to links; however for the sake of simplicity we only consider links in what follows, and ignore all technicalities related to symmetry factors.

Links can be projected onto a plane in such a way that they become 4-regular planar maps with decorations at each vertex representing under/over-passes. Links are always considered up to isotopy, which means in particular that the projected diagram is not unique. A famous theorem by Reidemeister asserts that two diagrams correspond to the same link iff they are related by a finite sequence of moves known as Reidemeister moves. However, in the case of alternating links, there is a simpler procedure. First, alternating diagrams can be reduced to diagrams having minimal crossing number by iteratively eliminating irrelevant crossings. Second, two such *reduced* alternating diagrams are topologically equivalent if and only if they are connected by a sequence of moves (see Fig. 1) known as *flypes*. This simplifies drastically the combinatorial treatment of alternating links and allows some exact enumerations to be performed, see Refs. 3–6.

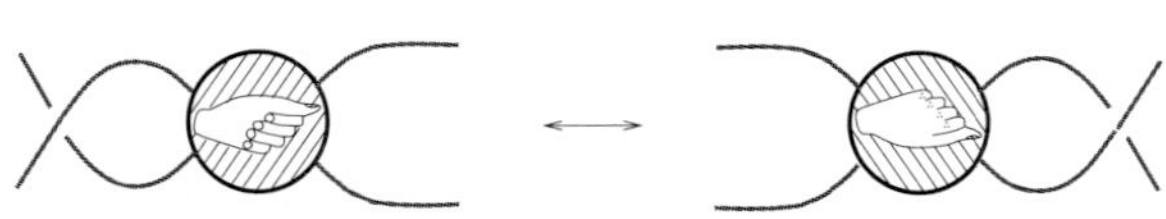

Fig. 1. A flype. The hand represents an arbitrary tangle.

Let us now describe the statistical lattice model, whose partition function reproduces the generating series that counts alternating links. First we define a simpler model that counts all diagrams: each configuration is simply a 4-regular planar map (which can be considered as a "random lattice"). Loops can then be drawn on the map in such a way that they cross each other at each vertex: this way we recover the connected components of the underlying alternating link (Fig. 2).

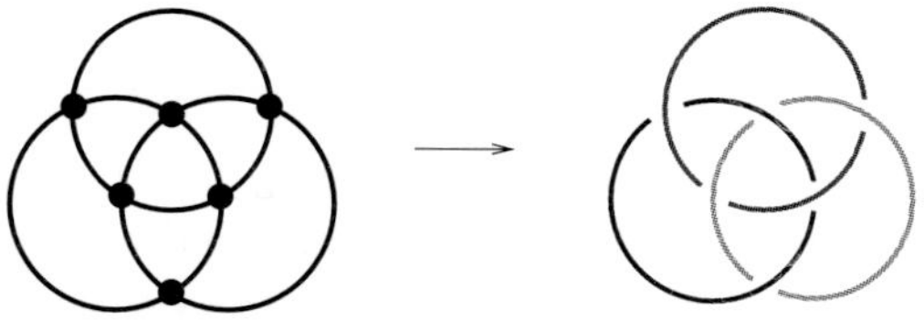

Fig. 2. 4-regular planar map and alternating link diagram.

The partition function is then given by

$$Z(g,n) = \sum_{\text{planar maps}} g^{\#\text{vertices}} n^{\#\text{loops}} \; ; \tag{1}$$

that is, each crossing has an associated weight of g, and each connected component a weight of n, where g and n are two parameters. Up to issues of symmetry factors, the partition function (1) coincides with the "large N free energy" defined in Ref. 6, and is the generating series of the number of alternating link diagrams.

However, in order to count each alternating link once, we need to restrict ourselves to reduced diagrams and to take into account the flyping equivalence. It is shown in Ref. 6 that this can be performed by replacing the former model with a more general model in which loops can either cross as before, or "avoid each other" at each vertex (Fig. 3). The partition function now takes the form

$$Z(g_1,g_2,n) = \sum_{\text{planar maps}} \sum_{\text{loopdrawings}} g_1^{\#\text{crossings}} g_2^{\#\text{tangencies}} n^{\#\text{loops}} \; . \tag{2}$$

Using a procedure similar to the renormalization procedure in quantum field theory, one can choose g_1 and g_2 as a function of a new parameter g (and of n) in such a way that $Z(g_1(g,n), g_2(g,n), n)$ is the desired generating series of the number of alternating links. The appearance of the extra terms in the partition function (i.e. that $g_1 \neq g$, $g_2 \neq 0$) is what is needed to cacnel the multiple occurrences of links due to topological equivalences; in

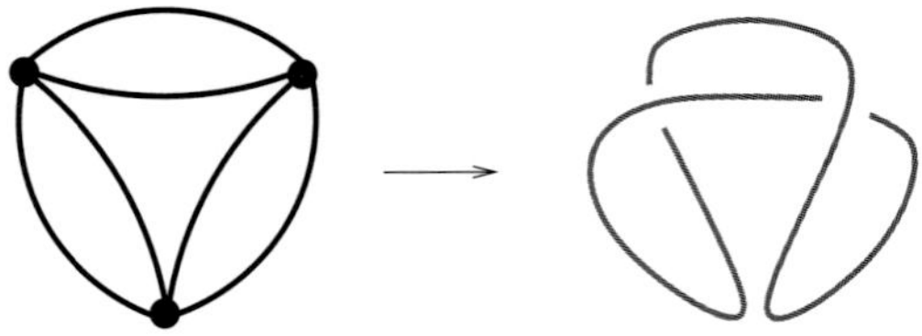

Fig. 3. 4-regular planar map and a possible loop drawing which contains both crossings and tangencies.

particular, the fact that we need both crossings and tangencies comes from the fact that on Fig. 1, the reconnections of the external lines inside the subtangle can be either in a crossing pattern or tangent pattern.

Let us call from now on $a_{k,p}$ the coefficient of this series, i.e. the number of alternating links with k components and p crossings; also, define the "partial" generating series

$$a_p(n) = \sum_{k=1}^{\infty} a_{k,p} n^k \,. \tag{3}$$

3. Some conjectures on asymptotic counting

The model described above belongs to a well-studied category of statistical mechanics models. They are usually referred to as "$O(n)$ models" since in one possible formulation in which n is supposed to be a positive integer, they display an $O(n)$ symmetry. A key question is to understand what kind of critical behavior, i.e. of long-range correlations, these models can develop. More precisely we are interested in the most *generic* behavior of a model with $O(n)$ symmetry. In a recent series of papers (cf Refs. 7–8), it was suggested that in the region $n < 2$ there is a (low temperature) phase in which the $O(n)$ symmetry is *spontaneously broken*, thus leading to a system of $n-1$ free fields in the long-range limit (so-called Goldstone bosons). This phase is stable to perturbations, so that it is likely to be the most generic critical phase.

Now a random planar map in which the number of vertices grows large can be assimilated to a continuous random surface, which suggests a connection to two-dimensional gravity. The latter theory has been exactly solved in Refs. 9–11, and the upshot is as follows: the large order asymptotic behavior of the generating series is indeed dictated by two-dimensional quantum gravity, and is of the form

$$a_p(n) \underset{p\to\infty}{\sim} e^{\sigma(n)\,p + (\gamma(n)-3)\,\log p + \kappa(n)} \tag{4}$$

where σ, κ are unspecified "non-universal" parameters, whereas the "universal" exponent γ is given by

$$\gamma = \frac{c-1-\sqrt{(1-c)(25-c)}}{12} \,. \tag{5}$$

In Eq. (5), the *central charge* c describes the critical behavior of the underlying statistical model (roughly, it counts the number of degrees of freedom); in the present case we conjecture that we are inside the spontaneously broken symmetry phase alluded to above, so that

$$c = n - 1 \,. \tag{6}$$

Note that by a Legendre transform (which takes care of the summation over k), Eq. (4) also provides the asymptotic behavior of $a_{k,p}$ as $k, p \to \infty$ in such a way that k/p remains finite.

One case is of special importance: when n is sent to zero, the factor n^k in Eq. (3) suppresses the contribution of links with multiple components, so that we find the conjectured behavior of knots:

$$a_{0,p} \underset{p\to\infty}{\sim} c\,\tau^p \cdot p^{-\frac{19+\sqrt{13}}{6}} \,. \tag{7}$$

4. Numerical checks

It is natural, in order to check the conjecture above, to try to compute the first few values of $a_{k,p}$. A procedure to carry this out has been implemented in Refs. 12–13. It uses the concept of a (variable size) transfer matrix. Usually in statistical mechanics a transfer matrix is an operator that describes discrete time evolution. Here the situation is somewhat special since the lattice itself is dynamical. We shall describe the simplest algorithm for a simple knot diagram; various extensions allow to take into account links, tangles, tangencies, etc.

The principle of the algorithm is to iterate a transfer matrix which "builds" the diagrams slice by slice by following the knot as it winds around itself. This is best explained by a picture (Fig. 4). The intermediate configurations of the system ("states") can be described as a set of several open lines which still need to be reconnected, as well as the date of their connections in the past. One line is marked as the "active line". The process at

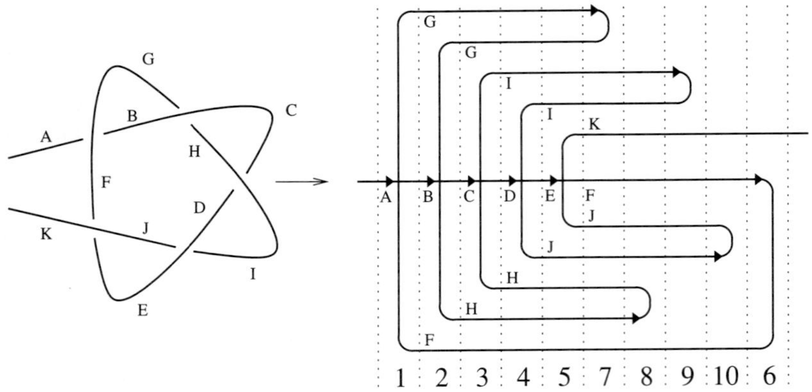

Fig. 4. A knot diagram and its corresponding "sliced" diagram. The steps are ordered by their number below the diagram (for planarity reasons one cannot simply order them from left to right). The dashed lines show the intermediate configurations. The active line is distinguished by an arrow. Steps 1–5 are creations of new crossings, while steps 6–10 are reconnections.

each step is that the "active line" can either reach a new crossing, thereby creating three new open lines, or reconnect to another existing open line.

It is important to note that we are not counting the diagrams one by one; only part of the information on the "past" of the knot diagram is needed to describe its "future", so that a single state can stand for many partial knot diagrams. This is encoded in the structure of the states on which the transfer matrix acts. Typically the space of states has a maximum dimension along the evolution of the order of 2.7^p, which also governs the computation time. This is to be compared with the total number of diagrams which is roughly 12^p. The transfer matrix sends each of these states onto the sum of all states which are obtained by adding a crossing or creating a reconnection.

This approach has been successful in the sense that it allowed to produce tables counting various alternating objects up to $p = 23$ crossings, which is much further than ever obtained before (some of the data are available at `http://ipnweb.in2p3.fr/lptms/membres/pzinn/table-tangle-diags`). However, it did not help with the determination of the critical exponent: although the general form (4) of the asymptotic expansion was confirmed, the estimates for γ had low accuracy, possibly due to strong finite-size corrections.

In Ref. 14, another approach was proposed: random sampling of large planar maps. In order to produce these maps one uses Schaeffer's bijection between decorated planar trees and planar maps. In the case of binary trees and 4-regular planar maps, the bijection is shown on Fig. 5; for more details see Refs. 15–16. Trees are very easy to create at random (with uniform probability) in linear time in the number of vertices. In practice we can generate, by this method, maps with up to $p \sim 10^7$ vertices.

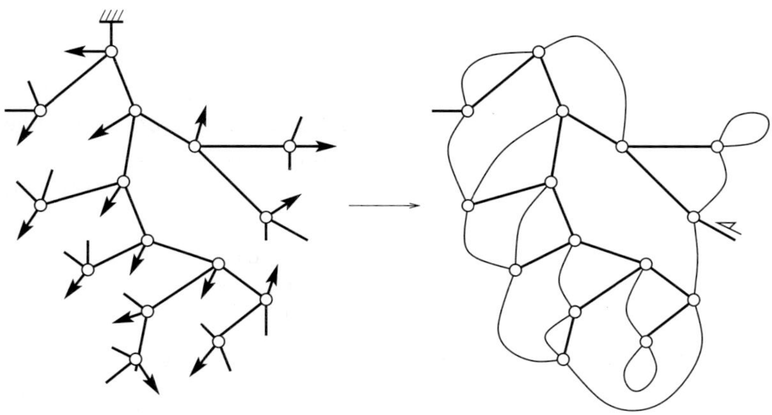

Fig. 5. Bijection from decorated rooted planar trees to 4-regular rooted planar maps.

Of course, one needs to compute a quantity out of these maps which somehow tests the conjecture (6). If we fix the number of crossings p, then we recall that the generating series $a_p(n)$ is given by Eq. (3). Since in the present situtation the probability law is uniform, this corresponds to the value $n = 1$. The simplest choice is to differentiate Eq. (4) with respect to n, assuming all parameters to depend smoothly on n, and take $n = 1$:

$$\frac{d}{dn} \log a_p(n)_{|n=1} = \sigma' p + \gamma' \log p + \kappa' + o(1) \ . \tag{8}$$

Furthermore, note that according to Eq. (3),

$$\frac{d}{dn} \log a_p(n)_{|n=1} = \frac{\sum_{k=1}^{\infty} k \, a_{k,p}}{\sum_{k=1}^{\infty} a_{k,p}} \equiv \langle k \rangle_p \ . \tag{9}$$

The result is a testable conjecture for the asymptotic behavior of the average number $\langle k \rangle_p$ of components of a random alternating link with given crossing number p. Note that since γ' is supposed to be universal it is probably unaffected by the renormalization procedure and we can sample alternating link diagrams directly without bothering to select reduced diagrams or implementing flypes.

According to Eqs. (4–5), $\gamma' = 3/10$. Simulations have been made for $p = 2^i$ where i ranges from 2 to 24 (the most up to date data can be found at `http://ipnweb.in2p3.fr/lptms/membres/pzinn/mc-planar-maps`). The fits performed in Ref. 14 on the data suggest a value of γ' in the range 0.29–0.33, which is compatible with the conjecture.

More work is needed in order to settle this issue. An interesting question is whether one can do simulations on large maps outside of $n = 1$, i.e. with non-uniform probability laws. Since no known algorithm directly produces maps with a probability proportional to $n^{\#\text{components}}$, one must use a dynamical process with appropriate moves which converges as fast as possible to this probability distribution. Some progress has been made in this direction (Ref. 17). Especially important would be the limit $n \to 0$, i.e. random generation of large knot diagrams (see also Ref. 22).

5. Virtual knot theory

Virtual knot theory is an extension of classical knot theory in which the basic objects, virtual knots and links, are defined as equivalence classes of *virtual diagrams*, that is 4-regular planar maps which are decorated with both classical and virtual crossings, under virtual Reidemeister moves, see Fig. 6 (and Ref. 18 for more details).

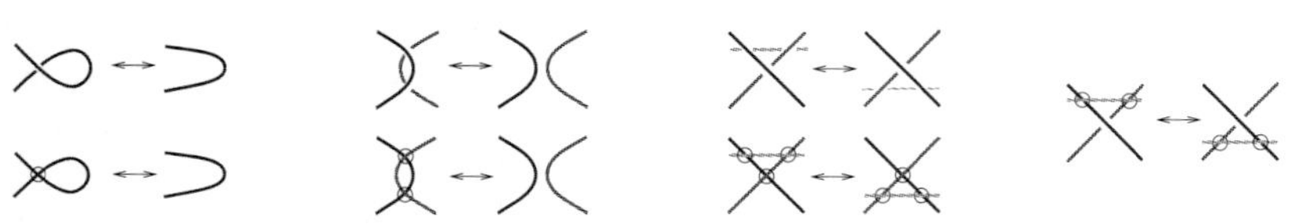

Fig. 6. Virtual Reidemeister moves. Virtual crossings are indicated by small circles.

One way to interpret virtual diagrams is that one is trying to force non-planar diagrams to be drawn in a planar way, hence the appearance of virtual crossings and virtual Reidemeister moves. Indeed virtual links turn out to be closely related to links drawn on thickened higher genus surfaces, see Refs. 19–21. In particular to each virtual link can be associated the (minimal) genus of the surface on which one can draw it.

Results on the enumeration of virtual alternating links and tangles should be a straightforward generalization of the classical case. In Ref. 21, some exact enumeration results were obtained under the assumption that the flyping equivalence generalizes in a specific way to virtual alternating links. They were tested using a computer generated database of virtual alternating links and tangles, available at `http://ipnweb.in2p3.fr/lptms/membres/pzinn/virtlinks`.

Furthermore, one can easily come up with a generalized conjecture for the asymptotic behavior of the number of virtual links: when the number of crossings p goes to infinity at fixed genus h,

$$a_p^{(h)}(n) \underset{p\to\infty}{\sim} \mathrm{e}^{\sigma(n)\,p + ((1-h)(\gamma(n)-2)-1)\,\log p + \kappa(n)} \tag{10}$$

(we recall that n is the parameter conjugate to the number of connected components k), where $\gamma(n)$ is still given by Eqs. (4–5).

References

1. J. Hoste, M. Thistlethwaite and J. Weeks, *The First 1,701,936 Knots*, *The Mathematical Intelligencer* 20 (1998) 33–48.
2. W.W. Menasco and M.B. Thistlethwaite, *The Tait Flyping Conjecture*, *Bull. Amer. Math. Soc.* 25 (1991) 403–412; *The Classification of Alternating Links*, *Ann. Math.* 138 (1993) 113–171.
3. C. Sundberg and M. Thistlethwaite, *The rate of Growth of the Number of Prime Alternating Links and Tangles*, *Pac. J. Math.* 182 (1998) 329–358.
4. P. Zinn-Justin, J.-B. Zuber, *Matrix Integrals and the Counting of Tangles and Links*, proceedings of the 11th International Conference on Formal Power Series and Algebraic Combinatorics, Barcelona June 1999; *Discr. Math.* 246 (2002),343, `arXiv:math-ph/9904019`.
5. P. Zinn-Justin, J.-B. Zuber, *On the Counting of Colored Tangles*, *J. Knot Theor. Ramif.* 9 (2000), 1127, `arXiv:math-ph/0002020`.
6. P. Zinn-Justin, *The General O(n) Quartic Matrix Model and its application to Counting Tangles and Links*, *Commun. Math. Phys.* 238 (2003), 287–304, `arXiv:math-ph/0106005`.
7. N. Read and H. Saleur, *Exact spectra of conformal supersymmetric nonlinear sigma models in two dimensions*, *Nucl. Phys.* B **613**, 409–444 (2001), `arXiv:hep-th/0106124`.
8. J.L. Jacobsen, N. Read and H. Saleur, *Dense loops, supersymmetry, and Goldstone phases in two dimensions*, *Phys. Rev. Lett.* 90 (2003) 090601, 4pp, `arXiv:cond-mat/0205033`.
9. V. G. Knizhnik, A. M. Polyakov and A. B. Zamolodchikov, *Fractal structure of 2D quantum gravity*, *Mod. Phys. Lett. A* 3, 819–826 (1988).
10. F. David, *Conformal field theories coupled to 2D gravity in the conformal gauge*, *Mod. Phys. Lett. A* 3, 1651–1656 (1988).
11. J. Distler and H. Kawai, *Conformal field theory and 2D quantum gravity*, *Nucl. Phys.* B 321, 509 (1989).
12. J.L. Jacobsen and P. Zinn-Justin, *A Transfer Matrix approach to the Enumeration of Knots*, *J. Knot Theor. Ramif.* 11 (2002), 739–758, `arXiv:math-ph/0102015`.

13. J.L. Jacobsen and P. Zinn-Justin, *A Transfer Matrix approach to the Enumeration of Colored Links*, *J. Knot Theor. Ramif.* 10 (2001), 1233–1267, `arXiv:math-ph/0104009`.
14. G. Schaeffer and P. Zinn-Justin, *On the Asymptotic Number of Plane Curves and Alternating Knots*, submitted to *J. Exp. Math.* , `arXiv:math-ph/0304034`.
15. G. Schaeffer, *Bijective census and random generation of Eulerian planar maps with prescribed vertex degrees*, *Electron. J. Combin.* **4** (1997), no. 1, Research Paper 20, 14 pp. (electronic).
16. G. Schaeffer, *Random sampling of large planar maps and convex polyhedra*, in *Annual ACM Symposium on Theory of Computing* (Atlanta, GA, 1999), 760–769 (electronic), ACM, New York, 1999.
17. J.L. Jacobsen and P. Zinn-Justin, work in progress.
18. L. Kauffman, *Virtual knot theory*, *Europ. J. Combin.* **20** (1999) 663-690, `math.GT/9811028`; *Detecting virtual knots*, Chicago preprint.
19. J.S. Carter, S. Kamada and M. Saito, *Stable equivalence of knots on surfaces and virtual knot cobordisms*, *J. Knot Theory and its Ramifications*, **11** (2002) 311–322, `arXiv:math.GT/0008118`.
20. G. Kuperberg, *What is a virtual link?*, *Algebr. Geom. Topol.* 3 (2003), 587–591, `arXiv:math.GT/0208039`.
21. P. Zinn-Justin and J.-B. Zuber, *Matrix Integrals and the Generation and Counting of Virtual Tangles and Links*, accepted for publication in *J. Knot Theor. Ramif.*, `arXiv:math-ph/0303049`.
22. Y. Diao, C. Ernst and U. Ziegler, *Generating Large Random Knot Projections*, present volume.

SERIES ON KNOTS AND EVERYTHING

Editor-in-charge: Louis H. Kauffman *(Univ. of Illinois, Chicago)*

The Series on Knots and Everything: is a book series polarized around the theory of knots. Volume 1 in the series is Louis H Kauffman's Knots and Physics.

One purpose of this series is to continue the exploration of many of the themes indicated in Volume 1. These themes reach out beyond knot theory into physics, mathematics, logic, linguistics, philosophy, biology and practical experience. All of these outreaches have relations with knot theory when knot theory is regarded as a pivot or meeting place for apparently separate ideas. Knots act as such a pivotal place. We do not fully understand why this is so. The series represents stages in the exploration of this nexus.

Details of the titles in this series to date give a picture of the enterprise.

Published:

Vol. 1: Knots and Physics (3rd Edition)
by L. H. Kauffman

Vol. 2: How Surfaces Intersect in Space — An Introduction to Topology (2nd Edition)
by J. S. Carter

Vol. 3: Quantum Topology
edited by L. H. Kauffman & R. A. Baadhio

Vol. 4: Gauge Fields, Knots and Gravity
by J. Baez & J. P. Muniain

Vol. 5: Gems, Computers and Attractors for 3-Manifolds
by S. Lins

Vol. 6: Knots and Applications
edited by L. H. Kauffman

Vol. 7: Random Knotting and Linking
edited by K. C. Millett & D. W. Sumners

Vol. 8: Symmetric Bends: How to Join Two Lengths of Cord
by R. E. Miles

Vol. 9: Combinatorial Physics
by T. Bastin & C. W. Kilmister

Vol. 10: Nonstandard Logics and Nonstandard Metrics in Physics
by W. M. Honig

Vol. 11: History and Science of Knots
edited by J. C. Turner & P. van de Griend

Vol. 12: Relativistic Reality: A Modern View
edited by J. D. Edmonds, Jr.

Vol. 13: Entropic Spacetime Theory
by J. Armel

Vol. 14: Diamond — A Paradox Logic
by N. S. Hellerstein

Vol. 15: Lectures at KNOTS '96
by S. Suzuki

Vol. 16: Delta — A Paradox Logic
by N. S. Hellerstein

Vol. 17: Hypercomplex Iterations — Distance Estimation and Higher Dimensional Fractals
by Y. Dang, L. H. Kauffman & D. Sandin

Vol. 19: Ideal Knots
by A. Stasiak, V. Katritch & L. H. Kauffman

Vol. 20: The Mystery of Knots — Computer Programming for Knot Tabulation
by C. N. Aneziris

Vol. 24: Knots in HELLAS '98 — Proceedings of the International Conference on Knot Theory and Its Ramifications
edited by C. McA Gordon, V. F. R. Jones, L. Kauffman, S. Lambropoulou & J. H. Przytycki

Vol. 25: Connections — The Geometric Bridge between Art and Science (2nd Edition)
by J. Kappraff

Vol. 26: Functorial Knot Theory — Categories of Tangles, Coherence, Categorical Deformations, and Topological Invariants
by David N. Yetter

Vol. 27: Bit-String Physics: A Finite and Discrete Approach to Natural Philosophy
by H. Pierre Noyes; edited by J. C. van den Berg

Vol. 28: Beyond Measure: A Guided Tour Through Nature, Myth, and Number
by J. Kappraff

Vol. 29: Quantum Invariants — A Study of Knots, 3-Manifolds, and Their Sets
by T. Ohtsuki

Vol. 30: Symmetry, Ornament and Modularity
by S. V. Jablan

Vol. 31: Mindsteps to the Cosmos
by G. S. Hawkins

Vol. 32: Algebraic Invariants of Links
by J. A. Hillman

Vol. 33: Energy of Knots and Conformal Geometry
by J. O'Hara

Vol. 34: Woods Hole Mathematics — Perspectives in Mathematics and Physics
edited by N. Tongring & R. C. Penner

Vol. 35: BIOS — A Study of Creation
by H. Sabelli

Vol. 36: Physical and Numerical Models in Knot Theory
edited by J. A. Calvo et al.